Mixed-Valence Compounds

NATO ADVANCED STUDY INSTITUTES SERIES

Proceedings of the Advanced Study Institute Programme, which aims at the dissemination of advanced knowledge and the formation of contacts among scientists from different countries

The series is published by an international board of publishers in conjunction with NATO Scientific Affairs Division

A	Life Sciences	Plenum Publishing Corporation
B	Physics	London and New York
C	Mathematical and Physical Sciences	D. Reidel Publishing Company Dordrecht, Boston and London
D	Behavioural and Social Sciences	Sijthoff & Noordhoff International Publishers
E	Applied Sciences	Alphen aan den Rijn and Germantown U.S.A.

Series C – Mathematical and Physical Sciences

Volume 58 – Mixed-Valence Compounds

Mixed-Valence Compounds

Theory and Applications in Chemistry, Physics, Geology, and Biology

Proceedings of the NATO Advanced Study Institute held at Oxford, England, September 9-21, 1979

edited by

DAVID B. BROWN
The University of Vermont, Department of Chemistry, Cook Physical Science Building, Burlington, Vermont, U.S.A.

D. Reidel Publishing Company

Dordrecht : Holland / Boston : U.S.A. / London : England

Published in cooperation with NATO Scientific Affairs Division

Library of Congress Cataloging in Publication Data

CIP

NATO Advanced Study Institute, Oxford, 1979.
Mixed-valence compounds.

(NATO advanced study institutes series : Series C, Mathematical and physical sciences ; v. 58)
Includes index.
1. Valence (Theoretical chemistry)–Congresses. 2. Chemical reaction, Conditions and laws of–Congresses. I. Brown, David Basset, 1943- II. Title. III. Series.
QD469.N37 1979 541.2'24 80-23749
ISBN-13: 978-94-009-9078-4 e-ISBN-13: 978-94-009-9076-0
DOI: 10.1007/978-94-009-9076-0

Published by D. Reidel Publishing Company
P.O. Box 17, 3300 AA Dordrecht, Holland

Sold and distributed in the U.S.A. and Canada
by Kluwer Boston Inc.,
190 Old Derby Street, Hingham, MA 02043, U.S.A.

In all other countries, sold and distributed
by Kluwer Academic Publishers Group,
P.O. Box 322, 3300 AH Dordrecht, Holland

D. Reidel Publishing Company is a member of the Kluwer Group

TABLE OF CONTENTS

PREFACE

It has been a decade since two seminal reviews demonstrated that mixed-valence compounds share many unique and fascinating features. The insight provided by those early works has promoted a great deal of both experimental and theoretical study. As a result of extensive efforts, our understanding of the bonding and properties of mixed-valence compounds has advanced substantially. There has been no comprehensive treatment of mixed-valence compounds since 1967, and the meeting convened at Oxford in September, 1979, provided a unique opportunity to examine the subject and its many ramifications.

Mixed-valence compounds play an important role in many fields. Although the major impact of the subject has been in chemistry, its importance has become increasingly clear in solid state physics, geology, and biology. Extensive interest and effort in the field of molecular metals has demonstrated that mixed-valency is a prerequisite for high electrical conductivity. The intense colors of many minerals have been shown to be due to mixed-valency, and the electron-transfer properties of certain mixed-valence metalloproteins are important in biological processes. Experts from all of these areas participated in this meeting, and the truly interdisciplinary nature of the subject made it a unique learning experience for all in attendance.

Several different types of papers are included in this volume. The subject as a whole is surveyed, and the descriptive chemistry of different types of materials is detailed. Several different theoretical approaches to mixed-valency are included, and the activity in this area - indeed, the significant controversy in this area - are indicative of the vigor of the field. Finally, several papers are concerned

D. B. Brown (ed.), Mixed-Valence Compounds, vii-viii.

with the application of specific experimental techniques to the study of mixed-valence compounds.

The Advanced Study Institute at Oxford, and consequently this monograph, would not have been possible without the generous financial support of NATO. Participation of some students from non-NATO nations was made possible by a grant from the U.S. Army Research Office - Europe. The warm hospitality of St. John's College, Oxford and the Inorganic Chemistry Laboratory provided an ideal location for the free exchange of ideas. The impressive and conscientious efforts of the lecturers helped to provide a coherent approach to the subject, and their efforts are appreciated. I am particularly grateful to the organizing committee - Peter Day, Andy Ludi and Tom Meyer - for their suggestions and support. Finally, I want to thank the participants at the meeting, whose enthusiastic involvement made it a truly exciting experience.

Burlington, Vermont David B. Brown
March, 1980

I. AN OVERVIEW OF MIXED-VALENCY

INTRODUCTION TO MIXED-VALENCE CHEMISTRY

P. Day

Oxford University, Inorganic Chemistry Laboratory, South Parks Road, Oxford, OX1 3QR, England

DEFINITIONS

To begin with, some definitions: 'mixed valency' is one of several names, such as 'mixed oxidation state' or 'non-integral oxidation state', used to describe inorganic or metal-organic compounds in which an element is present in more than one level of oxidation. Since the conventional definition of 'oxidation state' implies that we can define the number of valence shell electrons associated with each atom in a compound we shall exclude compounds in which covalency or electron delocalization between the constituent atoms is so great that we cannot judge, even approximately, how many valence shell electrons should be assigned to each kind of atom. Thus Fe_3O_4 comes within our compass but Cu_3Si does not. Without any doubt in Fe_3O_4 the most loosely bound electrons are mainly localized on the Fe atoms. The choice then is to say that each Fe has 5.33 d-electrons on average, or that one Fe per formula unit has 6 while two others have 5. The same argument applies to Prussian Blue, $Fe_4[Fe(CN)_6]_3 14H_2O$, or the Creutz-Taube ion $(NH_3)_5Ru(pyrazine)Ru(NH_3)_5^{5+}$.

Apart from compounds in which the chemical formula alone is sufficient evidence of mixed valency, there is a further category of substances having apparently integral oxidation numbers, but ones which would be highly unusual for the element in question. Typical examples are $Pt(NH_3)_2Cl_3$ and SbO_2. In neither case

D.B. Brown (ed.), Mixed-Valence Compounds, 3-24.

do we find properties such as paramagnetism, as anticipated for Pt(III) and Sb(IV), but instead there are two distinct kinds of cation site in the lattice, so we assign formal oxidation states Pt(II,IV) and Sb(III,V).

HISTORY

Mixed valence chemistry is as old as the hills: biotite, vivianite and many other minerals testify to this geological fact. Not much younger in the evolutionary timescale are the clusters of mixed valency Fe atoms and pairs of Cu atoms found in enzymes with electron transfer or nitrogen and oxygen-fixing functions. Arguably the first mixed valency compound made by man was Prussian Blue, prized as an ink and dyestuff, and the subject of early industrial espionage (1). In the latter nineteenth century many other such brightly coloured substances came to light in the Inorganic Chemistry Laboratories of Europe and the USA, often named in the fashion of the time after their inventors: Wells' Salt, Wolfram's Red Salt, etc. Theories, albeit couched in purely qualitative language, were soon developed to explain why these compounds had such striking colours. With the hindsight of quantum mechanics we know now that there are many different kinds of optical transition which can lead to light absorption in the visible, but one which has been investigated less than most of the others is the transfer of an electron from one metal ion to another. Yet this source of colouration was recognized first by Alfred Werner in 1896 (2) and stated explicitly by, among others, Zintl and Rausch in 1924 (3):

> "To produce colours ... which are not given by the sum of the individual colours of the components ... it is necessary that the distribution of oxidation states within the molecule can exchange under the influence of light so as to produce the light absorption and hence the colour. This situation occurs most frequently in inorganic chemistry when the same element is present in different valence states in the same molecule."

Probably the first attempts to relate mixed valency to the physical properties of inorganic substances in any quantitative way were those of Verwey and his colleagues at the Philips Research Laboratory. From 1937 to 1950 they prepared many examples of what they called 'controlled valency semiconductors' (4). These are mostly transition metal oxides in which a second oxidation state is introduced by doping the single valence lattice with an aliovalent closed shell cation of comparable ionic radius. Thus for instance the insulating perovskite $LaMnO_3$ becomes conducting when some of the La(III) is replaced by Sr(II), so that an equal amount of Mn(III) is oxidized to Mn(IV). Verwey et al. were also the first to note that rapid electron exchange between metal ions of different oxidation state was facilitated when they occupied sites of similar geometry (5). One of the neatest examples of this phenomenon is Fe_3O_4, in whose lattice Fe(II) and Fe(III) are both found in octahedral sites. Much more detail about the intricate story of Fe_3O_4 is given in the chapter by Goodenough.

Quite a different strand in the history of mixed valence chemistry from the solid state physicists' experiments on oxides arose from inorganic chemists' interest in kinetics and mechanisms of electron transfer reactions. In the 1960's it appeared that potentially a great deal of information relevant to inner- and outer-sphere oxidation-reduction processes could be obtained by examining thermal and Franck-Condon electron transfer (i.e. respectively conductivity and optical spectra) in mixed valence oligomers and solids (6). These strands came together with the publication of two reviews in 1967 (7,8). Solid state aspects of mixed valency received further substantial impetus from the discovery of one-dimensional molecular metals (9), while chemists became concerned with the redox states of metal cluster compounds (10), including the biological examples (11) and their inorganic analogues (12). Mineralogists, too, found mixed valency of interest (13) while potential applications of colour and conductivity in mixed valency compounds have appeared in electrochromism (14) and photochemistry (15). Table 1 summarizes some of the diverse fields to which mixed valency chemistry has contributed.

OCCURRENCE OF MIXED VALENCY

Whereabouts in the Periodic Table do we find mixed valency, and in what kind of compounds? Clearly variable oxidation state is a requirement so many examples are found among the d-block elements. Those f-block elements which occur in oxidation states other than +3, namely at the beginning and end of the

Table 1. Fields to which mixed valency has contributed

Chemistry		Ref
Oxidation-reduction	$[(NH_3)_5Ru(pz)Ru(NH_3)_5]^{5+}$	16
Electrochemistry	AgO	17
Metal-metal bonding	$Nb_6Cl_{12}^{2+}$	10
Photochemistry	$PW_{12}O_{40}^{6-}$	15
Dyes and pigments	Ru Red	18
Analytical chemistry	Prussian Blue	19
One dimensional compounds	Wolfram's Red Salt	20
Organometallics		21
Physics		
Hopping conduction	$Li_xNi_{1-x}O$	22
Metal-insulator transitions	V_nO_{2n-1}	23
Ordering of defects	Mo_nO_{3n-1}	24
Superconductivity	$LiTi_2O_4$; $BaBi_xPb_{1-x}O_3$	25
Ferromagnetism	$La_xSr_{1-x}MnO_3$	26
Molecular metals	$K_2Pt(CN)_4Br_{0.30}3H_2O$	27
Electrochromism	WO_{3-x}	14
Mineralogy		
Micas and asbestos	Biotite; crocidolite	28
Phosphates	Vivianite	29
Biology		
Electron transfer enzymes	Fe_4S_4 Ferredoxins	11
Oxygen binding enzymes	Cu Haemocyanines	30
Cancer chemotherapy	Pt Blues	31

lanthanides and on either side of Gd, also contribute examples, as do the earlier actinides. In the post-transition elements one finds electron configurations ns^2 as well as ns^0, and mixtures of these provide yet more instances. In all, nearly 40 elements form mixed valency compounds.

A representative selection of compounds and their oxidation states is listed in Table 2. Nearly all the 3d elements from mixed valent oxides while in the 4d and 5d series halides and coordination complexes, especially ammines, are widespread. A lot of mixed valence halides also occur in the B-subgroups. More detailed descriptions of representative examples of all these compounds are given in the chapter by Ludi.

CLASSIFICATION OF MIXED VALENCE COMPOUNDS

Among the compounds in Table 2 are metallic conductors and insulators, ferromagnets and antiferromagnets and materials which are black, white and all colours of the rainbow. Are there, then, any identifiable characteristics specifically associated with the presence of mixed valency? There must be, as we can see from the following example (32). Consider a lattice containing Ti(III) (d^1) and Ti(IV) (d^o) on two types of site A,B which may be crystallographically distinguishable by having different bond lengths, site symmetry, etc. If the A sites carry the excess electrons the ground state valence bond configuration is A(III)B(IV). The energy required to transfer the electron from A to B depends on the difference in crystal potential, ligand field, etc. between the sites because, if they were indistinguishable, no energy would be needed at all, since the ionization potential of Ti(III) is equal and opposite to the electron affinity of Ti(IV). In simple terms, if the electron can be transferred from site to site with no expenditure of energy we would have a metal, whilst if a large activation energy were needed the compound would be an insulator. Mixed valency properties are therefore related very closely to the similarity or difference between the two sites containing the ions of differing valency.

Class I

The pure valence bond configuration A(III)B(IV) would only be an accurate description of the ground state if the A and B sites were so different that the A(IV)B(III) configuration was of extremely high energy. Compounds of this kind would have properties which were essentially a superposition of those of A(III) and B(IV) taken separately. Mössbauer isomer shifts, core shell ionization energies or visible-ultraviolet electronic excitations localized on A will be just as in an

Table 2. Examples of mixed valency compounds

Ti	III,IV	oxides	Ti_nO_{2n-1}; $LiTi_2O_4$
		solutions	chloride, sulphate
V	III,IV	oxides	V_nO_{2n-1}; $Na_xV_2O_5$
	IV,V	oxyanions	$V_{10}O_{28}^{n-}$
Cr	II,III	halide	Cr_2F_5
	III,VI	oxides	Cr_5O_{12}; KCr_3O_8
Mn	II,III	oxide	Mn_3O_4
	III,IV	oxide	$La_xSr_{1-x}MnO_3$; $La_xCa_{2-x}MnO_4$
	IV,VII	oxide	$(H_3O)_2[Mn(MnO_4)_6]11H_2O$
Fe	II,III	oxide	Fe_3O_4
		halide	$Fe_2F_5.xH_2O$; KFe_2F_6
		complexes	Prussian Blue; $Fe_3O(OAc)_6(H_2O)_3^{n+}$
		organo-metallics	$(fulvalene)_2Fe_2^+$
		enzymes	Ferredoxins
Co	II,III	oxide	Co_3O_4
	III,IV	oxide	$La_xSr_{1-x}CoO_3$
		complexes	$(NH_3)_5CoO_2Co(NH_3)_5^{4+}$
Ni	II,III	oxide	$Li_xNi_{1-x}O$
		complexes	$Ni(diphenylglyoxime)_2(I_3)_{0.33}$
Cu	I,II	halide	$(Co(NH_3)_6)(CuCl_5)_x(CuCl_4)_{1-x}$
		sulphide	CuS; KCu_4S_3
		proteins	Caerwoplasmin
Nb	II,III	halides	$Nb_6X_{14}7H_2O$; Nb_3Br_8
	IV,V	oxide	$Nb_{3n+1}O_{8n-2}$
Ta	II,III	halide	Ta_6X_{15}
Mo	V,VI	oxide	Mo_nO_{3n-1}; $K_{0.26}MoO_3$
		oxyanions	$SiMo_{12}O_{40}^{n-}$
W	V,VI	oxide	W_nO_{3n-1}; Na_xWO_3
		oxyanions	$PW_{18}O_{62}^{n-}$
	II,III	halide	W_6Br_{14}
Tc	II,III	halide	$Tc_2Cl_8^{3-}$

Ru	II,III	complexes	$(NH_3)_5RuLRu(NH_3)_5^{n+}$; $Ru_3O(OAc)_6L_3^{n+}$
	III,IV	complexes	Ru Red
Os	II,III	complexes	$(NH_3)_5OsN_2Os(NH_3)_5^{5+}$
Rh	II,III	halide	Rh_6Cl_{14}
Ir	III,IV	sulphate	$Ir_3N(SO_4)_6(H_2O)_3^{4+}$
Pd	II,IV	halide	PdF_3
		complexes	$Pd(NH_3)_2Cl_3$; $Pd(bgd)_2I$
Pt	II,IV	oxide	$Na_xPt_3O_4$
		halide	$(PtA_4Cl_2)(PtA_4)Cl_4$
		complexes	$K_2Pt(CN)_4Br_{0.30}3H_2O$; $Pt(dpg)_2I$
Ag	O,I	fluoride	Ag_2F
		oxide	$Ag(Ag_6O_8)NO_3$
	I,III	oxide	AgO
Au	I,III	halide	$CsAuCl_3$
		complexes	$Au(dmg)_2(AuCl_2)$
Ga	I,III	halide	$GaCl_2$
		complexes	$(GaL_4)(GaCl_4)$
In	I,III	halide	$InCl_2$; In_4Cl_7; In_4Br_6
Tl	I,III	oxide	TlO; TlS
		halide	$TlCl_2$; Tl_4Cl_6
Sn	II,IV	aq. solution	$Sn_2Cl_6(H_2O)_2$
Pb	II,IV	oxide	Pb_3O_4
		halide	$Co(NH_3)_6PbCl_6$
Sb	III,V	oxide	Sb_2O_4
		halide	Cs_2SbCl_6
Bi	III,V	oxide	$BaBiO_3$
		halide	$Bi_{12}Cl_{14}$
I		polyiodides	I_n^-
Xe	VI,VIII	oxide	$K_4Xe_3O_{12}$

A(III) compound. The intervalence transition A(III) B(IV) → A(IV)B(III) will have such a high energy that it will not contribute to the visible colour of the compound. This kind of behaviour we call class I (7). Typical examples from Table 2 are $GaCl_2$ and $Cu(en)_2(CuBr_2)_2$. The environments of the two Ga sites in $GaCl_2$ are shown in Figure 1. Half the Ga atoms are surrounded by a tetrahedron of Cl atoms at 2.19 Å and the other half by an irregular dodecahedron of Cl at distances ranging from 3.2-3.3 Å. From the known co-ordination preferences of Ga(I) and Ga(III) it is clear that the tetrahedral site is Ga(III) and the dodecahedral one is Ga(I). $GaCl_2$ is colourless, diamagnetic and insulating.

Class II

In the next category (class II) we consider compounds in which the A and B sites, though crystallographically distinguishable, are rather similar. For instance they may both be octahedral, but with small differences in bond length or angle. Then the valence bond configuration A(IV)B(III) has an energy not too much greater than A(III)B(IV). If there exists a suitable perturbation matrix element to mix them the correct ground state wavefunction is not just A(III) B(IV) but a linear combination

$$(1-\alpha^2)^{\frac{1}{2}}(A(III)B(IV)) + \alpha(A(IV)D(III)) \qquad (1)$$

where α has been called the valence delocalization coefficient and the function (1) has been normalized. Note that the perturbation operator 'mixing' the two valence bond configurations must be totally symmetric under the operations of the point group of the local cluster or molecule so that should the A sites be surrounded by a number n of B sites we have to choose the appropriate linear combination function

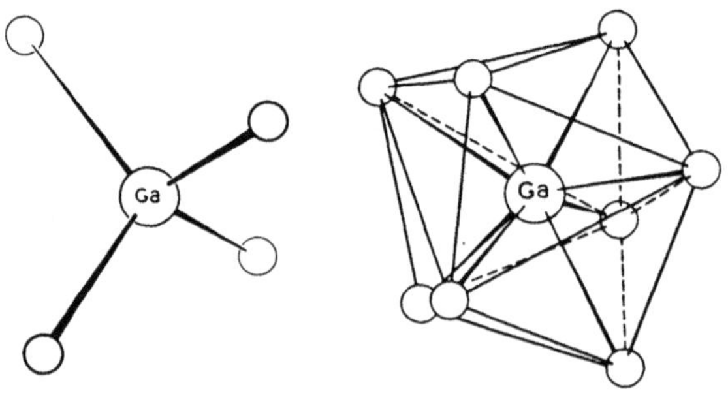

Figure 1. Coordination around Ga in $GaCl_2$.

$$\Psi_k = A(IV) \sum_j C_{kj} B_1(IV) B_2(IV) \ldots B_j(III) \ldots B_n(IV) \quad (2)$$

to mix with the zero-order ground state function

$$\Psi_o = A(III) B_1(IV) \ldots\ldots B_n(IV).$$

These arguments are worked out in more detail in (7).

Using perturbation theory, the magnitude of α depends on the energy (E_k) of I_k with respect to I_o and the off-diagonal matrix element (V) between them:

$$\alpha^2 = N^2 \left[1 - \left\{ 1 + \frac{2E_k^2 - 2E_k (E_k^2 + 4V^2)^{\frac{1}{2}} + 4V^2}{4V^2} \right\}^{-1} \right] \quad (3)$$

where $E_k = \langle \Psi_k | \mathcal{H} | \Psi_k \rangle$ and $V = \langle \Psi_o | \mathcal{H} | \Psi_k \rangle$.

If E_k is only a few electron volts optical transitions between the states mainly composed of Ψ_o and Ψ_k should be observed in the visible, near ultraviolet or near infrared, and it is this kind of excitation which gives so many mixed valency compounds their bright colours. Since α remains small, however, any excitations localized on either the A or B sites (for example a ligand field transition of A(III)) will appear very little shifted from their positions in the corresponding single valency compound, as in class I. On the other hand, if the vertical (Franck-Condon) intervalence electron transfer process only requires an eV or two, adiabatic (thermally activated) electron transfer should also take place, at a rate which, with luck, may span the time-scale of physical measurements like NMR or Mössbauer spectroscopy. If the compound is a discrete dimer or oligomer electron transfer only takes place within the molecule, but if it is a solid bulk charge migration may be detected as semiconductivity.

The properties of one of the most famous class II compounds, Prussian Blue, are described in detail in the chapter by Ludi. Another set of examples, on which a variety of physical measurements have been made, are the compounds containing octahedral SbX_6^- and SbX_6^{3-}, where X is a halogen. Structures of two such salts are shown in Figure 2. Depending on the cation they are either blue or red, colours which are not found in any other Sb compounds, and which result in each case

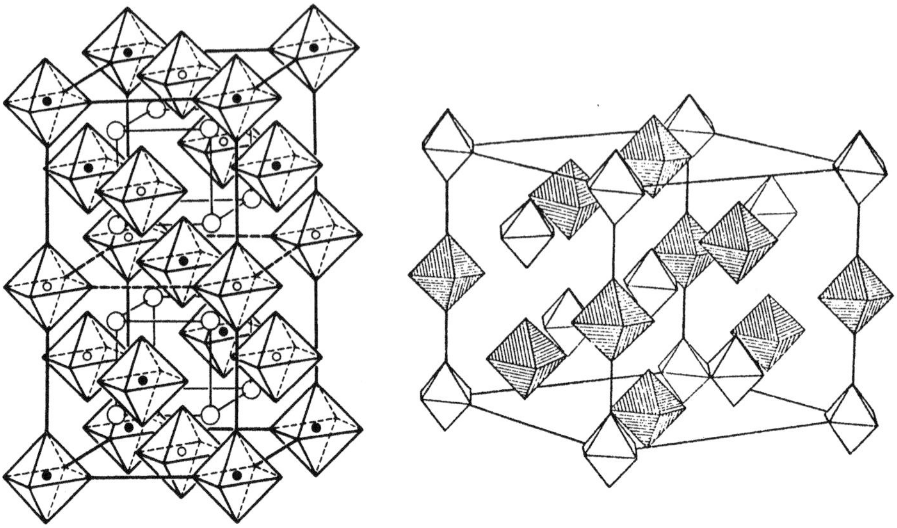

Figure 2. Crystal structures of (a) $(NH_4)_2SbBr_6$ and (b) $(C_3H_7NH_3)_4SbCl_6(Cl)_2$.

from single unstructured absorption bands about 5000 cm^{-1} wide at room temperature (33). The compounds are diamagnetic, and the difference in bond length between the two kinds of SbX_6 octahedra is 0.24 Å in the bromide (34) and 0.28 Å in the chloride (35). Both the far infrared spectra (36) and ^{121}Sb Mössbauer spectra (37) show separate peaks characteristic of Sb(III) and Sb(V), though the report (38) that the Mössbauer spectra broadened and split at higher temperature because of electron exchange was not confirmed by later measurements (39).

The electrical conductivity of hexachloroantimonates is low but one can easily verify that it results from the presence of mixed oxidation states by measuring it as a function of Sb concentration in a series of mixed crystals such as $Cs_2Sb_xSn_{1-x}Cl_6$. Figure 3 shows that the specific conductivity k is proportional to x^2 or to y in the series $Cs_2Sb_yIn_{0.5-y}Sb_{0.5}Cl_6$ where only the Sb(III) concentration is varied (40). Except in the most dilute solid solutions the semiconduction activation energy is constant (e.g. 0.76 ± 0.02 eV for Cs_2SbCl_6).

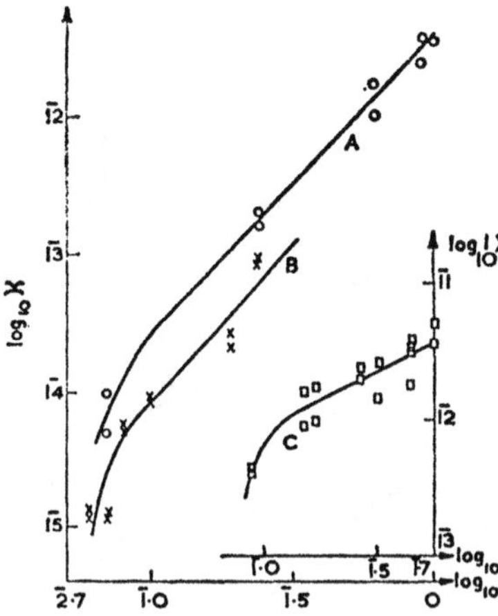

Figure 3. Concentration dependence of conductivity k in A, $Cs_2Sb_xSn_{1-x}Cl_6$; B, $(NH_4)_2Sb_xSn_{1-x}Cl_6$; C, $Cs_2Sb_yIn_{0.5-y}Sb_{0.5}Cl_6$ (40).

Optical spectra of the same solid solutions also serve to confirm that the intensity of the visible band is proportional to the product of the concentration [Sb(III)][Sb(V)], and single crystal spectra (Figure 4) enable one to evaluate the transition moment μ. It turns out to be very small (0.02 Å), corresponding to a small value of α (41). This is a common feature of class II compounds, which often appear dark in colour, not because the intervalence absorption is particularly intense, but because the effective concentration of the chromophore, in terms of pairwise interactions in the solid, is so large.

The intensity of the band relates to in eq. (1) in a very simple way. Writing the ground and excited states for a pair of mixed valence $SbCl_6$ ions as

$$\begin{aligned} \Psi_o &= (1-\alpha^2)^{\frac{1}{2}}(A(V)B(III)) + \alpha(A(IV)B(IV)) \\ \Psi_k &= \beta(A(V)B(III)) + (1-\beta^2)^{\frac{1}{2}}(A(IV)B(IV)) \end{aligned} \qquad (4)$$

The transition moment

$$\mu = \langle \Psi_o | e\underline{r} | \Psi_k \rangle = \tfrac{1}{2}e(\alpha+\beta)R \qquad (5)$$

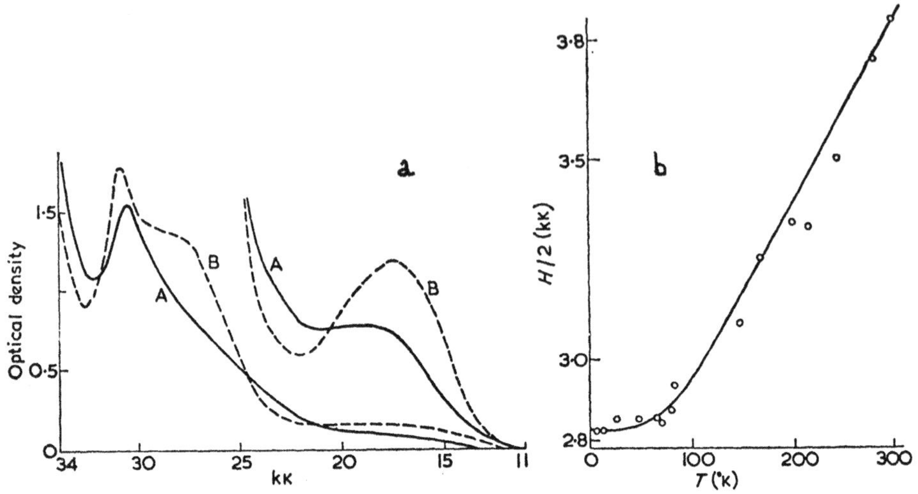

Figure 4. (a) Transmission spectra of two crystals of $(CH_3NH_3)_2Sb_xSn_{1-x}Cl_6$; A, 300 K; B, 6 K. (b) Variation of halfwidth of mixed valence band with temperature. The line is calculated from eq. (9) with $\omega = 210\ cm^{-1}$, s = 130 (41).

assuming zero differential overlap, where R is the distance between the two Sb. In Cs_2SbCl_6 R is 7.1 Å so $\frac{1}{2}(\alpha + \beta)$ is only 0.003.

Two other features of Figure 4 are of interest. First, quite typically for class II compounds, there is a transition in the near ultraviolet resulting from a local excitation on one of trapped valence ions. In the present case it is the $5s^2(^1S) \quad 5s^15p^1(^3P_1)$ transition of $SbCl_6^{3-}$. Second is the notable variation of the intervalence bandshape with temperature. At 6K the band is about 10% more intense than at room temperature and somewhat narrower. On the other hand (again typically for intervalence bands) it is still quite broad, and completely unstructured. In general, the shape of an absorption envelope $G(\nu)$ is the sum of terms involving the overlap integrals between all the vibrational wavefunctions χ_v of the ground state and of the excited state χ'_v, weighted, for temperatures above absolute zero, by a Boltzmann factor relating to the populations $P(v)$ of the χ_v levels (42). Then

$$G(\nu) = \sum_{v,v'} P(v) |\langle \chi'_{v'} | \chi_v \rangle|^2 \delta(\nu - (E_{v'} - E_v)/\hbar) \qquad (6)$$

On the assumption that the ground and excited state potential energy surfaces are harmonic:

$$\begin{aligned} E_v &= E_o(g) + h\omega_j(g)(v+\tfrac{1}{2}) \\ E_{v'} &= E_o(u) + h\omega_j(u)(v'+\tfrac{1}{2}) \end{aligned} \qquad (7)$$

If we make a further drastic assumption that all the vibration modes have the same frequency ω the summation in eq. (6) leads to a Gaussian band envelope with a halfwidth H given by

$$H^2 = 8(\ln 2)h^2\omega^2 S \coth(h\omega/2kT) \qquad (8)$$

where S is the 'Huang-Rhys factor', the ratio between the vibrational energy excited in the upper state and the energy of a single vibrational quantum:

$$S = (\tfrac{1}{2}\omega^2\Delta q^2/(h\omega) \qquad (9)$$

Figure 4(b) compares the halfwidth of the Sb(III,V) intervalence band as a function of temperature with a line calculated from eq. (8). Further elaborations of these formulae are given in the chapter by Schatz. The value of ω which fits the halfwidth variation (210 cm^{-1}) should be compared with the normal mode frequencies of the constituent single valence complex-ions. Totally symmetric stretching frequencies of $SbCl_6^-$ and $SbCl_6^{3-}$ are 329 and 267 cm^{-1} respectively (36), but at this stage the extent of involvement of other modes in the intervalence charge transfer process is unknown.

Class III

There are many compounds whose formulae indicate the presence of mixed valence but whose structures do not contain crystallographically distinguishable sites where the oxidation states might be localized. One of the first of this kind to be discovered was the cubic tungsten bronze series Na_xWO_3. Here the WO_3 matrix has the ReO_3 structure, with WO_6 octahedra sharing vertices, while Na occupy some of the cube centres. Within the range $0.4 < x < 0.9$ x can take any value. NMR measurements on the ^{23}Na nucleus show that the Na is present as Na^+ so electrons have been donated to the t_{2g} orbitals of W. Nevertheless, the W sites are all crystallographically equivalent so no distinction between W(VI) and W(V) is possible. In an equation like eq. (1) one would have to say that had its maximum value because E_k (eq. (3)) was zero. The bronzes are therefore metallic, their specific conductivity is proportional to x, and their optical and magnetic properties are typical of metals, with plasma reflectivity edges which move across the visible as the electron concentration varies, and Pauli conduction electron paramagnetism.

The bronzes have three-dimensional lattices but another series of class III compounds which has excited a lot of attention in the last ten years have chains of mixed valence metal atoms. These are the 'molecular metals', of which $K_2Pt(CN)_4Br_{0.30}3H_2O$ (KCP) is the prototype. In this case it is the Br^{-} which are non-stoichiometrically distributed in channels between the chains of square $Pt(CN)_4$ groups, stacked plane to plane. At room temperature the Pt atoms are equally spaced, but their average oxidation state is + 2.3. In becoming oxidized from +2 (d^8) electrons are removed from the z^2 orbitals so KCP is a metallic conductor along the Pt chains. When the electric vector of the incident light is parallel to the chains the optical properties are also those of a metal, but when perpendicular to the chains the crystals are transparent. A complication peculiar to one-dimensional conductors is that in theory they are unstable to lattice distortions which introduce a long wavelength periodicity into the chains. This has the effect of opening a gap at the Fermi surface, rendering the material semiconducting (43).

Not all class III compounds form infinite lattices. In finite clusters we no longer have the continuous spectrum of excitation as found in a metal, but a discrete set which still, however, bears no resemblance to the excitations of related single valence clusters. Good examples are the $M_6X_{12}^{n+}$ clusters with M = Nb, Ta and X - halogen, described in more detail by McCarley in this volume, or the carbonyl clusters described by Dahl. If the clusters are isolated from one another in the solid there cannot be any electron transfer between them, and the solid is an insulator, albeit very intensely coloured as a result of the intracluster transitions. When the interaction between the metal atoms is strong, the valence bond formalism we have used in our classification scheme no longer applies, and a molecular orbital approach is more appropriate. The intracluster electronic transitions are then seen simply as excitations from one MO configuration to another.

Table 3 collects some of the best known examples of each mixed valency class. Many more appear in later chapters.

THE ROLE OF BRIDGING LIGANDS

We have made a case for classifying the properties of mixed valence compounds by means of the parameter α. In turn α is determined by E_k and V (eq. (3)). E_k is presumed to relate to the difference in site potential around the sites of differing valency, and hence to the structure, but what about V? In particular, we have not considered how the presence or absence of bridging ligands affects the strength of interaction between the metal sites, yet it clearly must. In many class II compounds the metal-metal spacing is 5-6 Å, much too large for direct overlap.

Imagine two metal ion sites A and B with orbitals χ_A, χ_B separated by a ligand L whose highest occupied and lowest unoccupied orbitals are ϕ_L, ϕ_L^* (44). For simplicity all the orbitals are non-degenerate. Say the ground state configuration is $\chi_A^2\phi_L^2\chi_B^1$ with a determinental wavefunction

$$\Psi_o = |\chi_A\bar{\chi}_A\phi_L\bar{\phi}_L\chi_B| \tag{10}$$

The intervalence charge transfer configuration $\chi_A\phi_L^2\chi_B^2$ has a wavefunction

$$\Psi_1 = |\chi_A\phi_L\bar{\phi}_L\chi_B\bar{\chi}_B| \tag{11}$$

After interaction with the intervalence configuration

$$\begin{aligned} \Psi_G &= (1-\alpha)^{\frac{1}{2}}\Psi_o + \alpha\Psi_1 \\ \Psi_E &= (1-\beta^2)^{\frac{1}{2}}\Psi_1 + \beta\Psi_o \end{aligned} \tag{12}$$

as in eq. (4). Assuming zero differential overlap, $\langle\Psi_o|\mathcal{H}|\Psi_1\rangle \sim \langle\chi_A|\mathcal{H}|\chi_B\rangle$ which must be close to zero if χ_A and χ_B are widely separated. We know that in practice there has to be an interaction, because intervalence absorption bands have appreciable intensity even when χ_A, χ_B are far apart, it is necessary to

Table 3. Examples of mixed valence classes

Sites very different	Sites similar but distinguishable	Sites identical
$GaCl_2$	Eu_3S_4	Na_xWO_3
Sb_2O_4	Ti_4O_7	$K_2Pt(CN)_4Br_{0.30}3H_2O$
$Cu(en)_2CuBr_2$	$Fe_4[Fe(CN)_6]_3 14H_2O$	$K_{1.74}Pt(CN)_4 3H_2O$
	Cs_2SbCl_6	$Hg_{2.86}AsF_6$
		$Nb_6Cl_{12}^{2+}$
	Fe Fe	
		Gd_2Cl_3

invoke configurations in which electrons have been transferred to or from the bridging ligand. Although such 'local' charge transfer states usually have much higher energies than the intervalence exiited state, they are better able to mix with it because their configurations differ from the intervalence configuration in the occupancy of orbitals on neighbouring atoms. Thus we take transitions $\chi_A \rightarrow \phi_L^*$ and $\phi_L \rightarrow \chi_B$ with wavefunctions

$$\begin{aligned} \Psi_2 &= |\chi_A \phi_L \bar{\phi}_L \bar{\phi}_L^* \chi_B| \\ \Psi_3 &= |\chi_A \bar{\chi}_A \phi_L \chi_B \bar{\chi}_B| \end{aligned} \qquad (13)$$

By second-order perturbation theory we then find

$$\alpha = \sum_{i=2,3} \frac{\langle \Psi_o | \mathcal{H} | \Psi_i \rangle \langle \Psi_1 | \mathcal{H} | \Psi_i \rangle}{(E_1 - E_o)(E_i - E_o)}$$

$$\beta = \sum_{i=2,3} - \frac{\langle \Psi_o | \mathcal{H} | \Psi_i \rangle \langle \Psi_1 | \mathcal{H} | \Psi_i \rangle}{(E_1 - E_o)(E_i - E_1)} \qquad (14)$$

where $E_n = \langle \Psi_n | \mathcal{H} | \Psi_n \rangle$. In practice these expressions have to be evaluated by a mixture of empiricism and theory. The energies of the local charge transfer states can often be identified in the spectrum, either of the mixed valence species itself, or in related single valence species. The off-diagonal matrix elements reduce to metal-ligand resonance integrals. Calculations of this kind yield values of for Prussian Blue and the related dimer $(CN)_5Fe(CN)Fe(CN)_5^{6-}$ of about 0.1, in quite good agreement with the observed intensity of the intervalence band (44). They have also been used to compare a series of aromatic heterocycles as bridging groups in dimeric Ru(II,III) complexes (45). Direct measurement of α would be possible, in principle, by such techniques as polarised neutron scattering which measure spin transfer in the ground state. An attempt to do this with Prussian Blue (46) gave an upper limit of 0.1.

DYNAMICS OF ELECTRON TRANSFER

We have seen that in favourable cases it is

possible to calculate α, but to do so we have to assume some fixed molecular geometry. The model described so far is essentially a static one, taking no account of molecular vibrations. This is a grave defect because, for instance, if the A-L bond length was greater than L-B in the A-L-B example used in the last section, the 'excess' electron would come to rest on A, but if the other way round, then it would be found on B. In such a case we should draw two potential energy surfaces, one for the configuration $\chi_A^2\chi_B^1$ and the other for $\chi_A^1\chi_B^2$. Were the other ligands round A and B to be identical as, for example, in the Creutz-Taube ion $(NH_3)_5Ru(pyrazine)Ru(NH_3)_5^{5+}$, these two surfaces would be identical though displaced along A-L-B vibrational coordinate as in Figure 5 (47). On the other hand if E_k were very large, the minimum of one surface would be so much higher than the other that the vertical charge transfer energy, as observed in the optical spectrum, would be quite a good approximation to E_k. If a transition is to occur between one surface and the other they cannot be orthogonal. We have just discussed the mechanism for mixing the configurations $\chi_A^2\chi_B^1$ and $\chi_A^1\chi_B^2$ via local charge transfer configurations such as $\chi_A^1\phi_L^*\chi_B^1$ but for the present purpose we need only note that the effect of the mixing is to create an 'avoided crossing' at the point where the two surfaces intersect. The smallest vertical separation between the two new surfaces formed in this

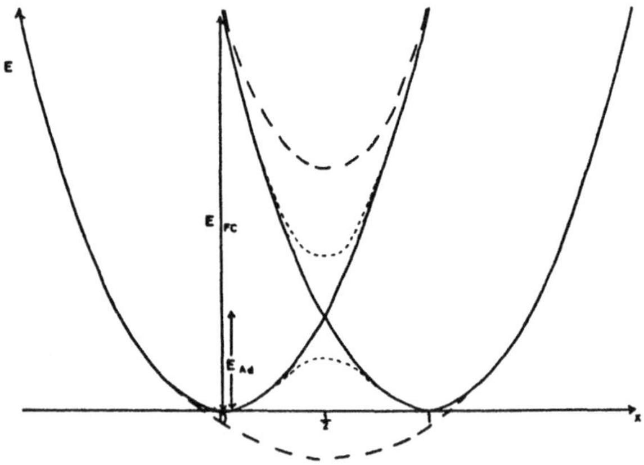

Figure 5. Potential energy surfaces for $\chi_A^1\chi_B^2$ and $\chi_A^2\chi_B^1$. The dotted lines are the surfaces when $V \ll E_{Ad}$ and the dashed lines when $V \gg E_{Ad}$ (47).

way is 2V but now, both V and E_k are functions of the vibrational as well as the electronic coordinates. This point is discussed in detail by Schatz but here we simply make a qualitative remark about the form of the resulting surfaces.

Defining the vibrational energy needed to make the A-L and L-B bond lengths equal as E_{Ad}, we see the ground state potential energy surface can take two limiting forms, depending on whether E_{Ad} is much greater than, or smaller than V. If the former (dotted line in Figure 5) the avoided crossing is a minor effect and minimum potential energy is close to the minima of the two original parabolae. In our nomenclature the compound would be in class II, with different A-L and L-B bond lengths. If $E_{Ad} \ll V$ on the other hand, the ground state potential energy surface would have only a single minimum half way between the two original ones (dashed line in Figure 5). Then the equilibrium ground state geometry is symmetrical, A and B are indistinguishable, and the compound would be in class III. Which of these cases applies to any particular complex is something which, for the moment, must be left to experiment.

CONCLUSION

This introductory chapter has tried to present some of the general features of mixed valency compounds, and to relate their very diverse physical properties to a categorization based on structure. Little detail has been given about any specific substances, or even about the theoretical models. What we have tried to do is to show the range of interest of the subject, to indicate that this widespread class of inorganic compounds poses subtle and demanding problems for theorists as well as experimentalists, and to suggest a qualitative picture which may aid future developments.

REFERENCES

1. H.M. Powell, Proc. Chem. Soc. 73 (1969).

2. A. Werner, Z. anorg. Chem. 12, 46 (1896).

3. E. Zintl and A. Rauch, Ber. Deut. Chem. Ges. 57, 1739 (1924).

4. E.J.W. Verwey, P.J. Haayman, F.C. Romaeijh and G.W. van Oosterhout, Philips. Res. Rep. 5, 173 (1950).

5. E.J.W. Verwey, P.J. Haayman and F.C. Romaeijh, J. Chem. Phys. 15, 181 (1947).

6. D. Culpin, P. Day, P.R. Edwards and R.J.P. Williams, Proc. Chem. Soc. 450 (1965).

7. M.B. Robin and P. Day, Adv. Inorg. Chem. and Radiochem. 10, 247 (1967).

8. N.S. Hush, Prog. Inorg. Chem. 8, 391 (1967).

9. K. Krogmann and H.D. Hausen, Z. anorg. Chem. 358, 67 (1968).

10. F.A. Cotton and T.E. Haas, Inorg. Chem. 3, 10 (1964).

11. L.H. Jensen, Ann. Rev. Biochem. 42, 461 (1974).

12. E.J. Laskowski, R.B. Frankel, W.O. Gillum, G.C. Papaefthymiou, J. Reneau, J. Ibers and R.H. Holm, J. Amer. Chem. Soc. 100, 5322 (1978).

13. G. Smith and R.G.J. Strens, 'Physics and Chemistry of Minerals and Rocks' ed. R.G.J. Strens, New York, Wiley, 1976, p. 583.

14. e.g. R.S. Crandell and B.W. Faughnan, Appl. Phys. Lett. 28, 95 (1976).

15. L. Chalkley, J. Opt. Soc. Amer. 44, 699 (1954).

16. C. Creutz and H. Taube, J. Amer. Chem. Soc. 91, 3988 (1969).

17. J.A. McMillan, Chem. Rev. 62, 65 (1962).

18. M.A. Joly, Compt. rend. 115, 1299 (1892).

19. E. Vogel, Textbook of Qualitative Inorganic Analysis, Longmans.

20. R.J.H. Clark, Ann. N.Y. Acad. Sci. 313, 672 (1978).

21. D.O. Cowan and F. Kaufman, J. Amer. Chem. Soc. 92, 219 (1970).

22. R.R. Heikes and W.D. Johnston, J. Chem. Phys. 26, 582 (1957).

23. S. Kachi, T. Takada and K. Kosuge, J. Phys. Soc. Jap. 18, 1839 (1963).

24. L. Kihlborg, Arkiv. Kemi. 21, 471 (1964).

25. D.C. Johnston, J. Low Temp. Phys. 25, 145 (1976); A.W. Sleight, Chem. Techn. 6, 468 (1976).

26. G.H. Jonker and J.H. van Santen, Physica 16, 337 (1950).

27. H.R. Zeller, Festkorperprobleme, 13, 31 (1973).

28. D.W. Robbins and R.G.J. Strens, Mineral. Mag. 38, 551 (1972).

29. P. Day, Endeavour 29, 45 (1970).

30. I.M. Klotz, G.H. Czerlinski and H.A. Fiess, J. Amer. Chem. Soc. 80, 2920 (1958).

31. B. Rosenberg, Cancer Chemother. Rep. 59, 589 (1975).

32. P. Day, 'Low-dimensional Cooperative Phenomena', ed. H.J. Keller, New York, Plenum Press, 1975, p. 191.

33. P. Day, Inorg. Chem. 2, 452 (1963).

34. S.L. Lawton and R.A. Jacobson, Inorg. Chem. 5, 743 (1966).

35. G. Birke, H.P. Latscha and H. Pritzkow, Z. Naturforsch. 31b, 1285 (1976).

36. T. Barrowcliffe, I.R. Beattie, P. Day and K. Livingstone, J. Chem. Soc. A 1810 (1967).

37. A.Y. Aleksandrov, S.P. Ionov, A.M. Pritchard and V.I. Goldanskii, J.E.T.P. Letters 13, 8 (1971).

38. A.Y. Aleksandrov, S.P. Ionov, D.A. Baltrunas and E.F. Makarov, J.E.T.P. Letters 16, 147 (1972).

39. G. Longworth and P. Day, Inorg. Nucl. Chem. Lett. 12, 451 (1976).

40. L. Atkinson and P. Day, J. Chem. Soc. (A) 2432 (1969).

41. L. Atkinson and P. Day, J. Chem. Soc. (A) 2423 (1969).

42. J.J. Markham, Rev. Mod. Phys. 31, 956 (1959).

43. See chapters in 'Low-dimensional Cooperative Phenomena', ed. H.J. Keller, New York, Plenum Press, 1975.

44. B. Mayoh and P. Day, J.C.S. Dalton Trans. 846 (1973).

45. D. Richardson and H. Taube, unpublished.

46. P. Day, F. Herren, A. Ludi, H.U. Güdel, F. Hulliger and D. Givord, Helv. Chim. Acta, in press.

47. B. Mayoh and P. Day, J. Amer. Chem. Soc. 94, 2885 (1972).

DESCRIPTIVE CHEMISTRY OF MIXED-VALENCE COMPOUNDS

A. Ludi

Institut für Anorganische Chemie, Universität Bern,
CH-3000 Bern 9, Switzerland.

Mixed valence chemistry is illustrated by the presentation of the structures of diverse typical compounds. In a phenomenological description of mixed valence compounds the optical spectra of binuclear complexes are discussed in terms of the model proposed by Hush. Examples of systems showing a wide variation in the range of electronic coupling between the metallic centers are presented. The results of a number of experimental studies of the representative pyrazine bridged ruthenium(II,III) dimer are summarized. Prussian Blue, $Fe_4[Fe(CN)_6]_3 \cdot xH_2O$, representing polynuclear crystalline compounds serves as a second characteristic example to demonstrate the close connection between spectroscopic and magnetic properties on one hand and the mixed valence interaction on the other hand.

INTRODUCTION

A phenomenological approach to define and describe a mixed valence compound simply uses the stoichiometric formula of the corresponding substances as the guiding criterion. The purpose of this chapter is to present a cursory description of mixed valence compounds of the general composition M_kL_ℓ. We introduce the restrictive condition that the ligands L are represented by ions or molecules having closed shell electronic configurations. The polynuclear mixed valence compound can then be considered to consist of classical or Werner complexes as composite structural elements. In other words, we restrict our discussion to those mixed valence systems where the constituent mononuclear units can

D.B. Brown (ed.), Mixed-Valence Compounds, 25-47.

be successfully described in terms of ionic bonding or ligand field theory. Compounds with extended covalent bonding, like the Fe_4S_4-clusters and intermetallic phases, will not be considered in this context. The conventional bookkeeping procedure may then be applied to derive the oxidation states of the k metal ions in M_kL_ℓ. A fractional oxidation state (e.g. $Fe_2(CN)_{11}^{6-}$, Pb_3O_4, Na_xWO_3) or a highly unusual oxidation state (e.g. $Pt(en)_2Br_3$, Rb_2SbCl_6, $CsAuCl_3$) is therefore a clear indication to label a compound as mixed valence. The occurrence of several stable oxidation states is an almost trivial condition for an element to form mixed valence compounds.

In order to study the properties of a substance induced by mixed valence we concentrate our attention on the following system

$$L_nM^M - L_b - M^NL'_m$$

This binuclear unit may exist as a stable dimeric complex, e.g. the Creutz-Taube ion $(NH_3)_5$RuNNRu$(NH_3)_5^{5+}$, or it may represent the essential structural element of a polynuclear crystal, e.g. Fe(III)-NC-Fe(II) in Prussian Blue. The superscripts M and N symbolize the two different oxidation states. Commonly a difference of one unit between M and N is observed, differences of two units are found in the B subgroups and at the end of the transition series exemplified by Pd and Pt compounds. The two coordination units may be identical or different depending on the nature of the bridging ligand L_b and on the nature and geometrical arrangement of the terminal ligands L and L'.

Consequently, the properties of a mixed valence compound can be related to the additive properties of the mononuclear constituents and to the effects originating in the electronic communication between the metal centers across the bridging ligand. The interplay of additive and cooperative effects thus determines the electronic structure of the compound and it is reflected in the Robin-Day classification scheme. The most obvious effect of the metal-metal interaction is the intense color of many mixed valence compounds, a phenomenon which cannot be attributed to a combination of the individual absorption spectra of the mononuclear moieties. Most of the early studies in this area of inorganic chemistry were actually initiated by this truly eye-catching property. Prior to 1967 when two comprehensive reviews were published[1,2] mixed valence compounds had not been studied systematically but were discovered and investigated in a rather haphazard way.

Structural data yields valuable information as to whether the static coordination geometries around the metal ions of a mixed valence compound are equivalent or not. Moreover, crystal and molecular structures of a few representative mixed valence compounds serve to illustrate the wide range of chemical classes covered by these substances. A coarse distinction is made into two groups, a first one comprising small molecular units and a second one consisting of polynuclear crystalline solids. The Creutz-Taube complex (Fig. 1) represents a whole series of binuclear complexes, its crystal structure[3] showing the two ruthenium sites to be equivalent. An example with clearly distinct metal sites is displayed by the structure of $[Mn_2O_2(bipy)_4]^{3+}$ [4] in Fig. 2. The structure of the trinuclear species

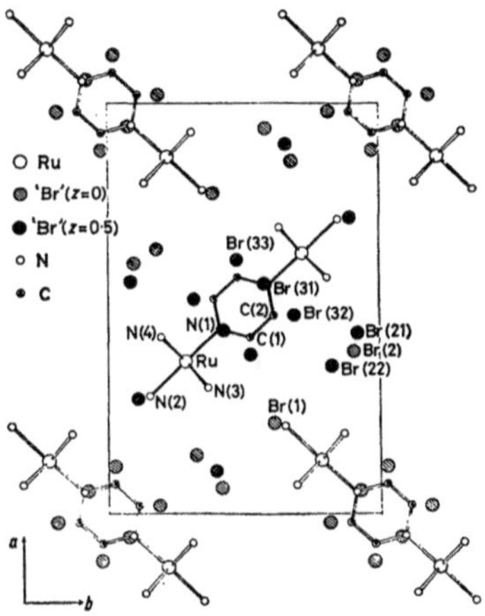

Figure 1. The crystal structure of $[(NH_3)_5RuN\bigcirc NRu(NH_3)_5]Br_{10/3}Cl_{5/3}\cdot 4H_2O$ (ref. 3).

$[Ru_3O_2(en)_2(NH_3)_{10}]Cl_6$, an analog of Ruthenium Red, has two identical terminal metal sites but which are distinct from the central one in the linear Ru-O-Ru-O-Ru arrangement (Fig. 3). A mixed valence complex with all three iridium ions being structurally equivalent is shown in Fig. 4 displaying the structure of the ion $[Ir_3N(SO_4)_6(H_2O)_3]^{4-}$.[6] Optical and Mössbauer data, however, have been interpreted as originating from a Class II mixed valence complex formulated as an Ir(III,IV,IV) salt.[7] Analogous geometries are attributed to various complexes, mostly carboxylates of Ru, Fe, and Cr, with an oxo-ion replacing the central nitrido ion.

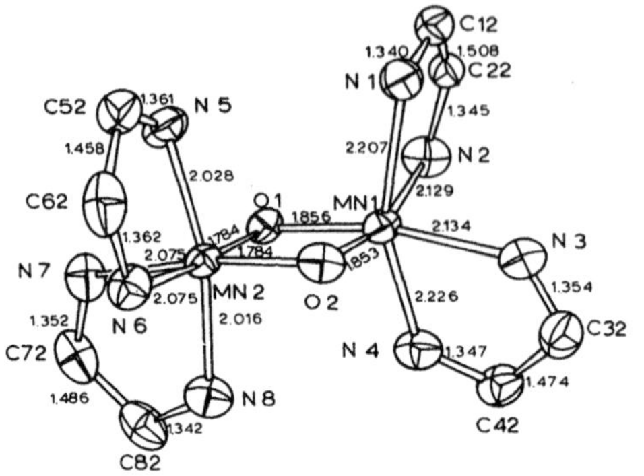

Figure 2. The structure of $[Mn_2O_2(bipy)_4]^{3+}$ (ref. 4).

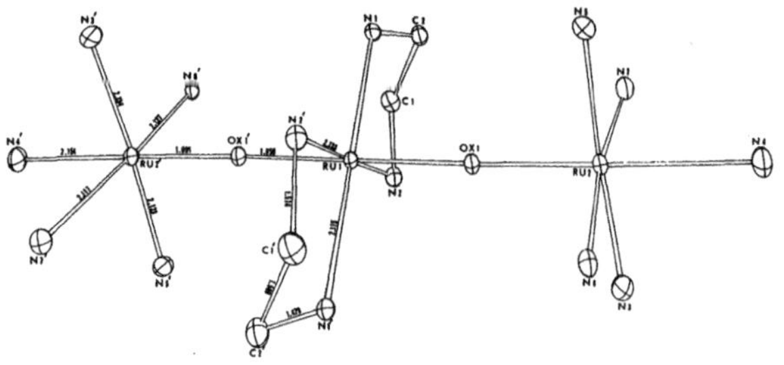

Figure 3. The structure of $[Ru_3O_2(en)_2(NH_3)_{10}]Cl_6$ (ref. 5).

During the past few years there has been considerable interest in the physical and chemical properties of one-dimensional compounds where again mixed valence compounds form an important group.[8] In the first example (Fig. 5), $Pt(NH_3)_2Br_3$ [9], the chain contains equal numbers of Pt(II) and Pt(IV) bridged by Br^- and consists of square planar (Pt(II)) and octahedral (Pt(IV)) coordination units alternating in a regular arrangement. A similar structural principle is realized for $M(I)AuCl_3$,[10] linear $Au(I)Cl_2$ units alternating with square planar $Au(III)Cl_4$ units. Direct metal-metal bonds occur in the structure of KCP, $K_2Pt(CN)_4Br_{0.3}\cdot 3H_2O$, the prototype of the highly anisotropic

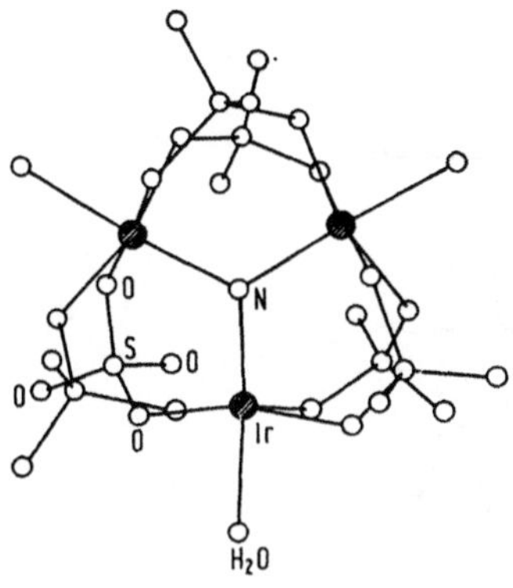

Figure 4. The structure of the ion $[Ir_3N(SO_4)_6(H_2O)_3]^{4-}$ (ref. 6).

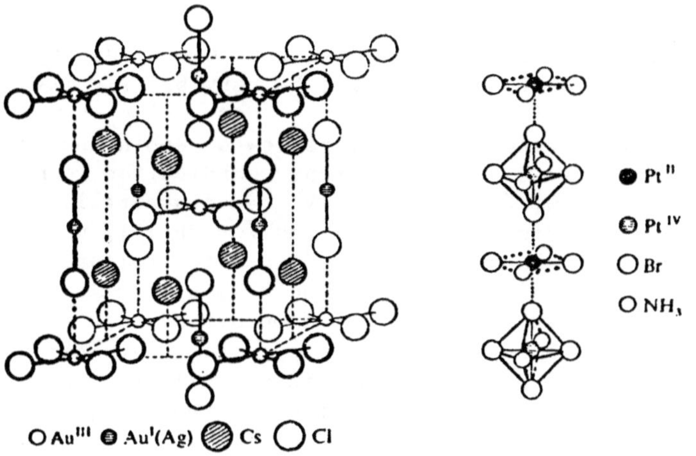

Figure 5. The chain structure of $Pt(NH_3)_2Br_3$ and $MAuCl_3$ (ref. 10).

conductors. The analytical formula does not conform to a simple Pt(II):Pt(IV) ratio but implies a partly oxidized system, common for a variety of cyano and oxalato complexes of Pt. The one-dimensional metallic behavior of these compounds is reflected by the short Pt-Pt distances along the chain, approaching the value found in Pt metal.

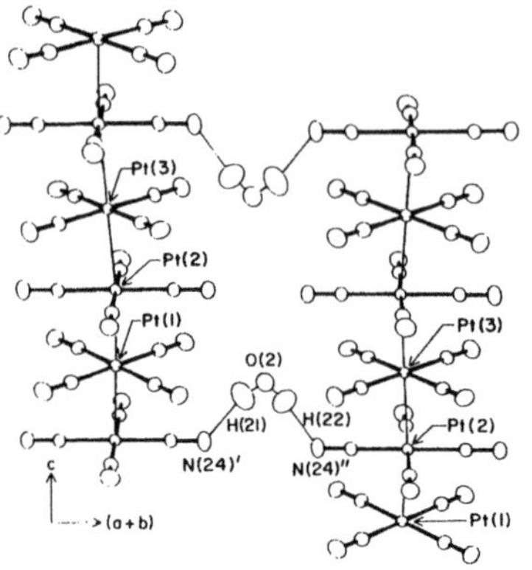

Figure 6. The chain structure of the partly oxidized platinum cyano complex $K_{1.75}Pt(CN)_4 \cdot 1.5H_2O$ (ref. 11).

The vast amount of mixed valence compounds crystallizing in three-dimensional lattices will be very briefly illustrated by a few arbitrarily chosen examples. For a comprehensive survey the reader is referred to the review by Robin and Day.[1] Two distinct metal sites with different coordination geometries are present in Pb_3O_4 (Fig. 7). Pb(IV) is octahedrally coordinated. Threefold coordination is realized for Pb(II), where three oxygen atoms form the basal plane of a distorted trigonal pyramid.[10] Octahedral coordination around both iron ions is observed in the structure of $Fe_2F_5 \cdot 2H_2O$.[12] The composition of the coordination environments, however, is different (Fig. 8). Fe2, the three-valent iron, is surrounded by six F^- while the coordination of the two-valent Fe1 is represented by trans-$FeF_4(H_2O)_2$.

We just skim the large group of mixed valence transition metal oxides by first mentioning the spinels (normal and inverse) Mn_3O_4, Co_3O_4, Fe_3O_4 as representing stoichiometric compounds (Fig. 9). Non-stoichiometric doped oxides are often related to either the NaCl structure, e.g. Li_xNiO or to the perovskite structure, e.g. the tungsten bronzes Na_xWO_3.

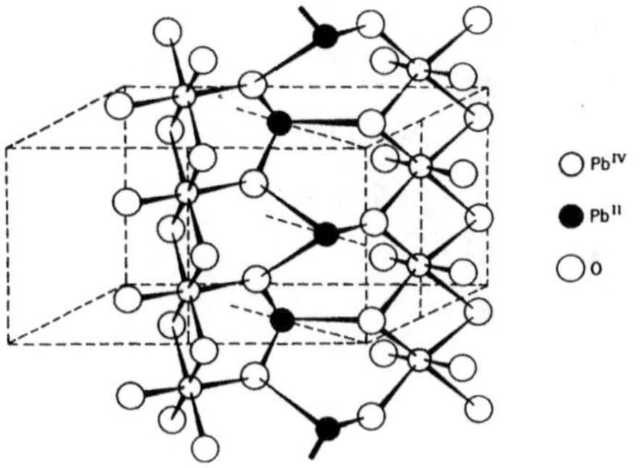

Figure 7. The crystal structure of Pb_3O_4 (ref. 2a).

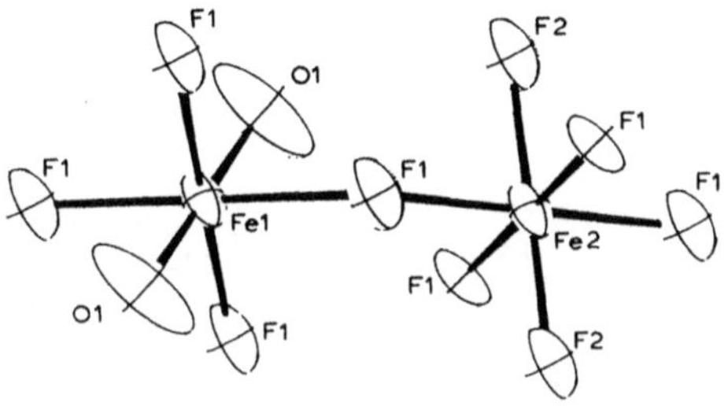

Figure 8. The coordination octahedra in $Fe_2F_5 \cdot 2H_2O$ (Fel: Fe(II), Fe2: Fe(III) (ref. 12).

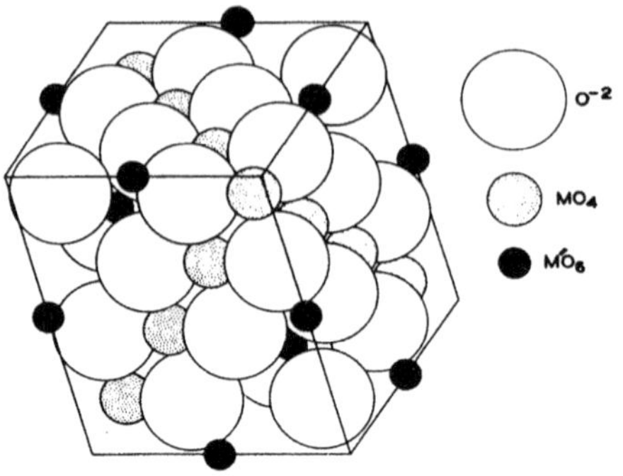

Figure 9. The spinel structure, $M(II)M(III)_2O_4$ (ref. 1).

BINUCLEAR SYSTEMS

In the wake of the first report of the pyrazine-bridged dimer by Creutz and Taube[13] a whole series of binuclear complexes $M\text{-}L_b\text{-}M'$ have been synthesized and described. The following chemical species have been employed as bridging ligands:

Cl^-, O^{2-}, N_2, CN^-

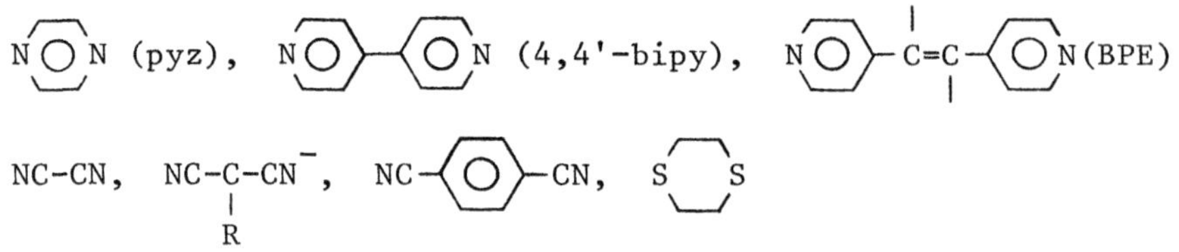

Prominent examples for the mononuclear moieties are represented by

$(NH_3)_5Ru(II,III)$[14], $(NH_3)_5Os(II,III)$[14], $(CN)_5Fe(II,III)$[15],

$(bipy)_2Mn(III,IV)$[16], $(bipy)_2ClRu(II,III)$[17].

We notice that the most frequent combination of oxidation states in these binuclear complexes is the [II-III] combination corresponding to the low-spin d^6-d^5 configurations.

Binuclear mixed valence complexes have attracted considerable attention for two reasons. First, they evidently represent the most simple mixed valence compounds to study the effects of electronic communication between the two metallic centers in a rather simple chemical environment. Secondly, they are the stable intermediates occurring in electron transfer processes of the inner sphere type.[14] Their synthesis can be achieved by direct combination of two appropriate mononuclear complexes provided one of them has a substitution labile coordination site like $(NH_3)_5RuH_2O^{2+}$ or $(CN)_5FeNH_3^{3-}$. Another preparative route is via one-electron reduction (oxidation) of the fully oxidized (reduced) binuclear complex. Finally the mixed valence species is generated by the following equilibrium reaction

$$M(III)\text{-}L_b\text{-}M(III) + M(II)\text{-}L_b\text{-}M(II) \rightleftarrows 2\ M(III)\text{-}L_b\text{-}M(II)$$

This equation which is usually written in a shorthand notation as

$$[3,3] + [2,2] \rightleftarrows 2\ [2,3]\ ;\quad K_c$$

is termed comproportionation equilibrium. Its constant K_c defines the stability of the mixed valence system with regard to its reduced and oxidized form. Conventionally the value of K_c is derived from measurements of the two reduction potentials, e.g. by cyclic voltammetry

$$[3,3] + e^- \rightarrow [2,3]\ ,\quad E_1$$
$$[2,3] + e^- \rightarrow [2,2]\ ,\quad E_2$$

For small E_1-E_2 differences cyclic voltammetry cannot be employed to produce reliable data. A spectrophotometric technique has been used successfully for such systems.[18] A wide spectrum of values of this constant has been reported:

	K_c	Ref.
$(CN)_5Fe\text{-}BPE\text{-}Fe(CN)_5^{5-}$	~4	15
$(NH_3)_5Ru\text{-}4,4'bipy\text{-}Ru(NH_3)_5^{5+}$	24	18
$(NH_3)_5Ru\text{-}pyz\text{-}Ru(NH_3)_5^{5+}$	$4x10^6$	14
$(NH_3)_5Ru\text{-}NC\text{-}C(R)\text{-}CN\text{-}Ru(NH_3)_5^{4+}$	10^{10}	14
$(phen)_2Mn(O)_2Mn(phen)_2^{3+}$	$4x10^{15}$	16
$(NH_3)_5Os\text{-}N_2\text{-}Os(NH_3)_5^{5+}$	10^{20}	14

For the preparative chemist the numerical value of K_c is a crucial factor for it determines the feasibility of isolating a pure solid mixed valence compound from its solution. Moreover, it has to be taken into account for the measurements of the extinction coefficients of optical absorption bands.

Intervalence Absorption

We have described an absorption band in the low energy region of the optical spectrum as the most conspicuous property of mixed valence compounds. This absorption, labelled intervalence or mixed valence band, is not observed in the spectra of mononuclear moieties and is attributed to an electron transfer from one metal site to the other.

$$[2,3] \xrightarrow{\nu_{IT}} [3,2]^*$$

This process generates a configuration [3,2]* with M(III) in the coordination environment of M(II) and vice versa for M(II). The excited species converts to the ground state configuration by vibrational relaxation. Selected data of intervalence spectra for symmetric as well as unsymmetric systems are presented in Table 1.

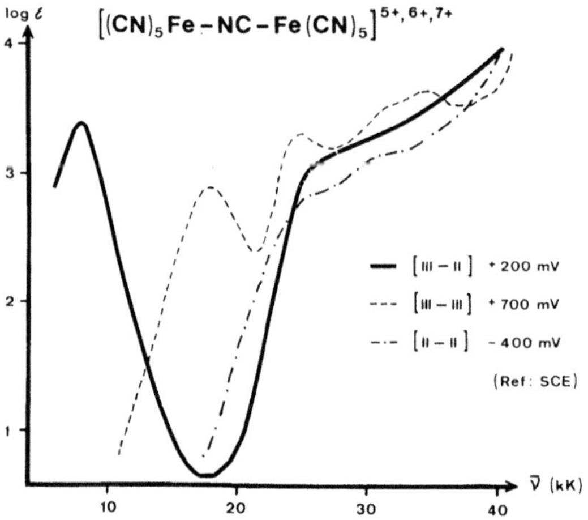

Figure 10. The absorption spectrum of $Fe_2(CN)_{11}^{n-}$ (ref. 24).

The intervalence absorption process has been discussed in a qualitative model proposed by Hush.[2b] Fig. 11 summarizes the essential features of this approach for a symmetric mixed valence species with weak metal-metal interaction (Class II). Assuming the harmonic approximation the two curves depict the potential energy as a function of one configurational coordinate. The corresponding vibrational mode may be envisaged as a combination of the metal-ligand stretching vibrations of the two mononuclear

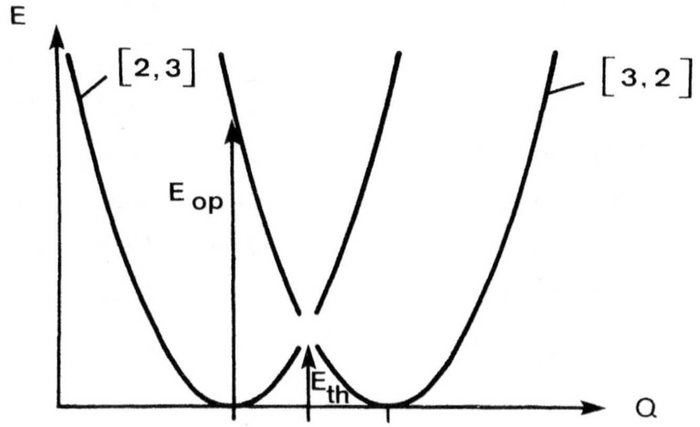

Figure 11. Potential energy diagram for intervalence absorption.

moieties. This model which has been recently refined and extended[19] has found widespread application also in the discussion of electron transfer processes.[2,20]

Applying the Franck-Condon principle, the vertical process indicated by E_{op} corresponds to the intervalence absorption. The energy barrier for thermal electron transfer is then $E_{th} = (1/4)E_{op}$. Since binuclear mixed valence systems are usually studied in solution the vertical energy E_{op} cannot be equated with the Franck-Condon term (E_{in}) but it contains also an energy E_{out} associated with the reorganization of the solvent environment to adapt to the electron transferred state.

$$E_{op} = E_{in} + E_{out} \tag{1}$$

From the solvent dependence of the second term[2b]

$$E_{out} = e^2 \left(\frac{1}{2a_1} + \frac{1}{2a_2} - \frac{1}{d}\right)\left(\frac{1}{n^2} - \frac{1}{D}\right) \tag{2}$$

the contribution E_{in} can be estimated. a_1 and a_2 are the radii of the two coordination spheres, d is the metal-metal distance, n and D are the refractive index and the static dielectric constant, respectively, of the solvent. Both the solvent and distance dependence of ν_{IT} have been investigated and a satisfactory agreement has been found.[27,28]

Table 1. Band Maxima ν_{IT} and Half-widths Δ for Intervalence Absorption in Binuclear Complexes (cm^{-1}).

Complex	ν_{IT}	Δ_{obs}	Δ_{cal}	Ref.
$(NH_3)_5Ru\text{-pyz-}Ru(NH_3)_5^{5+}$	6370	1500	3840	13
$(NH_3)_5Ru\text{-4,4'-bipy-}Ru(NH_3)_5^{5+}$	9700	5200	4730	18
$(NH_3)_5Ru\text{-pyz-}RuCl(bipy)_2^{4+}$	10400	5800	4900	21
$(bipy)_2ClRu\text{-pyz-}RuCl(bipy)_2^{3+}$	7700	4900	4220	22
$(NH_3)_5Ru\text{-NC-CN-}Ru(NH_3)_5^{5+}$	7000	1610	4020	23
$(bipy)_2Mn(\mu\text{-}O)_2Mn(bipy)_2^{3+}$	12000	4600	5260	16
$(CN)_5Fe\text{-pyz-}Fe(CN)_5^{5-}$	8300	4800	4400	15
$(CN)_5Fe\text{-CN-}Fe(CN)_5^{6-}$	7700	5100	4220	24
$(NH_3)_5Ru\text{-NC-}Fe(CN)_5^{-}$	10400	4600	4900	25
$(NH_3)_5Ru\text{-NC-}Os(CN)_5^{-}$	12600	6500	5400	25
Biferrocene	5260	3700	3500	26

A common feature of most intervalence bands is their large band width, typically around 5000 cm^{-1} (cf. Table 1). Within the configurational coordinate approach[29] the band half-width Δ is related to the energy $h\nu$ of the vibrational quantum participating in the Franck-Condon process according to

$$\Delta(T) = \Delta(0)\{\coth \frac{h\nu}{2kT}\}^{1/2} \tag{3}$$

$\Delta(T)$ and $\Delta(0)$ represent the half-width of the absorption band at T K and 0 K, the latter value being extrapolated from low temperature measurements. For the high temperature limit defined by $h\nu \ll kT$ the simple relation

$$\nu_{IT} = \frac{\Delta^2}{2310} \tag{4}$$

results for T = 300 K.[2b] An agreement, to about 10%, between Δ_{cal} and Δ_{obs} (Table 1) is usually taken as an indication that the Hush model is a satisfactory description of a mixed valence system. Since the relevant metal-ligand stretching vibrations are typically in the range 350-500 cm^{-1} the criterion for the high-temperature limit is obviously not valid. For the low temperature limit[29] - an equally poor approximation for actual vibrational frequencies - a similar agreement may be obtained. For $h\nu$ = 420 cm^{-1} the result is $\nu_{IT} = \Delta^2/2330$. Obviously neither limit can be considered to offer a good description. Rather the temperature dependence of Δ has to be measured to obtain an estimation of the vibrational energy involved in the intervalence transfer. Fig. 12 shows the data for the intervalence band of $(CN)_5Fe\text{-pyz-}Fe(CN)_5^{5-}$.[15] The best fit corresponds to $h\nu = 490\ cm^{-1}$. Clearly this figure is only a rough estimate but lies within the range of usual vibrational energies for metal-ligand stretching modes. Resonance Raman spectroscopy[30] is a powerful and elegant technique to determine this vibrational mode provided the intervalence band occurs in an accessible spectral range.

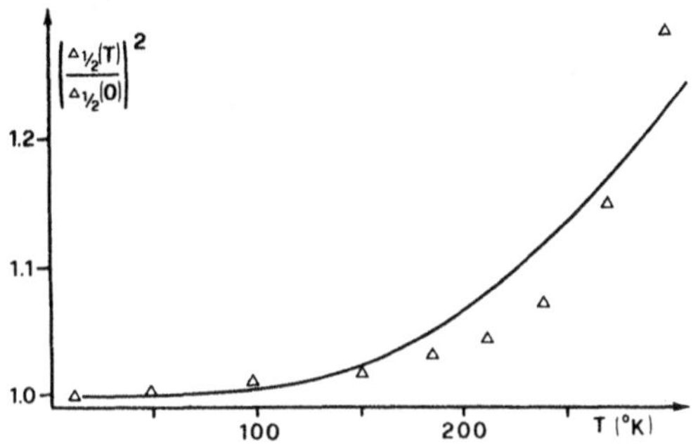

Figure 12. Temperature dependence of the half-width of the intervalence band of $(CN)_5Fe\text{-pyz-}Fe(CN)_5^{5-}$; solid line: theoretical curve for $h\nu = 490\ cm^{-1}$ (ref. 15).

The thermal energy of activation estimated by $E_{th} = E_{op}/4$ has been used to calculate the rate of electron transfer. A satisfactory agreement with corresponding data measured for closely related systems has been obtained.[26,27]

The integral intensity of the intervalence absorption can be used for an approximate calculation of the valence delocalization parameter α according to[1,2b]

$$\alpha^2 = \frac{4.24\cdot 10^{-4}\varepsilon_{max}\cdot\Delta}{\nu_{IT}\cdot d^2} \qquad (5)$$

ε_{max} and Δ are the extinction coefficient and the half-width respectively, of the intervalence band peaking at energy ν_{IT}, d is the metal-metal-distance. This equation assumes a Gaussian shape for the intervalence band. For the complexes listed in Table 1 the mixed valence parameter α is generally between 0.01 and 0.1.

The simple energy-coordinate diagram of Fig. 11 has to be modified to apply to a system with intermediate or strong coupling (Fig. 13). Within this model the classification of a compound

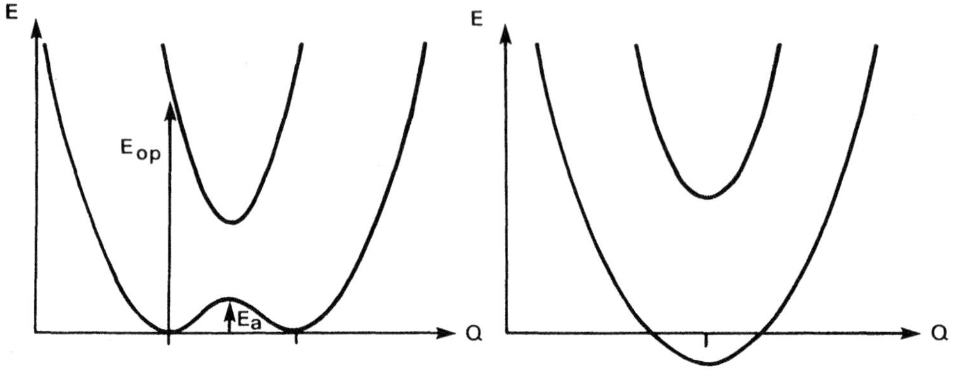

Figure 13. Energy diagram for intermediate and strong coupling.

as Class II or III is primarily determined by the magnitude of E_a in comparison to kT and hν. For the past decade many arguments have been collected to label the prototype of the binuclear systems, the Creutz-Taube complex, as either trapped valence or delocalized. Suffice it in this introductory chapter to present a brief summary of the essential studies:

Experimental technique	Assignment favored
UV-VIS-spectroscopy	localized[13]
Near infrared, solvent dependence, Δ	delocalized[13,14]
IR, Raman	controversial[13,31]
Mössbauer at 4.2 K	localized[32]
ESCA	controversial[33,34]
^{1}H NMR	delocalized[35]
EPR	localized[36]
Crystal structure	delocalized[3]
Model calculations	localized[37], borderline case[19]

This on first sight quite confusing situation will be understandable if we take the time scale of the specific spectroscopic technique into consideration to decide whether an electron transfer [2,3] ⇄ [3,2] is rapid or slow. The wide range of time scales (~10^{-5} to 10^{-14} sec) demonstrates that the issue localized versus delocalized has to be investigated in full cognizance of the limits imposed by the time scale of the particular experimental technique, a perhaps trivial albeit significant point.

PRUSSIAN BLUE

There doubtless does not exist a coordination compound with such a long history as Prussian Blue, $Fe_4[Fe(CN)_6]_3 \cdot xH_2O$ (x = 14-16), the earliest paper appearing at the beginning of the 18th century.[38] It also can be considered to be the prototype of mixed valence chemistry. A variety of studies concerning the assignment of oxidation states have been carried out; for a compilation of the literature see ref. 39. The long controversy whether this blue pigment is Turnbull's Blue, iron(II)hexacyanoferrate(III), or Prussian Blue, iron(III)hexacyanoferrate(II), has been unambiguously decided by a variety of physical measurements.[39] Regardless of the combination of oxidation states of the starting complexes, Fe^{z+}_{aq} and $Fe(CN)_6^{n-}$, the solid precipitate invariably is Prussian Blue, i.e. it contains the sequence Fe(III)-NC-Fe(II).[40]

The essential structural features of Prussian Blue have been derived from X-ray powder data more than forty years ago.[41] In a cubic face-centered lattice the metal sites are occupied alternatively by iron(II) and iron(III) bridged by cyanide in a regular array to produce a three-dimensional cubic network. Owing to its extremely low solubility, a common property of polynuclear transition metal cyanides, Prussian Blue has been studied only as a powder sample. The growth of single crystals suitable for X-ray work was achieved only rather recently.[40] The crystallographic investigation confirmed the cubic polymeric structure but revealed that 25% of the Fe(II) sites are vacant leading to an inherent structural disorder. Prussian Blue contains low-spin Fe(II) in C_6 octahedra and high-spin Fe(III) in a mixed nitrogen(NC)-oxygen(H_2O) environment of average composition $Fe(III)N_{4.5}O_{1.5}$. The structural role of the water molecule and the hydrogen bonds have been further elucidated by powder neutron diffraction.[42]

Robin[43] has assigned the visible absorption band responsible for the blue color to an intervalence transition. Since many of the properties typical for octahedral Fe(II) and Fe(III) have been detected in Prussian Blue by a variety of physical techniques this iron cyanide represents a typical Class II compound. The mixed valence behavior can be best described by considering the isolated cluster $Fe(II)(CN)_6Fe(III)_6$. The ground state is then given by

$$\psi_G = (1-\alpha^2)^{1/2}\psi_o + \alpha\psi_1 \tag{6}$$

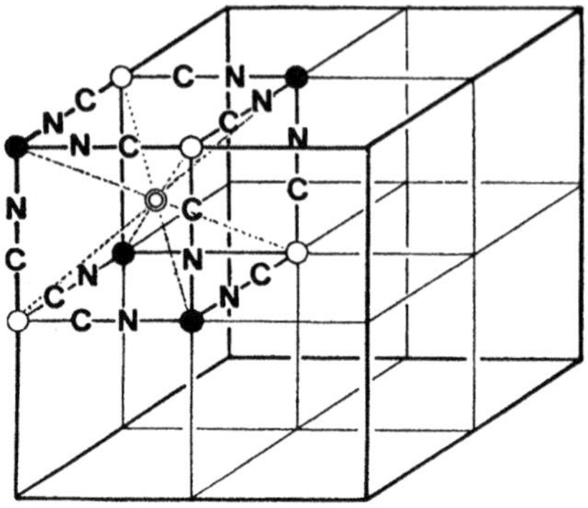

Figure 14. The unit cell of Prussian Blue; ○ Fe(II), "C-hole"; ● Fe(III), "(N,O)-hole".

In a shorthand notation we have $\psi_o = [C^6, N^{30}]$, symmetry A_{1g}, with C^i, N^j symbolizing the number of d-electrons in a C- and (N,O)-environment, respectively. Transfer of one electron from Fe(II) to Fe(III) produces $[C^5, N^{31}]$. This configuration leads to a state ψ_1 with symmetry A_{1g} which can mix with ψ_o and also to states ψ_i with symmetry T_{1u}. The intervalence band arises thus from the allowed transition

$$\psi_G(A_{1g}) \rightarrow \psi_i(T_{1u}).$$

Similar absorption bands attributed to homonuclear as well as heteronuclear intervalence transfer have been observed in a few other polynuclear transition metal cyanides.[44]

The temperature dependence of the intervalence band of Prussian Blue has also been evaluated using equation (3). The best fit of the observed half-widths to the theoretical curve corresponds to a value of 430 cm^{-1} for the vibrational mode associated with the intervalence transition. Again this frequency agrees with metal-ligand stretching modes of the corresponding mononuclear moieties.

The weak metal-metal interaction, typical for Class II behavior, is also responsible for the semiconductivity of Prussian

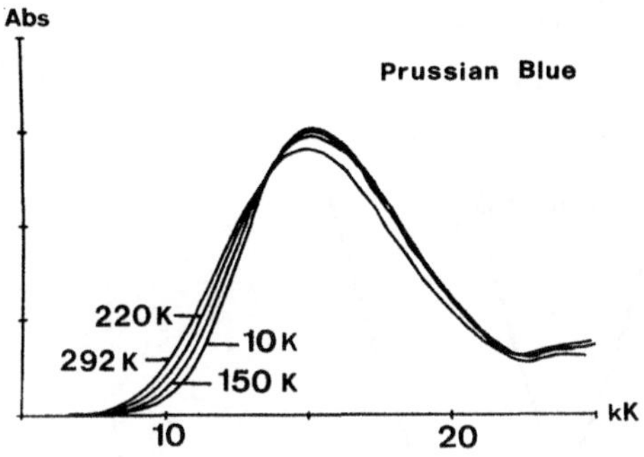

Figure 15. The intervalence band of Prussian Blue.

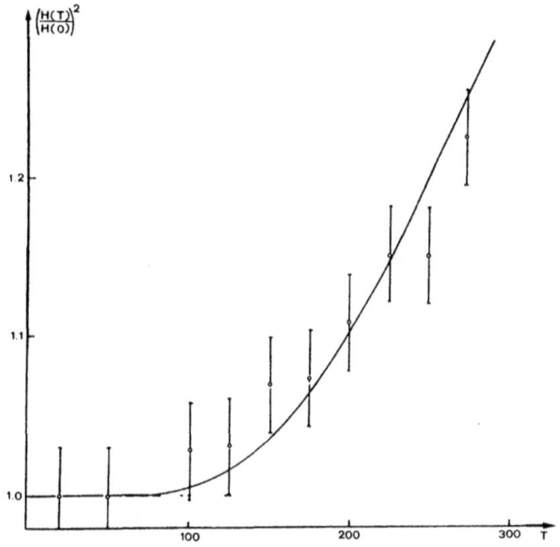

Figure 16. Temperature dependence of the half-width of the intervalence band of Prussian Blue; solid line: theoretical curve for $h\nu$ = 430 cm^{-1}.

Blue. In the Hush-diagram of Fig. 17, modified to account for the unsymmetric arrangement in Prussian Blue, E_{th} serves as a measure for the activation energy of conduction. Within the high-temperature limit (see above) the following simple relation has been derived[2b]

$$E_{th} = \frac{E_{op}^2}{4(E_{op}-E_o)} \tag{7}$$

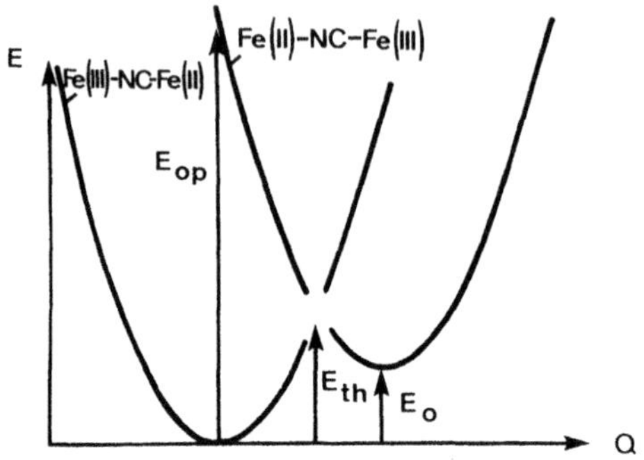

Figure 17. Potential energy-coordinate diagram for unsymmetric mixed valence systems.

A qualitative agreement was found between the values of E_{th} deduced from optical measurements and the observed activation energy of conduction, E_a.[45] Changes in the structure induced by dehydration from evacuation of the pressed powders and surface water films, however, have to be considered as severe limitations in measuring electric conductivities of Prussian Blue. Such difficulties were avoided in the studies of mixed valence halides where E_{th} from optical studies has been found to be consistent with the activation energy E_a of conduction.[46]

Table 2. Observed and Calculated Activation Energies (eV)[46]

	E_{op}(obs)	$E_{th}=\frac{1}{4}E_{op}$	E_a
Cs_2SbCl_6	2.30	0.57	0.76
Rb_2SbCl_6	2.43	0.61	0.87
Chlorocuprates(I,II)	2.18	0.55	0.73

Turning now to the magnetic properties of Prussian Blue we notice that the effective magnetic moment corresponds to high-spin Fe(III) and diamagnetic low-spin Fe(II).[47] Between 10 and 300 K perfect Curie-Weiss characteristics are observed corresponding to μ_{eff} = 5.98(2) B.M. per Fe(III). Magnetization measurements and magnetically perturbed Mössbauer spectra reveal ferromagnetic behavior at temperatures below a Curie point T_c of 5.5 ± 0.5 K.[48] The magnetic order has been further investigated by means of neutron diffraction. Magnetic intensity arising from

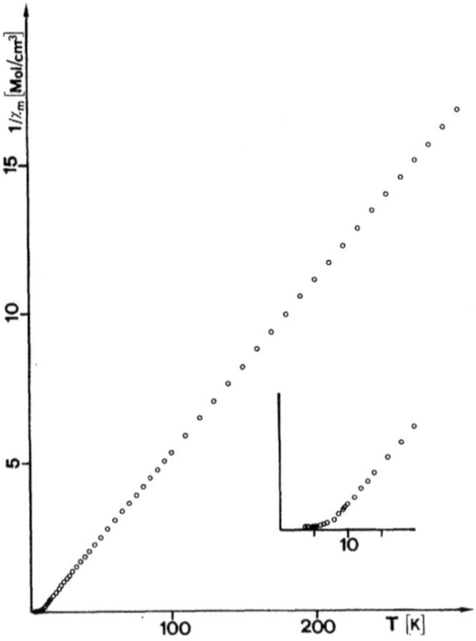

Figure 18. Magnetic susceptibility of Prussian Blue.

the magnetic moment of the neutron is superimposed on the nuclear Bragg reflections for Prussian Blue in its ferromagnetic state. The temperature dependence of this magnetic contribution follows in very good agreement the theoretical curve predicted by molecular field theory for S = 5/2. The resulting Curie temperature is T_c = 5.6 ± 0.1 K.[42] Ferromagnetism obviously requires a long range order of the spins of the Fe(III) ions. The two shortest Fe(III)-Fe(III) distances in the structure of Prussian Blue are 7.2 Å through space and 10.2 Å along the sequence Fe(III)-NC-Fe(II)-CN-Fe(III). A superexchange mechanism along this sequence is assumed to be the most plausible mechanism, the mixed valence interaction leading to a small spin transfer from Fe(III) to Fe(II)[49] which, to a first approximation, is diamagnetic. The mixed valence coefficient α and this small spin density at the Fe(II) site are thus closely related. An estimate of α has been made from the intensity of the intervalence band, measured in a colloidal suspension of Prussian Blue.[43] The result, $\alpha \simeq 0.1$, implies a localization of 99% for the optical electron at the Fe(II) sites. Calculations using a perturbation model predicted

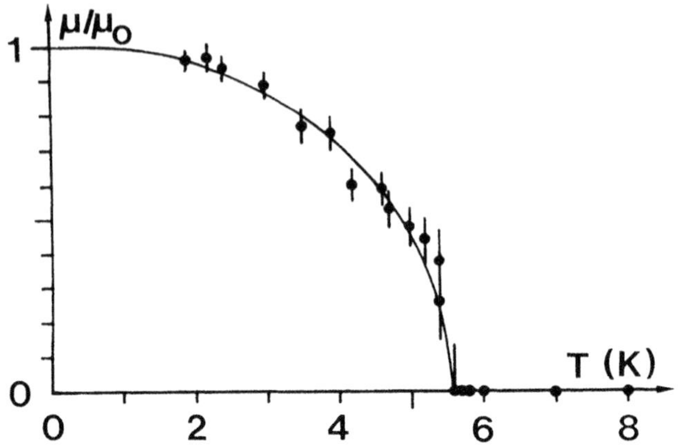

Figure 19. Temperature dependence of the ordered magnetic moment of Prussian Blue.

a value of 0.105 for α.[50] With these figures the Curie temperature of Prussian Blue has been calculated to be on the order of 10 K[49] in good agreement with the observed value.

A direct method, i.e. independent of approximations and assumptions of a given theoretical model, to determine spin densities in ferromagnetic solids is polarized neutron diffraction. Since Prussian Blue orders ferromagnetically at low temperatures spin density measurements are equivalent to a direct determination of the mixed valence parameter α. Such experiments on a powder sample have been evaluated to give an upper limit of about 5% for the spin delocalization in the electronic ground state of Prussian Blue.[51]

FINAL REMARKS

Our phenomenological presentation of mixed valence chemistry evidently has been quite sketchy and therefore selective. It certainly became apparent, however, that the scope of this particular branch of inorganic chemistry is broad indeed and offers many challenges for preparatively as well as theoretically inclined chemists. We just allude to three major topics of mixed valence chemistry as areas for future fruitful scientific activity.

So far the only really systematic investigation of binuclear complexes was confined to ruthenium compounds.[14] Homo- and heteronuclear systems containing other metals certainly deserve further attention. A methodical variation of the bridging ligand is expected to tune binuclear systems from the firmly trapped to the completely delocalized systems as has been demonstrated for specific examples by Taube.[14] The extension and refinement of theoretical models to explicitly include the specific properties of the bridging ligand has been started[19,50] and hopefully will make further advances.

There still exists a wide gap between small isolated molecular units like bi- and trinuclear complexes and crystalline solids with extended metal-metal-interactions. In this respect a first few encouraging steps have been made to stimulate further activity:

- μ-Pyrazine complexes up to six ruthenium centers[52]
- Ferrocene-ferricenium polymers[53]
- Ligand-bridged $[Ru_3O(OAc)_6L_3]$-clusters[54]

We may consider mixed valence chemistry as the conceptual link between coordination chemistry and solid state physics.

Finally, the concept of mixed valence chemistry offers a pathway to design and actually synthesize new compounds with certain optical, electric, and magnetic properties. Materials research in its broadest sense always depends on the conception and synthesis of interesting novel compounds.

References

(1) M.B. Robin and P. Day, Adv. Inorg. Chem. Radiochem. 10, 247 (1967).
(2) (a) G. Allen and N.S. Hush, Prog. Inorg. Chem. 8, 357 (1967). (b) N.S. Hush, Prog. Inorg. Chem. 8, 391 (1967).
(3) J.K. Beattie, N.S. Hush, P.R. Taylor, C.L. Raston, and A.H. White, J. Chem. Soc. Dalton 1977, 1121.
(4) P.M. Plaksin, R.C. Stoufer, M. Mathew, and G.J. Palenik, J. Am. Chem. Soc. 94, 2121 (1972).
(5) P.M. Smith, T. Fealey, J.E. Earley, and J.V. Silverton, Inorg. Chem. 10, 1943 (1971).
(6) M. Ciechanowicz, W.P. Griffith, D. Pawson, A.C. Skapski, and M.J. Cleare, Chem. Comm. 1971, 876.
(7) D.B. Brown, M.B. Robin, J.D.E. McIntyre, and W.F. Peck, Inorg. Chem. 9, 2315 (1970).
(8) J.S. Miller and A.J. Epstein, Prog. Inorg. Chem. 20, 1 (1976).

(9) J. Wallén, C. Brosset, and N.G. Vannerberg, Arkiv Kemi 18, 541 (1962).
(10) A.F. Wells, "Structural Inorganic Chemistry", Clarendon Press, Oxford, 1975.
(11) J.M. Williams, K.D. Keefer, D.M. Washecheck, and N.P. Enright, Inorg. Chem. 15, 2446 (1976).
(12) W. Hall, S. Kim, J. Zubieta, E.G. Walton, and D.B. Brown, Inorg. Chem. 16, 1884 (1977).
(13) C. Creutz and H. Taube, J. Am. Chem. Soc. 91, 3988 (1969); 95, 1086 (1973).
(14) H. Taube in "Synthesis and Properties of Low-Dimensional Materials", J.S. Miller and A.J. Epstein, eds., New York Academy of Sciences, Vol. 313, 1978, pp 481-495.
(15) F. Felix and A. Ludi, Inorg. Chem. 17, 1782 (1978).
(16) S.R. Cooper and M. Calvin, J. Am. Chem. Soc. 99, 6623 (1977).
(17) T.J. Meyer, Acc. Chem. Res. 11, 94 (1978).
(18) J.E. Sutton, P.M. Sutton, and H. Taube, Inorg. Chem. 18, 1017 (1979).
(19) S.B. Piepho, E.R. Krausz, and P.N. Schatz, J. Am. Chem. Soc. 100, 2996 (1978); 101, 2793 (1979).
(20) R.A. Marcus, Annu. Rev. Phys. Chem. 15, 155 (1964).
(21) R.W. Callahan, G.M. Brown, and T.J. Meyer, Inorg. Chem. 14, 1443 (1975).
(22) R.W. Callahan, F.R. Keene, T.J. Meyer, and D.J. Salmon, J. Am. Chem. Soc. 99, 1064 (1977).
(23) G.M. Tom and H. Taube, J. Am. Chem. Soc. 97, 5310 (1975).
(24) R. Glauser, U. Hauser, F. Herren, A. Ludi, P. Roder, E. Schmidt, H. Siegenthaler, and F. Wenk, J. Am. Chem. Soc. 95, 8457 (1973).
(25) A. Ludi, Chimia 26, 647 (1972).
(26) D.O. Cowan, C. Le Vanda, J. Park, and F. Kaufmann, Acc. Chem. Res. 6, 1 (1973).
(27) M.J. Powers, D.J. Salmon, R.W. Callahan, and T.J. Meyer, J. Am. Chem. Soc. 98, 6731 (1976).
(28) C. Creutz, Inorg. Chem. 17, 3723 (1978).
(29) B. Di Bartolo, "Optical Interactions in Solids", J. Wiley & Sons, Inc., New York, 1968.
(30) J.R. Campbell, R.J.H. Clark, and P.C. Turtle, Inorg. Chem. 17, 3622 (1978).
(31) T.C. Strekas and T.G. Spiro, Inorg. Chem. 15, 974 (1976).
(32) C. Creutz, M.L. Good, and S. Chandra, Inorg. Nucl. Chem. Lett. 9, 171 (1973).
(33) P. Citrin, J. Am. Chem. Soc. 95, 6472 (1973).
(34) J.K. Beattie, N.S. Hush, P.R. Taylor, Inorg. Chem. 15, 992 (1976).
(35) J. Elias and R.S. Drago, Inorg. Chem. 11, 415 (1972).
(36) B.C. Bunker, R.S. Drago, D.N. Hendrickson, R.M. Richman, and

S.L. Kessell, J. Am. Chem. Soc. 100, 3805 (1978).
(37) B. Mayoh and P. Day, J. Am. Chem. Soc. 94, 2885 (1972); Inorg. Chem. 13, 2273 (1974).
(38) "Miscellanea Berolinensia ad incrementum scientiarium", Berlin, 1710, pp. 377-379.
(39) A.G. Sharpe, "The Chemistry of Cyano Complexes of the Transition Metals", Academic Press, New York, N.Y. 1976, pp 99-126.
(40) H.J. Buser, D. Schwarzenbach, W. Petter, and A. Ludi, Inorg. Chem. 16, 2704 (1977).
(41) J.F. Keggin and F.D. Miles, Nature 137, 577 (1936).
(42) F. Herren, P. Fischer, A. Ludi, and W. Hälg, submitted for publication.
(43) M.B. Robin, Inorg. Chem. 1, 337 (1962).
(44) A. Ludi and H.U. Güdel, Struct. Bonding (Berlin) 14, 1 (1973).
(45) Y. Hara and S. Minomura, J. Chem. Phys. 61, 5339 (1975).
(46) L. Atkinson and P. Day, J. Chem. Soc. A, 1969, 2432.
(47) D. Davidson and L.A. Welo, J. Phys. Chem. 32, 1191 (1928).
(48) A. Ito, M. Suenaga, and K. Ono, J. Chem. Phys. 48, 3597 (1968).
(49) B. Mayoh and P. Day, J. Chem. Soc. Dalton 1976, 1483.
(50) B. Mayoh and P. Day, J. Chem. Soc. Dalton 1974, 846.
(51) P. Day, F. Herren, A. Ludi, H.U. Güdel, F. Hulliger, and D. Givord, submitted for publication.
(52) A. von Kameke, G.M. Tom, and H. Taube, Inorg. Chem. 17, 1790 (1978).
(53) D.O. Cowan, J. Park, C.U. Pittman, Y. Sasaki, T.K. Mukherjee, and N.A. Diamond, J. Am. Chem. Soc. 94, 5110 (1972).
(54) J.A. Baumann, S.T. Wilson, D.J. Salmon, P.L. Hood, and T.J. Meyer, J. Am. Chem. Soc. 101, 2916 (1979).

SYNTHETIC APPROACHES TO MIXED-VALENCE CHEMISTRY

David B. Brown and James T. Wrobleski*

Department of Chemistry, University of Vermont,
Burlington, Vermont 05405, USA

ABSTRACT

Several preparative routes to mixed-valence materials have been examined and the effect of the synthetic variables discussed. It has been shown that solubility is a dominant factor in the successful isolation of many such materials. Soluble metal cluster complexes are susceptible to sequential oxidation/reduction reactions which lead to mixed-valence clusters, and the importance and utility of electrochemical measurements in the preparation and understanding of these clusters is discussed. The preparation of mixed-valence polymers by solid-state reactions and by polymer growth techniques has been examined, and the types of behavior exhibited are shown to be explicable using arguments derived from the thermodynamics of solutions.

INTRODUCTION

The history of mixed-valence chemistry dates to the beginnings of coordination chemistry, since the first reported synthesis of a coordination compound involved a mixed-valence material. In 1704 Diesbach of Berlin, variously described as a chemist and an artist, prepared Prussian Blue by heating animal refuse and sodium carbonate together in an iron pot (1). A detailed preparation of Prussian Blue (2) and evidence demonstrating that it was an iron complex (3) were published in 1724. A short excerpt from that preparation (2) is reproduced below.

---

*Present address: Monsanto Company, St. Louis, Missouri.

D. B. Brown (ed.), Mixed-Valence Compounds, 49-71.

> Take 4 oz. of crude tartar and 4 oz. of dried crude nitre; powder them minutely and mix. Detonate them with charcoal, and you then have 4 oz. of extemporaneous alkali. While this salt is still hot it is finely powdered and 4 oz. of well-dried and finely powdered ox blood is added. . . .

Although modern synthetic work in mixed-valence chemistry rarely uses ox blood, the same basic process--serendipity or, depending on one's perspective, accident--is still responsible for the majority of the preparations of mixed-valence materials.

Because the ubiquity and importance of mixed-valence compounds were not generally recognized until 1967, when Robin and Day (4) and Hush (5) published their reviews of the area, it is not surprising that at that point there had been little systematic effort in the synthesis of such materials. Although this situation has changed in the past decade, few general treatments of the preparative aspects of mixed-valence chemistry have been published. Certain aspects of the field, such as the preparation of highly conducting one-dimensional solids and the design and synthesis of mixed-valence clusters, have been surveyed (6).

In this paper we hope to demonstrate that there are certain features common to the formation of stable or isolable mixed-valence compounds and that routes to such species need not be capricious. Our emphasis will not be so much on how to make particular compounds, but rather on the parameters which must be considered in the design of synthetic routes and on the competing reactions which may occur. A variety of factors, including the extent of interaction between adjacent sites, redox potential differences, solubility, and the availability of suitable precursors and oxidizing or reducing agents, dictate the success of an attempted preparation of a mixed-valence material, and the extent to which these factors can be controlled will be discussed. Following a brief discussion of systems where synthesis is either trivial or beyond the control of the synthesizer, we will concentrate on two types of mixed-valence material. First are those systems containing discrete molecular entities, clusters, where electrons can be added or subtracted sequentially and where, because solubility allows, the full range of synthetic techniques may be utilized. The second category includes polymeric systems which require incorporation of species of variable oxidation state within the growing polymer net or else solid-state transformations of a singly- to a mixed-valent species. The approach throughout this paper will be to seek generalizations which may prove to be useful while considering the synthesis of new mixed-valence materials.

We have made no effort to be comprehensive, or to survey the entire field of mixed-valence synthesis, for indeed it is still true that there are nearly as many different preparative routes to mixed-valence compounds as there are mixed-valence compounds.

PREPARATION BY PRECIPITATION

Many mixed-valence compounds have been isolated only because, under the conditions of a particular reaction, they are the least soluble species present. In an equilibrium involving singly- and mixed-valent species, it is not necessary (for synthetic purposes) that the mixed-valence complex be the thermodynamically most stable material, e.g., that it dominate the equilibrium. Rather, if the proper conditions--such as solvent, temperature, and counterion--prevail, then the equilibrium may often be distorted by precipitation of the desired component. One consequence of this is that a virtually unlimited number of mixed-valence compounds could probably be prepared with relatively little synthetic effort. A simple example of this is shown in eq 1:

$$Pt(NH_3)_4^{+2}(aq) + PtCl_6^{-2}(aq) \xrightarrow{H_2O} Pt^{II}(NH_3)_4Pt^{IV}Cl_6(s) \qquad (1)$$

Mixing aqueous solutions of the Pt(II) cation and the Pt(IV) anion leads to the immediate precipitation of the mixed-valence product. There is no apparent interaction among the sites in this compound, and it is probable that the properties of most materials prepared in this fashion will be relatively uninteresting, typically exhibiting this Robin and Day Class I behavior.

Not all mixed-valence materials prepared by simple precipitation reactions are themselves trivial, and many demonstrate significant interactions among the ions of different oxidation state. This will in general be true if Lewis acid-base interactions among the components are possible, and is particularly probable if ligand-bridged metal ions can be formed. A notable example of this is the formation of Prussian Blue from aqueous solutions of ferric and ferrocyanide ions (eq 2). The cation

$$Fe^{+3}(aq) + Fe(CN)_6^{-4}(aq) \rightarrow Fe_4[Fe(CN)_6]_3(s) \qquad (2)$$

present, $Fe(H_2O)_6^{+3}$, is labile to substitution, and hence behaves as a Lewis acid with respect to the Lewis base, the nitrogen atoms of the cyanide ligands in ferrocyanide. The resulting compound has Fe^{+2} and Fe^{+3} sites linked by cyanide bridges, and thus allows the intervalence transfer which is responsible for the characteristic color of Prussian Blue (7). As in the case of Prussian Blue, precipitation may be accompanied by substitution reactions, and other coincident reactions are possible. For example, this same Prussian Blue may be prepared by reaction of aqueous ferrous and ferricyanide ions. In this case, precipitation is accompanied not only by substitution but also by valence interchange.

The complexity of the equilibria which may be present in solution, and hence the difficulty in predicting or controlling the ultimate products, is clearly demonstrated by the reactions of manganese acetates in the presence of 2,2'-bipyridine (bipy) or pyridine (py). The controlled oxidation of $Mn^{II}(OAc)_2$ by permanganate in the presence of bipy gives the mixed-valence Mn(III,IV) ion, I. This same bipyridyl(III,IV) dimer is produced

$$[(bipy)_2\,Mn(\mu\text{-}O)_2Mn\,(bipy)_2]^{+3}$$

I

directly by the reaction of Mn^{III} with bipy at pH = 4.5. However, at low pH the dimer is itself unstable to disproportionation to $[(bipy)_2Mn^{III}(OH_2)(OH)]^{+2}$ and a Mn(IV,IV) dimer (8). In apparent contrast to this last result, reaction of bipy with the trimeric μ_3-oxo-bridged Mn(III) complex $[Mn_3O(CH_3CO_2)_6(H_2O)_3]^+$ in methanol, followed by acidification, gives I in nearly quantitative yields (9). Although these results are not readily predictable, they are all compatible with a fairly facile oxidation of Mn(III) to the mixed-valence Mn(III,IV) complex. The most surprising result, then, is that reaction of manganese(III) acetate, $Mn(H_2O)_2(CH_3CO_2)_3$ with pyridine gives <u>reduction</u> to the trimeric mixed-valence μ_3-oxo-bridged Mn(II,III,III) complex $Mn_3O(CH_3CO_2)_6(Py)_3$ (10). The apparently minor change of the ligands from bipy to py leads to seemingly very different stabilities for different oxidation states.

THE SOLUTION PREPARATION OF OLIGOMERIC MIXED-VALENCE COMPLEXES

Although the preceding examples would seem to suggest that the preparation of mixed-valence compounds is a rather arcane art, a large number of materials may be synthesized in a much more rational, and predictable, fashion. In particular, transformations involving soluble molecular species are not only frequently successful, but also offer great versatility in synthetic procedure, in architectural tinkering, and in the examination of chemical and physical properties. There has, consequently, been a great deal of successful effort directed toward the synthesis and study of mixed-valence organometallic clusters, metal-metal bonded clusters, and ligand-bridged oligomers.

If the mixed-valence cluster is stable, soluble, and structurally comparable to a singly-valent analog, then its synthesis becomes relatively trivial, involving the selection of suitable solvents and chemical or electrochemical redox agents. Fortunately, the great activity in inorganic and, particularly, organometallic chemistry during the past 25 years has produced vast numbers of metal clusters which are potential precursors to mixed-valence compounds. The more fundamental problem in designing a synthesis using one of these precursors is whether the mixed-valence product will be a distinct, stable material. Its existence, compared to the fully oxidized or fully reduced species, is dependent upon its stability toward disproportionation, which in turn is dependent upon Coulombic effects involving the species as a whole and also upon the extent and nature of the interaction between sites.

For simplicity, we will consider a generalized cluster, M_2, containing two metal ions which are susceptible to oxidation. Removal of a single electron to form the mixed-valence cluster M_2^+ occurs at potential E_1 and removal of a second electron to form the fully-oxidized complex M_2^{+2} occurs at potential E_2. (Because our interest is primarily in potential <u>differences</u>, we will generally use thermodynamic standard potentials, E°, and experimental half-wave potentials, $E_{½}$, interchangeably.) If the mixed-valence cluster is to be the dominant species at equilibrium, it is necessary that E_2 occur at a more positive potential than E_1 (e.g., that the standard electrode potential for the reduction $M_2^{+2} + e^- \rightarrow M_2^+$ be larger than the standard electrode potential for the reduction $M_2^+ + e^- \rightarrow M_2$). Under such conditions the equilibrium distribution of the various species is governed by the Nernst equation. Thus, the stability of the mixed-valence complex, M_2^+, with respect to disproportionation is given by:

$$2M_2^+ \overset{K_{dis}}{\rightleftharpoons} M_2^{+2} + M_2 \quad (3)$$

$$E = E_1 - E_2$$

$$\Delta G = -nFE = -RT \ln K_{dis}$$

$$\log K_{dis} \cong \frac{E}{0.059}$$

The behavior of a particular system is strongly dependent upon the value of E. Because there has been much confusion on this problem in the past, we will discuss, in a general way, the behavior in three regimes, corresponding to E positive, near zero, and negative. The discussion here will be confined to symmetrical clusters, M_2, where there is no difference in metal sites. The same considerations will apply to unsymmetrical clusters MM' as well, but in those cases the innate preference for oxidation at a particular site will generally cause an intrinsic potential difference corresponding to the case of E negative.

On purely electrostatic grounds, a difference in potentials for the loss of the first and second electrons is expected, since the second electron leaves behind an ion of greater positive charge than does the first. In general, then, the mixed-valence state is expected to have some stability in clusters where both fully reduced and fully oxidized forms exist. Added stability may exist as a result of interaction between the sites within a cluster. In a mixed valence system with trapped valences (X,Y), interaction with the valence interchange state (Y,X) provides a resonance or delocalization energy which stabilizes the mixed-valence state (11). The extent to which the excited state (Y,X) is mixed into the ground state (X,Y), and hence the extent of the resonance stabilization, is dependent upon the existence of suitable orbital pathways. Consequently, systems like biferrocene and $\{[(NH_3)_5Ru]_2(pyz)\}^{+5}$ (pyz = pyrazine, $C_4H_4N_2$), which have aromatic moieties connecting the metal ion sites, are reasonable candidates for stable mixed-valence systems. Indeed, for polyferrocenes each of the ferrocene centers is separately oxidizable to ferricinium at a distinct potential (12). Because the interaction between sites is potentially so strong, clusters with direct metal-metal interaction or with single-atom bridges are expected, and found, to yield a large number of mixed-valence materials. Thus, oxide-bridged clusters, such as the manganese-bipyridyl compounds discussed above and the μ-oxo-tetraphenylporphinatoiron(III,IV) clusters prepared by Hendrickson (13), are examples of readily accessible, stable mixed-valence clusters.

Because the success of the synthesis of a mixed-valence cluster from a singly-valent precursor is so strongly dependent upon the potentials of the redox equilibria involved, electrochemical methods of investigation have been particularly useful. Electrochemical techniques have been widely used as both probes and preparative tools in mixed-valence cluster chemistry. Their utility arises because the potentials of electron transfer processes may be examined directly, providing information regarding the existence and stability of the mixed-valence clusters.

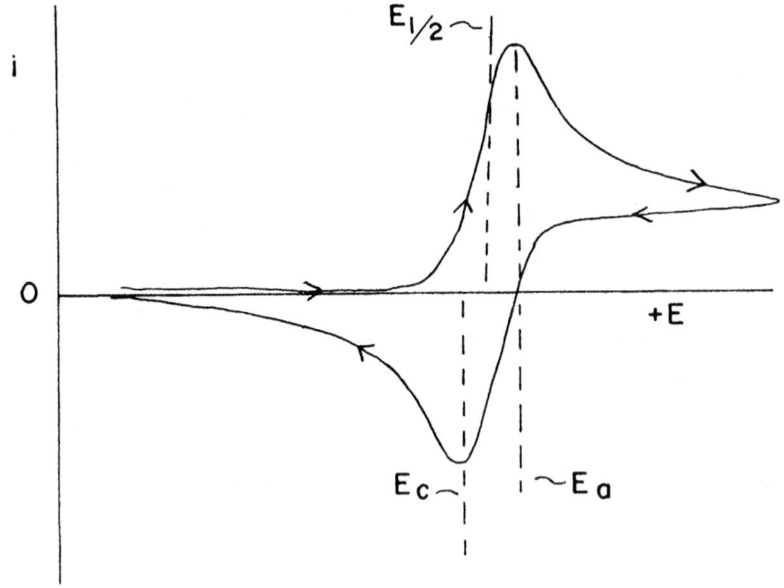

Figure 1. Typical cyclic voltammagram for a simple oxidation.

Although a great variety of electrochemical techniques are available, we will limit discussion here to cyclic voltammetry. Figure 1 shows a typical cyclic voltammagram (CV) for a simple oxidation. Over the course of the experiment the potential is swept in the direction indicated, first positive and then negative. In the case of a simple oxidation uncomplicated by preceding or succeeding chemistry, the current is controlled by a combination of thermodynamic (the Nernst equation) and kinetic (diffusion) factors. Although the mathematics leading to a curve of this shape can become rather complex, one feature of the curve is particularly important for our purposes. For a reversible, one-electron transfer, the peak-to-peak separation $E_a - E_c$ is 60 mV. (E_a is the potential at the current maximum in the anodic, or oxidation, wave and E_c is the corresponding potential for the cathodic, or reduction, wave observed upon current reversal.) In general, for an n-electron transfer process the peak separation is 60/n mV. Equivalent relationships involving n, the number of electrons transferred in a redox process, are available for many other electrochemical techniques. The implications of this relation with respect to the existence of mixed-valence species will be discussed below.

Figure 2 shows the cyclic voltammagram of biferrocene (12). There are two distinct one-electron processes, corresponding to

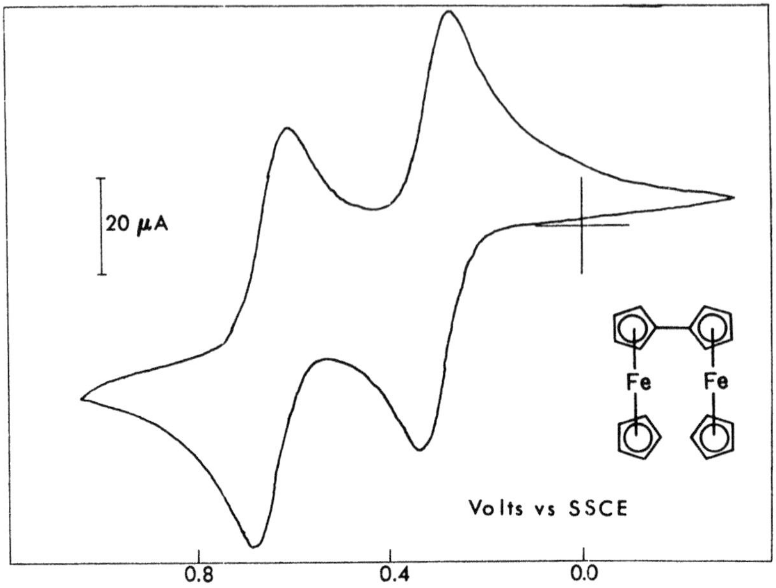

Figure 2. Cyclic voltammagram for the oxidation of biferrocene in 1:1 (v/v) CH_2Cl_2-CH_3CN containing 0.1 M. TBAH. Scan rate 200 mV/sec. (Courtesy of T. J. Meyer.)

oxidation of the ferrous compound [2,2] to the mixed-valence ion [2,3] and, at a more positive potential, oxidation of this species to the iron(III) complex [3,3]. The half-wave potentials of these two processes are 0.31 V and 0.65 V, respectively. As discussed previously, the equilibrium constant for the disproportionation of the mixed-valence ion can be calculated, and is found to be 1.7×10^{-6}. In this case, then, the mixed-valence state is a stable one, and its preparation is facile. In principle, its preparation may be affected using any oxidizing agent with a potential larger than that of the first wave (0.31 V here). If the oxidizing agent has a potential greater than that of the second wave, a stoichiometric amount must be added, otherwise an excess may be used. Of course, controlled potential electrolysis may also be used as a preparative technique. In the particular case of the biferrocene system, the mixed-valence complex has been prepared by oxidation with an excess of benzoquinone in the presence of picric acid (14). Direct synthetic advantage may be taken if the disproportionation equilibrium constant is small. Disproportionation of the Creutz and Taube ion, $\{[(NH_3)_5Ru]_2(pyz)\}^{+5}$, has K_{dis} ca. 10^{-6}, and the ion has been prepared by simply mixing stoichiometric quantities of the fully oxidized and fully reduced species (15).

As the strength of the interaction between sites becomes weaker, or as the distance between them becomes larger, the difference in electrode potentials will become smaller. In the limit of very large separation between sites, even the electrostatic contribution to the potential difference becomes negligible. Under these conditons, the isolation of the mixed-valence state becomes much more problematic. For example, the ion $[(bipy)_2ClRu(pyz)RuCl(bipy)_2]^{+2}$ ([2,2]) exhibits two one-electron cyclic voltammetry waves in which it is successively oxidized to the mixed-valence ion [2,3] and the fully oxidized [3,3] state (16). The potential difference for these two processes is only 0.12 V, and the disproportionation equilibrium (eq 4) has K = 0.01

$$2[2,3] \underset{}{\overset{K}{\rightleftharpoons}} [2,2] + [3,3] \qquad (4)$$

Although the mixed-valence ion is the dominant species, solutions of it at equilibrium contain ca. 10% of the fully oxidized and fully reduced forms.

As the interaction between sites becomes weaker still, the electrochemical behavior becomes more complex and the mixed-valence cluster ceases to dominate. For example, modifying the Creutz and Taube complex by substituting 4,4'-bipyridine in place of pyrazine leads to a cyclic voltammagram exhibiting one wave with a peak-to-peak separation of 114 mV. This was interpreted as being due to the superposition of two cyclic voltammetry waves separated by between 35 and 75 mV, corresponding to a fairly large disproportionation equilibrium (17). In the extreme of totally non-interacting centers, the microscopic redox potential for a given site is independent of the extent of oxidation of the other molecular sites, and the formal redox potentials and equilibrium constants are governed by statistical considerations. Recent theoretical treatment of this case has demonstrated that for non-interacting sites the cyclic voltammagram will have an identical shape to that of the corresponding species with only a single redox-active site (18). The curves will differ only in that the peak currents will be enhanced by the presence of multiple electroactive centers. As in the single-electron case, peak-to-peak separations should retain the theoretical limit of 60 mV.

A number of biferrocenes have been investigated using cyclic voltammetry, and the effect of bridging ligands upon the existence of discrete mixed-valence sites is clearly discernible (19). For species of the type II, separate one-electron waves are seen for the parent compound biferrocene and when X is small (CH_2) or electronically delocalized (C≡C). When X is large or unlikely to propagate an effective electronic interaction (X = Hg, CH_2CH_2)

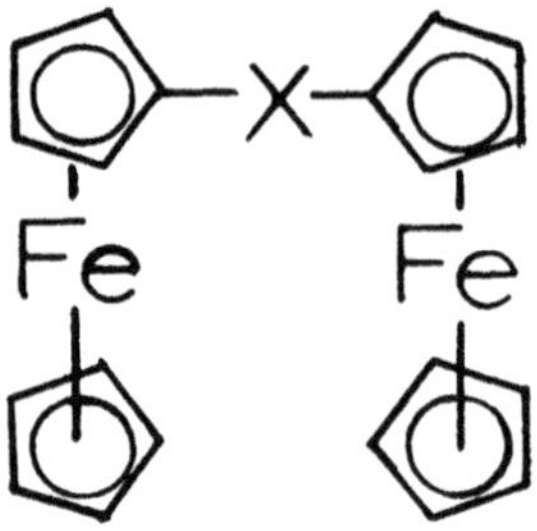

II

only a single wave is observed, with a peak current corresponding to the passage of two equivalents of electricity. Peak separations (and equivalent functions using other electrochemical techniques) were comparable for these systems and for ferrocene, demonstrating that two simultaneous one-electron processes, rather than a single two-electron process, were occurring. Similar behavior was observed in the electrochemistry of polyvinylferrocenes (18).

In these cases, where the difference in potentials for the two redox processes is zero, a statistical distribution of species prevails, and the equilibrium constant for disproportionation of the mixed-valence species is 0.25. This value arises because the mixed-valence ion has twice the probability of either of the two singly-valent species, analogous to the statistical factor in deprotonation of dicarboxylic acids. The important consequences of this value for the equilibrium constant have generally been overlooked. If one equivalent of oxidizing agent is added to a fully reduced two-metal cluster of this type, or if equal amounts of the fully reduced and fully oxidized forms are mixed, the resulting equilibrium mixture should contain the mixed-valence ion as the dominant (50%) species. Such a species should certainly be susceptible to spectroscopic characterization and may in fact be isolable. If a combination of solvent and counterion can be found with which the mixed-valence ion is the least soluble species, then in principle 100% yields of it may be attained. There have apparently been few attempts to isolate or examine mixed-valence complexes under such conditions. It should be emphasized, however, that the conclusion often reached--namely, that the observation of a single electrochemical wave implies that no stable mixed-valence species exists--is in general incorrect.

The final case to consider, and the sole exception to the last statement, is the case where $E_2^\circ < E_1^\circ$. This implies that, in an oxidation, it is easier to remove the second electron than the first. This is clearly contrary to electrostatic predictions, and can only occur if removal of the first electron leads to some electronic or molecular rearrangement which severely modifies the properties of the system. In such cases the disproportionation equilibrium is favored ($K > 0.25$) and the mixed-valent species is a minor component of the equilibrium. As in the previous case, cyclic voltammetry will exhibit a single wave, but the peak-to-peak separation will be only 30 mV (=60/n). It is this value which distinguishes a two-electron transfer from two simultaneous one-electron transfers. There are apparently few documented cases of this phenomenon. Fenton et al. (20) have recently reported the electrochemistry of a dicopper(II) macrocycle which undergoes two one-electron reductions with identical potentials (e.g., $E_2^\circ = E_1^\circ$). This implies a disproportionation equilibrium constant $K = 1$. Since this is larger than the statistical value of $K = 0.25$, it implies that the second electron is in fact easier (thermodynamically) to add than the first. Certain aryl substituted ethylenes exhibit a single two-electron oxidation wave, apparently as the result of the relief of steric strain by bond rotation following the initial oxidation (21).

Although this discussion has considered specifically only the ruthenium ammine dimers and the biferrocenes, it should be clear that the same considerations will apply to other soluble clusters. The synthesis of a mixed-valence cluster from its fully oxidized or reduced analog is primarily a matter of selecting suitable redox reagents and solvents.

THE PREPARATION OF POLYMERIC MIXED-VALENCE MATERIALS

Soluble mixed-valence complexes have a finite existence as distinct molecular, or ionic, entities. By contrast, the materials which will be discussed in this section exist only in the solid state. Dissolution, to the extent that it occurs, invariably involves either separation of the components of different formal oxidation state or redox processes which regenerate singly-valent species. Preparations of polymeric mixed-valence materials must then involve either solid-state reactions or polymer growth processes which incorporate the species of variable charge. Clearly, the precipitation reactions discussed above fall into this latter category. However, in many of the examples previously cited the driving force is the purely electrostatic interaction between, e.g., a Pt(II)-containing cation and a Pt(IV)-containing anion. In this section we will also consider the more complex cases--typified by $K_2Pt(CN)_4Br_{0.3}$--where ions of similar charge, but containing different oxidation states of a metal ion, aggregate through covalent bonding interactions.

A rather diverse group of mixed-valence materials may be

prepared by solid-state reaction techniques. Although solid-state syntheses are particularly applicable to the preparation of congruently-melting compounds such as bronzes, certain Magneli phases, mixed-valence perovskites and spinels, and vanadium-phosphate glasses, it is also possible to prepare, by solid-state reactions, mixed-valence coordination and organometallic complexes. It is instructive to list some conceivable synthetic techniques in order to gain an appreciation for the variety of processes occurring in the solid state which may be applied to the synthesis of mixed-valence compounds.

1. Binary or ternary transition-metal oxides or chalcogenides may be prepared by fusing together mixtures of upper- and lower-oxidation state compounds, as in eq 5. The facility of

$$\text{"FeO"} + Fe_2O_3 \rightarrow Fe_3O_4 \tag{5}$$

this reaction is demonstrated by the fact that it occurs even when the solid reactants are separated by an ionic filter, as shown in Figure 3. Oxide transport through the ionic ZrO_2 filter allows redox processes at both the ZrO_2/FeO and ZrO_2/Fe_2O_3 interfaces, with formation of the mixed valence Fe_3O_4 at each (22).

2. Thermal decomposition of singly-valent compounds often leads to mixed-valence species (23), as in eq 6. Occasionally,

$$2Ba(NO_3)_2 + 2Bi(NO_3)_3 \cdot 5H_2O \xrightarrow[1073°C]{air} Ba_2Bi^{+3}Bi^{+5}O_6 \tag{6}$$

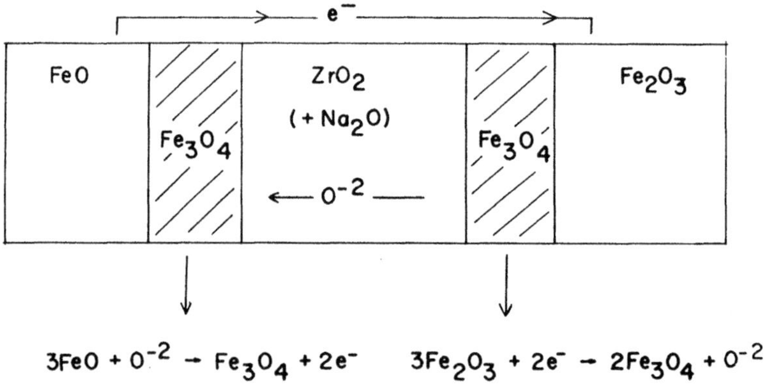

Figure 3. Solid state formation of Fe_3O_4 by oxide ion transport through an ionic filter. Completing the electrochemical cell by connecting the FeO and Fe_2O_3 ends allows the reaction to continue (22).

partial thermal decomposition at ambient conditions can lead to a mixture of singly- and mixed-valent materials. Thus, the electrical conductivity of $Cs_2BaCo^{II}(NO_2)_6$ is higher, by approximately five orders of magnitude, than that of analogs with other transition metals. This higher conductivity has been ascribed (24) to the generation of charge carriers by the solid state disproportionation of Co^{II} to Co^{I} and Co^{III}.

3. Photolysis of solids (using either ultraviolet or high-energy radiation) often involves electron-transfer processes which generate different oxidation states of an element. Notable examples of this are the color centers in many minerals and the blue color (due to colloidal sodium) of rock salt (25). This is, in general, a non-equilibrium situation in which the kinetic barrier to recombination or reversal is so high that long-lived mixed-valence materials are formed.

4. The reaction of a singly-valent coordination compound with an oxidizing or reducing atmosphere can produce mixed-valence polymers, as in eq 7. These partial oxidation processes

$$\text{Ni(II)(dimethylglyoxime)}_2 + I_2 \xrightarrow{200°C} \text{Ni(II,III)(dimethylglyoxime)}I_x \qquad (7)$$

have been used extensively in recent years, with activity in this area motivated largely by the current interest in the preparation of molecular metals (26).

The number and variety of materials which are, potentially, available by these routes is vast. Some useful preparative guidelines, based on the general reactions of solids, may be gleaned from several excellent monographs and texts (27). These references provide useful information on both theoretical (kinetic) and experimental (reactor design) considerations. They also are excellent sources for such germane topics as phase transformation kinetics and melting processes as well as gaseous diffusion into solids.

It should be emphasized that although the majority of solid-state syntheses of mixed-valence materials are based on stoichiometric considerations (i.e., eq 5), occasionally some surprising materials are obtained by attempting to synthesize mixed-valence compounds from "correct" (stoichiometric) mixtures of reactants. For example, although the Ba/Fe/S system contains a large number of compounds with "anticipated" Fe^{+2}/Fe^{+3} and Fe/S^{-2} ratios, two doubly-mixed-valence compounds $Ba_4Fe_2S_6[S_{2/3}(S_2)_{1/3}]$ and $Ba_{3.6}Al_{0.4}Fe_2S_6[S_{0.6}(S_2)_{0.4}]$ were prepared (28) by reacting BaS, Fe, and S powders (and Al foils) at 1100°C. The presence of both sulfide and disulfide anions in these materials (as in Covelline, "CuS" = $Cu_4^{+1}Cu_2^{+2}(S_2)_2S_2$) (29) is a reflection of the propensity of the chalcogenides to form oligomers at high temperature. On the other hand, KCu_4S_3 has been shown by single-crystal x-ray analysis to contain only S^{-2} anions, which suggests the formulation $KCu_3^{+1}Cu^{+2}S_3$ (30). Formation of polyanions is also observed in halogen chemistry, i.e., I_3^-, I_5^-, etc.

It is instructive to examine the thermodynamic factors involved in the existence of polymeric mixed-valence materials. Figure 4 contains a Born-Haber cycle for the formation of possible complexes between electron donor (D) and acceptor (A) molecules. Although as drawn Figure 4 is most appropriate for the consideration of organic complexes similar to those formed between tetrathiofulvalene (TTF) and tetracyanoquinodimethane (TCNQ), similar

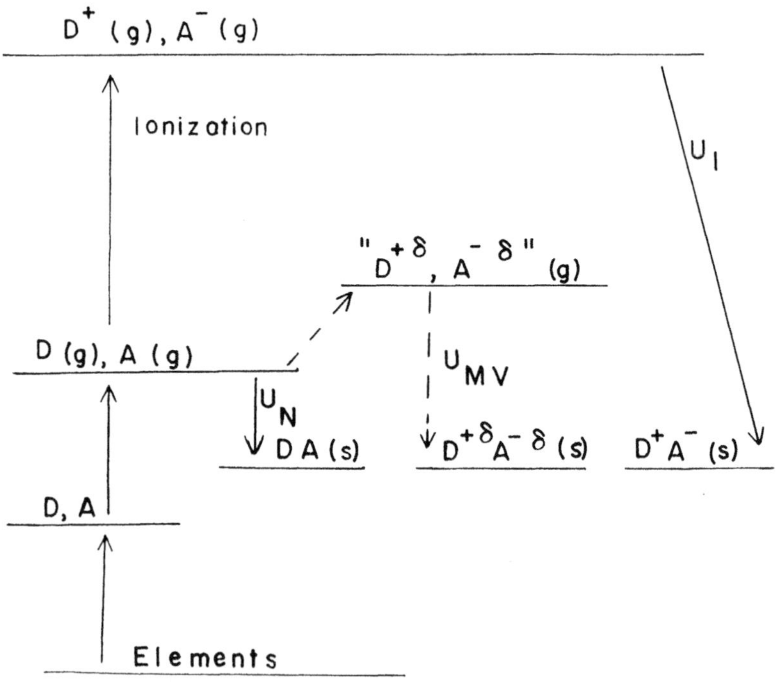

Figure 4. Born-Haber cycle for the interaction of donor (D) and acceptor (A) molecules, Calculations for the intermediate, mixed-valence complexes require formation of the imaginary, partially-ionized gaseous species.

cycles can be constructed for any material. The specific material which will form is determined by the relative magnitudes of the various ionization energies and binding energies, U. The ionization potential of D and the electron affinity of A (or quantities related to them) are, in general, measurable, and, in principle, the binding energies U can be calculated. However, unlike the simple ionic salts (e.g., NaCl) these calculations are distinctly non-trivial. The structures generally contain segregated stacks of $D^{(+)}$ and $A^{(-)}$, rather than alternating ions, so that covalent bonding terms must be considered, and simplifying assumptions made in other cases are not valid here. Nonetheless, by accounting for a variety of steric and electronic factors, calculations of U_N and U_I have been made (31), which agree reasonably well with experiment.

In the discussion which follows, we will assume that it is possible to calculate the free energy (rather than the enthalpy) of formation for a donor-acceptor complex with any arbitrary degree of charge transfer, using thermochemical schemes conceptually similar to that in Figure 4. It might then be possible to calculate curves such as those shown in Figure 5, which show the variation in free energy of formation as a function of the degree of charge transfer. With some slight modifications these curves are also appropriate for exhibiting the variation in free energy of reaction as a function of the extent of partial oxidation of a metal complex by some external reagent. (For the purposes of illustration we have arbitrarily chosen the energy of the complete charge-transfer state, D^+A^-, to be higher than that of DA.) Assuming that there are no structural phase changes on proceeding from the neutral to ionic extremes (a surprisingly reasonable assumption (32) in the case of materials analogous to TTF-TCNQ), a smooth variation in energy results. We stress that our interest here is in the general behavior of materials in mixed-valence states, and that our attempt to fit all of these cases into the framework of a single diagram presents obvious problems in precision. Nonetheless, three types of energy variation, represented by curves 1, 2, and 3 in Figure 5, should be possible, corresponding to the cases where enthalpy terms are respectively unimportant, repulsive, and attractive.

It should be noted that these curves are similar to those observed for the free energy of solutions of two components, and indeed arguments from the thermodynamics of solutions are applicable here. In the case of the partial oxidation reaction, as in the ideal solution, the free energy will be decreased by entropic factors. Thus, each added ion (X^-) can occupy a vast number of sites, and in so doing greatly increase the entropy of the system. Thus, in the case where no significant new enthalpy terms are

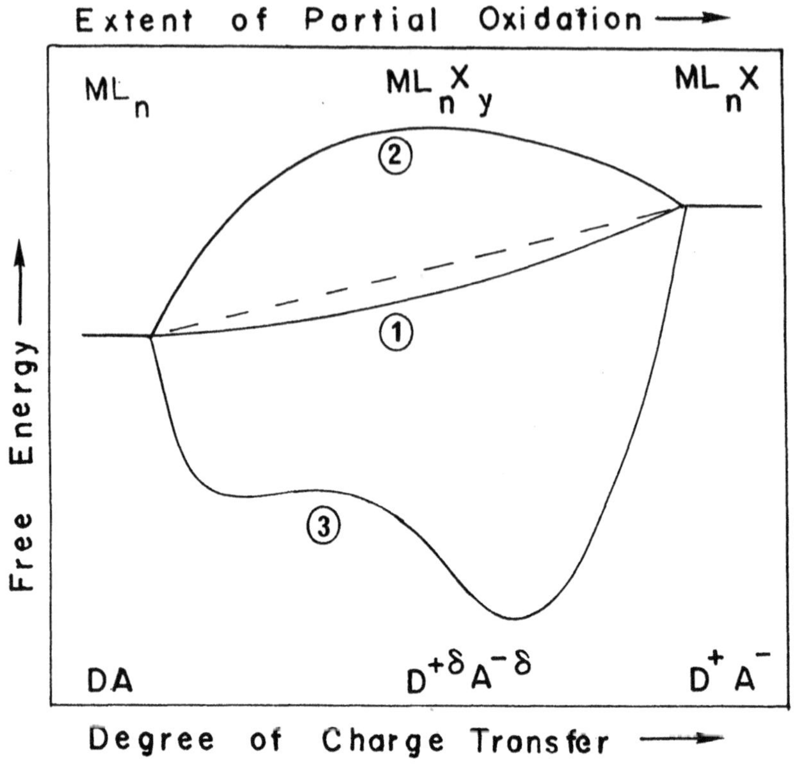

Figure 5. Variation in free energy as a function of the extent of charge transfer in a donor-acceptor complex or as a function of the degree of partial oxidation of a metal complex.

introduced, curves similar to 1 must prevail. Calculations show, however, that the entropy terms will be small, and that the overall behavior of a particular system is likely to be dominated by the enthalpy. Curve 1 represents the case where there is a monotonic increase in free energy as a function of the degree of intramolecular charge transfer. (For the ideal solution case, there should in fact be a shallow minimum near DA.) In the case of the donor-acceptor interaction this corresponds to the existence of a single stable form (the neutral compound as drawn here), and is certainly the most common situation. In the case of a partial oxidation reaction it corresponds to stoichiometric control, effectively the formation of a mixed-valence solid solution. This situation prevails, for example, in the doping of polyacetylenes with halogens (33). Thus, one can sequentially add iodine to solid $(CH)_x$ with incorporation of variable amounts of I dependent upon stoichiometry and conditions, and the physical properties of the product $[(CH)I_y]_x$ vary in a continuous fashion.

Another example of this behavior involves a case where mixed-valence polymer formation is simultaneous with crystallization. Where typical geometries are the same for different oxidation states of an element, co-crystallization--in effect solid solution formation--is feasible. This is exemplified by the preparation of hexachloroantimonates (III,V). Compounds of general formula $Cs_2Sb^{III}_yIn^{III}_{1/2-y}Sb^{V}_{1/2}Cl_6$ were prepared by co-crystallization of the appropriate amounts of the singly-valent components. In this fashion, complexes can be prepared in which the ratio of Sb^{III}/Sb^{V} is continuously variable from 0/1 to 1/1 (34).

For the donor-acceptor complex, curves 1 and 2 both predict the same result, either no or complete charge transfer. However, for the partial oxidation reaction curve 2 corresponds to instability of the mixed valence phase. Stoichiometric control can no longer lead to a mixed-valence material, since if prepared it would be unstable with respect to disproportionation to the unoxidized and fully oxidized extremes. Synthetic efforts in such cases will go unrewarded, and probably unreported as well.

The most interesting case, of course, is represented by curve 3, since here the mixed-valence state is most stable. Predictions of the energy at intermediate states are quite difficult. There is, for example, the complication that structural parameters, and hence the various Coulombic terms, will vary with the degree of charge-transfer. This has been demonstrated in the case of KCP analogs, where the Pt-Pt separation is related to the formal oxidation state of the platinum (35). Although detailed calculations of the energy at intermediate degrees of charge transfer have not yet been successful, it is clear that a major factor will be the ease of charge transfer <u>in the solid</u>. This in turn must be <u>related</u> to other measures of electron transfer in donor and acceptor molecules. Working from this perspective, and utilizing solution redox potentials, Torrance has been quite successful in rationalizing and predicting which combinations of donor and acceptor organic molecules will give incomplete charge transfer and, consequently, high electrical conductivity (32). As in curve 3 in Figure 5, it is possible that <u>several</u> relative minima exist, which makes synthetic control a blend of thermodynamics and stoichiometry. This is exemplified by the partial oxidation of nickel complexes of 1,4,5,8,9,12,13,16-octamethyltetrabenzporphyrin (OMTBP). I_2 oxidation of Ni(OMTBP) leads (36) to two distinct mixed valence phases, of composition Ni(OMTBP)$(I_3)_{0.35}$ and Ni(OMTBP)$(I_3)_{0.97}$.

Because the nature of these curves can never be known in detail, they are in themselves of little direct synthetic utility. Nonetheless, knowledge of the possibilities can prove to be useful in designing syntheses, as has been demonstrated for the donor-acceptor complexes by Torrance (32). Similar predictive ability should exist for the partial oxidation reactions as well. Thus, precursors which have available structural sites that can accommodate oxidants with little distortion and with only minor enthalpy changes should exhibit behavior approaching that of the ideal solution. Consequently, curves of type 1, and a range of mixed valence materials dictated by stoichiometry, should result. Alternatively, curves of type 2 or 3 are more likely if either the nature of the oxidant or structural considerations lead to some specific interaction between the complex and oxidant. Depending on the bonding or antibonding nature of that interaction--which may be predictable--the mixed-valence state will be either stabilized or destabilized.

It should be emphasized that in all of these cases, kinetic factors can obviate any generalizations based on thermodynamic factors. This is particularly important for reactions involving solids, because of the low reactant mobility in the solid state. Thus, many thermodynamically non-viable mixed-valence materials have undoubtedly been prepared, and, similarly, many potentially stable mixed-valence compounds are simply inaccessible because of synthetic (kinetic) limitations.

Preparative methods involving partial oxidation and subsequent crystallization have been particularly useful in preparing highly-conducting one-dimensional metal complexes such as KCP. The most convenient chemical preparation of this compound involves mixing aqueous solutions of $K_2Pt(CN)_4$ and $K_2Pt(CN)_4Br_2$ (or, equivalently, adding a sub-stoichiometric amount of Br_2 to aqueous $K_2Pt(CN)_4$). In this case the least soluble component is the mixed-valence KCP. Figure 6 demonstrates some of the chemical processes which are involved in a reaction such as the formation of KCP. It is clear that, given the large number of possibilities, it is fortunate that thermodynamics favors the mixed-valence compound here.

As a point of interest, it may be somewhat surprising that it is sometimes difficult to distinguish between single (homogeneous) phase and truly multiple (heterogeneous) phase syntheses of mixed-valence materials. For example, the partial oxidation of many "insoluble" stacked, square-planar $\underline{d}^8$ transition-metal complexes may be accomplished by slurrying the singly-valent compound in a solution of an oxidizing agent in an inert solvent. Although the reaction may in fact be heterogeneous, involving the

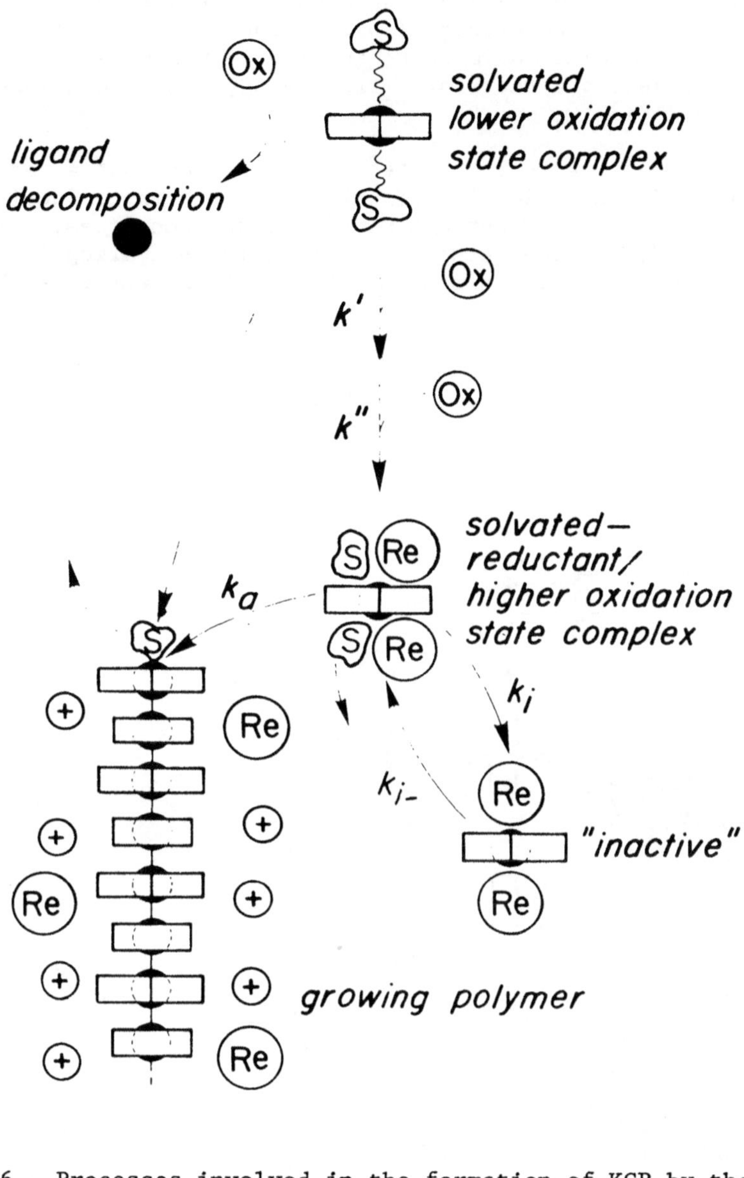

Figure 6. Processes involved in the formation of KCP by the partial oxidation reaction.

complex as the solid only, it is likely that in many cases product formation is the result of a complex series of equilibria. Thus, the reaction sequence may involve dissolution of the (insoluble) polymer into (soluble) monomers which are then rapidly oxidized by the solution-phase oxidizing agent. The fate of the oxidized monomer may then be either disproportionation or combination with unoxidized monomer to form the insoluble, mixed-valence polymer. Such processes are therefore strictly typical of a homogeneous-phase oxidation.

In recent years it has been demonstrated that a variation of this method leads cleanly and easily to certain conducting mixed-valence polymers. Electrochemical growth techniques, illustrated in Figure 7, are the methods of choice for preparing large single crystals of KCP. In this case, the electrode serves as both the

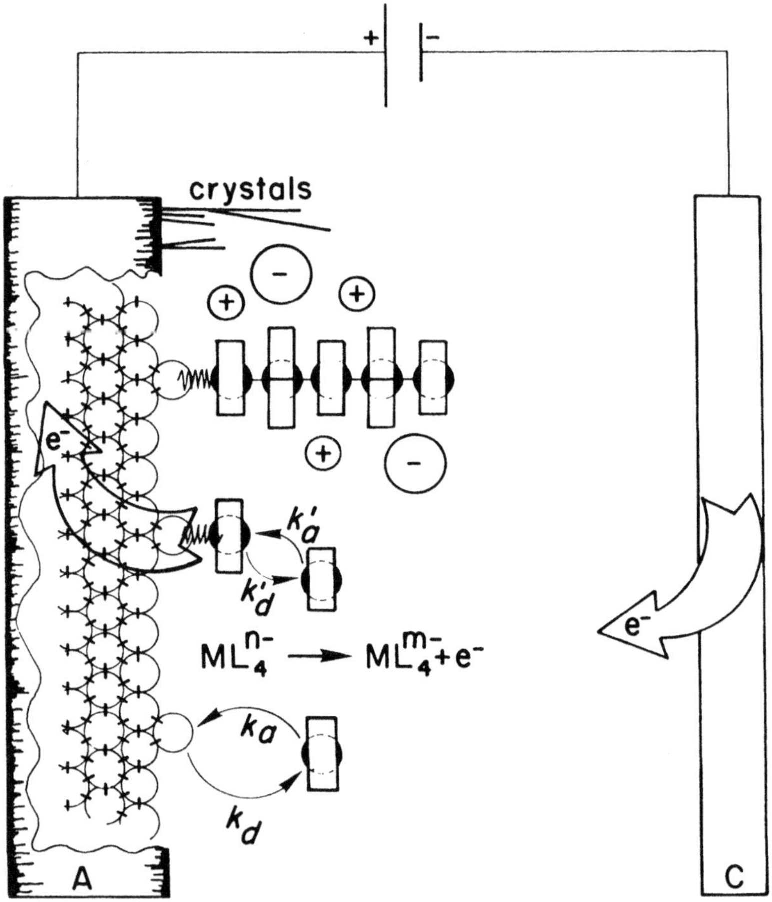

Figure 7. Electrochemical growth of a conducting, mixed-valence polymer.

oxidant and the nucleation site for crystal growth. Clearly, crystalline material will form only if the product is electronically conducting, since only then can electrons migrate from the crystal/solution interface.

Finally, it seems appropriate to emphasize that not all mixed-valence compounds need to be synthesized, since some are naturally occurring. A great number of mixed-valence minerals are known, most presumably forming during the slow crystallization processes which led to the formation of the earth's crust. Mixed-valence compounds are also formed in biological systems. Included here are the non-heme iron proteins which exhibit such a rich redox chemistry, and also the several animals--homing pigeons, honey bees, spirillum bacteria--which appear to possess a magnetic sense of direction as a result of their growth of single-domain crystals of magnetite, Fe_3O_4 (37).

ACKNOWLEDGEMENTS

We are particularly grateful to Mike Elliott for his careful reading of the manuscript and his valuable instruction in electrochemistry. In addition to reading the manuscript, Tom Meyer kindly provided the voltammagram of Figure 2. Some of our ideas on disproportionation equilibria were derived from a manuscript provided by Bob Gagne. Our efforts in mixed-valence syntheses are supported in part by the Office of Naval Research.

REFERENCES

1. Griffith, W.P.: 1962, Quart. Rev. 16, p. 188.

2. Woodward, J.: 1724, Phil. Trans. 33, p. 15. Translated in Powell, H.M.: 1959, Proc. Chem. Soc., p. 73.

3. Brown, J.: 1724, Phil. Trans. 33, p. 17.

4. Robin, M.B., and Day, P.: 1967, Adv. Inorg. Chem. Radiochem. 10, p. 247.

5. Allen, G.C., and Hush, N.S.: 1967, Prog. Inorg. Chem. 8, p. 357. Hush, N.S.: ibid., p. 430.

6. Miller, J.S.: 1978, Ann. N.Y. Acad. Sci. 313, p. 25. Meyer, T.J.: ibid., p. 496.

7. Robin, M.B.: 1962, Inorg. Chem. 1, p. 337.

8. Cooper, S.R., and Calvin, M.: 1977, J. Am. Chem. Soc. 99, p. 6623.

9. Uson, R., Riera, V., and Ciriano, M.A.: 1976, Trans. Met. Chem. 1, p. 98.

10. Baikie, A.R.E., Hursthouse, M.B., New, D.B., and Thornton, P.: 1978, J.C.S. Chem. Comm., p. 62.

11. Mayoh, B., and Day, P.: 1972, J. Am. Chem. Soc. 94, p. 2885.

12. Brown, G.M., Meyer, T.J., Cowan, D.O., LeVanda, C., Kaufman, F., Roling, P.V., and Rausch, M.D.: 1975, Inorg. Chem. 14, p. 506.

13. Wollmann, R.G., and Hendrickson, D.N.: 1977, Inorg. Chem. 16, p. 723.

14. Kaufman, F., and Cowan, D.O.: 1970, J. Am. Chem. Soc. 92, p. 6198.

15. Bunker, B.C., Drago, R.S., Hendrickson, D.N. Richman, R.M., and Kessell, S.L.: 1978, J. Am. Chem. Soc. 100, p. 3805.

16. Callahan, R.W., and Meyer, T.J.: 1976, Chem. Phys. Lett. 39, p. 82.

17. Tom, G.M., Creutz, C., and Taube, H.: 1974, J. Am. Chem. Soc. 96, p. 7827.

18. Flanagan, J.B., Margel, S., Bard, A.J., and Anson, F.C.: 1978, J. Am. Chem. Soc. 100, p. 4248.

19. Morrison, W.H., Jr., Krogsrud, S., and Hendrickson, D.N.: 1973, Inorg. Chem. 12, p. 1998. LeVanda, C, Bechgaard, K., and Cowan, D.O.: 1976, J. Org. Chem. 41, p. 2700.

20. Fenton, D.E., Schroeder, R.R., and Lintvedt, R.L.: 1978, J. Am. Chem. Soc. 100, p. 1931.

21. Phelps, J., and Bard, A.J.: 1976, J. Electroanal. Chem. 68, p. 313.

22. Schmalzried, H.: 1974, "Solid State Reactions," Verlag-Chemie, Weinheim, p. 179.

23. Cox, D.E., and Sleight, A.W.: 1979, Acta Cryst. B35, p. 1.

24. Barnes, J.C., and Thomas, J.M.: 1974, Inorg. Chim. Acta 9, p. 171.

25. Lehmann, G.: 1978, Angew. Chem. Int. Ed. Engl. 17, p. 89.

26. Hatfield, W.E., ed.: 1979, "Molecular Metals," Plenum Press, New York.

27. See, for example, A.K. Galwey, "Chemistry of Solids," Chapman and Hall, London, 1967, pp. 163-199; P. Hagenmuller, ed., "Preparative Methods in Solid State Chemistry," Academic Press, Inc., New York, 1972; and H. Schmalzried, "Solid State Reactions," Verlag-Chemie, Weinheim, 1974.

28. Hoggins, J.T., Rendon-Diazmiron, L.E., and Steinfink, H.: 1977, J. Solid State Chem. 21, p. 79.

29. Wells, A.F.: 1962, "Structural Inorganic Chemistry," 3rd edition, Oxford University Press, London, p. 883.

30. Brown, D.B., Zubieta, J.A., Valla, P.A., Wrobleski, J.T., Watt, T., Hatfield, W.E., and Day, P.: Inorg. Chem., submitted.

31. Metzger, R.M.: 1978, Ann. N.Y. Acad. Sci. 313, p. 145.

32. Torrance, J.B.: 1979, Acc. Chem. Res. 12, p. 79.

33. MacDiarmid, A.G., and Heeger, A.J.: 1979, "Molecular Metals," Hatfield, W.E., ed., Plenum Press, New York, p. 161.

34. Atkinson, L., and Day, P.: 1969, J. Chem. Soc. (A), p. 2423.

35. Reis, A.H., Jr., and Peterson, S.W.: 1978, Ann. N.Y. Acad. Sci. 313, p. 560.

36. Phillips, T.E., and Hoffman, B.M.: 1977, J. Am. Chem. Soc. 99, p. 7734.

37. Anon.: 1979, Mosaic 10, No. 3, p. 25.

II. MODELS FOR THE MIXED-VALENCE STATE

ELECTRON TRANSFER IN MIXED-VALENCE COMPOUNDS

Thomas J. Meyer

Department of Chemistry, University of North Carolina,
Chapel Hill, North Carolina 27514

Mixed-valence compounds where there is weak electronic coupling between sites provide perhaps the clearest experimental insight into electron transfer in solution. When combined with recent advances in electron transfer theory, they may help provide a nearly complete microscopic description of electron transfer in most chemical and biological systems.

Perhaps the most characteristic feature of many mixed-valence compounds is that a non-integral number of valence electrons are shared between an integral number of chemical sites. The development of an adequate description of the resulting electronic structure and associated vibrations, an issue which was first treated systematically by Robin and Day,[1] remains one of the fundamental themes in mixed-valence chemistry.[2,3] In some dimeric systems,[4] e.g.,

$(bpy)_2ClRu^{II}N$ 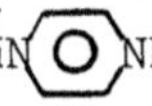 $NRu^{III}Cl(bpy)_2^{3+}$ (bpy is 2,2'-bipyridine),[5]

there is good experimental evidence that electronic coupling between sites is weak, that the resulting perturbation is insufficient to change the chemical and physical properties of the two sites appreciably, and that the excess electron is localized at one of the sites, at least on the time scale for transitions between molecular ($\sim 10^{-13}$–10^{-14}s) and even solvent ($\sim 10^{-11}$s) vibrational levels.

My attention will be devoted almost solely to such systems. The reason for interest is that they appear to offer perhaps the clearest experimental insight into electron transfer processes in

D.B. Brown (ed.), Mixed-Valence Compounds, 75-113.

solution. When combined with recent advances in theory, this insight may lead to a nearly complete microscopic description of electron transfer in most chemical and biological reactions.

In order to illustrate the value of mixed-valence compounds to the problem, consider first an "outer-sphere" reaction (eq. 1-3, py is pyridine),[6] in which a preassociation step (eq. 1, K_A),

$$\overset{II}{Ru}(NH_3)_5py^{2+} + \overset{III}{Ru}(NH_3)_5(py)^{3+} \underset{}{\overset{K_A}{\rightleftarrows}} [\overset{II}{Ru}(NH_3)_5py,\overset{III}{Ru}(NH_3)_5py]^{5+} \quad (1)$$

$$[\overset{II}{Ru}(NH_3)_5py,\overset{III}{Ru}(NH_3)_5py]^{5+} \xrightarrow{k_{et}} [\overset{III}{Ru}(NH_3)_5py,\overset{II}{Ru}(NH_3)_5py]^{5+} \quad (2)$$

$$[\overset{III}{Ru}(NH_3)_5py,\overset{II}{Ru}(NH_3)_5py]^{5+} \longrightarrow \overset{III}{Ru}(NH_3)_5py^{3+} + \overset{II}{Ru}(NH_3)_5py^{2+} \quad (3)$$

occurs prior to electron transfer (eq. 2, k_{et}). In most cases, k_{et} and K_A can not be separated by kinetic measurements, the observed rate constant is the product $k_{et}K_A$ (however, note ref. 7), and direct information about the electron transfer step is lost. Electron transfer in a mixed-valence complex, eq. 4,[8] is free

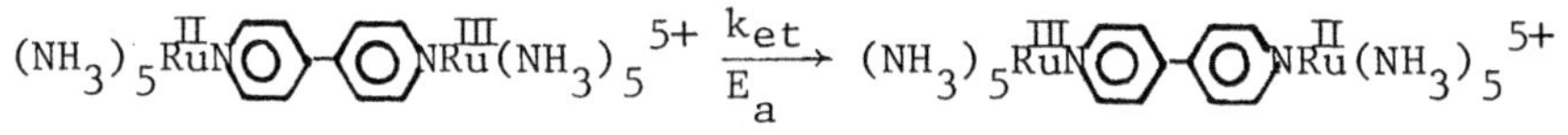 (4)

from such complications. Although the direct measurement of k_{et} has turned out to be an elusive and difficult problem, a point which will be taken up later, a series of spectral experiments (IR, epr, Raman, ...) can be carried out on the intact, odd electron system. Perhaps of most value is the appearance of optical or near IR absorption bands which can be assigned to the electromagnetically induced analog of eq. 4, (eq. 5). The properties of

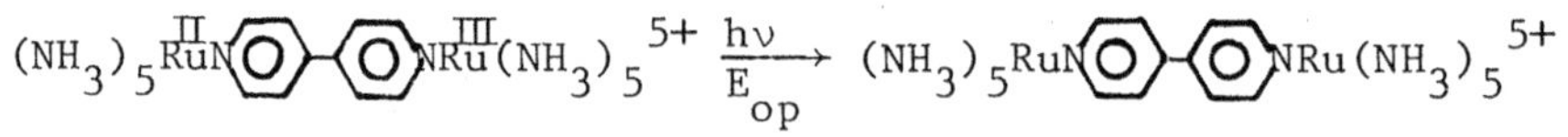 (5)

the transition in eq. 5, which has been termed Intervalence Transfer or Metal Metal Charge Transfer (MMCT), have been shown by Hush[9] and more recently by Hopfield[10] to be closely related to the properties of the thermally induced transition in eq. 4. In this account, the goal will be first to describe the current theoretical basis for understanding electron transfer reactions in solution and then to show how experimental results based on mixed-valence compounds can be used to support and extend the theory.

INTRODUCTION TO THE PROBLEM

The variations in rate constants in electron transfer reactions can be astonishing. Consider the self-exchange reactions between $Co(NH_3)_6^{3+}$ and $Co(NH_3)_6^{2+}$ and between $Ru(bpy)_3^{3+}$ and $Ru(bpy)_3^{2+}$

for which $k(65°C) \leqslant 10^{-9} \underline{M}^{-1} s^{-1}$ (11) and $k(25°) \sim 2 \times 10^{9} \underline{M}^{-1} s^{-1}$ (12). Even though $\Delta G^{o}=0$ for both cases and there is no need to form or break bonds in either net reaction, the rate variation is greater than 10^{18} given the temperature difference.

Before trying to account for such dramatic changes in reactivity, it is useful to consider their probable origins. For example, consider the self-exchange reaction between $Fe(H_2O)_6^{2+}$ and $Fe(H_2O)_6^{3+}$ which has been studied by isotopic tracer techniques.(13) Even if the reaction occurs by a simple outer-sphere mechanism, it must proceed through an association complex whose concentration at equilibrium may be low and will certainly vary with charge type and reactant size (eq. 1. A justification for assuming close contact between the reactants in the association complex will be developed later.). Electronic orbital overlap must occur between the electron donor and acceptor sites and that can vary widely between systems.

But the most important sources of variations in rate are usually the changes which occur in molecular and solvent structure with electron content or oxidation state. For the $Fe(H_2O)_6^{3+/2+}$ case, crystallographic studies show that average Fe-O bond lengths are longer by 0.14Å for the +2 ion than in the +3 ion (2.13 vs. 1.99Å).(14) For organic compounds, significant molecular distortions can occur following oxidation to radical cations or reduction to radical anions. The surrounding solvent also responds to changes in oxidation state. For ionic reactants, solvent molecules near the ion are highly polarized. As shown by the data below, the extent of polarization is highly dependent on ion charge.

	$\underline{\Delta H}$(15)
$Fe(H_2O)_6^{3+}(g) \longrightarrow Fe(H_2O)_6^{3+}(aq)$	-18.1 eV
$Fe(H_2O)_6^{2+}(g) \longrightarrow Fe(H_2O)_6^{2+}(aq)$	-4.7 eV

In fact, the very existence of an excess electron on only one of two possible redox sites creates a "trap" which tends to localize the electron. The origin of the trap is that the electron induces differences in molecular and solvent structure which must be overcome if electron transfer is to occur.

A SIMPLE CASE

A good place to begin our analysis is with electron transfer in a very simple mixed-valence compound, an atom and an ion in the gas phase, $A \cdots A^+ \longrightarrow A^+ \cdots A$. We will keep A and A^+ held at some fixed internuclear distance, R'. First consider the case where R' is much larger than the equilibrium internuclear separation R_e.

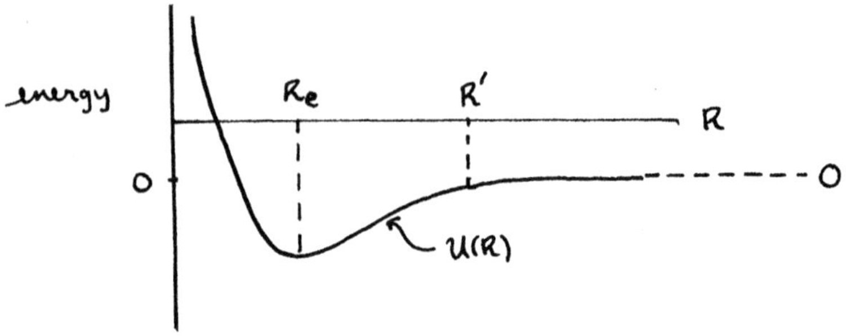

Figure 1. Plot of U(R) vs. internuclear coordinate (R) for the diatomic ion A_2^+.

Figure 1 shows a plot of potential energy for nuclear motion, U(R), vs. the internuclear coordinate, R. U(R) is the sum of the potential energy from internuclear repulsion (V_{NN}) and the potential (V_{el}) and kinetic (T_{el}) energies for electronic motion. In Figure 2 is shown schematically a plot of the potential energy solely for electronic motion at our fixed value of R, R'. For the simplest case of an A_2^+ ion, H_2^+, V_{el} is given by

$$V_{el} = -\frac{e^2}{r_a} - \frac{e^2}{r_b} \tag{6}$$

In eq. 6, r_a and r_b are the distances between the electron and the nuclei a and b whose coordinates are related by the internuclear separation, R'. From the figure, compared to the isolated

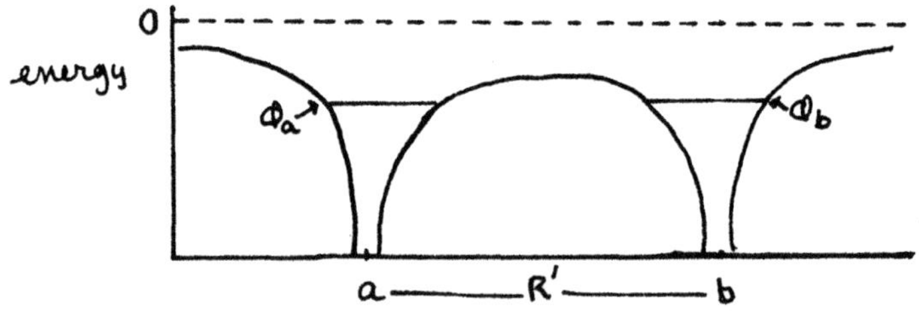

Figure 2. Schematic plot of V_{el} at $R' >> R_e$ for H_2^+(16)

atom energies which are indicated by the wave functions ϕ_a and ϕ_b, there is an increase in V_{el} in the internuclear region. At large, R, U(R) decreases in going from $R=\infty$ to $R=R'$ because the electronic kinetic energy, T_{el}, decreases faster than V_{el} increases.

Classically, an electron initially in ϕ_a would be trapped in ϕ_a but quantum mechanically, the region between nuclei is accessible by tunneling. When written in terms of the atomic Hamiltonian operators for each side, H_a+H_b, and the electron exchange operator, V, the Hamiltonian for the system becomes,

$$H = H_a + H_b + V \quad (7)$$

Taking into account the time dependent nature of the problem, a wave function for the system can be written as an expansion of the atomic wave functions ϕ_a and ϕ_b,

$$\psi(r,t) = c_a(t)\phi_a(r) + c_b(t)\phi_b(r+R') \quad (8)$$

where the expansion coefficients $c_a(t)$ and $c_b(t)$ are time dependent.[17,18] By inserting eq. 8 into the time dependent Schrodinger equation,

$$H\psi = i\hbar\delta\psi/\delta t \quad (9)$$

it is possible, after some manipulations, to derive a set of equations for the expansion coefficients.[18] Solving those equations for the case where the electron is initially at $\phi_a(r)$ at an energy, $E_a=E_b=E$, and using the result obtained in eq. 8, gives for $\psi(r,t)$,[18,19]

$$\psi(r,t)=[\cos(Vt/\hbar)\phi_a(r_a)-i\ \sin(Vt/\hbar)\phi_b(r_b)]\exp{-iEt/\hbar} \quad (10)$$

In eq. 10, the exponential term is the usual phase factor for the wave function and $V=V_{ab}$ is the "electron exchange matrix element" or "tunneling matrix element" which is strongly dependent on

$$V_{ab} = \langle\phi_a(r)|V|\phi_b(r+R')\rangle = V \quad (11)$$

R, $V_{ab}=V_{ab}(R)$.

By squaring $\psi(r,t)$ we obtain the probability distribution of the electron which gives an interesting result in the context of our electron transfer problem.

$$|\psi(r,t)|^2 = |\phi_a|^2\cos^2(Vt/\hbar) + |\phi_b|^2\sin^2(Vt/\hbar) \quad (12)$$

Eq. 12 shows that the probability distribution of the electron is oscillatory. If the electron is initially in ϕ_a, it will oscillate from ϕ_a to ϕ_b with an effective electron transfer frequency $2V_{ab}/h\ s^{-1}$.

Returning to the diagrams in Figures 1 and 2, as the internuclear separation decreases and approaches R_e (Fig. 1), the potential energy for electronic motion between the nuclei (Fig. 2) falls rapidly. In this region, the system becomes "delocalized" and an adequate description of electronic distribution is given by the usual stationary state molecular wave functions for the bonding and antibonding orbitals,

$$\psi(r) = \frac{1}{\sqrt{2}}(\phi_a \pm \phi_b).$$

VIBRATIONS AND COORDINATES

In the previous section, electron transfer occurred between two sites where there were discrete, well-defined energies. For a given electronic state of a molecule in solution, when both molecular and solvent vibrations are considered there is a near continuum of possible energies. A description of the system must now be made in terms of "average" energies and coordinates which depend on temperature and on the number of vibrational levels per energy interval because they determine the population statistics. The differences in such quantities as a function of electron content (oxidation state) provide an additional factor which must be taken into account if electron transfer is to occur.

In dealing with the problem of molecular vibrations quantum mechanically, it is illustrative to begin with the complete Hamiltonian operator for a molecule,

$$H = T_N + T_{el} + V_{NN} + V_{el} \qquad (13)$$

In eq. 13 T_N and T_{el} are the kinetic energy operators for nuclear and electronic motions, V_{NN} is the potential energy operator for nuclear-nuclear repulsions, and V_{el} is the potential energy operator for both electron-electron repulsions, and electron-nuclear attractions. Defining the molecular vibrations is greatly simplified by invoking the Born-Oppenheimer approximation. The Born-Oppenheimer approximation is based on the fact that the far heavier masses of the nuclei make nuclear motion slow relative to electron motion. During a cycle of electron motion, changes in nuclear configurations are negligible so that T_{NN} can be neglected and V_{NN} is essentially constant. It follows that for the separated electronic motion ($H_{el}=T_{el}+V_{el}$),

$$[H_{el}(r) + V_{NN}(r)]\psi_{el}(r) = U(r)\psi_{el}(r) \qquad (14)$$

In eq. 14, U(r) is the sum of the purely electronic energy E_{el} and V_{NN} both of which vary with changes in the nuclear coordinates r.

$$U(r) = E_{el}(r) + V_{NN}(r) \qquad (15)$$

Turning the argument around, on the time scale for a vibration, the electrons appear as an electron cloud or spring which adjusts rapidly to changes in r. Under these conditions U(r) is a smooth function of r and becomes the potential energy for nuclear motion. As a consequence, nuclear motion can be treated separately from electronic motion. Eq. 16 can be written for the separated nuclear motion ($H_N = T_N + U(r)$),

$$H_N\psi_N = E\psi_N \tag{16}$$

In eq. 16 E is the total energy since H_N contains operators for both the nuclear and electronic energies. Because of the separation of electronic and nuclear motions gained in eq. 14 and 16, the molecular wave function for the system can be adequately approximated by the product,

$$\psi(r_i, r_\alpha) = \psi_{el}(r_i, r_\alpha)\psi_{vib}(r_\alpha) \tag{17}$$

where r_i refers to electron and r_α to nuclear coordinates.

According to eq. 16, if it is to be determined how E varies with the vibrations of the system, it is necessary to evaluate U(r) as a function of r. This can be conveniently done by a Taylor's series expansion around the nuclear positions which are appropriate for the point of minimum or equilibrium potential energy ($U(r_e)$),

$$U(r)=U(r_e)+\sum_{i=1}^{3N}\left(\frac{\partial U}{\partial r}\right)_e(r_i-r_e)+\tfrac{1}{2}\sum_{i=1}^{3N}\sum_{j=1}^{3N}\left(\frac{\partial^2 U}{\partial r_i \partial r_j}\right)_e(r_i-r_e)(r_j-r_e)+\ldots \tag{18}$$

In the summations, N is the total number of atoms in the system being considered. The first derivative with respect to r is zero because $U(r_e)$ is a potential energy minimum. If terms higher than the quadratic one are neglected in the expansion, which means that anharmonicity effects are neglected, the expression for U(r) becomes,

$$U(r) = U(r_e) + \tfrac{1}{2}\sum_{i=1}^{3N}\sum_{j=1}^{3N}\left(\frac{\partial^2 U}{\partial r_i \partial r_j}\right)_e(r_i-r_e)(r_j-r_e) \tag{19}$$

It is much more convenient if eq. 19 is expressed in terms of the normal coordinates of the system, $Q_i=(Q-Q_e)_i$. They can be related to the usual cartesian coordinates ($r_i=x_1,y_1,z_1,x_2,\ldots$) by established procedures.[20] The use of normal coordinates has the advantage of simplifying both the nuclear kinetic and potential energy operators, as shown in eq. 20 for U, and it also factors the translational and rotational degrees of freedom out of the problem.

$$U(Q) = U(Q_e) + \tfrac{1}{2}\sum_{i=1}^{3N-6}\lambda_i Q_i^2 \qquad \left(\lambda_i=\left(\frac{\partial^2 U}{\partial Q_i^2}\right)_e\right) \tag{20}$$

Variations in the total potential energy of the system (minus the rotational and translational energies) with molecular vibrations can now be determined from the Schrodinger equation (eq. 16) using eq. 20 for the potential energy of nuclear motion. The answer is,

$$U = U(Q_e) + U_{vib} \tag{21}$$

$U(Q_e)$ is the sum of the electronic (E_{el}) and nuclear-nuclear repulsion (V_{NN}) energies at the equilibrium position where $Q=Q_e$. U_{vib} is the vibrational potential energy which is given by,

$$U_{vib} = \sum_i U_{vib,i} = \tfrac{1}{2}\sum_i \lambda_i Q_i^2 \tag{22}$$

Eq. 22 is of value to us because it gives an expression for the vibrational energy content of the system above that at equilibrium in terms of the molecular coordinates. We have begun to define the "trap" which localizes an electron in a mixed-valence compound.

By using $U_{vib,i}$ as the potential energy function, it is possible to determine the vibrational wave functions and quantized energy levels for each normal mode and therefore to establish a complete vibrational description for the system. The Hamiltonian operator for a single normal mode is given by,

$$H_{vib,i} = -\frac{\hbar^2}{2}\frac{\partial^2}{\partial Q_i^2} + \tfrac{1}{2}\lambda_i Q_i^2 \tag{23}$$

and for the complete vibrational Hamiltonian by,

$$H_{vib} = \sum_i H_{vib,i} \tag{24}$$

Since the vibrational Hamiltonian is a sum of terms each involving a single coordinate, it follows from the separation of variables theorem that,

$$\psi_{vib} = \prod_i \psi_{vib,i} \tag{25}$$

and that

$$E_{vib} = \sum_i E_{vib,i} \tag{26}$$

The vibrational wave functions $\psi_{vib,i}$, which are solutions to the Hamiltonian in eq. 23, are given by,

$$\psi_{vib,i}(Q_i) = \frac{1}{(2^{n_i} n_i!)}\left(\frac{\alpha_i}{\pi}\right)^{1/4} e^{-\alpha_i Q_i^2/2}\, H_{ni}(\alpha_i^{1/2} Q_i) \tag{27}$$

where, $\alpha=2\pi\nu/\hbar=\lambda_i^{1/2}/\hbar$. The vibrational energy associated with occupation of the n th level of the i th normal mode is,

$$E_{vib,i} = (n_i + \tfrac{1}{2})h\nu_i = (n_i + \tfrac{1}{2})\hbar\omega_i \tag{28}$$

and the total, quantized vibrational energy by,

$$E_{vib} = \sum_i (n_i + \tfrac{1}{2})\hbar\omega_i \tag{29}$$

Note that a whole series of normal modes may be found to be excited simultaneously in a molecule and that the total vibrational energy is given by the sum of $\hbar\omega_i(n_i+\frac{1}{2})$ for each mode.

In principle, we have now acquired detailed information about vibrational traps for electrons assuming that they consist of a series of harmonic oscillators. We have established a collection of normal modes which constitute the traps, the quantized energy levels for each mode, and through $\psi^2_{vib,i}$, the probability distribution of the nuclear coordinates in each level of each normal mode. Another factor that must be considered to help determine the rate of escape from the trap, is the population distribution within the set of vibrational levels as a function of temperature. From statistical mechanics,(21) the partition function for a harmonic oscillator normal mode is given by,

$$f_i = \sum_n \exp-(n_n\hbar\omega_i/kT) = \frac{1}{1-\exp(\hbar\omega_i/kT)} \tag{30}$$

The partition function in eq. 30 gives the population distribution in normal mode i from the n=0 level up. The fractional population in any particular level in a normal mode is given by,

$$\frac{\exp-(n_n\hbar\omega_i/kT)}{f_i}$$

A useful quantity for us is $\bar{n}_i$ the quantum number for the vibrational level at the statistical average since $(\bar{n}_i+\frac{1}{2})\hbar\omega_i$ gives the average vibrational energy content in normal mode <u>i</u>. $\bar{n}_i$ is given by,

$$\bar{n}_i = \frac{\exp-(\hbar\omega_i/kT)}{1-\exp-(\hbar\omega_i/kT)} = \frac{1}{\exp(\hbar\omega_i/kT)-1} \tag{31}$$

Turning back to the electron transfer problem, in order to define potential surfaces and energy-coordinate relationships and to deal with the mathematics involved, it is convenient to define a new set of "reduced" normal coordinates for the nuclei. The reduced coordinates are related to the usual normal coordinates by,

$$q_i = \frac{(Q-Q_e)_i}{<Q_i^2>^{1/2}} = \frac{Q_i}{<Q_i^2>^{1/2}} \tag{32}$$

Since the $<Q_i^2>^{1/2}$ are zero energy, root mean displacements, the q_i are dimensionless coordinates which describe nuclear motions in terms of fractional displacements from the mean at zero energy. Application of the Hellmann-Feynmann theorem (eq. 33)[16] to a normal vibration having effective force constant λ and reduced mass M gives for the variation of E with λ (Note eq. 23 for H_N, and eq. 29 for E where $\omega = (\lambda/M)^{1/2}$),

$$\frac{\partial E}{\partial \lambda} = <\psi_N|\frac{\partial H_N}{\partial \lambda}|\psi_N> \tag{33}$$

$$\frac{\partial((n+\frac{1}{2})\hbar(\lambda/M)^{1/2})}{\partial \lambda} = <\psi_N|\tfrac{1}{2}Q^2|\psi_N> \tag{34}$$

For the n=0 level, using $\lambda = M\omega^2$, gives,

$$<Q^2> = \frac{\hbar}{M\omega} \tag{35}$$

and for the specific normal mode i,

$$q_i = Q_i\left(\frac{M_i\omega_i}{\hbar}\right)^{1/2} \tag{36}$$

The potential energy of vibration associated with mode i is now given by,

$$U_{vib,i} = \tfrac{1}{2}\lambda_i Q_i^2 = \tfrac{1}{2}\hbar\omega_i q_i^2 \tag{37}$$

and the total vibrational potential energy by,

$$U_{vib} = \tfrac{1}{2}\sum_i \hbar\omega_i q_i^2 \tag{38}$$

POTENTIAL SURFACES AND ENERGY-COORDINATE RELATIONSHIPS

Having established both the variation in energy with vibrations at an electron trap site and a useful coordinate system, we can now turn to electron transfer in solution. The problem involves two different chemical sites and a coupling between them which can induce electron transfer. The process involved is a nonadiabatic transition between electronic states in the quantum mechanical sense because electronic coupling is weak. In the simplest case of a self-exchange reaction, the only difference between the reactant and product states is in the relative position of the excess electron.

$$Fe(H_2O)_6^{2+}, Fe(H_2O)_6^{3+} \longrightarrow Fe(H_2O)_6^{3+}, Fe(H_2O)_6^{2+} \quad (39)$$
$$\underline{(a)} \qquad\qquad\qquad \underline{(b)}$$

The molecular and solvent normal modes and frequencies are the same in both states, but even so, they are displaced relative to each other because of the shift in origin.

The latter point is illustrated in Figure 3 for the Fe^{3+}/Fe^{2+} case. It shows an energy-coordinate diagram for one of the normal modes of the system, q_i. The two complexes, $Fe(H_2O)_6^{3+}$ and $Fe(H_2O)_6^{2+}$, are taken to be at their equilibrium interreactant distance in an association complex (eq. 1). The particular normal mode shown is intended to represent a symmetric breathing mode at Fe^{2+}. Because of weak electronic coupling, the different molecular vibrations at Fe^{2+} and Fe^{3+} can be regarded as localized. The choice of a solvent vibration would have been equally appropriate but a solvent mode would be coupled simultaneously to both reactant sites.

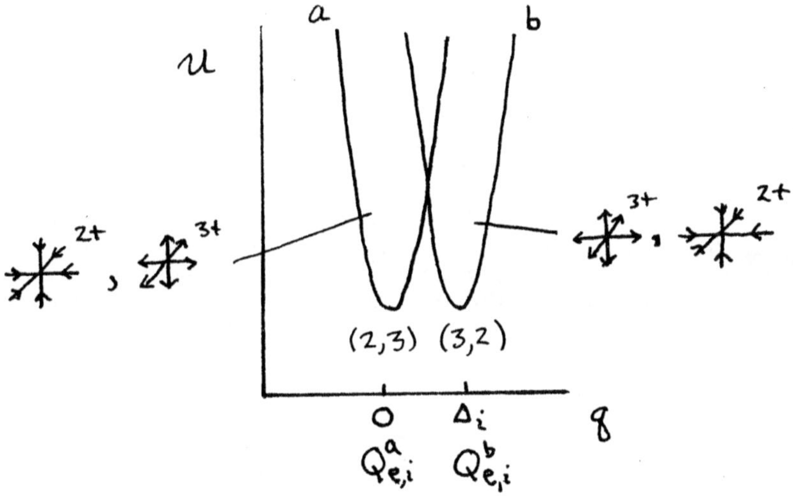

Figure 3. Energy-coordinate diagram for electron transfer between $Fe(H_2O)_6^{2+}$ and $Fe(H_2O)_6^{3+}$. The vibrational coordinate is that for the symmetric iron-ligand breathing mode at Fe^{2+}.

The vibrational potential energy associated with normal $\underline{i}$ in state $\underline{a}$ is given by,

$$U^a_{vib,i} = \tfrac{1}{2}\hbar\omega_i q_i^2 \quad (40)$$

The individual quantized energy levels (eq. 28) are not shown in the Figure.

In state b, the normal mode equations are identical in form but displaced from the origin for a. The coordinates in b are related to those in a by the displacement coordinate, Δ_i. Δ_i is the difference in Q_i values for the two states at their equilibrium positions divided by $<Q_i^2>^{\frac{1}{2}}$ for state a.

$$\Delta_i = \frac{Q^a_{e,i}-Q^b_{e,i}}{<(Q^a_{e,i})^2>^{\frac{1}{2}}} = (Q^a_{e,i}-Q^b_{e,i})\left(\frac{M_i\omega_i}{\hbar}\right)^{\frac{1}{2}} \quad (41)$$

The advantage in defining Δ_i is that it allows any coordinate in state b to be related to the origin in state a by a simple transformation,

$$q^b_i = q^a_i - \Delta_i \quad (42)$$

It follows directly that the vibrational energy in state b, where the oxidation states are reversed, is given by,

$$U^b_{vib,i} = \tfrac{1}{2}\hbar\omega_i(q_i-\Delta_i)^2 = \tfrac{1}{2}\hbar\omega_i q_i^2 - \hbar\omega_i q_i \Delta_i + \tfrac{1}{2}\hbar\omega_i\Delta_i^2 \quad (43)$$

$$U^b_{vib,i} = U^a_{vib,i} + E_{m,i} - \hbar\omega_i q_i \Delta_i \quad (44)$$

Eq. 44 follows from eq. 43 by noting that, $U^a_{vib,i} = \frac{1}{2}\hbar\omega_i q_i^2$ and defining $E_{m,i} = \frac{1}{2}\hbar\omega_i\Delta_i^2$. $E_{m,i}$ is the vibrational energy associated with curve b when it lies directly above the equilibrium coordinate for curve a (note Figure 4 below). The apparent coupling term, $-\hbar\omega_i q_i \Delta_i$, is simply a consequence of the displacement between curves a and b.

Equations like 42, 43, and 44 can be derived for all of the molecular and solvent vibrations in states a and b and the total vibrational potential energy functions are,

$$U^a_{vib} = \tfrac{1}{2}\sum_i \hbar\omega_i q_i^2 \quad (45)$$

$$U^b_{vib} = \tfrac{1}{2}\sum_i \hbar_i\omega_i(q_i-\Delta_i)^2 = U^a_{vib} + E_m - \sum_i \hbar\omega_i\Delta_i q_i \quad (46)$$

where E_m is defined as,

$$E_m = \tfrac{1}{2}\sum_i \hbar\omega_i\Delta_i^2 \quad (47)$$

Any accounting of a transition between states a and b must consider the possible roles of all of the molecular and solvent vibrations. Although the problem is quantized, it is useful to consider the classical limit where the quantum spacings, $\hbar\omega_i$, are small compared to kT and, $U_{vib} = E_{vib}$. As shown in Figure 4, the classical limit allows some important energy relationships to be visualized. Figure 4 is a generalized, energy-coordinate diagram which is intended to show schematically the effect of all of the vibrations of the system.

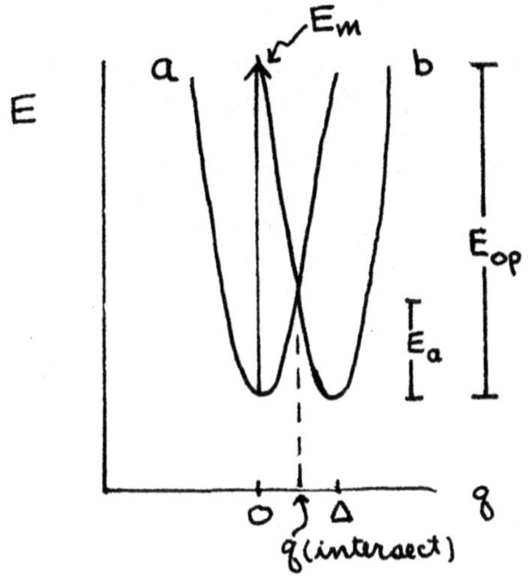

Figure 4. Generalized energy coordinate diagram in the classical limit.

The first relationship of value is that the vertical energy of state b above state a in its equilibrium nuclear configuration is given by,

$$E_m = \tfrac{1}{2}\sum_i \hbar\omega_i \Delta_i^2 = E_{op} \tag{48}$$

Eq. 48 is of interest in the context of symmetrical mixed-valence ions since E_m corresponds to the vertical or optical energy (E_{op}) of an Intervalence Transfer absorption band. The second relationship of interest pertains to the intersection region between surfaces which defines the classical energy of activation for thermal electron transfer. At the intersection region, $E^a_{vib} = E^b_{vib}$ and from eq. 45 and 46, it follows that for each normal mode in the intersection region,

$$q_i(\text{intersection}) = \Delta_i/2 \tag{49}$$

The classical energy of activation for the electron transfer reaction from state a to b is then given by,

$$E_a = E^a_{vib}(\text{intersection}) = \tfrac{1}{2}\sum_i \hbar\omega_i\left(\frac{\Delta_i}{2}\right)^2 \tag{50}$$

which immediately leads to the third equation of interest,

$$E_a = \tfrac{1}{4}E_{op} = \tfrac{1}{4}E_m \tag{51}$$

For a symmetrical mixed-valence ion, the energy gaps for the optical and thermal processes are related by a simple factor of 4.

For the more general case of electron transfer in an unsymmetrical system, e.g., $Ru(NH_3)_6^{2+}$,$Fe(H_2O)_6^{3+} \rightarrow Ru(NH_3)_6^{3+}$, $Fe(H_2O)_6^{2+}$, the situation is far more complicated because, in general, ΔG and ΔE are non-zero and the loss of chemical symmetry means that the normal modes are different in the initial and final states. General vibrational energy-coordinate relationships have been developed for such cases[22,23] but the problem is made far simpler by assuming that the frequencies and normal modes are the same in both the initial and final states. With that assumption, and the fact that ΔE between the two states is given by,

$$\Delta E = E^b_{vib}(Q^b_e) - E^a_{vib}(Q^a_e) \tag{52}$$

a series of useful energy-coordinate relationships can also be developed for the unsymmetrical case (Note Fig. 5).

In state a, the vibrational potential energy is,

$$U^a_{vib} = \tfrac{1}{2}\sum_i \hbar\omega_i q_i^2 \tag{53}$$

For state b, taking into account both the displacement in origin and the difference in internal energy between a and b, gives,

$$U^b_{vib} = \tfrac{1}{2}\sum_i \hbar\omega_i(q_i-\Delta_i)^2 + \Delta E = U^a_{vib} + \tfrac{1}{2}\sum_i \hbar\omega_i\Delta_i^2 + \Delta E - \sum_i \hbar\omega_i q_i\Delta_i \tag{54}$$

$$E_m = \tfrac{1}{2}\sum_i \hbar\omega_i\Delta_i^2$$

The situation is illustrated, once again in the classical limit where $U^a_{vib} = E^a_{vib}$, in Figure 5.

In the intersection region between surfaces, $E^a_{vib} = E^b_{vib}$, and it follows from eq. 53 and 54 that for each normal mode in the intersection region,

$$\hbar\omega_i\Delta_i q_i = \Delta E + \tfrac{1}{2}\hbar\omega_i\Delta_i^2 \tag{55}$$

so that

$$q_i(\text{intersect}) = \frac{\Delta_i}{2}(1+\Delta E/\tfrac{1}{2}\hbar\omega_i\Delta_i^2) = \frac{\Delta_i}{2}(1+\Delta E/E_{m,i}) \tag{56}$$

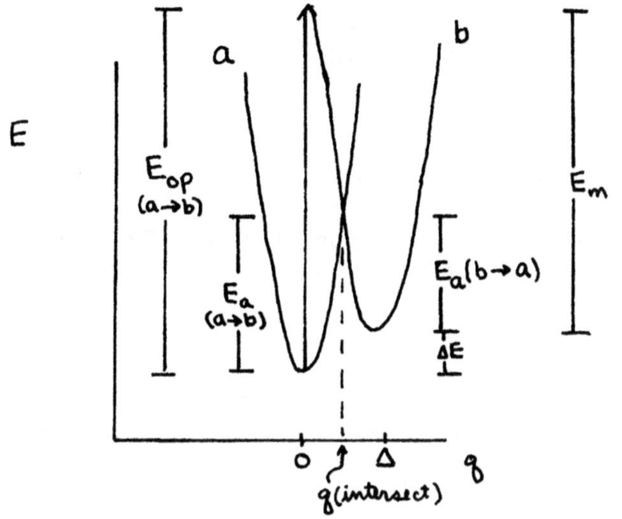

Figure 5. Energy-coordinate diagram for unsymmetrical electron transfer, classical limit.

The classical activation energy for the transition from state a to b is given by,

$$E_a(a\to b) = \tfrac{1}{2}\sum_i \hbar\omega_i q_i^2(\text{intersect}) = \tfrac{1}{2}\sum_i \frac{\hbar\omega_i\Delta_i^2}{4}(1+\Delta E/E_{m,i})^2 \qquad (57)$$

and,

$$E_a(a\to b) = \frac{E_m}{4}(1+\frac{\Delta E}{E_m})^2 = \frac{(E_m+\Delta E)^2}{4E_m} \qquad (58)$$

The optical transition from a→b is the quantity of most interest in unsymmetrical mixed-valence dimers because it corresponds to the energy of the IT band. Since the vertical energy between surfaces at $Q^a{}_e$ is given by,

$$E_{op} = E_{op}(a\to b) = \Sigma_i \hbar\omega_i\Delta_i^2 + \Delta E = E_m + \Delta E \qquad (59)$$

it follows that the optical and thermal processes are related by,

$$E_a(a\to b) = \frac{(E_{op}^2)}{4(E_{op}-\Delta E)} \qquad (60)$$

For the reverse optical and thermal reactions, it is easily shown that,

$$E_a(b\rightarrow a) = \frac{E_m}{4}(1-\frac{\Delta E}{E_m})^2 = \frac{(E_m-\Delta E)^2}{4E_m} \quad (61)$$

and that,

$$E_{op}(b\rightarrow a) = E_m - \Delta E \quad (62)$$

All of these useful relationships (eq. 50,51,58,59) were derived earlier by Hush although in his nomenclature, $E_m=\chi$ and $\Delta E = E_o$.[9]

ELECTRON TRANSFER TRANSITIONS

In the previous section, potential energy surfaces were developed which described the role of solvent vibrations in electron transfer. In the classical limit, the energies of thermal and optical processes were shown to be directly related. However, the actual electronic transition has not yet been considered nor has the quantum mechanical view of the role of molecular and solvent vibrations.

The underlying theory in this area extends back to the early work of Kubo and others on electron transfer and related processes in the solid state,[(24,25)] extends through the early work on chemical systems by Levich and Dogonadze and their coworkers,[(26)] and more recently to work by a number of authors including important contributions by Jortner and coworkers[(27,28)] and by Duke.[(18,29)] Electron transfer theory has been exhaustively reviewed by Schmidt.[(30)] The results described here will be based largely on the work by Jortner and Duke and their coworkers.

With the molecular and solvent vibrational potential surfaces described, the nature of the vibrational trap holding back the electron has been defined. The rate of escape of the electron from the trap can be treated by time-dependent perturbation theory.[(17,30)] There are clear similarities in the treatment with the procedure adopted for the H_2^+ case in an earlier section. The Hamiltonian operator (note eq. 7) is,

$$H = H_o + V \quad (63)$$

H_o is the Hamiltonian operator for the unperturbed system and V is the electron exchange operator which is the perturbation. With the perturbation, the system is no longer in an eigenstate but a new, time-dependent wave function, $\psi(t)$, can be defined by an expansion of the eigenfunctions of the unperturbed system. The expansion is valid if the expansion coefficients are time dependent,

$$\psi(t) = \sum_n C_n(t)\psi_n^o \exp-(iE_n^o t/\hbar) = \sum_n C_n(t)\psi_n(t) \tag{64}$$

In the expansion, ψ_n^o is the time-independent or stationary part of $\psi_n(t)$, E_n^o is the energy or eigenvalue, and $\exp-(iE_n^o t/\hbar)$ is the phase factor. The expansion coefficients squared, $|C_n(t)|^2$, give the probability of finding the system in the nth state following the perturbation.

The goal is to derive equations for the rate of change of $|C_n(t)|^2$ with time since they will describe the rate of transition from initial to final states. The procedure followed is to insert the expression for $\psi(t)$ in eq. 64 into the time-dependent Schrodinger equation (eq. 9), which after some trickery gives,[17,30]

$$i\hbar \frac{\partial C_n(t)}{\partial t} = \sum_k \langle\psi_n^o|V|\psi_k^o\rangle \exp(i\omega_{nk}t)C_k(0) \tag{65}$$

In eq. 65, $\langle\psi_n^o|V|\psi_k^o\rangle = V_{nk}$, is the electron exchange matrix element. The exponential term is the ratio of the phase factors for ψ_n and ψ_k,

$$\exp(i\omega_{nk}t) = [\exp-(iE_n^o t/\hbar)]^{-1} \exp-(iE_k^o t/\hbar) \tag{66}$$

$$\hbar\omega_{nk} = E_n^o - E_k^o$$

$C_k(0)$ is the coefficient of eigenstate k at t=0. If we focus our attention initially on transitions only from a single vibrational level <u>j</u> in our initial electronic state <u>a</u>, then, $C_k(0)=1(k=j)$; $C_k(0)=0(k\neq j)$ and eq. 65 becomes

$$i\hbar \frac{\partial C_n(t)}{\partial t} = V_{nj}\exp(i\omega_{nj}t) \tag{67}$$

Eq. 67 is only valid for times near zero where $C_n(t) << C_j(t)$. It is easily integrated to give,

$$C_n(t) = \frac{V_{nj}}{\hbar\omega_{nj}}(1-\exp(i\omega_{nj}t)) \tag{68}$$

and it follows that,

$$|C_n(t)|^2 = \frac{2V_{nj}^2}{\hbar^2} \frac{(1-\cos\omega_{nj}t)}{\omega_{nj}^2} \tag{69}$$

Eq. 69 gives the probability that a transition from initial level <u>j</u> to final level <u>n</u> has occurred after time <u>t</u>. It is an oscillating function and it is important to note that its amplitude has a pronounced peak at $\omega_{nj}=0$. Only transitions for which, $E_n^o \approx E_j^o$, are expected to make an important contribution. The total transition probability from j must take into account transitions to all possible vibrational levels in the final state, which leads to eq. 70,

$$\sum_n |C_n(t)|^2 = \frac{2}{\hbar^2} \sum_n V_{nj}^2 \frac{(1-\cos\omega_{nj}t)}{\omega_{nj}^2} \tag{70}$$

If the number of vibrational levels in the final state is sufficiently large, it can be described by a density of states function, $\rho_f(E^o_n)$. The summation in eq. 70 can then be replaced by an integral,

$$\sum_n |C_n(t)|^2 = \frac{2}{\hbar^2} \int |V_{nj}|^2 \frac{1-\cos\omega_{nj}t}{\omega_{nj}^2} \rho(E_n^o) dE_n^o \tag{71}$$

In order to calculate a transition rate, the time rate of change of the probability functions, $|C_n(t)|^2$, must be determined. When all possible transitions from level j in state a to state b are included, eq. 72 is the result.

$$W_{aj} = \frac{d}{dt} \sum_n |C_n(t)|^2 = \frac{2}{\hbar^2} \int |V_{nj}|^2 \frac{\sin\omega_{nj}t}{\omega_{nj}} \rho_f(E_n^o) dE_n^o \tag{72}$$

Assuming that $|V_{nj}|^2$ and $\rho_f(E_n^o)$ are reasonably constant near E_n^o, they can be taken out of the integral giving,

$$W_{aj} = \frac{2}{\hbar} |V_{nj}|^2 \rho_f(E_n^o) \int_{-\infty}^{+\infty} \frac{\sin\omega_{nj}t}{\omega_{nj}} d\omega_{nj} \tag{73}$$

or,

$$W_{aj} = \frac{d}{dt} \sum_n |C_n(t)|^2 = \frac{2\pi}{\hbar} |\langle\psi_n^o|V|\psi_j^o\rangle|^2 \rho_f(E_n^o) \tag{74}$$

The result in eq. 74 is the famous Fermi "Golden Rule". It gives the rate of transition from vibrational level j in state a to all of the vibrational levels in state b in terms of two quantities: (a) the density of vibrational levels in the final state with energies near j, $E^o_{aj} \approx E^o_{bn}$; and (b) the electron exchange matrix element which includes both the vibrational and electronic parts of the problem.

If the Born-Oppenheimer approximation holds, the nuclear and electronic wave functions can be separated, $\psi = \psi_{el}\psi_{vib}$ (eq. 17). The wave functions for the initial and final states can be written as,

$$\psi_n^{o} = \psi_b^{o}\chi_{bn}^{o}$$

$$\psi_j^{o} = \psi_a^{o}\chi_{aj}^{o}$$

where, for example, χ_{bn}^{o} is the vibrational wave function for vibrational level n in state b and χ_b^{o} is the electronic wave function. The electronic part of the electron exchange matrix element is only weakly dependent on nuclear coordinates; the electronic and vibrational contributions in the Golden Rule expression can be separated. The electron exchange matrix element becomes the product of a purely electronic matrix element and a vibrational overlap term,

$$\langle\psi_b^{o}\chi_{bn}^{o}|V|\psi_a^{o}\chi_{aj}^{o}\rangle \approx \langle\psi_b^{o}|V|\psi_a^{o}\rangle\langle\chi_{bn}^{o}|\chi_{aj}^{o}\rangle \tag{75}$$

This assumption, the Franck-Condon approximation, is commonly made in optical transitions. The transition rate through vibrational level j now becomes,

$$W_{aj} = \frac{2\pi}{\hbar}V_{ab}^{2}\langle\chi_{bn}^{o}|\chi_{aj}^{o}\rangle^{2}\rho_f(E_n^{o}=E_j^{o}) \tag{76}$$

In eq. 76, $V_{ab}=V_{ba}=\langle\psi_b^{o}|V|\psi_a^{o}\rangle$.

In order to calculate the total transition rate from state a to state b, transitions from all of the vibrational levels in a must be considered. Each level has a characteristic transition rate, W_{aj}, and a total accounting is possible by taking the product of the transition rate through each level and the statistical probability that the level is occupied,

$$W_a = \frac{\sum_j W_{aj}\exp-(E_{aj}^{o}/kT)}{\sum_j \exp-(E_{aj}^{o}/kT)}$$

$$= \frac{2\pi}{\hbar}V_{ab}^{2}\,\frac{\sum_j \exp-(E_{aj}^{o}/kT)\langle\chi_{bn}|\chi_{aj}\rangle^{2}\rho_f(E_{bn}^{o}=E_{aj}^{o})}{\sum_j \exp-(E_{aj}^{o}/kT)} \tag{77}$$

W_a is a thermally averaged, time dependent transition probability between the vibrational manifolds of states a and b. Strictly speaking, it is not a rate constant but rather a time dependent transition probability at a fixed interreactant distance, d. In general, W_a is a function of d, but for us d is a fixed quantity, either the interreactant distance across a ligand bridge or within an association complex, and it follows that,

$$W_a = k_{et} \tag{78}$$

The forms of the single level (eq. 76) and total transition probabilities (eq. 77) are revealing in terms of factors at the miscroscopic level which are involved in determining the transition rate. Those factors are: (1) From the V_{ab} term, there must be orbital overlap in order for electron transfer to occur. The assumption is made here that V_{ab} is a constant and the same for all vibrational levels. (2) Vibrational overlap between the wave functions for the initial and final levels must be nonzero. The larger the overlap relative to other overlaps, the more important is the contribution from $\underline{j}\rightarrow\underline{n}$ compared to other transitions. (3) The higher the density of levels in the acceptor manifold at the same energy as the initial level, $E_{aj}^{o}=E_{bn}^{o}$, the more important the transition through level $\underline{j}$. (4) The importance of transitions through $\underline{j}$ depends on the population in $\underline{j}$ and it is assumed that the population can be calculated using equilibrium statistical mechanics. The total transition through all levels will clearly depend on how the system is distributed throughout the levels.

Some of the points raised here are illustrated in Fig. 6 where two different sets of $\underline{j}\rightarrow\underline{n}$ transitions are considered.

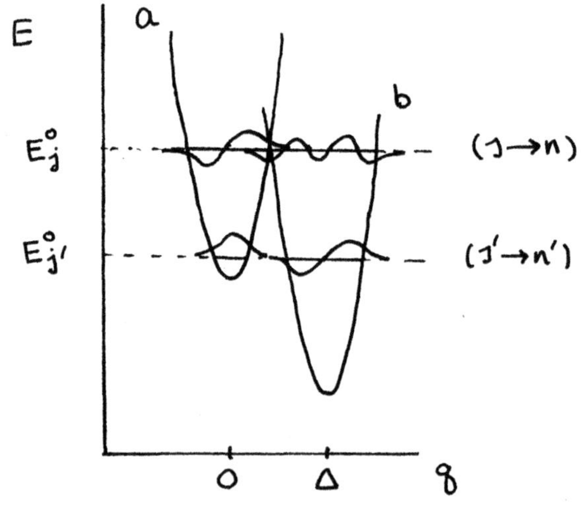

Figure 6. Energy coordinate diagram showing schematically the importance of vibrational overlap.

For both cases, the energy of the initial and final levels is the same ($E_{aj}^{o}=E_{bn}^{o}$). A schematic attempt has been made to illustrate the amplitudes of the wave functions involved. Overlap between the aj and bn levels is reasonably good. When vibrational level aj is populated, the probability of finding the system with nuclear coordinates near the intersection region between the potential energy curves and even beyond is relatively good. The uncertainty in position is of course, associated with the quantum mechanical nature of the problem. Another consequence of the dispersion in nuclear coordinates is the overlap region between χ_{aj}^{o} and χ_{bn}^{o}. In this region, the nuclear coordinates associated with having the electron on either redox site are the same. In the overlap region, there is no longer a vibrational distinction between states a and b. Because of V_{ab}, an electronic perturbation exists which allows electron transfer to occur. The net effect is the nonadiabatic transition between states a and b through level j.

The transition aj'→bn' is far less favorable. Because of the displacement between the potential surfaces, and the fact that the amplitudes of the vibrational wave functions die off quickly past the potential energy curve, vibrational overlap is small. At room temperature, where population statistics in levels near the intersection region are high, the contribution to the a→b transition rate through aj'→bn' is negligible. However, at low temperatures, the transition from a→b, although slow, will be dominated by transitions like aj'→bn' and the apparent activation energy will fall to zero. The transition between j' and n' is referred to as a "nuclear tunneling" process although the origin of the tunneling is the same as for transitions in the intersection region, overlap between vibrational wave functions.

Returning to the total transition rate, convenient theoretical techniques have been developed for evaluating eq. 77.[25,29,30] Although the details will not be developed here, it is useful to cite the results obtained in certain limiting cases.

In one limit, electron-vibration interactions are strong, E_m(eq. 47)>>$h\omega_j$; the quantum spacings for the vibrations are small compared to the energy gaps involved for optical and thermal electron transfer. In this limit, which is certainly appropriate for at least solvent vibrations, the transition probability is given by a semiclassical expression,(18,26-29)

$$W_a = \frac{2\pi V_{ab}^2}{\hbar}\left(\frac{\pi}{\sigma_{ab}^2}\right)^{\frac{1}{2}}\exp-\left[\frac{(E_m+\Delta E)^2}{4\sigma_{ab}^2}\right] \qquad (79)$$

The terms V_{ab}, E_m, and ΔE have all been encountered before. E_m is the vibrational energy of state b when it lies directly above the equilibrium coordinate of state a (q_j=0). The term σ_{ab}^2 is defined as,

$$\sigma_{ab}^2 = \tfrac{1}{2}\sum_j E_{m,j}(\hbar\omega_j)(2\bar{n}_j+1) = \sum_j E_{m,j}(\hbar\omega_j)(\bar{n}_j+\tfrac{1}{2}) \qquad (80)$$

The summation is taken over all of the vibrations involved in the transition and $\hbar(\omega_j)(\bar{n}_j+\frac{1}{2})$ is the average thermal energy in the jth mode. σ^2 is a mean square deviation function which is a measure of the sharpness of the vibrational distribution involved in the transition. It contains information about the number of levels per energy interval; the density of levels. Eq. 80 can also be written in terms of the hyberbolic cotangent function, coth $\hbar\omega_j/2kT$,

$$\sigma_{ab}^2 = \tfrac{1}{2}\sum_j E_{m,j}(\hbar\omega_j)\coth(\hbar\omega_j/2kT) \tag{81}$$

which follows from the definition of the coth function,

$$\coth u = \frac{e^u+e^{-u}}{e^u-e^{-u}} \tag{82}$$

and eq. 31 for $\bar{n}$.

If we are dealing only with solvent vibrations where $\hbar\omega \ll kT$, coth $(\hbar\omega/2kT) \rightarrow 2kT/\hbar\omega$, and σ_{ab}^2 becomes,

$$\sigma_{ab}^2 = 4E_m kT \tag{83}$$

The transition probability is now,

$$W_a = k_{et} = \frac{2\pi V_{ab}^2}{\hbar}\left(\frac{\pi}{kTE_m}\right)^{\frac{1}{2}} \exp-\frac{(E_m+\Delta E)^2}{4E_m kT} \tag{84}$$

According to eq. 84, if the quantum spacings between vibrational levels are small compared to kT, the transition from state a to state b will occur completely through vibrational levels at the intersection region where vibrational overlap is maximized. The thermal energy of activation is given by,

$$E_a = \frac{E_m}{4}\left(1 + \frac{\Delta E}{E_m}\right)^2 = \frac{(E_m+\Delta E)^2}{4E_m} \tag{84a}$$

At this point, it is useful to recall the assumptions made in the derivation of eq. 84. They include: (1) Weak electronic coupling. (2) Harmonic oscillator vibrations. (3) The same normal modes and frequencies in reactant and product states. (4) Strong electron-vibration interactions. (5) Vibrational quantum spacings which are small compared to kT.

TREATMENT OF MOLECULAR AND SOLVENT VIBRATIONS

In the previous section, time dependent perturbation theory was used to derive a general expression for the rate of electron transfer in solution. The expression was evaluated at the high temperature limit, $\hbar\omega << kT$, which gave eq. 84. In this limit, the exponential term in eq. 84 is identical to the expression for the classical energy of activation (eq. 58).

Since at room temperature, $kT \sim 200$ cm^{-1}, the assumption that $\hbar\omega << kT$ is valid for vibrations in polar solvents where $\hbar\omega \sim 1$-10 cm^{-1}. However, for molecular vibrations where $\hbar\omega$ is generally in the range 200-3500 cm^{-1}, the assumption clearly breaks down.

Because of the large difference in frequency ranges and vibrational lifetimes ($\sim 10^{-11}$s vs. $\sim 10^{-14}$s) for the two different types of vibrations, they are usually treated as separate problems. The high frequency molecular modes are treated as uncoupled vibrations at the two redox sites and the solvent vibrational modes as a collective property of both redox sites.

Both types of vibrations can be incorporated into the quantum mechanical treatment described in the previous section. The results of the analysis have been discussed by Jortner(27b) in a series of temperature ranges: (1) at very low temperatures, thermal activation to the intersection region can not occur. Electron transfer occurs by nuclear "tunneling" through vibrational levels low in the vibrational manifold (Note Figure 6); $E_a \sim 0$. (2) At very high temperatures where $kT >> \hbar\langle\omega_i\rangle >> \hbar\langle\omega_o\rangle$, both molecular and solvent vibrations are in the high temperature limit. $\langle\omega_i\rangle$ is the average frequency for the molecular vibrations (inner coordination sphere vibrations for metal complexes). $\langle\omega_o\rangle$ is the average frequency for the solvent vibrations (outer coordination sphere vibrations for metal complexes). Eq. 84 is appropriate for calculating the electron transfer rate constant. E_a is given by eq. 58. E_m includes the sum of contributions from both molecular and solvent vibrations, $E_m = \chi_i + \chi_o$. (c) In an intermediate temperature range, which is the most common case for electron transfer reactions, $\hbar\langle\omega_i\rangle > kT >> \hbar\langle\omega_o\rangle$. The solvent is behaving classically and can be treated using the high temperature limit. However, the molecular vibrations can not be treated classically and the contribution from each vibration must be taken into account. Nuclear tunneling occurs in concert with the usual background solvent contribution at the intersection region between surfaces. When taken together the two contributions represent the total vibrational distribution leading to electron transfer.

$$W_a = \frac{V_{ab}^2}{\hbar}(\exp(-S))\left(\frac{\pi}{\chi_o kT}\right)^{\frac{1}{2}} \sum_{n=o}^{\infty} \exp-\left[\frac{(\Delta E + \chi_o + n\hbar\langle\omega_i\rangle)^2}{4\chi_o kT}\right]\frac{S^n}{n!} \quad (85)$$

In the intermediate temperature region the transition probability is written in terms of a summation over the individual molecular vibrations (eq. 85). In eq. 85, S is the average electron-vibration coupling strength for the molecular vibrations (eq. 86) and n is the vibrational quantum number. In this temperature region the energy of activation is given by eq. 85a.

$$E_a > \frac{(\chi_o+\Delta E)^2}{4\chi_o} \tag{85a}$$

$$S = \tfrac{1}{2}\sum_j \Delta_j^{\,2} = \tfrac{1}{2}\frac{\chi_i}{\hbar\langle\omega_i\rangle} \tag{86}$$

The specific evaluation of the tunneling contribution from the molecular vibrations is obviously a difficult problem. However, work of this kind has appeared[27,28,31-33] including a recent paper on the $Co(NH_3)_6^{3+/2+}$ self-exchange reaction.[34]

For our purposes, which include largely spin-paired d^6-d^5 mixed-valence cases, and for many self-exchange reactions including both organic, e.g., p-$Me_2NC_6H_4NMe_2^{+/0}$,[35] and inorganic, e.g., $Ru(NH_3)_6^{3+/2+}$,[36] examples, the role of molecular vibrations may be small. For the self-exchange reactions, reasonable estimates for self-exchange rate constants can be obtained simply by assuming that solvent vibrations are the origin of the electron trap ($E_a \sim \chi_o/4$). In some cases crystal structure analyses show that only minor changes in structure occur between complexes in different oxidation states and, at best, only a small contribution from molecular vibrations is expected.[37]

Even if molecular vibrations enter the problem in a significant way, the number of vibrations involved may be small. Examples of an important type of molecular vibration for metal complexes are low frequency, symmetric metal-ligand breathing modes. For such cases, the summation in eq. 85 becomes a single term and classical methods for estimating the contribution from χ_i are available.[38]

A number of treatments have been given for the role of solvent vibrations in electron transfer taking advantage of the classical nature of the problem. They include the non-equilibrium polarization approach of Marcus,[39] a thermochemical argument developed by Hush,[40] and the polaron approach of Levich and Dogonadze.[26,41] In the latter approach, the solvent is approximated as a collection of harmonic oscillators of total energy,

$$E = \frac{\hbar\omega_o}{2}\sum_k(p_k^2 + q_k^2) \tag{87}$$

The p_k are the reduced momentum coordinates

$$p_k = \frac{(P-P_e)_k}{\langle P_k{}^2\rangle^{\frac{1}{2}}} = (P-P_e)_k(M_k\hbar\omega_k)^{-\frac{1}{2}} \tag{88}$$

defined by analogy with our earlier definition of the reduced normal coordinates, q_j (eq. 36). The frequency in eq. 88 is the average solvent frequency $\langle\omega_o\rangle$; it is taken to be 10^{-11} s^{-1} so that $\hbar\langle\omega_o\rangle \sim 1$ cm^{-1}. When eq. 87 is used in quantum mechanical operator form and applied to the problem of electron transfer, the contribution to the vibrational trap from solvent vibrations is given by,

$$E_s = \frac{\hbar\langle\omega_o\rangle}{2}\sum_k(q_{kb}{}^o - q_{ka}{}^o)^2 \tag{89}$$

The $\underline{q}^o$ values refer to equilibrium solvent coordinates before and after electron transfer has occurred. In the macroscopic continuum approximation, eq. 89 becomes,

$$E_s = \tfrac{1}{2}\left(\frac{1}{D_{op}} - \frac{1}{D_s}\right)\int(\vec{D}_f - \vec{D}_i)^2 dV \tag{90}$$

In eq. 90, D_{op} and D_s are the optical and static dielectric constants of the medium. The quantities $\vec{D}_f$ and $\vec{D}_i$ are the dielectric displacement vectors associated with the charge distributions in the final (f) and initial (i) states. The integration is performed over the entire volume of the system. From classical electrostatics, for a charge distribution in a medium, the actual electric field produced, $\vec{E}$, is related to the dielectric displacement, and the polarization induced in the medium $\vec{P}$, by

$$\vec{E} = \vec{D} - 4\pi\vec{P} \tag{91}$$

and to the dielectric displacement by,

$$\vec{D} = D_s\vec{E} \tag{92}$$

$\vec{D}$ is then the effective electric field exerted by the charge distribution in the absence of the medium. For an electron transfer process, the charge distributions before and after electron transfer are,

$$Z_1e,\ Z_2e \rightarrow Z_1e+\Delta e\ ,Z_2e-\Delta e \tag{93}$$

where Δe is the amount of charge transferred. It will be discussed below. Since for a given charge distribution Ze, $\vec{D}$ is given by,

$$\vec{D} = \frac{Ze}{r^2} \tag{94}$$

it follows that,

$$(\vec{D}_f - \vec{D}_i) = \frac{\Delta e}{r_1^{\,2}} - \frac{\Delta e}{r_2^{\,2}} \tag{95}$$

$$(\vec{D}_f - \vec{D}_i)^2 = \Delta e^2 [(\frac{1}{r_1})^4 + (\frac{1}{r_2})^4 - (\frac{2}{r_1^{\,2} r_2^{\,2}})] \tag{96}$$

where r_1 and r_2 are the vector distances from the centers of the two charge distributions Z_1e and Z_2e. Assuming that the redox sites are spherical, the integration of eq. 90 gives,

$$E_s = \frac{(\Delta e)^2}{2} (\frac{1}{D_{op}} - \frac{1}{D_s}) [\int_{a_1}^{\infty} (\frac{1}{r_1^{\,4}}) dV + \int_{a_2}^{\infty} (\frac{1}{r_2^{\,4}}) dV + 2\int_{d}^{\infty} \frac{1}{r_1^{\,2} r_2^{\,2}} dV]$$

$$= (\Delta e)^2 (\frac{1}{2a_1} + \frac{1}{2a_2} - \frac{1}{d}) (\frac{1}{D_{op}} - \frac{1}{D_s}) \tag{97}$$

For the first two terms, the integrations are carried out from the surfaces of the spheres throughout the volume of the solution, and for the third term, from the interreactant separation, d. If the reactants are in close contact $d = a_1 + a_2$.

The proper value to choose for Δe depends on the nature of the electron transfer process. The expression for χ_o gives the solvent vibrational contribution to the electron transfer barrier or trap as a macroscopic approximation. As such it represents the energy difference between those points on the potential energy surface where the system is initially at thermal equilibrium and then vertically above it corresponding to the optical transition in Figures 4 and 5. The following cases follow where e is the unit electron charge: (1) For optical electron transfer where the transition between surfaces is a vertical process,

$$E_m = \tfrac{1}{2}\sum_i h\omega_i q_i^{\,2} = \chi_i + \chi_o$$

$\Delta e = e$ and eq. 97 becomes,

$$E_s = \chi_o = e^2 (\frac{1}{2a_1} + \frac{1}{2a_2} - \frac{1}{d}) (\frac{1}{D_{op}} - \frac{1}{D_s}) \tag{98}$$

(2) For a self-exchange reaction, $E_a = E_m/4$ from eq. 51. It follows that $E_s = \chi_o/4$ and that $\Delta e = \tfrac{1}{2}e$. (3) For a chemically unsymmetrical electron transfer, E_a is defined by the intersection between

surfaces (eq. 58),

$$E_a = \frac{E_m}{4}\left(1 + \frac{\Delta E}{E_m}\right)^2 = \frac{\chi_i+\chi_o}{4}\left(1 + \frac{\Delta E}{E_m}\right)^2$$

and since,

$$E_s = \frac{\chi_o}{4}\left(1 + \frac{\Delta E}{E_m}\right)^2$$

it follows that,

$$\Delta e = \frac{e}{2}\left(1 + \frac{\Delta E}{E_m}\right).$$

The integrated result in eq. 97 is not exact since the integration fails to account for the volume of the ions. However, the correction involved appears to be small.(42) A more realistic approach to the integration of eq. 90 for cases like mixed-valence dimers has been proposed(43,44) based on earlier work by Kirkwood and Westheimer on the calculation of solvation energies for ellipsoidal shapes.

COMPARISONS BETWEEN CLASSICAL AND QUANTUM MECHANICAL RESULTS

In previous sections, expressions have been derived for electron transfer rates and energy barriers or traps based on molecular and solvent vibrations. When treated quantum mechanically, the result at ambient temperatures is mathematically complex (eq. 85) because of a nuclear tunneling contribution from the molecular vibrational levels. In the high temperature limit, $\hbar\omega << kT$, which is appropriate for solvent vibrations, the classical (eq. 58) and quantum mechanical (eq. 84) equations for the energy barrier are similar. The question might be realistically asked, why bother with the far more complicated quantum mechanical development?

The answer comes from experiment where limitations in the classical treatment can lead to incorrect predictions and to a loss of insight. Specific examples include the following: (1) In terms of the experimentally observed Arrhenius equation,

$$k_{et} = A\exp{-E_a/RT} \qquad (99)$$

the quantum mechanical treatment gives exact expressions for both the energy of activation (eq. 84) and for the pre-exponential term,

$$A = \frac{V_{ab}^2}{\hbar}\left(\frac{\pi}{\chi kT}\right)^{\frac{1}{2}} \qquad (100)$$

The expressions are based on detailed microscopic models, are not arbitrary, and as discussed later, are experimentally testable. In semiclassical treatments the pre-exponential term enters the problem as a product of the probability of a surface crossing, using a Landau-Zener approach, and a frequency term which is usually somewhat arbitrarily chosen to be the frequency for a low energy vibrational mode like diffusion or collision within a solvent cage. (2) There are reaction domains where the classical treatment simply breaks down. One is at low temperatures where $E_a \sim 0$ and electron transfer occurs by nuclear tunneling between reactant and product surfaces.[27b,45] A second occurs for electron transfer reactions which are sufficiently favored energetically that they actually become excited state processes (Figure 7). In such cases electron transfer by vibrational transitions through the

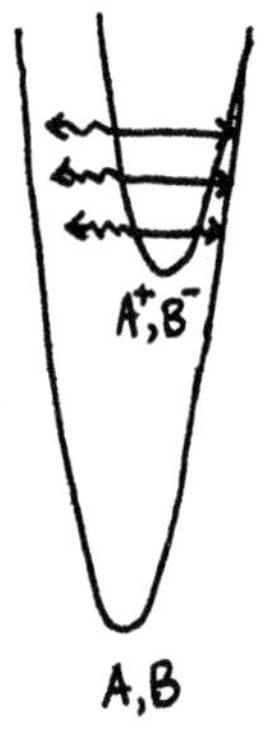

Figure 7. Energy coordinate diagram for a highly favored reaction.

intersection region are not the only efficient channels in terms of overlap because the reactant vibrational surface is actually within the product surface.[31-33,46] (3) In the quantum mechanical treatment no resort is made to an artificial intermediate state as in semiclassical theories based on transition states. The nature of the process arises quite naturally given the quantum mechanical description of the vibrations in the system and the equilibrium thermal distribution throughout the vibrational levels.

ELECTRON TRANSFER IN MIXED-VALENCE COMPOUNDS

The major theme to this point has been the development of a theoretical basis for understanding electron transfer. The obvious next step is experimental; the design of systems and measurements which test and elaborate on the theoretical predictions. The most obvious experimental approach is through kinetics, but electron transfer reactions are, in general, complex mechanistically. The interpretation of rate data in terms of the electron transfer step can be difficult and inferential at best. Even for simple outer-sphere reactions, e.g., eq. 1-3, the observed rate constant is a product of constants and direct information about electron transfer is not obtainable. There are exceptional cases based on Co(III) amine complexes where the electron transfer step can be isolated. Examples are known both of inner-sphere reactions based on complexes like(47-49)

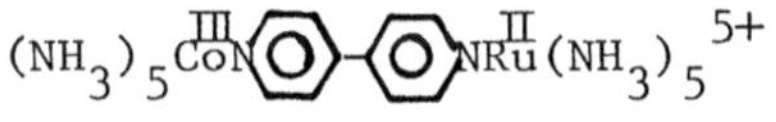

and outer-sphere reactions based on ion-pairs like,[7a]

$$[\overset{III}{Co}(NH_3)_6, \overset{II}{Fe}(CN)_6]^-$$

However, in general, it is a nontrivial situation to arrange experimentally.

Mixed-valence compounds have a distinct advantage. Spectroscopic studies can be carried out on the intact systems while electron transfer is occurring and there are no competing chemical processes to interfere. The most valuable spectroscopic observations to date have been based on IT absorption bands in solution. It should be noted that the experimental results discussed below were obtained on systems in which electronic coupling is weak. In fact, a quantitative statement can be made about the extent of coupling. The electron exchange matrix element, V_{ab}, can be estimated from the integrated intensity of an IT band.(1,9) The results presented here are based on chemical systems for which $V_{ab}<0.1$ eV.

Vibrational Barriers from Band Energies

In the high temperature limit, E_{op} and the thermal activation energy should be explicitly related (eq. 51 and 60). The relationship between E_{op} and E_a ($E_{op} = 4E_a$) is particularly simple for symmetrical mixed-valence compounds. The observation of IT bands for compounds like the biferrocenium ion, $[(C_5H_5)Fe(C_5H_4-C_5H_5)Fe-(C_5H_5)]^+$ (4a,50,51) and the 4,4'-bipyridine-bridged dimer in eq. 5, allows an immediate calculation to be made of E_a for the thermally activated, intramolecular process (e.g., eq. 101 where Fc is the ferrocenyl group.).

$$Fe\text{-}C{\equiv}C\text{-}Fc^{+} \xrightarrow[E_a]{k_{et}} Fc^{+}\text{-}C{\equiv}C\text{-}Fc \qquad (101)$$

Although rate constant data are not available for the intramolecular reactions, comparisons can be made with related outer-sphere self-exchange reactions. As an example, the intramolecular reaction in eq. 101 is closely related in terms of reactant dimensions to electron transfer between Fc and Fc^+ within an association complex, $Fc,Fc^+ \rightarrow Fc^+,Fc$. From the IT band energy for the dimer in eq. 101, E_a=5 kcal/mole, and from kinetics data for the Fc/Fc^+ self-exchange,(52) $E_a \sim 4$ kcal/mole, showing that reasonable agreement can be obtained.

Even more striking are the consequences of the predictions made in eq. 98 concerning the contribution of solvent vibrations. The total vibrational barrier or trap for a symmetrical mixed-valence compound is,

$$E_{op} = 4E_a = \chi_o + \chi_i \qquad (102)$$

and by including eq. 98, it follows that,

$$E_{op} = \chi_i + e^2\left(\frac{1}{2a_1} + \frac{1}{2a_2} - \frac{1}{d}\right)\left(\frac{1}{D_{op}} - \frac{1}{D_s}\right) \qquad (103)$$

According to eq. 103, χ_o can be calculated in any solvent for which D_s and D_{op} are known. A plot of E_{op} vs. $(1/D_{op}-1/D_s)$ should be linear and have as its intercept χ_i. Thus, as first noted by Taube and coworkers,[8a] numerical values for the molecular and solvent contributions to the vibrational trap should be obtainable from IT band measurements in a series of solvents.

The predicted linear dependence of E_{op} with solvent dielectric properties has been observed for several mixed-valence compounds.(5,8,51,53) The theoretically predicted slopes of the plots (eq. 103) depend on the structures of the compounds and can be calculated. When certain corrections are included, the agreement between observed and calculated slopes can be quite reasonable

One potential complication is that χ_i determined from solvent dependence studies includes the sum of all contributions to E_{op} which are insensitive to changes in the medium. In addition to molecular vibrations, other factors may be important. One is that in low symmetry cases or where spin-orbit coupling is important, the IT band intensity could be dominated by transitions other than the one of lowest energy,(8b) e.g.,

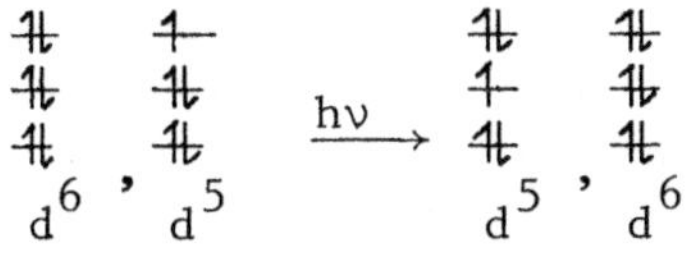

If zero field splitting and/or spin-orbit coupling are large, the contribution of such an effect in χ_i could be significant.

One of the few examples available of a solvent dependence study on an unsymmetrical mixed-valence compound was based on the dimer in eq. 104.(54)

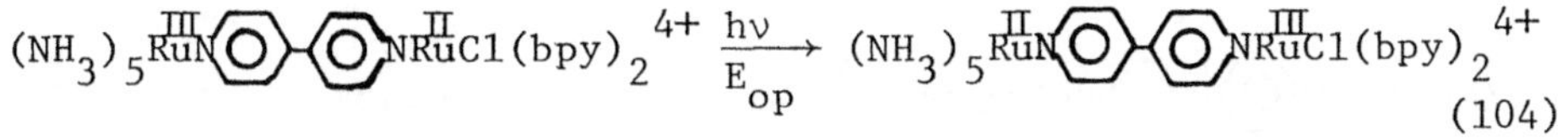

(104)

For an unsymmetrical compound, the initial product of the optical excitation is a vibrationally excited "redox isomer" of the system (Note Fig. 5). There are complications in interpreting the data obtained from such compounds. One is that the background theory was developed assuming equal frequencies for molecular and solvent vibrations in the initial and final states. In a dimer where there are significant chemical differences between the two sites, the assumption is expected to break down. A second complication is that the calculation of E_a depends both on E_{op} and ΔE (Note eq. 105 which is eq. 58 with $E_m = \chi_\cdot$),

$$E_a = \frac{\chi}{4}\left(1 + \frac{\Delta E}{\chi}\right)^2 \tag{105}$$

In addition to E_{op}, temperature dependent redox potential measurements are necessary to estimate ΔE before E_a can be calculated.

For an unsymmetrical dimer, E_{op} is predicted to vary with solvent dielectric as in eq. 106,

$$E_{op} = E_m + \Delta E = (\chi_i + \Delta E) + e^2\left(\frac{1}{2a_1} + \frac{1}{2a_a} - \frac{1}{d}\right)\left(\frac{1}{D_{op}} - \frac{1}{D_s}\right) \tag{106}$$

which follows from eq. 59 and 98. For the mixed ammine-bipyridine dimer, a plot of E_{op} vs. $(1/D_{op}-1/D_s)$ was linear although the linear correlation was less precise than for the symmetrical dimers.(54)

A number of χ_i and χ_o values obtained from solvent dependent E_{op} data are collected in Table 1.

Table 1. χ_i and χ_o values at room temperature in CH_3CN.

Compound	χ_i(eV)	χ_o(eV)	Ref.
$(NH_3)_5Ru$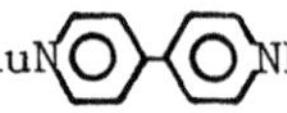$Ru(NH_3)_5{}^{5+}$	0.43	0.67	8
$(bpy)_2ClRu$$RuCl(bpy)_2{}^{3+}$	0.73	0.26	5
$(C_5H_5)Fe(C_5H_4-C_5H_4)Fe(C_5H_4)^+$	0.44	0.69	51
$(NH_3)_5Ru$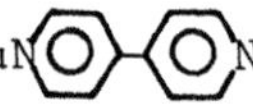$RuCl(bpy)_2{}^{4+}$	0.56	0.43	54

Distance Dependence

The availability of the series of closely related mixed-valence ions, $(bpy)_2ClRu(L)RuCl(bpy)_2{}^{3+}$ (L =

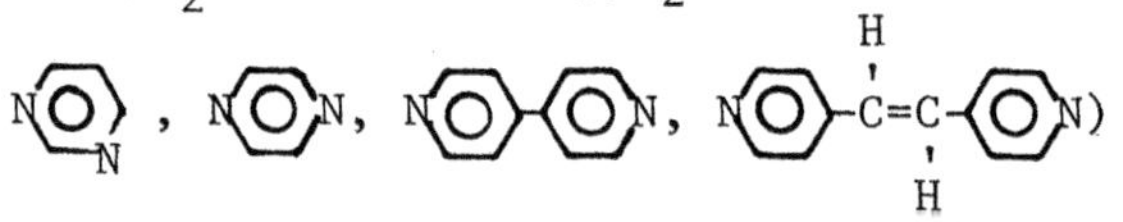

allowed an interesting experimental observation to be made concerning the role of inter-reactant distance, d, in optical electron transfer, and by inference, in outer-sphere electron transfer as well.[53] The predicted dependence of E_{op} on d is given in eq. 107 which is a rearranged version of eq. 103.

$$E_{op} = [\chi_i + e^2(\frac{1}{2a_1} + \frac{1}{2a_2})(\frac{1}{D_{op}} - \frac{1}{D_s})] - \frac{e^2}{d}(\frac{1}{D_{op}} - \frac{1}{D_s}) \qquad (107)$$

Eq. 107 predicts that as the interreactant separation increases, E_{op}, and E_a, should both increase. There is an energetic benefit from close contact in an electron transfer reaction associated with a minimization in the solvent vibrational contribution to the electron trap.

The dependence of E_{op} on $-1/d$ predicted by eq. 107 has been observed both for the ruthenium dimers[53] and for a series of mixed-valence biferrocenium ions.[51] For the ruthenium dimers, good agreement was obtained between experimental and theoretically derived slopes and intercepts although an empirical correction was necessary to take into account deviations from the model used to derive the solvent dependence result in eq. 97.

The results of the distance dependence experiment have broader implications. During electron transfer, a dependence on d is expected only in the solvent term. Solvent vibrations are coupled to both redox sites while the molecular vibrations at the separate redox sites are uncoupled. However, for an outer-sphere reaction, an additional d-dependent term enters the problem through the equilibrium constant for association prior to electron transfer (eq. 1). The form of the dependence can be seen from the Eigen-Fuoss equation,[55]

$$k = \frac{4\pi N_o d^3}{3000} \exp-[\frac{Z_A Z_B e^2}{dD_s RT}] \tag{108}$$

where Z_A and Z_B are the charges on the reactants and N_o is Avogadro's number. When the distance dependences in eq. 107 and 108 are combined, the energetic advantage of close contact remains.[53] This is true even for like-charged ions where the exponential term in eq. 108 is repulsive.

The preexponential term in the theoretical expression for k_{et} (eq. 84) is also distance dependent and in such a way that k_{et} is also benefitted at close contact distances. The effect exists both through minimization of χ_o in the $(\chi kT)^{-\frac{1}{2}}$ term but also in V_{ab}. V_{ab} is expected to be strongly dependent on d because overlap between the electron donor and acceptor wave functions will decrease exponentially with d.

Outer-Sphere Intervalence Transfer

From examples cited so far, IT absorption bands are a powerful experimental tool for studying electron transfer reactions. At first glance, the approach appears to be limited since it involves a special, relatively uncommon type of compound where an intact chemical link spans adjacent redox sites and provides an "orbital pathway" for electron transfer.

A first point to be considered is the role of the bridging ligand. In a complex like

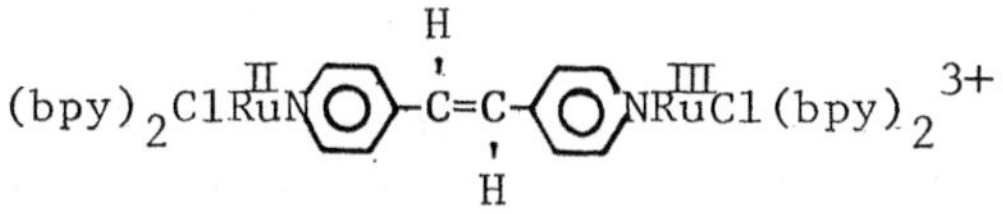

the bridging ligand clearly plays an active role in the electron conduction mechanism. Mixing of dπ(Ru(II)) character with the π* system of the bridging ligand "carries" the Ru(II) electronic wave function to the Ru(III) site enhancing orbital overlap. However, IT bands are also observed in complexes like

$$(NH_3)_5\overset{II}{Ru}N\bigcirc\!-CH_2-CH_2-\bigcirc\!N\overset{III}{Ru}(NH_3)_5{}^{5+} \quad (56)$$

and

$$[(bpy)_2Cl\overset{II}{Ru}(Ph_2PCH_2PPh_2)\overset{III}{Ru}Cl(bpy)_2]^{3+} \quad (57)$$

where the interaction between sites may be through-space and the role of the bridging ligand is simply to hold the two redox sites in the same molecular region.

An even more interesting observation is that in aqueous solution, MMCT (Metal Metal Charge Transfer) bands can be observed within ion-pairs as, for example, in,[58]

$$[\overset{III}{Ru}(NH_3)_5py,\overset{II}{Fe}(CN)_6]^- \xrightarrow{h\nu} [\overset{II}{Ru}(NH_3)_5py,\overset{III}{Fe}(CN)_6]^- \tag{109}$$

The appearance of such bands is encouraging from two points of view. It suggests that it may be possible to observe similar transitions in a variety of chemical materials. If so, information can be gained immediately about electron transfer properties from the properties of the absorption bands. It also allows the spectroscopic advantage enjoyed by mixed-valence compounds to be extended to outer-sphere reactions. The results obtained can be revealing. Analysis of the MMCT band for the ion-pair shows that the value for the electron exchange matrix element ($V_{ab}\sim 0.016$ eV) is close in magnitude to the values for some of the ligand-bridged complexes of the type,

$$(bpy)_2Cl\overset{II}{Ru}(L)\overset{III}{Ru}Cl(bpy)_2{}^{3+}$$

The inference can be drawn that through-space, outer-sphere overlap between donor and acceptor wave functions can be appreciable by the IT experiment, and made quantitative.

Calculation of k_{et} from IT Bands

One of the most useful results of the quantum mechanically derived equations is their explicit form. At least in the high temperature limit, the terms which appear in equations like 84 can be evaluated from the properties of IT bands. As mentioned above, V_{ab} can be estimated from band intensity measurements and $E_m = \chi$ from band energies. The results of a series of such calculations using eq. 84 and observed IT band intensities and energies are summarized in Table 2.[59]

Table 2. Calculated k_{et} values in acetonitrile at 23° [a]

Ion	E_{op}(eV)	V_{ab}(eV)	k(s^{-1})
	0.953	0.049	$2.4x10^{10}$
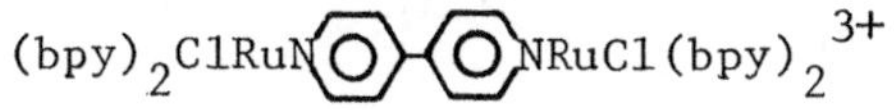	1.26	∿0.02	$\sim 2x10^{8}$
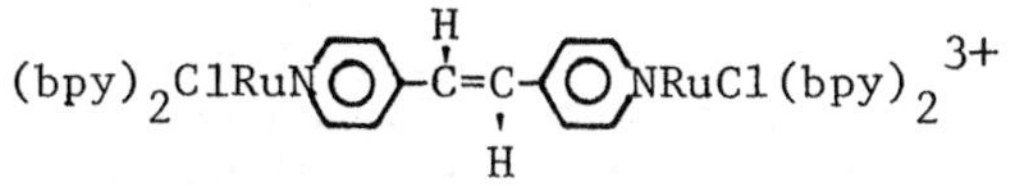	1.34	∿0.02	$\sim 8x10^{7}$

[a]Values taken from ref. 59

The numerical values obtained for k_{et} appear to be reasonable and are in good agreement with what could be expected based on the results obtained for the $Ru(bpy)_2(py)Cl^{2+/+}$ outer-sphere self-exchange reaction.(5,53)

However, the agreement may be partly fortuitous. For the compounds, solvent dependence studies show that χ_i is an appreciable fraction of E_{op} and the use of the high temperature limit approximation is questionable. In the future, a more detailed spectroscopic attack on the complexes may lead to a sufficiently detailed view of the molecular vibrations that their effects can be brought into the problem in a quantitative way using appropriate equations in the intermediate temperature region.(27)

FUTURE EXPERIMENTS

The spectral experiments based on the observation of IT or MMCT bands are of obvious value in understanding electron transfer reactions. Although far easier to perform, the experiment appears to give greater insight into the microscopic nature of the process than the results of simple rate measurements.

The ability to assess electron transfer properties by spectral measurements will hopefully be extended, and extended in several ways. One way is to new chemical systems including organometallic compounds and to compounds like the metalloporphyrins which have relevance as models in biological electron transfer. There is also the prospect of using the concepts developed here and applying them in a more general way to spectral transitions. Specific examples include the low energy absorption bands found in organic donor-acceptor complexes and absorption bands arising from metal to ligand and ligand to metal charge transfer transitions.

Additional spectroscopic studies will also be revealing. The key is to obtain detailed information about the molecular vibrations which are coupled to electron transfer and the obvious experiment is Resonance Raman in the IT band region.[60] Combining a knowledge of vibrational frequencies and normal modes from the Raman experiment with χ_i from solvent dependence measurements should define the molecular vibrations in sufficient detail to test the background theory critically.

The last experiment to mention may be the best of all, at least when combined with IT band measurements. For an unsymmetrical dimer like the mixed ammine-bipyridine complex, either direct photolysis into the IT band (eq. 104) or photolysis into an upper state followed by radiationless decay, may be a means of populating the thermodynamically unfavorable redox isomer. If the time resolution of the experiment is sufficiently rapid (ns-ps), the decay process back to the stable isomer should be observable,

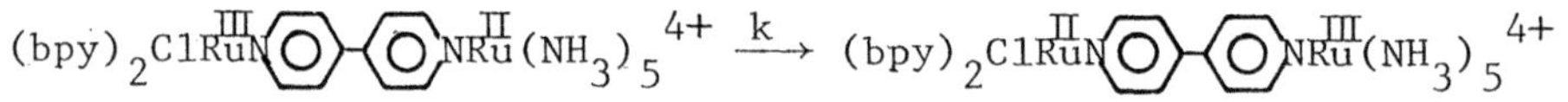

(110)

In fact, the results of such an experiment have been reported recently by Creutz and Sutin and their coworkers.[61] The combination of a rate constant for the relaxation process in eq. 110 and the information obtainable for the reverse process by spectral studies on the stable redox isomer, should provide a nearly complete picture experimentally. Hopefully, the execution of such experiments, including measurements at low temperatures to search for quantum effects, will become routine in the near future.

Acknowledgements

I would like to acknowledge the value of useful conversations with Jeff Curtis, Pat Sullivan and especially with Ed Kober in writing this manuscript and the Alfred P. Sloan Foundation for financial support.

REFERENCES

(1) M. B. Robin and P. Day, Adv. Inorg. Chem. Radiochem., 10, pp. 247 (1967).
(2) N. S. Hush, Chem. Phys., 10, pp. 361 (1975).
(3) P. N. Schatz, J. Am. Chem. Soc., 100, pp. 2996 (1978); Chem. Phys. Lett., 55, pp. 539 (1978).
(4)(a) D. O. Cowan, C. LeVanda, J. Park, and F. Kaufman, Acc.

Chem. Res. 6, pp. 1 (1973); (b) T. J. Meyer, Accts. Chem. Res., 11, pp. 94 (1978); (c) T. J. Meyer, Ann. N.Y. Acad. Sci., 313, pp. 496 (1978); (d) H. Taube, Ann. N.Y. Acad. Sci., 313, pp. 481 (1978).
(5)(a) R. W. Callahan, F. R. Keene, T. J. Meyer, and D. J. Salmon, J. Am. Chem. Soc., 99, pp. 1064 (1977); (b) R. W. Callahan and T. J. Meyer, Chem. Phys. Lett., 39, pp. 82 (1976).
(6) G. M. Brown and N. Sutin, J. Am. Chem. Soc., 101, pp. 883 (1979).
(7)(a) A. J. Miralles, R. E. Armstrong, and A. Haim, J. Am. Chem. Soc., 99, pp. 1416 (1977); (b) D. Gaswick and A. Haim, J. Am. Chem. Soc., 93, pp. 7347 (1976).
(8)(a) G. M. Tom, C. Creutz, and H. Taube, J. Am. Chem. Soc., 96, pp. 7827 (1974); (b) C. Creutz, Inorg. Chem., 17, pp. 3723 (1978).
(9)(a) N. S. Hush, Prog. Inorg. Chem., 8, pp. 391 (1967); (b) Electrochim. Acta, 13, pp. 1005 (1968).
(10)(a) J. J. Hopfield, Proc. Natl. Acad. Sci., USA, 21, pp. 3640 (1974); (b) J. J. Hopfield, Biophys. J., 18, pp. 311 (1977); (c) M. J. Potasek and J. J. Hopfield, Proc. Natl. Acad. Sci., USA, 74, pp. 229, 3817 (1977).
(11) D. R. Stranks, Discuss. Faraday Soc., 29, pp. 73 (1960); N. S. Birader and D. R. Stranks, Trans. Faraday Soc., 58, pp. 2421 (1963).
(12) F. R. Keene, R. C. Young, and T. J. Meyer, J. Am. Chem. Soc., 99, pp. 2468 (1977).
(13) J. Silverman and R. W. Dodson, J. Phys. Chem., 56, pp. 846 (1952).
(14) N. J. Hair and J. K. Beattie, Inorg. Chem., 16, pp. 245 (1977).
(15) F. Basolo and R. G. Pearson, "Mechanisms of Inorganic Reactions", John Wiley and Sons, New York, 1967, Chapter 2.
(16) I. N. Levine, "Quantum Chemistry", 2nd Ed., Allyn and Bacon, Inc., Boston, 1974, Chapter 14.
(17) E. Merzbacher, "Quantum Mechanics", Second Ed., John Wiley and Sons, New York, NY, 1970, Chapter 18.
(18) C. B. Duke, Proc. of the Conference on Tunneling in Biological Systems, B. Chance, Ed., in press.
(19) W. Kauzmann, "Quantum Chemistry", Academic Press, New York, NY, 1957, Chapter 14.
(20) E. B. Wilson, Jr., J. C. Decius, and P. C. Cross, "Molecular Vibrations", McGraw-Hill, New York, 1966.
(21) N. Davidson, "Statistical Mechanics", McGraw-Hill, New York, 1962, Chapters 6, 8, 11.
(22) R. C. O'Rourke, Phys. Rev., 91, pp. 265 (1953).
(23) S. H. Lin, J. Chem. Phys., 44, pp. 3759 (1966).
(24) R. Kubo and Y. Toyozawa, Prog. Theor. Phys. (Osaka), 13, pp. 161 (1955).
(25)(a) M. Lax, J. Chem. Phys., 20, pp. 1752 (1952); (b) K. Huang and A. Rhys, Proc. Phys. Soc., Ser. A, 204, pp. 413 (1951).
(26) V. G. Levich in "Physical Chemistry: An Advanced Treatise",

Vol. 9B, H. Eyring, D. Henderson, and W. Jost, Eds., Academic Press, New York, 1970; Adv. Electrochem. Electrochem. Eng. 4, pp. 249 (1966); (b) R. R. Dogonadze in "Reactions of Molecules at Electrodes", Ed. by N. S. Hush, Wiley-Interscience, London, 1971.
(27)(a) N. R. Kestner, J. Logan, and J. Jortner, J. Phys. Chem., 78, pp. 2148 (1974); (b) J. Jortner, J. Chem. Phys., 64, pp. 4860 (1976).
(28) J. Ulstrup and J. Jortner, J. Chem. Phys., 63, pp. 4358 (1975).
(29) T. F. Soules and C. B. Duke, Phys. Rev., B3, pp. 262 (1971).
(30) P. P. Schmidt, Chemical Soc., Specialist Periodical Report, Electrochemistry, Vol. 5, Chapter 2, 1974; ibid., Vol. 6, Chapter 4, 1975.
(31) S. Efrima and M. Bixon, Chem. Phys., 13, pp. 447 (1976).
(32) W. Schmickler, Ber. Bunsenges. Phys. Chem., 77, pp. 991 (1973).
(33) S. F. Fischer and R. P. Van Duyne, Chem. Phys., 26, pp. 9 (1977).
(34) E. Buhks, M. Bixon, J. Jortner, and G. Navon, Inorg. Chem., 18, pp. 2014 (1971).
(35) B. A. Kowert, L. Marcoux, and A. J. Bard, J. Am. Chem. Soc., 94, pp. 5538 (1972).
(36) T. J. Meyer and H. Taube, Inorg. Chem., 7, pp. 2369 (1968).
(37) H. C. Stynes and J. A. Ibers, Inorg. Chem., 10, pp. 2304 (1971).
(38) N. Sutin in "Inorganic Biochemistry", Vol. 2, G. L. Eichhorn, Ed., Elsevier, New York, N.Y., 1973, pp. 611.
(39) R. A. Marcus, J. Chem. Phys., 24, pp. 966 (1956); 43, pp. 679 (1965).
(40) N. S. Hush, Trans. Faraday Soc., 57, pp. 557 (1961).
(41) R. R. Dogonadze, Ber. Bunsenges. Phys. Chem., 75, pp. 628 (1971).
(42) Yu. I. Kharats, Sov. Electrochem., 9, pp. 845 (1973).
(43) R. D. Cannon, Chem. Phys. Lett., 49, pp. 299 (1977); Adv. Inorg. Chem. Radiochem., 21, pp. 179 (1978).
(44) E. D. German, Chem. Phys. Lett., 64, pp. 295 (1979).
(45)(a) D. De Vault and B. Chance, Biophys. J., 6, pp. 825 (1966); (b) D. De Vault, J. H. Parkes, and B. Chance, Nature (London), 215, pp. 642 (1967).
(46) J. K. Nagle, W. J. Dressick, and T. J. Meyer, J. Am. Chem. Soc., 101, pp. 3983 (1979).
(47) H. Fischer, G. M. Tom, and H. Taube, J. Am. Chem. Soc., 98, pp. 5512 (1976).
(48) A. Haim, Acc. Chem. Res., 8, pp. 264 (1975).
(49) D. Gaswick and A. Haim, J. Am. Chem. Soc., 96, pp. 7845 (1974).
(50) W. H. Morrison, Jr., and D. N. Hendrickson, J. Chem. Phys., 59, pp. 380 (1973); Inorg. Chem., 14, 2331 (1975).
(51) M. J. Powers and T. J. Meyer, J. Am. Chem. Soc., 100, pp. 4393 (1978).

(52) E. S. Yang, M.-S. Chan, and A. C. Wahl, J. Phys. Chem., 79, pp. 2049 (1975).
(53)(a) M. J. Powers, D. J. Salmon, R. W. Callahan, and T. J. Meyer, J. Am. Chem. Soc., 98, pp. 6731 (1976); (b) M. J. Powers and T. J. Meyer, J. Am. Chem. Soc., in press.
(54) M. J. Powers, R. W. Callahan, D. J. Salmon, and T. J. Meyer, Inorg. Chem., 15, pp. 1457 (1976).
(55)(a) R. M. Fuoss, J. Am. Chem. Soc., 80, pp. 5059 (1958); (b) M. Eigen, Z. Physik. Chem. [N.F.], 1, pp. 176 (1954).
(56) K. Rieder and H. Taube, J. Am. Chem. Soc., 99, pp. 7891 (1977).
(57) B. P. Sullivan and T. J. Meyer, Inorg. Chem., in press.
(58) J. C. Curtis and T. J. Meyer, J. Am. Chem. Soc., 100, pp. 6294 (1978); J. Curtis, work in progress.
(59) T. J. Meyer, Chem. Phys. Lett., 64, pp. 417 (1979).
(60) R. J. H. Clark and P. C. Turtle, Inorg. Chem., 17, pp. 2526 (1978); J. Chem. Soc. Dalton, pp. 1622 (1978).
(61) C. Creutz, P. Kroger, T. Matsubara, T. L. Netzel and N. Sutin, J. Am. Chem. Soc., 101, pp. 5442 (1979).

A VIBRONIC COUPLING MODEL FOR MIXED-VALENCE COMPOUNDS AND ITS APPLICATION TO REAL SYSTEMS

Paul N. Schatz

Department of Chemistry
University of Virginia
Charlottesville, Virginia 22901
U.S.A.

ABSTRACT: A recent vibronic coupling model proposed by Piepho, Krausz and Schatz is described, and detailed derivation is given of the secular equations which permit an exact solution. A variety of applications of the model are then presented. These include the use of model parameters to predict intervalence band contours and vice versa, the prediction of low energy tunneling transitions, and a refinement of the Robin and Day classification scheme. The calculation of magnetic moments and susceptibilities in mixed-valence and spin-crossover systems is also discussed briefly. Finally, the model is used to calculate electron transfer rates. A method is presented which should be valid even when the electronic coupling is strong. Thus transer rates are calculated which have a much greater range of validity than is the case in the usual treatments.

I. INTRODUCTION

This article describes a simple vibronic coupling model for mixed-valence compounds recently proposed by Piepho, Krausz and Schatz (PKS model (1)). The model leads to a set of secular equations which may be readily solved to any desired accuracy yielding explicit eigenfunctions and eigenvalues for the ground vibronic manifold. These quantities in turn may be used to calculate experimental observables associated with this manifold. The paper starts by outlining the model in detail and then discusses its application in calculating intervalence band profiles (1,2), tunneling transitions (3), degrees of delocalization and valence trapping (1,2), magnetic susceptibilities (4) and electron transfer kinetics (5).

D.B. Brown (ed.), Mixed-Valence Compounds, 115-150.

II. PKS MODEL; SYMMETRICAL CASE (A = B)

We proceed in a rather intuitive manner placing emphasis on the physical content of the model. A rigorous treatment may be found in Reference 1.

A mixed-valence system consists of two coupled subunits, A and B. In the symmetrical case (A = B), one or more symmetry operations exist which carry each nucleus in A into a symmetrically equivalent partner in B. Examples of such systems are the Creutz and Taube (C & T) complex (6),

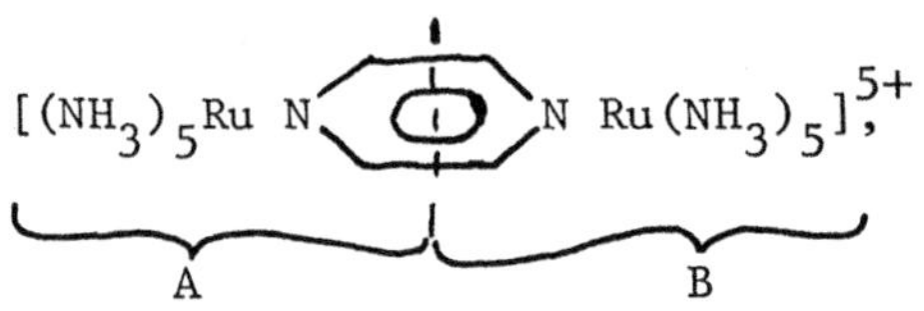

and the benzene dimer cation, $(C_6H_6)_2^+$ (7). We use the former to illustrate the theory.

In the C & T complex, we may formally assign Ru A the oxidation state 2+ and Ru B the oxidation state 3+, *or* vice versa. If we assume that there is no interaction (coupling) between the two subunits A and B, then electronic wave functions for the two equivalent assignments of oxidation states can be written,

$$\psi_a = \psi_M^A \psi_N^B, \quad \psi_b = \psi_N^A \psi_M^B \tag{1}$$

where M and N respectively designate the 3+ and 2+ oxidation states of Ru. Corresponding vibrational wave functions for each subunit can be obtained by expanding the potential energy in a power series in the normal coordinates followed by solution of the vibrational wave equation. Keeping terms through quadratic (the harmonic approximation) yields vibrational eigenfunctions composed of products of harmonic oscillator wave functions (8). However, if it is assumed that the subunits have the same point group symmetry in both oxidation states (M and N)--an assumption made throughout this treatment--then one finds to a first approximation that one only need consider the totally symmetric normal coordinates (1). Assuming that there is just one such on each subunit, designated respectively Q_A and Q_B, the vibrational potential energy of subunit A in oxidation state M can be written (in the harmonic approximation)

$$W_M^A(Q_A) = W_M^0 + \ell_M^A Q_A + \tfrac{1}{2} K_M^A Q_A^2 \tag{2}$$

with analogous expressions for W_N^A, W_M^B, W_N^B, where $W_M^0 \equiv W_M^A(Q_A = 0) = W_M^B(Q_B = 0)$. Eq (2) is a parabola with curvature K_M^A

whose minimum occurs at $(-\ell_M^A/k_M^A)$ with value $W_M^O - \frac{1}{2}(\ell_M^A)^2/k_M^A$.

Let us write the Hamiltonian operator for subunit A,

$$H^A = H_{el}^A + T_n^A \tag{3}$$

where T_n^A is the *nuclear* kinetic energy operator. Then

$$H_{el}^A \psi_M^A = W_M^A(Q_A)\psi_M^A$$

$$H_{el}^A \psi_N^A = W_N^A(Q_A)\psi_N^A \tag{4}$$

with exactly analogous equations applying if all A's are changed to B's. Eq (4) is the electronic Schrödinger equation (with the internuclear repulsion terms included in H_{el}^A). It follows immediately that

$$(H_{el}^A + H_{el}^B)\psi_a = (W_M^A(Q_A) + W_N^B(Q_B))\psi_a \equiv W_a\psi_a \tag{5}$$

with an analogous equation for ψ_b. We define the zero of energy by $W_N^O + W_M^O = 0$, and arbitrarily assume (as is the case in the C & T complex) that the lower oxidation state (N) is lower in energy. We choose coordinate origins by setting Q_A and Q_B to zero at the minima of W_N^A and W_N^B respectively (so that $\ell_N^A = \ell_N^B = 0$). Setting $\ell_M^A = \ell_M^B \equiv \ell$, $k_M^A = k_M^B \equiv k_M$, $k_N^A = k_N^B \equiv k_N$, and using eq (2) and its analogs, we get,

$$W_a = \ell Q_A + \tfrac{1}{2}k_M Q_A^2 + \tfrac{1}{2}k_N Q_B^2$$

$$W_b = \ell Q_B + \tfrac{1}{2}k_M Q_B^2 + \tfrac{1}{2}k_N Q_A^2 \tag{6}$$

It is now very useful to introduce the new coordinates,

$$Q_{\pm} = (1/\sqrt{2})(Q_A \pm Q_B) \tag{7}$$

Eq (6) then becomes,

$$W_a = (1/\sqrt{2})\ell Q_- + \tfrac{1}{4}(k_N + k_M)Q_-^2$$
$$+ \tfrac{1}{2}(k_M - k_N)Q_+Q_- + (1/\sqrt{2})\ell Q_+ + \tfrac{1}{4}(k_N + k_M)Q_+^2 \tag{8}$$

and W_b is obtained if $(-Q_-)$ is substituted for Q_- in eq (8). *We know make the approximation*, $k_M = k_N$. This is often a reasonable assumption, and it enormously simplifies the model, both conceptually and computationally. (We briefly discuss later the case, $k_M \neq k_N$.) With this assumption of equal force constants in both oxidation states, we obtain (setting $k_M = k_N \equiv k$)

$$W_a = (1/\sqrt{2})\ell Q_- + \tfrac{1}{2}kQ_-^2 + (1/\sqrt{2})\ell Q_+ + \tfrac{1}{2}kQ_+^2$$

$$W_b = (-1/\sqrt{2})\ell Q_- + \tfrac{1}{2}kQ_-^2 + (1/\sqrt{2})\ell Q_+ + \tfrac{1}{2}kQ_+^2 \qquad (9)$$

The cross term in Q_+Q_- has vanished and so the problem is separable with respect to Q_+ and Q_-. In Q_+ space, W_a and W_b give identical parabolas whose minima occur at $Q_+ = (-\ell/k\sqrt{2})$. We find later that Q_+ does not enter the calculation of spectroscopic or other properties of interest to us (when $k_M = k_N$). *The crucial coordinate is* Q_-. Defining the dimensionless variables,

$$q = 2\pi(\nu_-/h)^{\frac{1}{2}}Q_-$$

$$\lambda = (8\pi^2 h\nu_-^3)^{-\frac{1}{2}}\ell \qquad (10)$$

where $\nu_- = (2\pi)^{-1}\sqrt{k}$, is the fundamental vibrational frequency associated with normal coordinate Q_- (or q), eq (9) *in q space* becomes

$$W_a/h\nu_- = \lambda q + \tfrac{1}{2}q^2$$

$$W_b/h\nu_- = -\lambda q + \tfrac{1}{2}q^2 \qquad (11)$$

The physical significance of q (or Q_-) can be understood by reference to Fig. 1 using the C & T complex as an example. The lines from centers A and B depict the Ru-N interatomic distances assuming that the RuN_6 moiety is octahedral. It should be remembered from our earlier discussion that λ directly measures the difference, in the two oxidation states, of the equilibrium displacement along the totally symmetric normal coordinate (and is thus proportional to the difference in the Ru-N equilibrium distance in the 2+ and 3+ oxidation states in the C & T example). It can in fact be shown for octahedral subunits that this difference (Δr) is $\lambda(h\nu_-/3m_L k)^{\frac{1}{2}}$ where m_L is the mass of a ligand atom.

Since we have permitted no interaction between subunits A and B, the vibronic wave functions for electronic states ψ_a and ψ_b will be,

$$\phi_a = \psi_a\chi_{na}$$

$$\phi_b = \psi_b\chi_{nb} \qquad (12)$$

where the χ are defined by

$$[T_n(q) + W_a(q)]\chi_{na}(q) = E_{na}\chi_{na}(q)$$

$$[T_n(q) + W_b(q)]\chi_{nb}(q) = E_{nb}\chi_{nb}(q) \qquad (13)$$

with $T_n(q)$ the kinetic energy operator associated with coordinate q.

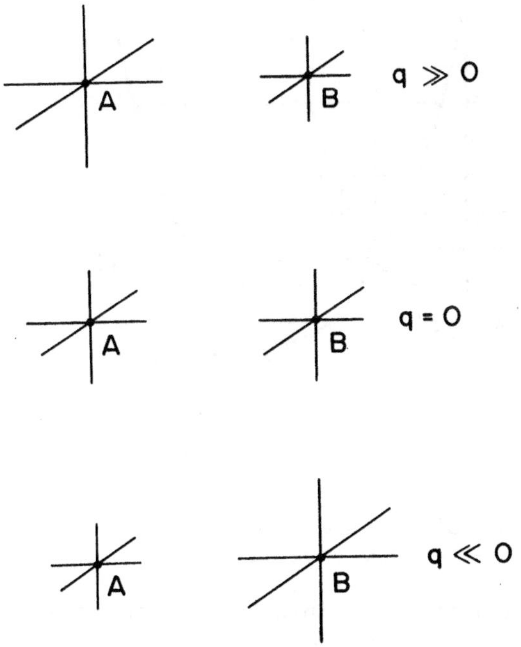

Figure 1. The crucial vibrational coordinate.

Since, by eq (11) W_a and W_b are harmonic potentials (i.e., are parabolas), χ_{na} and χ_{nb} are harmonic oscillator functions centered at $-\lambda$ and λ respectively; the vibronic eigenvalues are,

$$E_{n_a,n_b}/h\nu_- = -\lambda^2/2 + (n_{a,b} + \tfrac{1}{2}) \qquad (14)$$

with n_a and n_b = 0,1,2.... These vibrational levels are shown in Fig. 2.

Suppose that the system is in state $\psi_a(= \psi_M^A\psi_N^B)$ and we wish to compute the probability (or intensity) of an optical transition to state $\psi_b(= \psi_N^A\psi_M^B)$. Classically, this corresponds (in the C & T complex) to a transition in which an electron hops from Ru_B^{2+} to Ru_A^{3+} (a so-called intervalence transition) producing a final state consisting of Ru_B^{3+} and Ru_A^{2+}. (We show later that in fact this transition occurs with zero intensity if there is

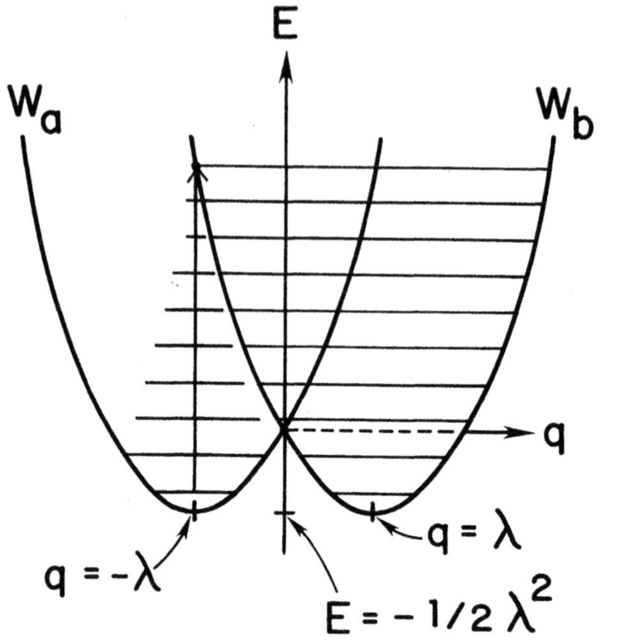

Figure 2. The potential surfaces W_a and W_b with their associated vibrational energy levels.

no coupling between A and B). This transition is shown by the vertical (Franck-Condon) arrow in Fig. 2.

The first point to note is that the final electronic state (ψ_b) is vibrationally excited ($n_b = 9$ in the diagram). This corresponds physically to the fact that the electronic motion (transition) is fast compared to the vibrational motion, and so immediately *after* the transition, Ru_A^{2+} finds itself in the equilibrium environment (Ru-N distances) of a Ru^{3+} and vice versa for Ru_B^{3+}. It is seen that the degree of vibrational excitation is proportional to λ, and this is physically reasonable since we know from the earlier discussion that λ is a direct measure of the difference in equilibrium displacement along the totally symmetric normal coordinate in the two oxidation states. We describe the fact that the electronic transition produces a vibrational change (i.e., that the electronic and vibrational excitations are coupled) by saying that the system is *vibronically coupled*. Since the magnitude of this coupling is proportional to λ, we refer to the latter as the vibronic (or electron-phonon) coupling parameter. Note that as $\lambda \to 0$, the two curves (potential surfaces) in Fig. 2 move into coincidence, and the transition energy goes to zero. Note also that when $\lambda \neq 0$, we have (at q=0) a Jahn-Teller-like situation, i.e., the two electronic states are degenerate and two equivalent, *lower energy* configurations can be reached by moving a distance $\pm\lambda$ along q.

The system becomes *much* more interesting if we let the two subunits (A and B) interact. Let us use the symbol V^{AB} to express the interaction (e.g., electrostatic) of each particle in A with every particle in B, and vice versa. We now show that V^{AB} couples (mixes) the two electronic states, ψ_a and ψ_b. If we write,

$$\psi_k = c_a\psi_a + c_b\psi_b \tag{15}$$

application of the linear variation method yields the secular equations

$$\sum_{j=a,b} c_j(H_{ij} - S_{ij}W_k) = 0, \qquad \begin{array}{l} i = a,b \\ k = 1,2 \end{array} \tag{16}$$

and the W_k are the two roots of the secular determinant,

$$|H_{ab} - S_{ab}W_k| = 0 \tag{17}$$

where $H_{ab} = \langle\psi_a|H_{el}|\psi_b\rangle$, $S_{ab} = \langle\psi_a|\psi_b\rangle$ and $H_{el} = H_{el}^A + H_{el}^B + V^{AB}$. Using eq (5) and its analog, we obtain, $H_{aa} = W_a + \langle\psi_a|V^{AB}|\psi_a\rangle$, $H_{bb} = W_b + \langle\psi_b|V^{AB}|\psi_b\rangle$. $\langle\psi_a|\psi_b\rangle$ is proportional to the overlap integral of orbitals centered on A and B. Since these centers are assumed to be relatively far apart, we make the approximation,

$$\langle\psi_a|\psi_b\rangle = 0 \tag{18}$$

and so in eq (17), $S_{ab} = \delta_{ab}$, $H_{ab} = H_{ba} = \langle\psi_a|V^{AB}|\psi_b\rangle$. V^{AB} is a complicated function of the electronic *and* vibrational (normal) coordinates of both A and B. *We approximate it by the value it has when all nuclei are held fixed at their equilibrium configurations (including q=0)* and thus write

$$V^{AB} = V_0^{AB} \tag{19}$$

where V_0^{AB} is a function only of electronic coordinates. $\langle\psi_a|V_0^{AB}|\psi_a\rangle$ ($= \langle\psi_b|V_0^{AB}|\psi_b\rangle$ by symmetry) is a constant term along the diagonal in eq (17) and may be dropped. We define a parameter ε by,

$$(h\nu_-)\varepsilon = \langle\psi_a|V_0^{AB}|\psi_b\rangle \tag{20}$$

We call ε the electronic coupling parameter becasue it is determined by purely electronic quantities and, as we see just below, because it couples ψ_a and ψ_b (since $\varepsilon \sim H_{ab}$). *The heart of the model is the assumption that* ε *is small compared to electronic binding energies within the subunits.* This is reflected in eq (15) by the implicit assumption that one need only use the ground electronic state functions, ψ_a and ψ_b, to describe the situation where $\varepsilon \neq 0$, i.e., first order perturbation theory is adequate. Using eqs (11), (18) and (20), eq (17) becomes (with W_k in units of $h\nu_-$),

$$\begin{vmatrix} \lambda q + \tfrac{1}{2}q^2 - W_k & \varepsilon \\ \varepsilon & -\lambda q + \tfrac{1}{2}q^2 - W_k \end{vmatrix} = 0 \qquad (21)$$

and the roots of the resulting quadratic are

$$W_{1,2} = (q^2/2) \mp \sqrt{\varepsilon^2 + \lambda^2 q^2} \qquad (22)$$

The corresponding electronic wave functions are

$$\psi_1 = -(1/N\sqrt{2})[(\varepsilon + \lambda q - \sqrt{\varepsilon^2 + \lambda^2 q^2})\psi_a + (\varepsilon - \lambda q - \sqrt{\varepsilon^2 + \lambda^2 q^2})\psi_b]$$

$$\psi_2 = (1/N\sqrt{2})[-(\varepsilon - \lambda q - \sqrt{\varepsilon^2 + \lambda^2 q^2})\psi_a + (\varepsilon + \lambda q - \sqrt{\varepsilon^2 + \lambda^2 q^2})\psi_b] \qquad (23)$$

where $N = [\lambda^2 q^2 + (\varepsilon - \sqrt{\varepsilon^2 + \lambda^2 q^2})^2]^{\frac{1}{2}}$. (It is important to keep in mind in this and subsequent equations that we are working in a one-dimensional (q) space. Thus for example, we use eq (11) for W_a and W_b, *NOT* eq (9), i.e., the coordinate Q_+ has been dropped.) Eq (23) clearly exhibits the coupling (mixing) of ψ_a and ψ_b when ε is "turned on". In addition, it is seen that in general ψ_1 *and* ψ_2 *are explicit and sensitive functions of q.* Physically this is so because when ε is turned on, the mixing of ψ_a and ψ_b is (to first order) directly proportional to ε and inversely proportional to the energy difference, $\Delta W \equiv |W_a - W_b|$. But ΔW is a strong function of q because *the potential surfaces* W_a *and* W_b *are not parallel* (Fig. 2).

Inspection of eq (23) shows that the q- dependence of ψ_1 and ψ_2 vanishes only in two limiting cases, namely, $\varepsilon=0$ or $\lambda=0$. This follows in the former case because there is no perturbation mixing ψ_a and ψ_b, and eq (23) reduces to $\psi_1 = \psi_b$, $\psi_2 = \psi_a$. It happens in the latter case ($\lambda=0$, $\varepsilon \neq 0$) because *the two potential surfaces are parallel*--in fact by eq (22), they are identical parabolas vertically displaced by $2|\varepsilon|$. With a little rearranging, eq (23) can be shown in that case to reduce to $\psi_1 = (1/\sqrt{2})(\psi_a + \psi_b)$, $\psi_2 = (1/\sqrt{2})(\psi_a - \psi_b)$.

The chemically interesting cases arise when ε *and* λ are *simultaneously nonzero.* Fig. 3 shows the potential surfaces for a varity of cases (including the unsymmetrical ($A \neq B$) case to be discussed later).

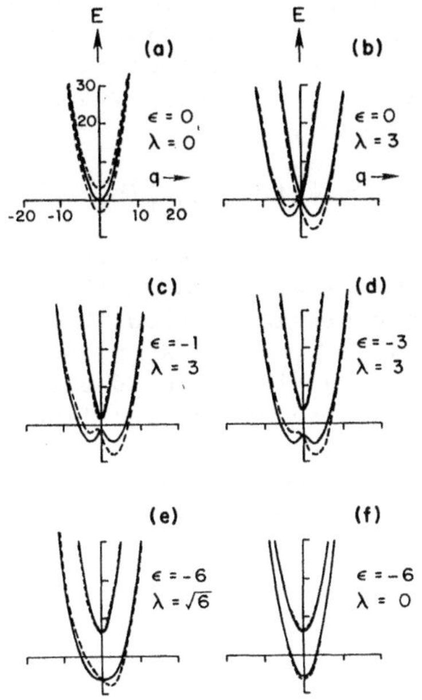

Figure 3. Potential energy surfaces (in units of $h\nu_-$) in q space according to eq (22) (for W=0) and its generalization (1,2) (for $W \neq 0$) for a range of values of the parameters ε and λ. The solid curves apply for W=0 (the A=B case), and the dashed curves apply for W=3.

It follows from eq (22) that the height of the barrier on the lower potential surface is given by

$$E_a/h\nu_- = \tfrac{1}{2}\lambda^2 - |\varepsilon| + \varepsilon^2/2\lambda^2; \quad |\varepsilon| \lesssim \lambda^2 \tag{24}$$

E_a is sometimes referred to as the activation energy for thermal electron transfer.

Associated with the potential surfaces (Fig. 3) will be a complete set of vibronic energy levels and eigenfunctions. It is clearly of great interest to determine these quantities since, by quantum mechanical principles, one can then calculate, within the context of the model, observables associated with this ground vibronic manifold, including the spectroscopic properties of the intervalence band, which is a distinctive feature of mixed-valence systems. In fact, this turns out to be a surprisingly easy task, and was accomplished many years ago by Fulton and Gouterman in their treatment of the excited states of symmetrical dimers (9).

An obvious method of proceeding, since the electronic functions are known (eq (23)), is to associate with each of them a complete set of vibrational functions, say $\chi_{1,n_1}(q)$ and $\chi_{2,n_2}(q)$, where n_1, $n_2 = 1,2\ldots\infty$. A general *vibronic* wave function analogous to eq (15) is then,

$$\Psi_k(r,q) = \psi_1(r,q)\chi_{1,k}(q) + \psi_2(r,q)\chi_{2,k}(q) \quad (25)$$

where $k = 1,2\ldots\infty$, and r symbolizes all the electronic coordinates implicit in the functions ψ_M and ψ_N (eq (1)). Formally, we may again use the linear variation method, this time to determine the unknown "coefficients" $\chi_{1,k}$ and $\chi_{2,k}$ *at each value of q*. However, whereas in eqs (16) and (17) we used only the electronic Hamiltonian (since we were holding the nuclei fixed), we must now use the full Hamiltonian (H), i.e., we must include the nuclear kinetic energy operators (T_n) since we are allowing nuclear motion. Putting it another way, we are now solving the *dynamic problem*. Recalling again that we are working in q- space, we may write,

$$H = H_{el} + T_n(q) \quad (26)$$

where $T_n(q)$ is the kinetic energy operator for vibrational coordinate q. But by definition, H_{el} is diagonal in ψ_1, and ψ_2, and so eq (25) leads to the secular equations,

$$\sum_{j=1,2} [(W_i - E_k)\delta_{ij} + \langle\psi_i|T_n(q)|\psi_j\rangle]\chi_{j,k} = 0;$$
$$i = 1,2;\ k = 1,\ldots,\infty \quad (27)$$

where $\langle\psi_i|T_n(q)|\psi_j\rangle\chi_{j,k} \equiv \int \psi_i^*(r,q)T_n(q)\psi_j(r,q)\chi_{j,k}(q)d\tau_{el}$, and E_k are the vibronic eigenvalues. If ψ_i and ψ_j are *not* strong functions of q (i.e., if *either* λ or ε=0) then

$$T_n(q)\psi_j\chi_{j,k} = \psi_j T_n(q)\chi_{j,k} \quad (28)$$

In that case, eqs (27) and (25) become respectively (since $\langle\psi_i|\psi_j\rangle = \delta_{ij}$,

$$[W_i(q) + T_n(q)]\chi_{i,k}(q) = E_k\chi_{i,k}(q);$$
$$i = 1,2;\ k = 1,\ldots,\infty \quad (29)$$

$$\Psi_k^{(i)}(r,q) = \psi_i(r,q)\chi_{i,k}(q);$$
$$i = 1,2;\ k = 1,\ldots,\infty \quad (30)$$

Eqs (29) and (30) are the familiar results of the Born-Oppenheimer approximation. In particular, ψ_i is assumed to be at most a slowly varying function of q so that eq (28) is a good approximation. (In the crude adiabatic approximation, ψ_i would be set equal to its value at q=0.) The vibrational function $\chi_{i,k}$ is determined *only* by the W_i potential surface. *Physically, the nuclear motion is confined to that surface alone.* Thus we see that *the Born-Oppenheimer approximation results if the nuclear kinetic energy operator is assumed to commute with the electronic functions,* ψ_j (see eq (28)). Furthermore, since under these circumstances (ε or λ=0) the surface is harmonic in our treatment, the $\chi_{i,k}$ are simply harmonic oscillator functions, and the dynamic problem has been solved.

The situation changes profoundly if the electronic functions depend explicity (and strongly) on q (ε and λ *both* $\neq$ 0) since eq (28) no longer applies. Thus $T_n(q)$ must be applied to ψ_j (i.e., the function in eq (23) must be differentiated twice with respect to q) *before* the integration over electronic coordinates can be performed in $\langle\psi_i|T_n(q)|\psi_j\rangle$, eq (27). Under these circumstances this matrix element is *not* diagonal in ψ_i and ψ_j and so both $\chi_{1,k}$ and $\chi_{2,k}$ appear in *both* (i=1,2) equations (eq (27)). Thus $\chi_{1,k}$ and $\chi_{2,k}$ are determined by *both* potential surfaces ($W_1(q)$ and $W_2(q)$), i.e., the vibronic functions $\psi_1\chi_{1,k}$ and $\psi_2\chi_{2,k}$ are coupled (mixed) by the nuclear motion, and thus the Born-Oppenheimer approximation (eq (30)) fails and must be replaced by eq (25).

Physically, this means that the *nuclear motion cannot be confined to a single potential surface.* Aside from the complicated nature of the $\chi_{i,k}$ in this case (they are *not* harmonic oscillator functions), ψ_1 and ψ_2 are not convenient in spectroscopic applications because their q dependence *invalidates* the Franck-Condon approximation ($\langle\psi_i\chi_i|\vec{m}_{el}|\psi_j\chi_j\rangle = \langle\psi_i|\vec{m}_{el}|\psi_j\rangle^\circ \langle\chi_i|\chi_j\rangle$, where $\vec{m}_{el}$ is the electronic part of the electric dipole moment operator, and superscript zero indicates that the nuclear coordinates are held fixed at equilibrium).

In such circumstances, *it is much more convenient to use electronic wave functions which have no explicit q dependence,* and so we return to the functions ψ_a and ψ_b. To take advantage of the interchange symmetry (A=B), we use the functions $\psi_\pm$ defined by

$$\psi_\pm = (1/\sqrt{2})(\psi_a \pm \psi_b) \tag{31}$$

If we substitute ψ_1 and ψ_2 from eq (23) into eq (25), the latter assumes the form,

$$\Phi_\nu(r,q) = \psi_+(r)\chi_{+,\nu}(q) + \psi_-(r)\chi_{-,\nu}(q) \tag{32}$$

where the χ_+ and χ_- are (q-dependent) linear combinations of the χ_1 and χ_2, and we use Φ_ν in place of Ψ_k to emphasize the change in electronic basis. Proceeding as before (eqs (25)-(27)), eq (32) leads to the secular equations,

$$\sum_{j=+,-} [\langle\psi_i|H_{el} + T_n|\psi_j\rangle - \langle\psi_i|\psi_j\rangle E_\nu]\chi_{j,\nu} = 0;$$
$$i = +,-; \; \nu = 1,2,\ldots\infty \tag{33}$$

From the definition of $\psi_\pm$, $\langle\psi_i|\psi_j\rangle = \delta_{ij}$, but $\langle\psi_i|H_{el}|\psi_j\rangle$ is not diagonal. We now assume that ψ_a and ψ_b are at most slowly varying functions of q. In particular, this requires that the subunit electronic state $(\psi_M^A, \psi_N^A, \psi_M^B, \psi_N^B)$ *not* show Jahn-Teller effects. We may then assume that $T_n(q)$ commutes with ψ_a and ψ_b (and, therefore, with $\psi_\pm$), and it follows immediately that $\langle\psi_i|T_n|\psi_j\rangle\chi_{j,\nu} = \delta_{ij}T_n\chi_{j,\nu}$. The matrix elements of $\langle\psi_i|H_{el}|\psi_j\rangle$ (i,j = a,b) aredisplayed in eq (21), so it is an elementary matter to compute the corresponding matrix elements for i,j = +,-. Eq (33) thus becomes,

$$(T_n(q) + \tfrac{1}{2}q^2 + \varepsilon - E_\nu)\chi_{+,\nu}(q) + \lambda q\chi_{-,\nu}(q) = 0$$
$$\lambda q\chi_{+,\nu}(q) + (T_n(q) + \tfrac{1}{2}q^2 - \varepsilon - E_\nu)\chi_{-,\nu}(q) = 0$$
$$\nu = 1,2,\ldots,\infty \tag{34}$$

where E_ν is in units of $h\nu_-$.

If this pair of *coupled equations* can be solved for each value of ν, i.e., if $\chi_{+,\nu}$, $\chi_{-,\nu}$ (and E_ν) can be determined, then eq (32) gives the complete set of *vibronic* eigenfunctions of the ground manifold. As Fulton and Gouterman have shown (9), this is easily accomplished if $\chi_{+,\nu}$ and $\chi_{-,\nu}$ are expanded in *the complete orthonormal set* of harmonic oscillator functions $\chi_n(q)$ $(n=0,1,2,\ldots,\infty)$, i.e., *harmonic oscillator functions in coordinate q are used as the vibrational basis.* Thus

$$\chi_{+,\nu} = \sum_{n=0}^{\infty} C_{\nu,n}\chi_n; \quad \chi_{-,\nu} = \sum_{n=0}^{\infty} C'_{\nu,n}\chi_n \tag{35}$$

and eq (32) assumes the form,

$$\Phi_\nu(r,q) = \psi_+ \sum_{n=0}^{\infty} C_{\nu,n}\chi_n + \psi_- \sum_{n=0}^{\infty} C'_{\nu,n}\chi_n \tag{36}$$

Because of the symmetry (A=B), group theoretic principles permit us to classify the Φ_ν in accord with their behavior under interchange; Φ^- and Φ^+ designate respectively vibronic functions which do or do not change sign when the nuclei are interchanged. (Note again that this classification is with respect to the nuclear configuration q=0.) Since $\psi_+ \to \psi_+$, $\psi_- \to -\psi_-$ and $q \to (-q)$ (and thus $\chi_n(-q) \to (-1)^n \chi_n(q)$) under an interchange operation, it follows that eq (36) can be written,

$$\Phi_\nu^+ = \psi_+ \sum_{n=0,2,4,\ldots}^{\infty} r_{\nu n}\chi_n + \psi_- \sum_{n=1,3,5,\ldots}^{\infty} r_{\nu n}\chi_n$$

$$\Phi_\nu^- = \psi_+ \sum_{n=1,3,5,\ldots}^{\infty} s_{\nu n}\chi_n + \psi_- \sum_{n=0,2,4,\ldots}^{\infty} s_{\nu n}\chi_n \tag{37}$$

Our task is thus to calculate the coefficients $r_{\nu n}$ and $s_{\nu n}$ for each value of ν. Let us first do this for the $r_{\nu n}$. We note that

$$(T_n + \tfrac{1}{2}q^2)\chi_n = (n + \tfrac{1}{2})\chi_n$$

$$\langle\chi_n|q|\chi_{n+1}\rangle = \langle\chi_{n+1}|q|\chi_n\rangle = ((n+1)/2)^{\frac{1}{2}} \tag{38}$$

That is, *the* χ_n *are eigenfunctions of* $(T_n + 1/2\ q^2)$ with eigenvalues $(n + 1/2)$ in units of $h\nu_-$; this of course is why they are such a convenient basis. The second of eqs (38) is a well known property of harmonic oscillator functions. In view of the first of eqs (37), substituting eq (35) into eq (34) gives,

$$(T_n(q) + \tfrac{1}{2}q^2 + \varepsilon - E_\nu^+) \sum_{n,\text{even}} r_{\nu n}\chi_n + \lambda q \sum_{n,\text{odd}} r_{\nu n}\chi_n = 0$$

$$\lambda q \sum_{n,\text{even}} r_{\nu n}\chi_n + (T_n(q) + \tfrac{1}{2}q^2 - \varepsilon - E_\nu^+) \sum_{n,\text{odd}} r_{\nu n}\chi_n = 0 \tag{39}$$

We now multiply eqs (39) by each χ_n in turn $(n=0,1,2,\ldots,\infty)$ and integrate over q. Using eq (38), the following set of secular equations is obtained;

$$\sum_{n=0}^{\infty} r_{\nu n}(H_{mn} - \delta_{mn}E_\nu^+) = 0; \qquad m = 0,1,2,\ldots \quad \nu = 0,1,2,\ldots$$

with

$$H_{mn} = \lambda[(\sqrt{m/2})\ \delta_{m,n+1} + (\sqrt{(m+1)/2}\ \delta_{m,n-1}] + (m + \tfrac{1}{2} + (-1)^m\varepsilon)\delta_{m,n} \tag{40}$$

The corresponding secular determinant yields the eigenvalues, E^+_ν. Proceeding in exactly the same way for the second of eqs (37), the analogous set of equations for the $s_{\nu n}$ are obtained by making the substitutions, $r_{\nu n} \to s_{\nu n}$, $\varepsilon \to -\varepsilon$, $E^+_\nu \to E^-_\nu$ in eq (40).

Thus the solution to the dynamic problem has been reduced to a very simple matrix diagonalization. It is true that this matrix is infinite, *but drastic truncation is possible without significant loss of accuracy*. For example, to calculate the intervalence band contour (see later), a basis of 40 quanta ($n=0,1,\ldots 39$ in eq (40)) is usually entirely satisfactory. This requires for example, only a few seconds of central processor time on a CDC Cyber 172. Furthermore, if there is any doubt, the basis size can be increased to see if significant changes occur.

An excellent, concise discussion of electronic potential surfaces and associated nuclear motions may be found in Reference 10.

III. PKS MODEL; UNSYMMETRICAL CASE ($A \neq B$)

The symmetrical case is relatively uncommon in practice. More usual is the situation in which the two centers contain either different atoms, or the identical atoms are located in different environments. For example, $[(NC)_5\text{-Ru-C-N-Fe-}(CN)_5]^{5+}$ and $[(NC)_5\text{-Fe-C-N-Fe-}(CN)_5]^{5+}$ illustrate these two cases. Such systems can be treated by the same basic vibronic coupling model if some additional simplifications are made and if the energies of ψ_a and ψ_b are not too disparate. We outline this case here.

Returning to the beginning of Section II, equations analogous to eqs (1)-(5) still apply, but now $W_M^A(Q_A = 0) \equiv W_M^{\circ A} \neq W_M^{\circ B}$ and $W_N^{\circ A} \neq W_N^{\circ B}$. In addition, while we again define coordinate origins by $\ell_N^A = 0 = \ell_N^B$, it is no longer a symmetry requirement that $\ell_M^A = \ell_M^B$, $k_M^A = k_M^B$ or $k_N^A = k_N^B$. These equalities are the *ADDITIONAL assumptions required* in the unsymmetrical case to again allow the degree of freedom in coordinate Q_+ to be ignored. With these assumptions, W_a and W_b in eq (6) et seq. contain the additional constant terms, $(W_M^{\circ A} + W_N^{\circ B})$ and $(W_N^{\circ A} + W_M^{\circ B})$ respectively and it is again assumed that $k_M = k_N \equiv k$. Finally, following eq (19), it is no longer true that $\langle\psi_a|V^{AB}|\psi_a\rangle \equiv W_\circ^{AB}(A)$ is equal to $\langle\psi_b|V^{AB}|\psi_b\rangle \equiv W_\circ^{AB}(B)$. Proceeding exactly as before, one finds that eqs (20)-(23) and (25)-(36) and (38) continue to apply provided only that the zero of energy is now defined by,

$$W_M^{\circ A} + W_N^{\circ B} + W_M^{\circ B} + W_N^{\circ A} + W_\circ^{AB}(A) + W_\circ^{AB}(B) = 0,$$

AND the term λq is replaced everywhere by the term $\lambda q+W$ where

$$(h\nu_-)W = \tfrac{1}{2}[(W_M^{\circ A} + W_N^{\circ B}) - (W_N^{\circ A} + W_M^{\circ B}) + W_\circ^{AB}(A) - W_\circ^{AB}(B)] \qquad (41)$$

(An *analytical expression* analogous to eq (24) cannot be written when $W \neq 0$.)

W is thus the one additional parameter introduced in the unsymmetrical case and clearly goes to zero when A=B. Since the interchange symmetry has been lost, eq (37) is replaced by,

$$\Phi_\nu = \sum_{n=0}^{\infty} (\psi_+ r_{\nu n} \chi_n + \psi_- r'_{\nu n} \chi_n) \tag{42}$$

The secular equations corresponding to eq (42) are found to be,

$$\sum_{n=0}^{\infty} r_{\nu n} (H_{mn} - \delta_{mn} E_\nu) + \sum_{n=0}^{\infty} r'_{\nu n} H''_{mn} = 0$$

$$\sum_{n=0}^{\infty} r'_{\nu n} (H'_{mn} - \delta_{mn} E_\nu) + \sum_{n=0}^{\infty} r_{\nu n} H''_{mn} = 0$$

$$H_{mn} = (m + \tfrac{1}{2} + \varepsilon)\delta_{mn}; \; H'_{mn} = (m + \tfrac{1}{2} - \varepsilon)\delta_{mn},$$

$$H''_{mn} = \lambda[(\sqrt{m/2})\, \delta_{m,n+1} + (\sqrt{(m+1)/2}\, \delta_{m,n-1}] + W\delta_{mn}$$

$$m = 0,1,2,\ldots; \quad \nu = 0,1,2,\ldots \tag{43}$$

It is necessary to diagonalize a single block twice the size of the two separate blocks (corresponding to Φ_ν^+ and Φ_ν^-) of the symmetrical case.

IV. APPLICATIONS

The previous two sections present and solve a simple vibronic coupling model for mixed-valence systems. Since the complete set of vibronic eigenfunctions for the ground manifold is obtained, one can calculate observable properties associated with this manifold. Comparisons of such calculations with experimental results provide tests of the theory, and theoretical predictions of the model should give additional insight into such systems. In this section, we discuss several applications and predictions of the model. In Section V we use the model to calculate the electron transfer rate in symmetrical (A=B) mixed-valence systems.

A. Intervalence Band Contours and the Determination of ε, λ and W

It is clear from the discussion in Sections II and III that the complete ground vibronic manifold is determined by the values of ε, λ and W. We show here how the intervalence band contour may be estimated from these parameters, or vice versa.

In the symmetrical case, symmetry requirements allow only $\pm \rightarrow \mp$ transitions, and the dipole strength (intensity) of a vibronic line, arising from a collection of randomly oriented molecules, is given by,

$$D(\nu' \rightarrow \nu) = \sum_{\gamma=x,y,z} \frac{(N_{\nu'} - N_\nu)}{N} |\langle\Phi^+_{\nu'}|m_\gamma|\Phi^-_\nu\rangle|^2 \tag{44}$$

where $N_{\nu'}/N$ is the fractional population in state ν'. $N_{\nu'} = \exp(-E_{\nu'}/kT)$ and $N = \Sigma_{\nu'}N_{\nu'}$. m_γ is the γth component of the molecule-fixed electric dipole operator, $\vec{m} = \Sigma_i e_i \vec{r}_i$. Let us first rewrite eq (37) in the form of eq (32);

$$\Phi^+_{\nu'} = \psi_+\chi^+_{+,\nu'} + \psi_-\chi^+_{-,\nu'}$$

$$\Phi^-_\nu = \psi_+\chi^-_{+,\nu} + \psi_-\chi^-_{-,\nu} \tag{45}$$

Now the electric dipole operator in eq (44) may be written,

$$m_\gamma = m^e_\gamma + m^n_\gamma \tag{46}$$

where e and n respectively denote the electronic and nuclear contributions. Denoting $\langle\Phi^+_{\nu'}|m_\gamma|\Phi^-_\nu\rangle \equiv TD_\gamma = \langle\Phi^+_{\nu'}|m^e_\gamma + m^n_\gamma|\Phi^-_\nu\rangle = TD_\gamma(e) + TD_\gamma(n)$, let us examine the various contributions. Using eq (45),

$$TD_\gamma(e) = \langle\psi_+\chi^+_{+,\nu'} + \psi_-\chi^+_{-,\nu'}|m^e_\gamma|\psi_+\chi^-_{+,\nu} + \psi_-\chi^-_{-,\nu}\rangle \tag{47}$$

We now make the Franck-Condon approximation, i.e., we assume that since ψ_+ and ψ_- are slowly varying functions of q, the integrals *over electronic wave functions* can be evaluated holding the nuclei fixed at equilibrium (including q=0). Then the vibrational functions can be factored out and noting that (see eq (37) $\langle\chi^\pm_{+,\nu'}|\chi^\mp_{+,\nu}\rangle = \langle\chi^+_{-,\nu'}|\chi^-_{-,\nu}\rangle = 0$, eq (47) becomes,

$$TD_\gamma(e) = \langle\psi_+|m^e_\gamma|\psi_-\rangle^o\langle\chi^+_{+,\nu'}|\chi^-_{-,\nu}\rangle + \langle\psi_-|m^e_\gamma|\psi_+\rangle^o\langle\chi^+_{-,\nu'}|\chi^-_{+,\nu}\rangle \tag{48}$$

where superscript zero indicates that the matrix element is evaluated at q=0. (Note that this treatment would not work using the strongly q-dependent functions, ψ_1 and ψ_2 (eq (23)), since the Franck-Condon approximation would be unjustified.

At this point we can see clearly why the vibrational motion in normal coordinate Q_+ (or in fact *any* of the multitude of other normal coordinates) plays no significant role in the model (recall discussion below eq (9)). In the harmonic approximation, inclusion of Q_+ and all other vibrational motions (except q) would multiply each of the two terms in eq (48) by a Franck-Condon overlap factor of the form $\langle\chi_{i',n'}(Qi)|\chi_{i,n}(Qi)\rangle$ for each such normal coordinate. But with the assumption of equal force constants in both oxidation states ($k_M = k_N$), the harmonic oscillator functions, $\chi(Q_+)$, on potential surface W_a and W_b are members of the same identical set of orthonormal functions since the two surfaces are identical in Q_+ space (eq (9)). Thus via eq (31) and using the orthonormality of these vibrational functions, it follows that $\langle\chi'_{n'}(Q_+)|\chi_n(Q_+)\rangle = 0$ unless n' = n. Thus *no change can occur in the vibrational quantum number associated with* Q_+ in the course of the transition. Precisely the same argument applies for all the other *(non-totally symmetric)* normal coordinates (except q) because we have assumed (following eq (1)) that *the subunits have the same point group symmetry (and force constants) in both oxidation states.*

We assume that $\psi_\pm$ and m^e_γ are real, and we can, therefore, write,

$$\begin{aligned}\langle\psi_+|m^e_\gamma|\psi_-\rangle^\circ &= \langle\psi_-|m^e_\gamma|\psi_+\rangle^\circ\\ &= \tfrac{1}{2}[\langle\psi_a|m^e_\gamma|\psi_a\rangle^\circ - \langle\psi_b|m^e_\gamma|\psi_b\rangle^\circ]\\ &= \tfrac{1}{2}[\langle\psi^A_M\psi^B_N|m^e_\gamma|\psi^A_M\psi^B_N\rangle^\circ - \langle\psi^A_N\psi^B_M|m^e_\gamma|\psi^A_N\psi^B_M\rangle^\circ]\end{aligned} \tag{49}$$

$m_\gamma(=m_x,m_y,m_z$ where $m_x = \Sigma_i e_i x_i$, etc.) is defined with respect to an origin midway between A and B, and must be expressed in terms of the subunit centers (which are chosen to lie along the z axis):

$$m^e_x = m^{eA}_x,\quad m^e_y = m^{eA}_y,\quad m^e_z = m^{eA}_z - (R/2)\sum_i e_i$$

$$m^e_x = m^{eB}_x,\quad m^e_y = m^{eB}_y,\quad m^e_z = m^{eB}_z + (R/2)\sum_i e_i \tag{50}$$

where R is the distance between subunit centers and e_i is the charge of the ith electron. We now substitute eq (50) into eq (49) *and assume that the subunit centers have inversion symmetry.* This assumption has the immediate consequence that all m^{eA}_γ and m^{eB}_γ (γ=x,y,z) make zero contribution (parity selection rule) and so for example,

$$\begin{aligned}\langle\psi^A_M\psi^B_N|m^e_z|\psi^A_M\psi^B_N\rangle^\circ &= \langle\psi^A_M|m^{eA}_z - (R/2)\sum_{i=1}^{m} e_i|\psi^A_M\rangle^\circ +\\ &\quad\ \langle\psi^B_N|m^{eB}_z + (R/2)\sum_{i=1}^{n} e_i|\psi^B_N\rangle^\circ = \frac{(n-m)}{2}\,eR;\\ \langle\psi^A_N\psi^B_M|m^e_z|\psi^A_N\psi^B_M\rangle^\circ &= \frac{(m-n)}{2}\,eR\end{aligned} \tag{51}$$

where n and m are respectively the number of electrons associated with oxidation states N and M. The corresponding terms for m_x^e and m_y^e are zero. Assuming $(n-m) = 1$, substituting eq (37) into eq (48) and using eqs (49)-(51), the final result is

$$D^e(\nu' \rightarrow \nu) = \left(\frac{N_{\nu'} - N_{\nu}}{N}\right) S^2_{\nu'\nu} \; |\langle\psi_+|m_z^e|\psi_-\rangle^\circ|^2 \quad (52)$$

where $|\langle\psi_+|m_z^e|\psi_-\rangle^\circ|^2 = e^2R^2/4$, and

$$S_{\nu'\nu} = \sum_n r_{\nu',n} s_{\nu n} \quad (53)$$

We see that the transition in eq (52) is z-polarized.

Let us now consider $TD_\gamma(n)$. Eq (47) applies if we simply replace e everywhere by n. If we make the Frank-Condon approximation and recall that $\langle\psi_+|\psi_-\rangle^\circ = 0 = \langle\chi^{\pm}_{+,\nu'}|\chi^{\mp}_{+,\nu}\rangle = \langle\chi^{\pm}_{-,\nu'}|\chi^{\mp}_{-,\nu}\rangle$, the result is,

$$TD_\gamma(n)(\text{Franck-Condon}) = 0 \quad (54)$$

If instead we expand the integrals over electronic wave functions (eq (47) for *both* m_γ^e and m_γ^n) in a power series in normal coordinates retaining only linear terms, the result is,

$$TD_\gamma{}' = \left(\frac{\partial m_\gamma^+}{\partial q}\right)_\circ \langle\chi^+_{+,\nu'}|q|\chi^-_{+,\nu}\rangle + \left(\frac{\partial m_\gamma^-}{\partial q}\right)_\circ \langle\chi^+_{-,\nu'}|q|\chi^-_{-,\nu}\rangle \quad (55)$$

where $m_\gamma^\pm \equiv \langle\psi_\pm|m_\gamma^e + m_\gamma^n|\psi_\pm\rangle$, and the subscript zero indicates that the derivative is evaluated at $q = 0$ holding all other normal coordinates at equilibrium. Making the one-center approximation and assuming simple additivity of the dipole derivatives on the two centers,

$$TD_\gamma{}' = \left(\frac{1}{2C}\right)\left[\left(\frac{\partial m_\gamma^A}{\partial Q_A}\right)_M + \left(\frac{\partial m_\gamma^A}{\partial Q_A}\right)_N\right]_\circ \left[\langle\chi^+_{+,\nu'}|q|\chi^-_{+,\nu}\rangle + \langle\chi^+_{-,\nu'}|q|\chi^-_{-,\nu}\rangle\right] \quad (56)$$

where the subscripts M and N indicate the oxidation states for which the dipole derivatives are evaluated. $C = \pi\sqrt{2}(\nu\text{-}/h)^{\frac{1}{2}}$. (Because of the symmetry, B may be substituted for A.)

The dipole derivatives are non-zero if the totally symmetric normal coordinate Q_A (and Q_B) is associated with an infrared active fundamental. In the case of octahedral subunits (and in fact in any case where the subunits are non-polar), such derivatives are zero. However, if the subunits are polar, as for example in the case of the diketone anion radical discussed by Schroeder and Mazur (11), a contribution arises from eq (56). In the diketone case, Q_A and Q_B represent carbonyl stretching vibrations which *are* infrared active.

The dipole derivatives in eq (56) may often be estimated from known absolute infrared intensity data, and it may be anticipated in general that $TD_\gamma{}'$ will be substantially smaller than $TD_\gamma(e)$ (eq (48)). However, it should be noted that the vibrational factors are entirely different in the two cases. These factors can of course be explicitly calculated by the model to help assess the relative contributions of $TD_\gamma(e)$ and $TD_\gamma{}'$ in specific cases.

When $W \neq 0$ ($A \neq B$), the $\pm \rightarrow \mp$ selection rule no longer applies, but eq (52) continues to hold if eq (53) is replaced by,

$$S_{\nu'\nu} = \sum_{n=0}^{\infty} (r_{\nu'n} s'_{\nu n} + r'_{\nu'n} s_{\nu n}) \tag{53a}$$

where the transition is from a state $\Phi_{\nu'}$ (eq (42) with $\nu = \nu'$) to a state Φ_ν (eq (42) with $r_{\nu n}$, $r'_{\nu n}$ replaced respectively by $s_{\nu n}$, $s'_{\nu n}$). In the unsymmetrical case, eq (54) still applies and eq (56) may be generalized in a straightforward manner.

The central role of the parameters ε, λ and W has already been stressed. These quantities together with ν_- define the complete set of eigenfunctions and eigenvalues of the ground vibronic manifold. To extract these parameters from a known band contour, one proceeds by a trial and error fitting procedure, guessing values of ε, λ (and W if $A \neq B$) and diagonalizing the secular equations (eq (40) or (43)). The intervalence band is the totality of lines, $\nu' \rightarrow \nu$, eqs (52) and (56) (but see next subsection). Associating a gaussian lineshape with each transition, an intervalence band contour can be synthesized for comparison with experiment. The widths of such gaussians must be chosen in an *ad hoc* manner. We have previously done this (1,2) assuming the same width for all lines.

The trial and error procedure, though potentially tedious, can be accomplished with remarkable speed and economy using simple interactive graphics. In a few-minutes, one can quite thoroughly explore parameter space to ascertain the uniqueness and parameter sensitivity of various fits. Helpful guidance is furnished by previous analyses of band contours (1) and moments (2), the latter of which cover a very wide range of possible parameters. Application of this procedure to the C & T complex (1), $[(bpy)_2ClRu\text{-}pyr\text{-}RuCl(bpy)_2]^{3+}$ (3), and $(C_6H_6)_2^+$ (2) have been discussed previously. It appears possible to fit the first and third of these with a remarkable degree of uniqueness, the fit of the C & T complex being shown in Fig. 4.

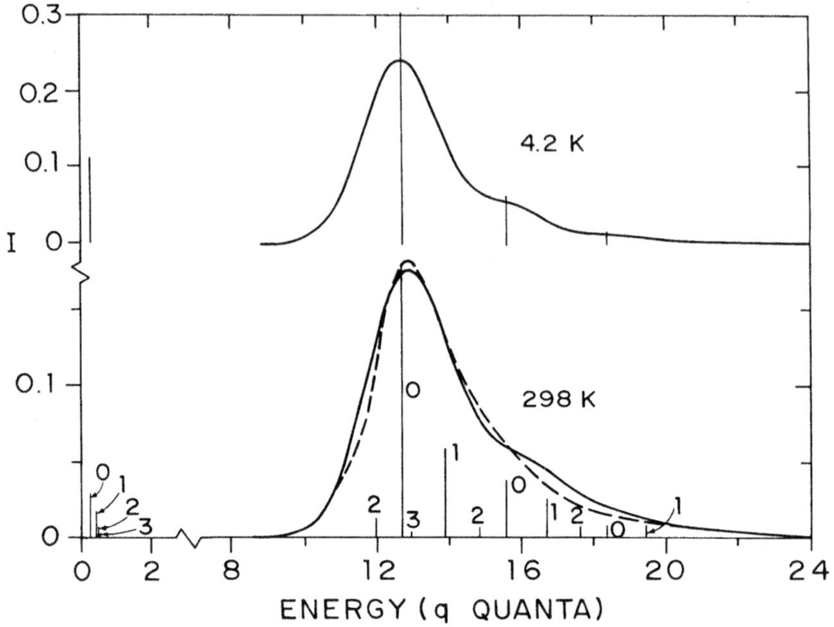

Figure 4. Intervalence band (dashed curve) of the Cruetz and Taube complex at room temperature (from Ref. (6)) scaled in intensity to the "best fit" of the model (stick spectrum and solid curve) which uses $\varepsilon = -6$, $\lambda = 2.7$. The corresponding model spectrum at 4.2 K is shown directly above on a vertical scale compressed by a factor of 2. The calculated height of each stick is $S^2_{\nu'\nu}/2$ scaled by the appropriate normalized population factor. The solid curves are synthesized from gaussians taking all widths at half-height to be 2.4 q quanta. The labels 0-3 designate transitions from the ground and first three hot vibronic levels respectively. These latter three are calculated to lie respectively 131, 362, and 613 cm^{-1} above the ground vibronic level. The energy of a q quantum is $\nu_- = 500$ cm^{-1}.

As a general rule (2), increasing λ and/or W decreases the band intensity, moves the band to higher energy and increases its width. Increasing $|\varepsilon|$ increases the band intensity, moves the band to higher energy and decreases its width.

B. Infrared "Tunneling" Transitions

If both $|\varepsilon|$ and λ^2 are greater than unity, one can quite cleanly divide transitions into two types; the first occur between the two potential surfaces and give rise to the intervalence band just discussed. But in addition, low energy transitions are predicted between neighboring pairs of levels in the lower potential surface (3). Both cases are illustrated in Fig. 5. Virtually no discussion of this latter possibility appears

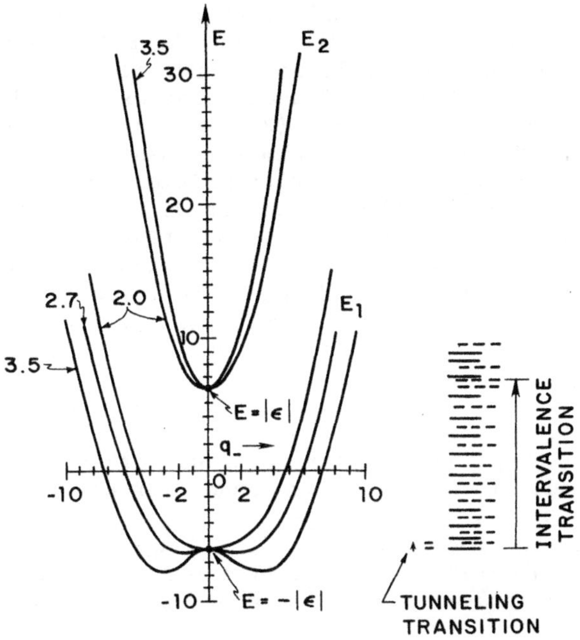

Figure 5. Upper (E_2) and lower (E_1) potential surfaces (in units of q quanta) for $\varepsilon = -6$, $\lambda = 2.0, 2.7, 3.5$. E_2 is omitted for $\lambda = 2.7$. The energies of the lower lying vibronic levels for $\lambda = 2.7$ are shown to the right; the solid lines have symmetry (+) and the dashed lines symmetry (-). The long vertical arrow shows the strongest single intervalence transition in the low-temperature limit for $\lambda = 2.7$. Tunneling transitions should occur primarily between successive low-lying (populated) vibronic levels. The short vertical arrow shows the one from the ground vibronic level.

in the literature, and this writer knows of only one reported observation of such a transition, namely that in the diketone anion radical mentioned previously (11). These predicted "tunneling" transitions are shown in Fig. 4 for the C & T complex. It has been pointed out that these transitions can be very intense and should be accessible to observation (3). By a similar analysis, the tunneling transition for $(C_6H_6)_2^+$ is predicted to occur at ~265 cm^{-1} with an integrated intensity about 40% of the intervalence band which occurs in the 900 nm region. Physically, these transitions have the character of a vibrational motion, the electronic charge adjusting adiabatically to the vibration in q, whereas the transitions between the surfaces are vibronic i.e., occur over a short interval compared to a vibrational cycle.

Observation (or otherwise) of these tunneling transitions constitutes a very important test of the PKS model (1) which claims to be able to predict their location and intensity from a knowledge of ε, λ and W (and ν_-) deduced, for example, by fitting the intervalence band. Krausz (27) has recently searched for these transitions in the C & T complex using a docosane mull at 4.2 °K. No evidence for such transitions was found out to a frequency of 20 cm^{-1} in the far infrared. This result can only be reasonably accounted for if λ is substantially smaller than 2.7. In that event, the energy of such transitions increases (to ν_- as $\lambda \to 0$) and the intensity drops sharply (to zero as $\lambda \to 0$). If this explanation is correct, then the C & T complex is much more delocalized than indicated in our analysis (see subsection C), and the high energy shoulder in the intervalence band must be explained in some other way since the band becomes symmetrical as $\lambda \to 0$. In fact, the shoulder may be explicable when the effects of spin-orbit coupling and tetragonal distortion on the subcenters are considered. These points are under further investigation.

Applying the formula, $\Delta r = \lambda(h\nu_-/3m_L k)^{\frac{1}{2}}$, to the C & T complex, one calculates a value, $\Delta r = 0.1$ Å, as the difference in Ru-N bond distance in the 2+ and 3+ oxidation states if λ = 2.7.

C. Probability Distribution in Configuration (q) Space

Having ε, λ and W and thereby the ground vibronic manifold eigenfunctions, one can get an illuminating picture of the mean oxidation state (or degree of valence trapping) of the subunit centers. To do this, one squares the vibronic eigenfunction and integrates over electronic coordinates. Thus

$$P_\nu(q) \equiv \int |\Phi_\nu|^2 d\tau_{el} \qquad (57)$$

and substituting from eq (42) noting that $\langle\psi_+|\psi_-\rangle = \delta_{+,-}$,

$$P_\nu(q) = \sum_{n,n'} \chi_n(q)\chi_{n'}(q)(r_{\nu n}r_{\nu n'} + r'_{\nu n}r'_{\nu n'}) \qquad (58)$$

The $r_{\nu n}$, etc. are obtained from the diagonalization and the χ_n are the well known harmonic oscillator functions whose higher members are easily generated from the recursion relation for the

Hermite polynomials. Thus eq (58) is easily evaluated by computer as a function of q. Plots of two cases are shown in Fig. 6 (W=0) and Fig. 7 (W=0.3).

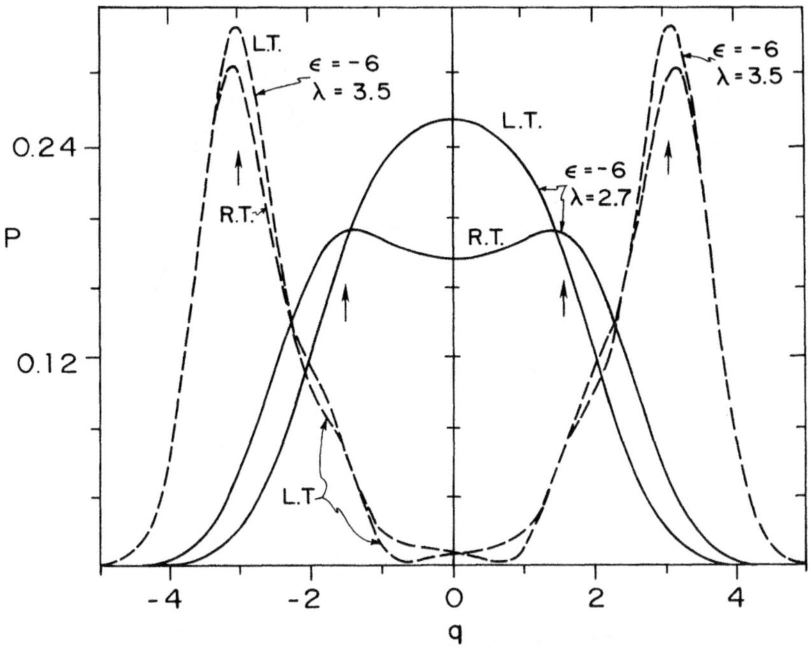

Figure 6. The calculated probability distributions in q space at room temperature (RT) and in the low-temperature limit (LT) for $\varepsilon = -6$, $\lambda = 2.7$ (solid curves) and $\varepsilon = -6$, $\lambda = 3.5$ (dashed curves). The vertical arrows designate the location of the calculated minima in the lower potential surface ($q = \pm 1.53$ and ± 3.05, respectively).

If more than one vibronic level is populated, Boltzmann factors are applied in the usual way;

$$P(q) = \sum_{\nu} P_{\nu}(q) e^{-E_{\nu}/kT} / \sum_{\nu} e^{-E_{\nu}/kT} \tag{59}$$

In the case of the C & T complex ($\varepsilon = -6$, $\lambda = 2.7$, $W = 0$), Fig. 6 shows that the probability function in q-space (P(q)) is predicted to have a single maximum at $q = 0$ at low temperatures. This means that the most probable configuration is one in which both subunits are equivalent and are intermediate between the 2+ and 3+ oxidation states. This is the *untrapped* or *delocalized* (also sometimes referred to as the $2\frac{1}{2} \leftrightarrow 2\frac{1}{2}$) case. On the other

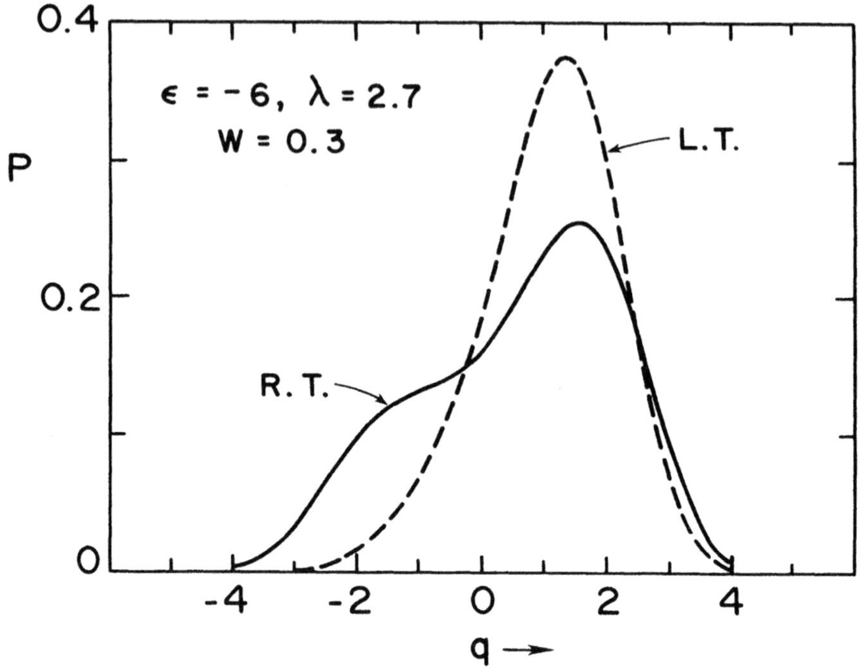

Figure 7. P(q) for the unsymmetrical case: $\varepsilon = -6$, $\lambda = 2.7$, $W = 0.3$. Notation as in Fig. 6.

hand, for the parameters, $\varepsilon = -6$, $\lambda = 3.5$, $W = 0$ of Fig. 6, there are two configurations of maximum probability each corresponding roughly to the situation in which one Ru is in the 3+ and the other is in the 2+ state. This is the *trapped* or *localized* (or 2-3 ↔ 3-2) case. Distinct Ru(II) and Ru(III) species are present.

It is interesting to note (Fig. 6) that as the temperature is raised, a slight degree of trapping is predicted for the C & T complex. This reflects the fact that the C & T parameters ($\varepsilon = -6$, $\lambda = 2.7$) represent a borderline case between trapped and untrapped. In fact, the barrier on the lower potential surface is calculated to be 57 cm^{-1} whereas the calculated lowest vibronic level lies about 30 cm^{-1} *above* the top of the barrier. Thus while using the existence of a barrier on the lower potential surface (E_a of eq (24)) as a criterion for trapping is qualitatively reasonable, a plot of P(q) gives a *quantitative description*. As would be expected, a non-zero W favors trapping and the predominance of one of the two electronic configurations (ψ_a over ψ_b or vice versa). This is illustrated in Fig. 7.

It is also possible to develop a quantitative description of the *degree* of delocalization by integrating $|\Phi_\nu|^2$ over vibrational coordinate q. This and related matters are discussed in some detail in Ref. 2, Section V.

D. The Classification of Mixed-Valence Systems

The Robin and Day classification scheme proposed in 1967 (12) stimulated much interest in mixed-valence systems. This classification was based primarily on a parameter E_ℓ which is roughly equivalent to our W. This scheme can be refined in a natural way by also including the parameters ε and λ (2). Thus compounds in classes I, II, and III (12) correspond respectively to the cases $|\varepsilon| \ll (\lambda^2 + W)$, $|\varepsilon| \lesssim (\lambda^2 + W)$ and $|\varepsilon| > (\lambda^2 + W)$. Class I compounds are strongly localized (valence trapped); the intervalence band tends to high energy and large width and is of low or negligible intensity. Class III compounds are completely delocalized, the intervalence band tends to small width and high intensity. Class II compounds are intermediate; there is some delocalization, but valences are still distinguishable. This requires a small barrier in the lower potential surface. The class II and Class III cases are best distinguished by a probability plot, P(q), the occurrence of a double maximum distinguishing the former from the latter. Further discussion of the present and Robin and Day schemes may be found in Ref. 2, Section VII.

E. Unequal Force Constants ($k_M \neq k_N$)

If the assumption of equal force constants in the two oxidation states is dropped, the normal coordinates Q_+ and Q_- are coupled (see eq (8)), i.e., the potential surfaces W_a and W_b contain cross terms of the form $(k_M - k_N)Q_+Q_-$. The consequence is that the coordinates Q_+ and Q_- and their associated motions cannot be separated. The system becomes a two-dimensional (pseudo) Jahn-Teller problem. The resulting vibronic matrix for the A=B case is given in Ref. (1), eq (51). This problem can be solved by the same techniques described in Section II, but the vibrational basis required for reasonable convergence increases greatly. Such calculations are now in progress.

The qualitative effect of unequal force constants is that simultaneous transitions in both Q_+ and Q_- are allowed, and the intervalence band contains additional transitions, their importance being proportional to the ratio, $|(k_M - k_N)/(k_M + k_N)|$.

F. Magnetic Moments and Susceptibilities in Mixed-Valence and Spin-Crossover Systems

As a first approximation, we might expect that the magnetic

moment of a mixed-valence system will simply be the average of the magnetic moments of the two subunits. We examine here how much the electronic (ε) and vibronic (λ) coupling in the PKS model modify this expectation both in mixed-valence and in the related spin-crossover systems.

For the electronic basis states, $\psi_\pm$, making the one-center approximation,

$$<\psi_\pm|\mu_\alpha|\psi_\pm> = \tfrac{1}{2}(\mu_{aa} + \mu_{bb}) \equiv \bar{\mu}$$
$$<\psi_\pm|\mu_\alpha|\psi_\mp> = \tfrac{1}{2}(\mu_{aa} - \mu_{bb}) \equiv \Delta\bar{\mu} \qquad (60)$$

where μ_α is a component of the magnetic moment operator and $\mu_{aa} = <\psi_a|\mu_\alpha|\psi_a>$, etc. Then considering first the symmetrical case and using eq (37),

$$<\Phi_\nu^+|\mu_\alpha|\Phi_\nu^+> = <\psi_+|\mu_\alpha|\psi_+> \sum_{n,n'\,\mathrm{even}} r_{\nu n} r_{\nu n'} <\chi_n|\chi_{n'}>$$
$$+ <\psi_-|\mu_\alpha|\psi_-> \sum_{n,n'\,\mathrm{odd}} r_{\nu n} r_{\nu n'} <\chi_n|\chi_{n'}> \qquad (61)$$

and recalling that $<\chi_n|\chi_{n'}> = \delta_{nn'}$, $<\Phi_\nu^+|\Phi_\nu^+> = 1 = \Sigma_n r_{\nu n}^2$ and using (60),

$$<\Phi_\nu^+|\mu_\alpha|\Phi_\nu^+> = \bar{\mu} \qquad (62)$$

In exactly the same way,

$$<\Phi_\nu^-|\mu_\alpha|\Phi_\nu^-> = \bar{\mu} \qquad (63)$$

Also,

$$<\Phi_\nu^+|\mu_\alpha|\Phi_{\nu'}^-> = \Delta\bar{\mu} \sum_n r_{\nu n} s_{\nu' n} = 0 \qquad (64)$$

where the same method has been used noting that $<\Phi_\nu^+|\Phi_{\nu'}^-> = 0 = \Sigma_n r_{\nu n} s_{\nu' n}$. It also follows that

$$<\Phi_\nu^\pm|\mu_\alpha|\Phi_{\nu'}^\pm> = 0 \qquad (65)$$

when $\nu \neq \nu'$. Thus we conclude that the simple averaging of magnetic moments mentioned above is in no way modified by the PKS model in the symmetrical case. Magnetic susceptibilities can be calculated treating the two subunits independently.

For the unsymmetrical ($W \neq 0$) case, we proceed in the same way, but using eq (42). The results are,

$$<\Phi_\nu|\mu_\alpha|\Phi_\nu> = \bar{\mu} + 2\Delta\bar{\mu} \sum_n r_{\nu n} r'_{\nu n}$$

$$\langle\Phi_\nu|\mu_\alpha|\Phi_{\nu'}\rangle = \Delta\bar{\mu}\, S_{\nu\nu'} \qquad (66)$$

where $S_{\nu\nu'}$ is defined in eq (53a). In this case we see that vibronic effects do appear through the term $\Delta\bar{\mu}$ which measures the *difference* in magnetic moments for state $\psi_a = \psi_M^A\psi_N^B$ *vs* state $\psi_b = \psi_N^A\psi_M^B$. Thus vibronic effects will modify the magnetic susceptibility in the unsymmetrical case.

The PKS model can also be applied to spin crossover systems which formally are quite similar to the $W\neq0$ case just discussed. In a spin crossover system, the two potential surfaces in Fig. 2 would represent *two different spin states* of a single-center metal complex, for example an iron (III) complex. However, in general, the two surfaces would have different energy zeros ($W\neq0$). Spin-orbit states on the two surfaces belonging to the same irreducible representation would then be connected by a spin-orbit coupling matrix element *which is formally identified with the parameter* ε. λ plays the same role as before, and the eigenfunctions and eigenvalues can be determined by solving essentially the same secular equations previously discussed (eq (43)).

Vibronic effects appear which are directly analogous to those in eq (66), and they do have an effect on the calculated magnetic susceptibility. However, for the systems which show crossover effects experimentally, the spin-orbit coupling matrix elements (ε) are small, and so vibronic effects are calculated to have only a small effect on the observed susceptibilities.

V. ELECTRON TRANSFER KINETICS (15)

Electron transfer between interacting potentials plays an important role in many physical problems. For localized mixed-valence systems, the lower potential surface exhibits a barrier of height E_a (see eq (24)), and hence it is meaningful to ask the following question: if the system starts out in one of the two wells--clearly a *non*-stationary state--how much time will elapse before it is in the other well? We first define this problem more precisely, and then we propose a solution which has a much greater range of validity than the usual treatments.

A. Definition of Problem

Let us return to our initial treatment, Section II, and consider a symmetrical (A=B) mixed-valence system in which there is no interaction between the subunits. This situation is illustrated in Fig. 2, the potential surfaces are defined by eq (11), and the vibronic eigenstates are given in eq (12). If the system is in vibronic state ϕ_{na} (a stationary state) at time t=0, and if there is neither interaction between the subunits nor any external perturbation, then the probability of the system undergoing transition to a state ϕ_{n_b} is zero. (Indeed, this is approximately true even in the presence of electromagnetic radiation of the required frequency because the transition dipole (TD) is ≈ 0. Thus $TD_\gamma(e) = \langle\phi_{n_a}|m_\gamma^e|\phi_{n_b}\rangle \approx \langle\psi_a|m^e|\psi_b\rangle^o \langle\chi_{n_a}|\chi_{n_b}\rangle \approx 0$ in the one center approximation in view of eq (18).)

Let us now suppose that at t=0 the subunits are allowed to interact. We express this interaction by the parameter ε and assume that it is turned on instantaneously, i.e., for $t<0$, $\varepsilon=0$; for $t\geq 0$, $|\varepsilon|$ = constant $\neq 0$. Then as t increases from zero, electron transfer between the subunits will occur, or more precisely, q will change from a negative to a positive value (Fig. 2). Since the occupied vibronic levels are usually below the top of the barrier, the transition from one well to the other is a tunneling process. It is the kinetics of this process that we wish to examine.

B. General Approach

Several methods of treating this problem exist in the literature. One of the oldest and most widely used is the method of beats (MOB) (16,17). In this method, one solves the Schrödinger time-dependent equation after the perturbation (ε in our case) has been turned on. One finds that tunneling to the other well has occurred after a time t_s given by,

$$t_s = h/2(E_1 - E_0) \qquad (67)$$

where E_0 and E_1 are the pair of levels which arise when application of the (small) perturbation splits the double degeneracy (Fig. 2) of the unperturbed ground state levels. Another widely used method regards the tunneling as a radiationless transition between two interacting manifolds. The transfer rate can then be calculated using the Fermi Golden Rule (18-22) provided the perturbation is small (also a requirement in the MOB). However, Weiner (23) has argued recently that the method of beats gives rates that are much too high when interactions of the system with a heat bath are significant during time t_s (eq (67)). This would be the expected situation, for example, for a mixed-valence system in solution.

In the treatment presented here, we use an equation given by Weiner (23) which relates the transfer rate to the energy splitting produced by the perturbation. Our rate expression reduces to the Fermi Golden Rule result (22) when the perturbation is small, but in view of Weiner's analysis, it should also remain valid for much larger values of the perturbation parameter, $|\varepsilon|$, in which case it predicts much higher rates than the perturbation expression given, for example, in Ref. 22.

C. Calculation of the Transfer Rate

Using semiclassical methods, it can be shown (see Ref. 23, for example) that the transmission function, D, can be written,

$$D = \pi^2 \left(\frac{E_1 - E_0}{h\nu}\right)^2 \tag{68}$$

where ν is the classical frequency of oscillation in a single well and E_1 and E_0 are defined as in eq (67). By the *definition* of D, the integrated probability of transfer through the barrier (P) in time t is given by,

$$P = \nu D t \tag{69}$$

Weiner (23) has recently shown that a relation analogous to this can be obtained *without* using semiclassical arguments by making a decomposition of the *exact stationary states of the perturbed system* into scattering states. His relation is written,

$$K_n = \pi^2 \left(\frac{E_{n+1} - E_n}{h\nu_n}\right)^2 \tag{70}$$

where K_n is the Weiner transmission coefficient analogous to D, and ν_n is the classical frequency associated with either of the exact states, E_n or E_{n+1}. Eq (70), however, applies for any split pair of low-lying (compared to E_a) states provided the splitting is not too large compared to $h\nu_n$.

The transition probability per sec (i.e., the transfer rate) is given by

$$P_n = \nu_n K_n \tag{71}$$

and the thermally averaged transfer rate is

$$\overline{P} = \sum_{n=0}^{\infty} P_n e^{-E_n/kT} \Big/ \sum_{n=0}^{\infty} e^{-E_n/kT} \tag{72}$$

In the PKS model, all E_n are calculated "exactly". Using our earlier notation and assuming that ν_n has the constant value, ν_-, eq (70) becomes,

$$K_\nu = \pi^2 \left(\frac{E_\nu^+ - E_\nu^-}{h\nu_-}\right)^2 \tag{73}$$

Substituting this into eq (72),

$$\overline{P} = \frac{\pi^2 \nu_- \sum_{\nu=0}^{\infty} (\Delta E_\nu)^2 (e^{-E_\nu^+/kT} + e^{-E_\nu^-/kT})}{\sum_{\nu=0}^{\infty} (e^{-E_\nu^+/kT} + e^{-E_\nu^-/kT})} \tag{74}$$

where $\Delta E_\nu \equiv (E_\nu^+ - E_\nu^-)/h\nu_-$. The pairs of Boltzmann factors in eq (74) take account of the fact that the two members of the split pair in eq (70) or eq (73) have different populations. Eq (74) is our general result for the transfer rate. Since $E_\nu^\pm$ and, therefore, ΔE_ν can be calculated from the PKS model, $\overline{P}$ can be calculated for any values of ε and λ provided ΔE_ν is not too large compared to 1. We now investigate the implications of this criterion in more detail.

Let us first consider the case of weak electronic coupling, i.e., $|\varepsilon| \ll \lambda^2$. In this case, Fulton and Gouterman (9) calculated an explicit expression for the vibronic energy levels using first order perturbation theory:

$$E_\nu^\pm/h\nu_- = (\nu + \tfrac{1}{2} - \tfrac{1}{2}\lambda^2) \pm (-1)^\nu \varepsilon e^{-\lambda^2} L_\nu^{(o)}(2\lambda^2) \tag{75}$$

where $L_\nu^{(o)}$ is the associated Laguerre polynomial of order zero and degree ν. If this expression is substituted into eq (74) and higher order terms in the exponentials are dropped, one obtains

$$\overline{P} = 4\pi^2\nu_- \frac{\sum_{n=0}^{\infty} \varepsilon^2 e^{-2\lambda^2} [L_n^{(o)}(2\lambda^2)]^2 e^{-nh\nu_-/kT}}{\sum_{n=0}^{\infty} e^{-nh\nu_-/kT}} \tag{76}$$

The summation in the numerator can be performed (22), and the result is,

$$\overline{P} = 4\pi^2\nu_-\varepsilon^2 \exp(-2\lambda^2 - Ze^{-h\nu_-/2kT}) I_o(Z) \tag{77}$$

where $I_o(Z)$ is the zeroth order modified Bessel function and

$$Z = 4\lambda^2 e^{-h\nu_-/2kT}/(1 - e^{-h\nu_-/kT}) \tag{78}$$

This is precisely the result obtained by Laplante and Siebrand (22) using the Fermi Golden Rule. Thus eq (74) reduces to a well established result in the weak electronic coupling limit.

To calculate $\overline{P}$ more generally, we now follow Weiner's argument and use eq (74) simply substituting for $E_\nu^\pm$ the "exact" values obtained from the PKS diagonalization (eq (40)). The results as a function of temperature for $\lambda=3$ are shown by the solid lines in Fig. 8 for a wide range of ε values. For comparison, the results based on eq (76), applicable for small $|\varepsilon|$, are shown by the dash-triple dot curves. As expected, both agree in the range of small $|\varepsilon|$. But as $|\varepsilon|$ gets larger, the results diverge markedly, especially at low temperature. Precisely where eq (74) begins to fail badly is not clear, but Weiner's analysis suggests that its range of validity should be *much greater* than

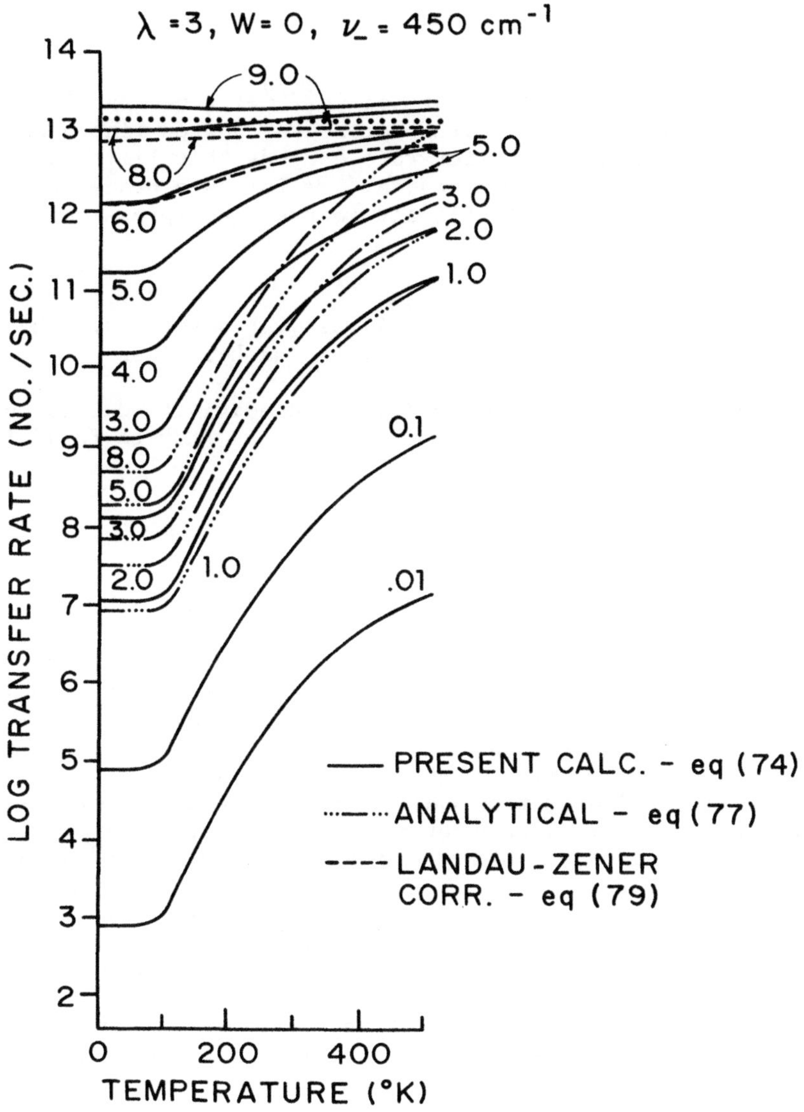

Figure 8. log of the transfer rate versus absolute temperature as a function of the electronic coupling (ε) for the case $\lambda = 3$, $W = 0$, $\nu_- = 450\ cm^{-1}$. The values of $|\varepsilon|$ are appended to the curves. The solid curves are the results of the present treatment using eq (74); the analytical results use eq (77) and apply for small $|\varepsilon|$. The dotted horizontal line at the top is the limiting transfer rate for a single harmonic potential well with classical frequency $\nu_- = 450\ cm^{-1}$.

for eq (77). Additional discussion of this and related points will be found in Ref. 15.

Finally we note (15) that for $|\varepsilon|/\lambda^2 \to 1$, the transfer rate should simply approach ν_- (the horizontal dotted line near the top of Fig. 8). Our transfer rate exceeds this limit and hence needs modification in this region. We follow a procedure originally due to Zener (24-26), and we use the expression,

$$P(LZ) = \nu_-(1 - e^{-\bar{P}/\nu_-}) \tag{79}$$

where $\bar{P}$ is calculated from eq (74) as just described. The results using P(LZ) are shown in Fig. 8 by the dashed lines.

Further discussion of eq (74), including its dependence on ν_-, ε, λ and W and some of its applications, may be found in Ref 15.

VI. CONCLUSIONS

The vibronic coupling model (PKS model) provides a precise quantum mechanical interpretation of the potential surfaces (Fig. 3) employed in many discussions of mixed-valence systems. In particular, we have shown that the nuclear motion (dynamical problem) can be described in a rigorous and unified manner over the whole range of vibronic (λ) and electronic (ε) coupling parameters. In the localized (or static) limit ($|\varepsilon|/(\lambda^2 + W)$ small), the nuclear motion can be ignored, and a semiclassical treatment (2) will yield a number of widely used formulas due to Hush. The loss of validity of these formulas as $|\varepsilon|/(\lambda^2 + W)$ increases has been discussed elsewhere in explicit and quantitative detail (2). In that discussion, it was also shown that familiar properties and concepts--e.g., the degree of delocalization, valence trapping, electron transfer rates, and the Robin and Day classification scheme--which previously were amenable to analysis only for strongly localized systems, can in fact be generalized in a perfectly natural way as the electronic coupling increases.

We suggest a two-fold role for our model. First, it provides a unified conceptual framework (albeit a much simplified one!) within which to view mixed-valence systems. The transition from the static to the dynamic case is smooth and continuous, and the previous theoretical emphasis on the static limit appears in a clearer perspective. In addition, hopefully, stimulus is provided to recognize, study and prepare additional systems where dynamic effects are important.

Secondly, *to the extent that the model approximates real systems,* it has important predictive and correlative powers. For example, one could deduce the parameters ε and λ from spectroscopic measurement of an intervalence band contour (Section IV. A) and then use these parameters to calculate the electron transfer rate of the system (Section V). This in turn not only gives important information about electron dynamics but is immediately relevant to the question of whether the system is localized or delocalized, and on what time scale. Furthermore, the parameter λ can be quantitatively related to bond length differences in the two subunit oxidation states, a point which can be checked against and correlated with other available information. Clearly, the usefulness of the model in such respects can be assessed only through systematic tests against experimental data. In this regard, attempts to observe the predicted far infrared tunneling transitions (Section IV. B) would be of great value. The existence of such transitions is a clear requirement of the theory, and their location and intensity would provide a critical test of the model.

Finally, we note some obvious extensions and applications. One of considerable current interest is the calculation of resonance Raman spectra in mixed-valence systems *when the excitation is within the contour of the mixed valence band.* Again since we have the eigenfunctions of the complete ground vibronic manifold, it is straightforward to make such calculations (13). Resonance Raman measurements of this type have in fact been made by Clark and coworkers (28), primarily on II-IV linear chain systems. Remarkably sharp, well defined (and often very) long progressions are observed. *Such progressions should occur in the q-mode, and hence the spacings in the progression should correspond in energy to the successive vibronic energies calculated by our model* which should also be able to predict the observed relative intensity pattern. Thus an important potential test of the model is available. To make a direct comparison with the Clark *et al.* results, it is necessary for us both to extend the model to infinite linear chains, and to treat the case in which neighboring metal atoms differ by two in oxidation number. These extensions as well as resonance Raman calculations are now in progress. It also appears possible to use the model to calculate the electrical conductivity of one-dimensional mixed-valence chains.

Another extension of great importance is the treatment of electron transfer kinetics (Section V) *in the unsymmetrical* ($W \neq 0$) *case.* Such work is also in progress.

Other obvious extensions include treatment of the case of unequal force constants (Section IV. E) and an examination of the related matter of anharmonic effects. The latter will couple the q mode to other molecular (and solvent) modes and is probably largely responsible for the smooth, unresolved contours generally observed in intervalence bands. In addition, the model as presently formulated takes no account of the solvent; this very important matter clearly deserves attention. Finally, it would be interesting to extend the model to the case where the two centers have different point group symmetries.

ACKNOWLEDGMENTS

This paper has summarized the joint contributions of several workers. My interest in these systems was strongly stimulated by the low temperature optical and MCD measurements made by Drs. E.R. Krausz and H. Güdel on thin foils containing mixed-valence ions prepared by the latter worker. Dr. S.B. Piepho made *the* major theoretical contribution by appreciating the relevance of the earlier Fulton and Gouterman work and by connecting, applying and extending that work to mixed-valence systems and to the analysis of intervalence band contours. Dr. K.Y. Wong has played the major role in connecting the model to earlier theoretical work and in applying it to various problems, particularly to the treatment of electron transfer kinetics. E.R. Krausz has made many contributions and has been stubbornly insistent that experimental facts be heeded. I should also like to acknowledge a very helpful conversation with Professor J.H. Weiner.

The work was supported by the National Science Foundation under grants CHE74-17833 and CHE77-08311.

REFERENCES

1. Piepho, S.B.; Krausz, E.R.; and Schatz, P.N.: 1978, J. Am. Chem. Soc. 100, pp. 2996.

2. Wong, K.Y.; Schatz, P.N.; and Piepho, S.B.: 1979, J. Am. Chem. Soc. 101, pp. 2793.

3. Schatz, P.N.; Piepho, S.B.; and Krausz, E.R.: 1978, Chem. Phys. Lett. 55, pp. 539.

4. Wong, K.Y.; Piepho, S.B.; and Schatz, P.N.--to be published.

5. Wong, K.Y.; and Schatz, P.N.--to be published.

6. Creutz, C.; and Taube, H.: 1969, J. Am. Chem. Soc. 91, pp. 3988; Creutz, C.; and Taube, H.; 1973, J. Am. Chem. Soc. 95, pp. 1086.

7. Badger, B.; and Brocklehurst, B.: 1969, Trans. Far. Soc. 65, pp. 2582.

8. Wilson, E.B., Jr.; Decius, J.C.; and Cross, P.C.: 1955, *Molecular Vibrations*, McGraw-Hill, Chapter 3.

9. Fulton, R.L.; and Gouterman, M.: 1961, J. Chem. Phys. 35, pp. 1059; 1964, 41, pp. 2280.

10. Stephens, P.J.: 1976, Adv. Chem. Phys., Ed. by I. Prigogine and S.A. Rice (Wiley) 35, pp. 197, Appendix I.

11. Schroeder, A.H.; and Mazur S.: 1978, J. Am. Chem. Soc. 100, pp. 7339.

12. Robin M.B.; and Day, P.: 1967, Adv. Inorg. Radiochem. 10, pp. 247.

13. Wong, K.Y.: unpublished work.

14. See Reference 4.

15. See Reference 5.

16. Eyring, H.; Walter, J.; and Kimball, G.E.: 1944, *Quantum Chemistry*, Wiley, pp. 310.

17. Cribb, P.H.; Nordholm, S.; and Hush, N.S.: 1978, Chem. Phys. 29, pp. 31.

18. *Radiationless Processes in Molecules and Crystals*, 1976, Ed. by F.K. Fong, Springer, Berlin.

19. Hopfield, J.J.: 1974, Proc. Nat. Acad. Sci. U.S.A. 71, pp. 3640.

20. Freed, K.F.; and Fong, F.K.: 1975, J. Chem. Phys. 63, pp. 2890.

21. (a) Jortner, J.: 1976, J. Chem. Phys. 64, pp. 4860; (b) Ulstrup, J.; and Jortner, J.: 1975, J. Chem. Phys. 63, pp. 4358--and references therein.

22. Laplante, J.P., and Siebrand, W.: 1978, Chem. Phys. Lett. 59, pp. 433.

23. Weiner, J.H.: 1978, J. Chem. Phys. 69, pp. 4743; Weiner, J.H.: 1978, J. Chem. Phys. 68, pp. 2492.

24. Landau, L.D.; and Lifshitz, E.M.: 1965, *Quantum Mechanics*, Pergamon Press, Oxford, pp. 327-329.

25. Zener, C.: 1932, Proc. Roy. Soc. A137, pp. 696.

26. Soules, T.F., and Duke, C.B.: 1971, Phys. Rev. B, 3, pp. 262.

27. Krausz, E.R.: private communication.

28. Clark, R.J.H.; and Stewart, B.: 1979, *The Resonance Raman Effect-Review of the Theory and of Applications in Inorganic Chemistry,* Structure and Bonding Ed. by J.D. Dunitz et al. (Springer Verlag) 36, pp. 68-71.

ELECTRON DELOCALIZATION, STRUCTURE AND DYNAMICS IN MIXED-VALENCE SYSTEMS

N.S. Hush

Department of Theoretical Chemistry
The University of Sydney, N.S.W. 2006
Australia.

1. INTRODUCTION

Mixed-valence systems may exhibit the property of electron localization, which distinguishes them sharply from molecules, complex ions or solid state compounds usually encountered in chemistry. Where electron localization or, as it is often termed, valence trapping, occurs, it is possible to distinguish between subsystems which differ from each other by the presence of one or more valence electrons. Thus, in the solid state, complex ions of Pb(II) and Pb(IV) can be simultaneously identified in compounds of the type $Rh(NH_3)_6PbCl_6$, or Fe(II) and Fe(III) in many naturally-occurring silicate minerals.

The interactions between the ionic centres can, however, be so strong that all electrons in the system are completely delocalized. If so, the mixed-valence system has a stationary ground state. This contrasts with trapped-valence systems, in which the ground state is non-stationary, so that electron transfer between ion centres can occur.

These are the two extreme situations, between which there is the possibility of a wide range of systems exhibiting weak to fairly strong interionic interaction. A useful classification has been provided by Robin and Day[1], in an extensive review of mixed-valence systems. The reviews of Allen and Hush[2] and Hush[3] also discuss these systems, together with new experimental material.

The differences in structure are governed primarily by two factors. For an effective two-site system, these are the inter-

D.B. Brown (ed.), Mixed-Valence Compounds, 151-188.

site electronic coupling or transfer integral, and the magnitude of coupling of the electron and phonon manifolds. The relative magnitude of these quantities determines the nature of the ground state: if the ratio of the electronic coupling integral to the electron-phonon coupling energy is very small, the ground state of the mixed-valence system will be anticipated to exhibit localization. If it is very large, a delocalized ground state is expected. We note also that the absolute value of these terms will strongly influence structural and dynamic properties, so that they should in principle be accessible to measurement.

In the following sections, concepts relating to delocalization, and the problems of interpretation of specific physical measurements, will be discussed. We shall also be concerned with the adequacy of models employed in interpretation of the electronic spectra of delocalized complexes, and finally with recent developments in the theory of double-well exchange dynamics. We would not be able to carry the discussion very far without the knowledge provided by the very large number of dimeric metal complexes of bifunctional ligands prepared and studied in the highly important work of Taube and co-workers[4] and of Meyer and co-workers[5]. It is as well at this point to refer to the precursor of such complexes, the Creutz-Taube ion[6]:

which has the structure illustrated in Fig. 1. This serves (or so I shall argue) as the paradigm of delocalized mixed-valence ions.

Electron localization also occurs, of course, in other than mixed-valence systems: the electronic states of disordered one-dimensional solids are all trapped[7] and in metals the phenomenon of Anderson localization[8] is well-known. An interesting general review of the more chemical aspects of such systems, also with reference to mixed-valence systems, has been given by Haas.[9]

2. ELECTRON COUPLING INTEGRAL

We discuss briefly first the electronic coupling (transfer) integral. Let us suppose that two electronic wavefunctions ϕ_a and ϕ_b have origins (or centroids) at $\underline{X}_a$ and $\underline{X}_b$ respectively. The electronic coupling integral between ϕ_a and ϕ_b is $J(\underline{X})$, expressed as

$$J(\underline{X}) = \langle\psi_a(\underline{X}_a\underline{x}_a)|H'|\psi_b(\underline{X}_b\underline{x}_b)\rangle \quad (1)$$

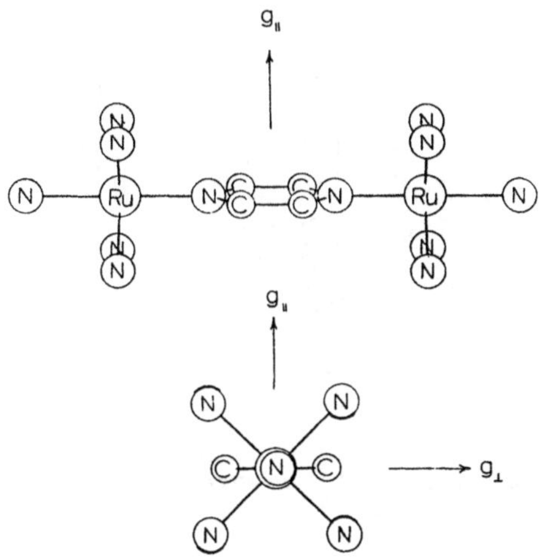

Figure 1. Two perspectives of the framework of the Creutz-Taube ion, from X-ray measurements.[36] [Arrows indicate directions of e.s.r. g-tensor components].[38]

where $\underline{X}$, $\underline{x}$ designate electronic and nuclear coordinates respectively, and H' is a perturbing Hamiltonian appropriate to the system. The value of $J(\underline{X})$ will be expected initially to decrease with increasing separation between centroids a and b, but since ϕ_a and ϕ_b will in general be molecular rather than atomic functions, the dependence of $J(\underline{X})$ on nuclear coordinates will in turn be functions of the detailed electronic structure of the participating subsystems.

Perhaps the simplest, but nevertheless very important, situation is that in which two transition-metal complex ions of different formal valence interact weakly in an 'outer-sphere' configuration. For example, we could consider the interaction between two hexaquo ions $M(II)(H_2O)_6^{2+}$ and $M(III)(H_2O)_6^{3+}$. If, in the independent orbital representation, the transferring electron is of type e_g^*, maximum overlap between donor and acceptor orbitals will occur in the configuration shown in Fig. 2, where the ligand octahedra are in apical contact. Such a configuration could well be the most probable one for transfer in an ionizing solvent, in which the ions may freely mutually orient.

Designating the relevant orbitals of the two complex ions as $\phi(e_g^*)_a$ and $\phi(e_g^*)_b$, we may write

$$\phi(e_g^*)_a = c(d_{x^2-y^2}(a)) - \frac{1}{2}(1-c^2)^{\frac{1}{2}}(\sigma_1+\sigma_2+\sigma_3+\sigma_4), \quad (2)$$

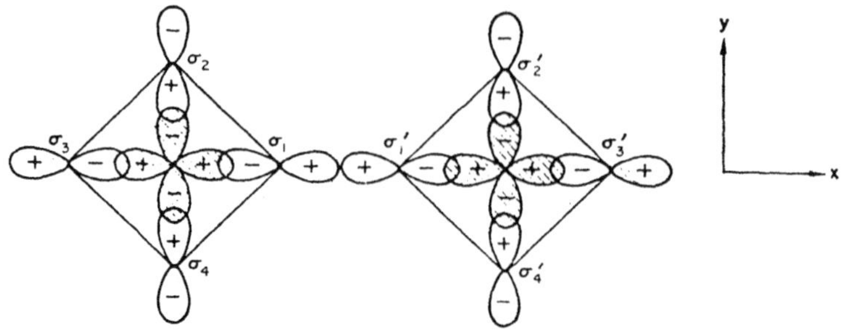

Figure 2. σ-overlap of octahedrally co-ordinated transition ions in apical contact.[14c]

where $\frac{1}{4}(1-c^2)$ measures the delocalization of d-electron density into each of the appropriate oxygen σ-orbitals; $\phi(e_g^*)_b$ is similarly defined. A one-electron calculation of $J(\underline{X})$ for this system was carried out[10] using the value of c appropriate to the Cr(III)/Cr(II) aquo system, ignoring Jahn-Teller distortions. The results are shown in Fig. 3. The abscissa in this figure is the distance between the oxygen nuclei of the apical ligands.

It is clear from Fig. 3 that $J(\underline{X})$, in this typical case, is a sensitive function of the separation of nearest-neighbour ligand centres, and much less so of the total distance between ion centres. It is also a less sensitive function of the amplitude of the electronic wavefunction at the ligand centre. At the outer-sphere close contact separation, assumed to be twice the radius of a water molecule (2.8 Å), the value of $|J|$ is calculated to be 0.023 eV. Thus the calculated electronic energy gap between upper and lower potential surface is $|2J| = 0.046$ eV in a transition-state configuration.

It is evident from this treatment that considerable care has to be exercised when the 'distance' of electron transfer is discussed. While the separation between ion centres may be large, the wavefunction of the transferring electron will in general be spread out over the metal-ligand framework, and the magnitude of the transfer integral in the most probable configuration for exchange will depend critically on near-neighbour ligand overlap and interaction. Thus, for example, estimates of large electron-transfer distances in cytochrome C/oxidant exchange could be interpreted in this way, without necessarily invoking - as is sometimes done - long-range electron tunnelling mechanism for the process (cf. ref. 11).

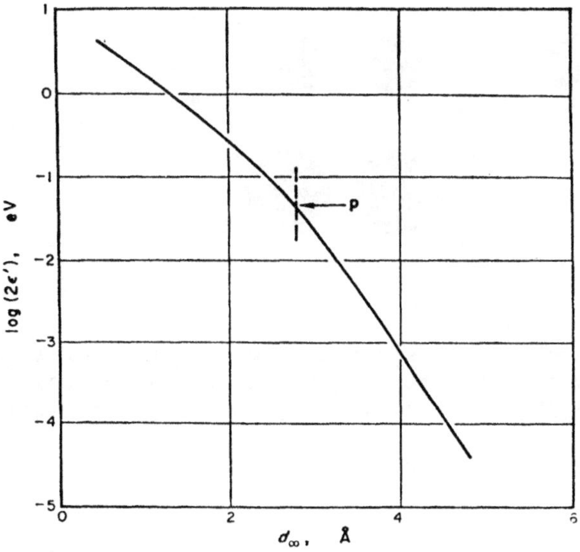

Figure 3. The electronic coupling integral as a function of distance d_{OO} of adjacent oxygen atoms for octahedrally co-ordinated transition ions with c^2=0.7, appropriate to the Cr^{2+}-Cr^{3+} couple. [2E'≡2|J| in eV, the separation of upper and lower surfaces]. The point 'p' indicates close-contact separation at which d_{OO} equals the diameter of a water molecule.[14c]

In the above example, we have discussed a mixed-valence ion solution, in which interaction occurs only during the lifetime of a collision. In solids or complexes the geometry is of course fixed. Where there is weak interaction between distinguishable complex ions, similar general considerations apply, except that the configuration may not be the most favourable one for an electron transfer of a particular symmetry type. The situation is somewhat different in bridged complexes, which are analogous to 'inner-sphere' collision/transition state complexes in solution electron transfer processes. As a typical example, we consider the Creutz-Taube ion[6] shown in Fig. 1. In this ion, electronic interaction between the two metal centres will occur primarily via the π-electron system of the ligands and the 4dπ ruthenium orbitals. For a delocalized structure (either ground state or transition state), a one-electron calculation[12] taking into account only the pyrazine π-orbitals, leads to the following forms (Fig. 4a) for the highest occupied ($2b_{2u}$) and half-occupied levels:

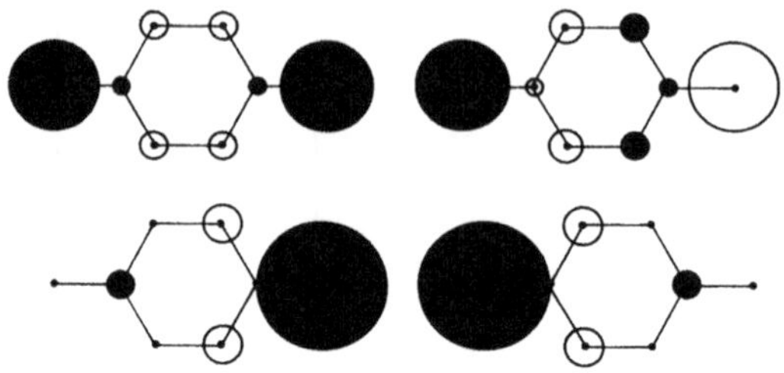

Figure 4. One-electron representation of pπ/dπ orbitals involved in coupling in Creutz-Taube ion.[12]
(a) $2b_{2u}$ (doubly filled) and $2b_{3g}$ (half-filled).
(b) Localized orbitals $2^{-\frac{1}{2}}(2b_{2u} \pm 2b_{3g})$.

What are the analogues of the localized orbitals ϕ_a and ϕ_b of the outer-sphere complexes in such a structure? In the independent-electron model, these will be the sum and difference of the above in-phase and out-of-phase molecular orbitals; i.e., $2b_{2u} \pm 2b_{3g}$. The form of the localized orbitals is thus as shown in Fig. 4b. One-electron levels[12] of the pπ/dπ system are shown in Fig. 5. The significance of 'localization' here is that in either orbital the probability density of an electron is localized at one ruthenium plus a set of ligand centres at all of which the probability density of an electron in the second orbital is nearly zero. We notice however that, as in the outer-sphere case, the orbitals ϕ_a and ϕ_b are not simply atomic, but are spread out over a molecular framework: here, and in other similar mixed-valence complexes containing symmetrical bridging ligands with even-membered rings, the essential features of the localized orbitals can be described in terms of perturbed non-bonding orbitals. We also note that the separation of ϕ_a and ϕ_b is not simply that of the metal nuclei: as in the general outer-sphere case, $J(\underline{X})$ is a function of both metal and ligand nuclear coordinates.

3. ADIABATIC ELECTRON TRANSFER

We have referred to outer-sphere electron transfer between ions in ionizing solvents. For ions of the same charge type, these are of the kind:

$$a^{z_1} + b^{z_2} \rightarrow a^{z_1-1} + b^{z_2+1} \quad . \tag{3}$$

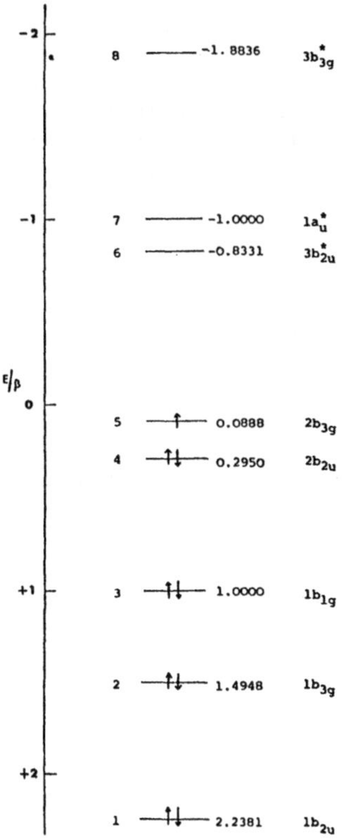

Figure 5. One-electron pπ/dπ levels of Creutz-Taube ion. Energies in units of carbon-carbon resonance integral.[12]

Historically, theoretical interest in solution transfer preceded that in mixed-valence systems. One can see, however, that if a and b are the same, the above reaction is a symmetrical self-exchange, and the precursor collision complex preceding electron transfer is a short-lived mixed-valence complex. It is natural, then, that the concepts employed in interpreting thermal and optical properties of mixed-valence systems grew from work on kinetics of homogeneous electron exchange and the exchange between ions in solution and an electrode. A brief comment on the role of the concept of delocalization in this work will thus be useful.

Outer-sphere electron transfer (homogeneous and heterogeneous) was interpreted in the high-temperature limit, as essentially adiabatic processes in theories put forward independently by

Marcus[13] and by Hush[10,14,15]. Marcus' treatment was based directly on the properties of the energy surface, while that of Hush was a simplified density functional approach, in which activation energy was calculated as a function of the formal charge distribution. In the latter approach, the electronic wavefunction Ψ_s at any points on the reaction path is expressed as:

$$\Psi_s = c_{1s}\phi_a + c_{2s}\phi_b \quad . \tag{4}$$

The transferring electron probability density is

$$\Psi^*\Psi = c_1^2\phi_a^*\phi_a + c_2^2\phi_b^*\phi_b + 2c_1c_2\phi_a^*\phi_b \quad . \tag{5}$$

This corresponds to a probability density c_1^2 associated with kernel a, c_2^2 with b, and an overlap density $2c_1c_2$ in the internuclear region. The weak-interaction condition is that Jc_1c_2 is of the order of k_BT in the transition-state region. To the accuracy to which this can be neglected, the free energy of activation for reaction is expressed as:

$$\Delta G^{\ddagger} = w_{ab}^r + \alpha(1-\alpha)\chi_{ab} + \alpha\Delta G_o \quad . \tag{6}$$

In this expression, w_{ab}^r is the free energy change in bringing the reactants (r) to the pre-transfer collision separation, and the probability density (c_1^2) associated with the electron acceptor a in the transition state is designated by α. The term ΔG_o is the overall free energy change. The quantity χ_{ab} represents approximately the change in intramolecular energy plus energy of interaction with the dielectric, in the vertical (optical) transition a + b(e) → a(e) + b, and is related to corresponding quantities for the self-exchange transfers as

$$\chi_{ab} \simeq \tfrac{1}{2}(\chi_{aa} + \chi_{bb}) \quad . \tag{7}$$

It is assumed that reaction is confined to the lower energy surface. Where the avoided crossing permits this, the transition-state value of α is found from differentiation of eq. (4) to be $\alpha = \tfrac{1}{2}(1 + (\Delta G_o/\chi_{ab}))$ and that consequently

$$\Delta G^{\ddagger} = w^{r}_{ab} + \frac{\chi_{ab}}{4} (1 + \frac{\Delta G_o}{\chi_{ab}})^2 , \qquad (8)$$

$$= w^{r}_{ab} + \alpha^2 \chi_{ab} .$$

Since the maximum value that α may take is 1, the above expression holds for $\Delta G_o < \chi_{ab}$. For higher endothermicities, the free energy of activation reduces simply to $w^{r}_{ab} = \Delta G_o$. The minimum value for α is 0, and so the quadratic expression is not applicable for reactions more exothermic than $-\chi_{ab}$, below which the activation energy reduces to w^{r}_{ab}. This very simple approach permits correlation of a great deal of kinetic information. In the interval $-\chi_{ab} < \Delta G_o < \chi_{ab}$, the quadratic expression is essentially identical with that of Marcus. Outside this interval, however, the vanishing of the activation energy for highly exoenergetic processes and its convergence towards ΔG_o (apart from w^r) for highly endoenergetic processes emerge naturally from the charge density approach and are in general agreement with experiment. The equality of the transition state charge density parameter and Marcus' parameter $|m|$ in the region of agreement has a simple significance[15]. These concepts are clearly also applicable, with modification, to mixed-valence systems. The formal concept of the distribution of charge density in the activated state can be given a rigorous interpretation[16].

The high-temperature rate constant can be expressed as

$$k = \kappa \frac{k_B T}{h} \exp-(\Delta G^{\ddagger} / k_B T) , \qquad (9)$$

where κ is the transmission coefficient, given by[10]

$$\kappa = 1 - \exp-(\eta) \; ; \quad \eta = \frac{2\pi J^2}{\hbar\omega_o} \left[\frac{\pi}{\chi k_B T}\right]^{\frac{1}{2}} . \qquad (10)$$

For $\eta \ll 1$, this reduces to

$$\kappa = \eta . \qquad (11)$$

In this expression, ω_o is the frequency of the effective phonon mode coupled to the electron motion. It should be noted that in perturbation approaches to small polaron hopping rates and analogous processes, which are often taken as models for mixed-valence systems, it is implicitly or explicity assumed that $\eta \ll 1$. If this does not hold, these expressions are not applicable. It is likely that J is sufficiently large in many mixed-valence complexes for this to be so, so that some caution is necessary in using small polaron models. Further aspects of this are discussed in the final section.

4. INTERVALENCE TRANSFER

The theory of the optical properties of trapped-valence systems, and in particular of intervalence transfer[3,17] will be mentioned in other papers at this meeting and will not be discussed in any detail here. We may note that in general the band characteristics can be correlated remarkably well, that solvent dependence, in general terms, is in conformity with the theory, as is indicated particularly by the extensive work of Taube and of Meyer and co-workers[4,5,18]. The correlation of rate constant for thermal transfer and optical properties predicted by this approach also appears (in the high-temperature limit) to be essentially correct. It would be useful to have more data on comparatively simple systems so that such comparisons could be pursued in more detail. An interesting example is the mixed-valence sulphide Eu_3S_4, for which, unusually, two independent measurements (Mössbauer[19] and conductivity[20]) yield an activation energy which correlates very well with the energy and bandwidth of an intervalence transition observed by Dyke[21]; subsequent work on the temperature-dependence of the reflectance spectrum (Wong and Hush[22]) yielded a value for the effective coupling frequency 430 ± 30 cm^{-1}. Single crystal work on this compound would be valuable.

5. A MODEL HAMILTONIAN FOR DELOCALIZED COMPLEXES

We return to the particular case of binuclear mixed-valence complexes. Examples of these are iron and ruthenium dimers of such types as

(i) $[(bipy)_2ClRuN$  $NRuCl(bipy)_2]^{3+}$ (ref. 5)

(ii) $[(NC)_5Fe\ CN\ Fe(CN)_5]^{6-}$ (ref. 23)

as well as the Creutz-Taube complex in Fig. 1. As in that species, the Ru ions in (i) have formal oxidation states II and III, as also do the iron ions in (ii).

For the purpose of a general overview of electronic properties, we use here a model potential introduced previously[24] in order to discuss the criteria for delocalization in such mixed-valence systems. In this, we simplify the problem by assuming that the electron is coupled in either localized state to a single vibrational mode. We continue to assume that the electronic problem can be reduced to consideration of only one orbital in each localized state.

Two orthonormal electron wavefunctions ϕ_a and ϕ_b are again assumed, with origins at X_a and X_b. An electron in ϕ_a or ϕ_b is linearly coupled to a doubly degenerate vibration with force constant k (frequency ω_o). The Hamiltonian for the system is then, with $\hbar=1$:

$$H = (n_a+n_b)E_{oa}+n_b\Delta+J(c_b^{\dagger}c_a+c_a^{\dagger}c_b)$$

$$+ \frac{\omega_o}{4} \sum_{i=a,b} [\dot{Q}_i^2+Q_i^2] + \sum_{i=a,b} gn_iQ_i \quad . \qquad (12)$$

In eq. (12), E_{oa} is the energy of an electron in ϕ_a, and Δ is the difference $E_{ob}-E_{oa}$; Δ is of course zero for a symmetrical complex. The electron occupation number operator for site i is n_i. The electronic coupling integral, as before, is J, taken here to be a constant quantity; $c_i^{\dagger}$ and c_i are site electron creation and annihilation operators respectively ($n_i=c_i^{\dagger}c_i$). The linear electron-phonon coupling constant is denoted by g, and Q_i is the dimensionless normal mode coordinate for mode ω_o at site i.

In evaluating the intramolecular contribution to the electron vibrational coupling energy χ in a symmetrical electron transfer between hexaquo ions, the unsymmetrical coupling of totally symmetric 'breathing' modes of the aquo ions was considered[10] (cf. fig. 6). In the present model, we are also concerned with the unsymmetrical coupling of the two normal modes - i.e., with the difference $Q_2 - Q_1$. Denoting the unsymmetrical (u) and symmetrical (s) combinations as

$$q_s = \frac{1}{\sqrt{2}} (Q_1 + Q_2) - 2\sqrt{2}\ g/\omega$$

$$q_u = \frac{1}{\sqrt{2}} (Q_1 - Q_2) \quad , \qquad (13)$$

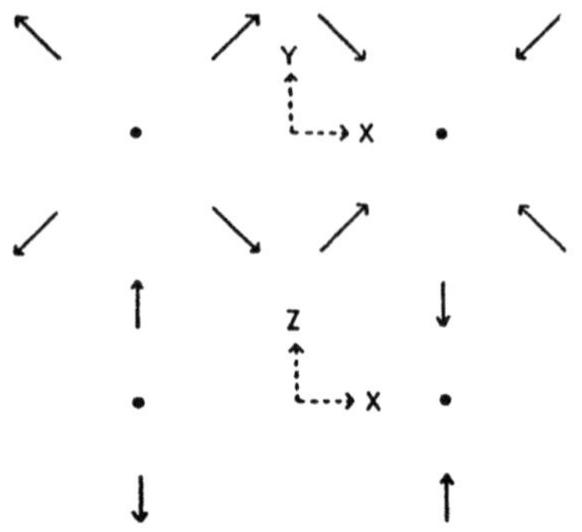

Figure 6. Out-of-phase combination of breathing modes of two hexaquo ions as intramolecular electron-vibrational coupling mode in outer-sphere electron transfer [Dots represent metal nuclei].[14d]

the Hamiltonian can be rewritten as $\tilde{H}$, where

$$\tilde{H} = (n_a+n_b)E_{oa}+n_b\Delta+ J(c_b^\dagger c_a+c_a^\dagger c_b)-g^2/\omega_o + \frac{\omega_o}{4}[\dot{q}_s^2+q_s^2] + \frac{\omega_o}{4}[\dot{q}_u+q_u^2] + \frac{g}{\sqrt{2}}(n_a-n_b)q_u \; . \tag{14}$$

This is evidently a convenient formulation, as the equations involving the symmetrical vibration are now decoupled from those involving the unsymmetrical mode. The former are ignored. The potential energy of the electron coupled to the unsymmetrical mode q_u is:

$$V_\pm = \frac{1}{2}\Delta + y^2 \pm [J^2 + (\frac{1}{2}\Delta-y)^2]^{\frac{1}{2}} \tag{15}$$

where the dimensionless parameter $y=q_u(2g/k)^{-1}$ and the energy unit is taken to be $2g^2/k$. The reason for this last choice is that $2g^2/k$ is, for a symmetrical trapped-valence complex, the electron-vibrational coupling energy approximately equal to $h\nu_{max}$ of the intervalence transfer band. This is the quantity denoted earlier by χ.

For a symmetrical complex, with $\Delta=0$, eq. (15) reduce to

$$V_{\pm} = y^2 \pm [J^2 + y^2]^{\frac{1}{2}}. \tag{16}$$

This provides us with a criterion for delocalization. If the magnitude J of the electronic coupling integral is less than $\frac{1}{2}\chi$, the model corresponds to a trapped-valence state (Fig. 7a); there are two equal vibrational wells on the lower surface separated by a barrier at y=0 the height of which approaches χ as $|J|\to 0$. This is the weak-interaction case. The electron will transfer back and forth between sites a and b at a rate governed by J, k and g and the mechanism of coupling of the system to a heat bath.

The criterion for the appearance of two minima on the lower surface is thus:

$$|J| < \tfrac{1}{2}\chi \quad . \tag{17}$$

(Note that $\frac{1}{2}\chi$ is twice the barrier height for J=0). We denote the dimensionless ratio $\frac{1}{2}\chi/|J|$ as Θ. Then for localized complexes, $\Theta<1$. Where $\Theta>1$, a single minimum is found on the lower surface, and the complex is predicted to be delocalized in the ground state. (Fig. 7b). The value of Θ for the Creutz-Taube

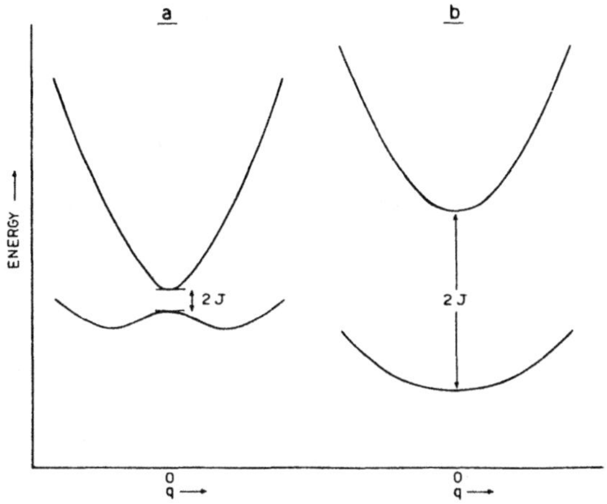

Figure 7. Potential surfaces for (a) weak-coupling ($|J|<g^2/k$) and (b) strong coupling ($|J|>g^2k$) as function of normal coordinate q.[24]

ion proposed by Taylor, Beattie and Hush[25] corresponds to a quite strongly delocalized complex.

An intermediate case arises in the presumably rare instance of a potential for which the barrier at y=0 is very small and of the order of one quantum of the unsymmetrical vibration. Temperature-dependent localization would then be observed.

The adiabatic Born-Oppenheimer solutions of the Schrödinger equation with the Hamiltonian for coupling with the unsymmetrical mode are, for the electronic wavefunctions:

$$\psi_\pm = c_a^\pm \phi_a \pm c_b^\pm \phi_b \tag{18}$$

$$c_a^\pm / c_b^\pm = |J| / \{\tfrac{1}{2}\Delta - y \pm [J^2 + (\tfrac{1}{2}\Delta - y)^2]^{\frac{1}{2}}\} \quad .$$

It was pointed out in ref. (24) that the formalism used here was chosen to draw attention to the formal identity of the problem with that of the pseudo-Jahn-Teller effect. It is also possible to solve the full vibronic problem, in which the nuclear kinetic energy operator is diagonalized. This is particularly simple to do for this model potential; the eigenvalue matrix elements were first given, for a different application, by Merrifield[26]. The vibronic coefficient matrix is of course related to that of eq. (18) by a canonical transformation. Such a vibronic approach has recently been proposed by Schatz and co-workers[27,28] and is discussed below in § 10.

6. X-RAY PHOTOEMISSION SPECTROSCOPY OF DELOCALIZED COMPLEXES

It has proved difficult to draw unequivocal conclusions about the extent of delocalization from the results of a number of different types of physical measurement on mixed-valence systems. It might be thought that X-ray photoemission spectroscopy would be an exception to this. It is known that the binding energies of core electrons are a function of the effective charge on the atom, so that, for example, the $3d_{5/2}$ photoionizations from Ru(II) and Ru(III) compounds show easily measurable differences in binding energy. Theoretical calculation of this charge-dependence requires the use of very good quality wavefunctions: however, for our present purpose we require to know only the experimental values.

Since the time-scale of the photoelectron experiment is much shorter than that of an electron transfer process, it may seem reasonable to conclude that if two distinguishable peaks are observed for, say, a $3d_{5/2}$ ionization in a mixed valence Ru(II)/

Ru(III) complex, with energies corresponding approximately to those observed in single-valence Ru(II) and Ru(III) compounds of analogous structure, then the mixed-valence complex has a trapped-valence ground state. For a delocalized complex, a single peak at the average energy would similarly be predicted. Until recently, these assumptions have been fairly generally held to be correct.

On closer examination, however,[24] one notices that this assumes that the valence electrons, and in particular the odd electron in the half-filled valence sub-shell, play only a spectator role in the core ionization. Focusing attention, for simplicity, on the odd valence electron, for which, as before, the localized orbitals at sites a and b are ϕ_a and ϕ_b, it is clear that the energy E_o of the localized orbital will depend on whether or not a core electron has been removed. Suppose that the electron occupies the orbital at a. The core binding energy from this ion is designated by BE(occ). Similarly, the core binding energy for the case that a level ϕ is empty is BE(unocc). These are therefore the binding energies for a particular core ionization from Ru(II) and Ru(III) respectively. Let the binding energy difference be given by δ, where

$$\delta = \text{BE (unocc)} - \text{BE (occ)}. \qquad (19)$$

Then, since the core hole is localized at either a or b, and we assume that coupling of the two Ru centres other than through the orbitals can be neglected, the orbital energy levels E_{oa} and E_{ob} are no longer equivalent, but are related by:

$$E_{oa} - E_{ob} = \delta.$$

Thus the core-photoionized state of such a delocalized mixed-valence complex will no longer have a symmetrical charge distribution.

In effect, the core-ionized system has a strong electric field acting on the valence electrons, which must result in polarization. The potential energies of the lower and upper electronic states as a function of the dimensionless distance parameter y are given by eq. (15) of the last section, with $\delta = \Delta$. For a delocalized mixed-valence complex, there will be, on these assumptions, not one core photoionization line at the average energy $\frac{1}{2}$(BE(occ) + BE(unocc)), but two, at energies:

$$BE(1) = [J^2 + (\tfrac{1}{2}\Delta)^2]^{\frac{1}{2}} + J$$

$$BE(2) = -[J^2 + (\tfrac{1}{2}\Delta)^2]^{\frac{1}{2}} + J \quad . \tag{21}$$

This is illustrated in Fig. 8. As above, energies are expressed in units of the electron-vibrational coupling energy χ.

The odd-electron electronic wave function for the core-ionized complex is given by

$$\psi_{\pm}^{\text{core ionized}} = r_a^{\pm}\phi_a + r_b^{\pm}\phi_b \tag{22}$$

where the coefficients r_a, r_b are given by eq. (15). An increase of the electron coupling integral J with respect to the binding energy difference should cause the ratio of intensities of the two lines to approach unity, and their energies to approach those of the separated ions, as in a trapped-valence situation. Specifically, the ratio of the intensities I of the peaks at low and high binding energies are given by:

$$\frac{I(\text{low BE})}{I(\text{high BE})} = \frac{1+(1+\alpha^2)^{\frac{1}{2}}+\alpha}{1+(1+\alpha^2)^{\frac{1}{2}}-\alpha} \tag{23}$$

with energies

$$BE_{\pm} = \tfrac{1}{2} \; [BE(\text{unocc}) + BE(\text{occ})] + \tfrac{1}{2}\Delta[\alpha\pm(1+\alpha^2)^{\frac{1}{2}}] \tag{24}$$

In these expressions, α is the dimensionless parameter $|J|/\tfrac{1}{2}\Delta$.

It is interesting to apply these expressions to the Creutz-Taube complex, for which $3d_{5/2}$ X-ray photoelectron data have been given by Citrin[29], together with spectra of oxidised and reduced species. The experimental value of Δ is 2.70 eV. From the electronic absorption spectrum, the energy of the intervalence band yields the value 0.4 eV for $|J|$, assuming that the complex is delocalized. Inserting these values into eqs. (23) and (24), we obtain the results shown in Fig. 9. These show two peaks separated by 2.82 eV. The electronic wavefunctions for the odd electron in the core-photoionized state are then

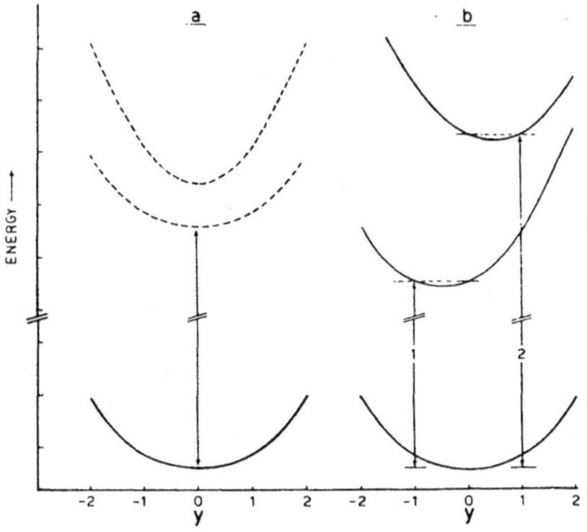

Figure 8. Effect of electron relaxation on core level photoionization for complex with delocalized ground state. Electron relaxation in photoionized state ignored in (a), taken into account (b). In hypothetical case (a)(Koopmans theorem case) only one transition is allowed at average binding energy. In case (b), two transitions are allowed, the two upper highly polarized photoionized states having the valence electron strongly localized near one nucleus. Transition energies are close to those of separate ions BE(occ) and BE(unocc) for this choice of parameters. [J=0.4 eV, Δ=2.70 eV]. One vertical division corresponds to 1 eV if vibrational coupling energy $2g^2/k$ is set equal to 0.5 eV. Abscissa is dimensionless coordinate y.[24]

$$\psi_+^{\text{core ionized}} = 0.9896\ \phi_a + 0.1435\ \phi_b$$

$$\psi_-^{\text{core ionized}} = 0.1435\ \phi_a - 0.9896\ \phi_b\ . \qquad (25)$$

This corresponds to 98% localization of the electron at site a on the lower potential energy surface and 2% at a on the upper surface.

The XPS spectrum of the tosylate salt* of the Creutz-Taube ion[29] in fact exhibits two $3d_{5/2}$ peaks, rather than one intermediate one, the energies being consistent with the above analysis for a delocalized system. This was, quite naturally,

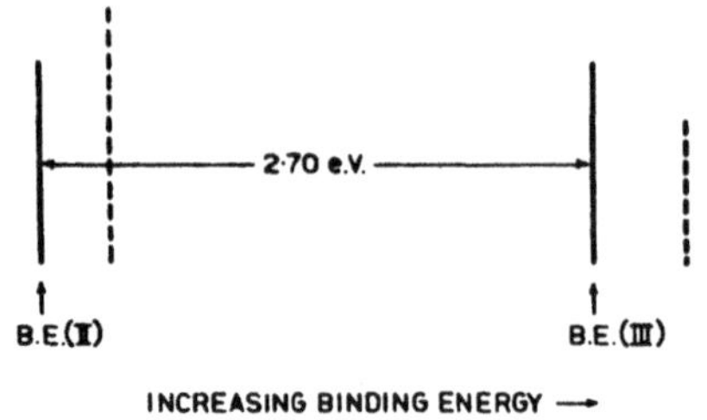

Figure 9. Calculated XPS spectrum. Solid lines: inner shell binding energies and relative intensities for M(II) and M(III) ions with δ=2.70 eV. Broken lines: energies and intensities for M(II)...M(III) complex with delocalized ground state ($|J|$=0.4eV) and δ=2.70 eV. [eqs. (23) and (24)].[24]

initially interpreted as evidence for a trapped-valence ground state. It would be very desirable to have accurate intensity data for this and similar ions for which there is good independent evidence of a delocalized ground state, in order to test the proposed relationship[23]. It should be noted that it is assumed that the ions interact via only one orbital on each metal centre; is there is more extensive interaction, this single-orbital model if inappropriate. A recent discussion[30] of the XPS spectrum of a binuclear copper complex in which the metal-metal interaction occurs via many orbitals in terms of the above model thus requires modification.

Considerable theoretical interest has been growing in the last few years in distortions in the ground states of core-ionized symmetrical systems of a more general type. Thus the O(1s) hole state of CO_2^+ is calculated to have broken symmetry, with one bond longer than the other, and the analysis of the vibrational envelope for the transition is in agreement with this[31,32]. As in the mixed-valence systems, a non-totally symmetric vibrational mode is principally involved.

7. ELECTRONIC SPECTRA OF DELOCALIZED COMPLEXES

When the interaction between localized states is too large to be treated as a small perturbation, the perturbation approach on which the general expressions for intervalence transfer in trapped-valence systems are based is no longer applicable. I shall discuss here the characteristics of the mixed-valence band of a completely delocalized symmetric dimeric complex.

The first analysis of the mixed-valence band of a delocalized complex - actually, the Creutz-Taube ion - was proposed by Taylor,

Beattie and Hush[25]. The basic assumptions of the model were those of §5, above. Thus a single vibrational mode ω_o is associated with each of the two localized sites, and the Hamiltonian (14) applies. Decoupling of the symmetric combination of normal modes yields the Hamiltonian (15), which is a function of only one non-totally symmetric mode; the potential energy corresponding to this is given by eq. (16) for a symmetric system.

For a delocalized system, with $\Theta<1$, the upper and lower potential surfaces both have single minima undisplaced from $y=0$. The quadratic force constants of the lower and upper surfaces are given approximately by:

$$k_- \simeq k_o(1-\Theta)$$

$$k_+ \simeq k_o(1+\Theta) \quad . \tag{26}$$

The maximum frequency of absorption at zero temperature is given by

$$h\nu_{max} = 2|J| + \tfrac{1}{2}(h\omega^+ - h\omega^-) \quad . \tag{27}$$

Whereas in the trapped-valence case, we may easily determine χ from the position of the band maximum, but have to analyse band characteristics in order to estimate J, in the delocalized systems the situation is exactly reversed. An estimate of χ requires an analysis of the second and possibly higher moments of the absorption envelope.

The second central moment μ_2 of an absorption envelope, which is the most accessible, can be written as:

$$\mu_2 = \mu_2(\Delta) + \mu_2(\Gamma) + \mu_2(LW). \tag{28}$$

The first term is the contribution from horizontal displacement of equilibrium positions in upper and lower states, which dominates in the trapped-valence case. (Δ should not be confused with the energy term in eq. (12)). The second results from frequency dispersion, and is a function of the difference of force constants ($\Gamma=k_+-k_-$). The last term is the linewidth, assumed temperature-independent.

In terms of the model proposed, $\mu_2(\Delta)$ is zero, and the bandwidth results from the frequency dispersion plus a contribution from linewidth. It is immediately evident that a much narrower bandwidth is expected for the delocalized complex. As can be seen by inspection of Fig. 7b, transitions from the turning points of the zero-point vibration of the lower surface fall very close to the corresponding turning points on the upper surface. This is qualitatively in accordance with the fact that the half-width of the Creutz-Taube complex (1.35kK in D_2O at 300K) is only a fraction of that of trapped-valence diruthenium complexes.

In order to fit the data to experiment, one may (in adiabatic Born-Oppenheimer approach) calculate the vibrational eigenvectors for a range of values of parameters, directly calculate the moments of the absorption envelope and fit to experiment. Alternatively, since the model can only be an approximate one, one may choose to employ the approximate quadratic force constants (eq. (26)) and express $\mu_2(\Gamma)$ as:

$$\mu_2(\Gamma) = \tfrac{1}{2}\hbar^2\omega_o^2 \left[\frac{\Theta}{(1-\Theta)^{\frac{1}{2}}} \coth(\hbar\omega_o(1-\Theta)^{\frac{1}{2}}/2k_BT\right] . \qquad (29)$$

The latter course was taken by Taylor et al. to fit their measured temperature dependence of the reflectance spectrum of the chlorotetrabromide salt of the complex. This yielded a value of $\Theta \sim 0.6$. Together with their infra-red data, this result left little doubt that the Creutz-Taube complex is indeed delocalized, although the analysis is clearly open to later refinement. The potential energies corresponding to a value of Θ very close to this are shown in Fig. 10.

This problem has also recently been examined by Schatz and co-workers in interesting publications, in which the Creutz-Taube ion is also specifically considered[27,28]. Schatz et al. adopt the model of §5 for the delocalized system with three parameters equivalent to the J, χ and ω_o of our analysis. The innovation these authors make is to solve the vibronic problem, using the equivalent of the Merrifield eigenmatrix, and to fit to the observed lineshape. Their analysis for the Creutz-Taube ion yields the value Θ=1.215, with ω_o set at 500 cm^{-1}. The potential surfaces corresponding to this choice of Θ and ω_o are shown in Fig. 11. It can be seen that the proposed lower surface is extremely flat and has a slight barrier (57 cm^{-1}), corresponding to the fact that Θ is greater than 1. It is concluded by Schatz et al. that the complex is almost delocalized but shows a small temperature-dependent localization.

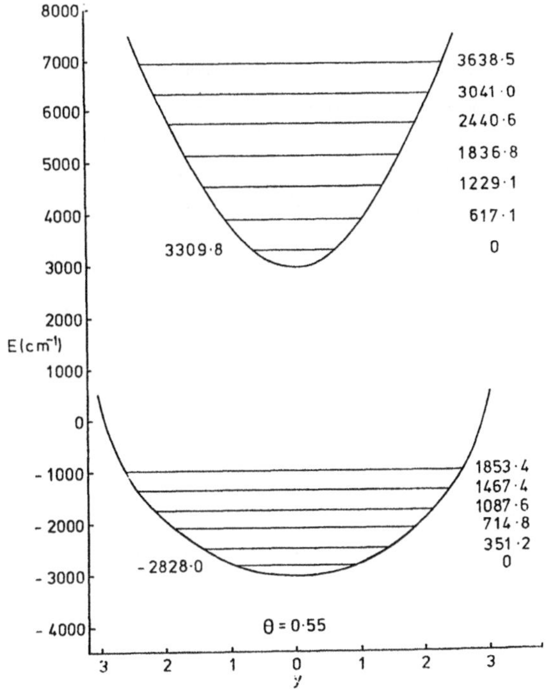

Figure 10. Potentials for lower and upper surfaces in Creutz-Taube ion, as function of dimensionless distance coordinate y, with $|J|$=0.4 eV, ω_o=500 cm^{-1} and Θ=0.55. Adiabatic vibrational eigenvalues indicated at right.[12]

Before considering the striking difference in the potentials of Figs. 10 and 11 we should ask whether the model of §5 is in fact adequate for this particular application. The answer is that it is not. An important physical effect has not yet been taken into account. This is simply that when an electron is (as in a delocalized complex) excited from a bonding to a less or anti-bonding orbital the average bond lengths will change, and in general will increase. This means that totally symmetric modes of the complex will also be excited in the mixed-valence transition, increasingly so as the bond strength increases. A simple Hückel-type calculation[12] for the π-electron structure of the Creutz-Taube complex yields for the quantity $(\Sigma_{\mu\nu}\delta P_{\mu\nu})^2$ the value 0.255 for the Ru-N bonds and 0.047 for the ring bonds, where $P_{\mu\nu}$ is the π bond order between centres μ and ν. This can give an order-of magnitude estimate of the energy excited in symmetric vibrational modes in the mixed-valence transition.

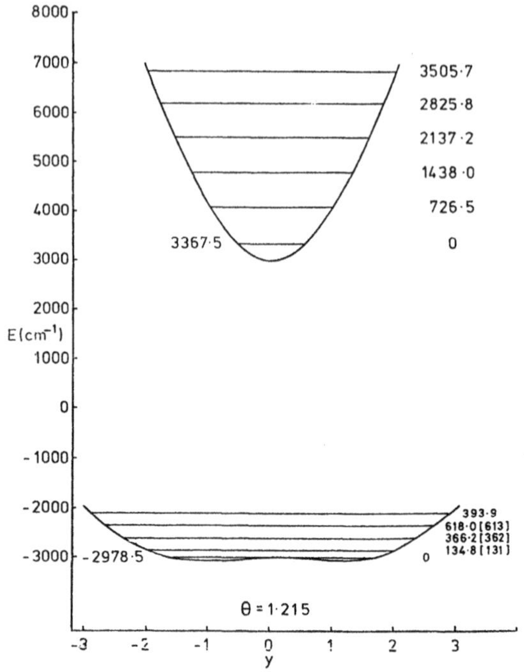

Figure 11. As for Fig. 10, but with Θ=1.215 (value proposed by Schatz et al.[27]). Numbers in square brackets are corresponding vibronic levels given in ref. 27.[12]

Unfortunately, it is not easy to accommodate this effect in a model potential without introducing details of the electronic structure of the complex, although it is intuitively obvious that its importance will increase with decreasing J. However, one may fit the data for the absolute value of the second moment of the absorption and its temperature dependence assuming also a contribution from $\mu_2(\Delta)$, where the energy of coupling to a symmetric mode, assumed also to have the frequency ω_o, is $S\hbar\omega_o$. Fig. 12 shows an absorption envelope computed[12] in the adiabatic Born-Oppenheimer approach with Θ=0.55, ω_o=500 cm^{-1}, and S=0.62 and a linewidth 1043 cm^{-1}; this results from a convolution of the symmetric and asymmetric mode progressions with the linewidth. The value of ω_o is chosen to facilitate comparison with the calculations of refs. (27,28).

This is a quite reasonable fit. It is to be noted that only a very small average coupling of an excited-state symmetric mode (in this case just over half a quantum) produces a contribution $\mu_2(\Delta)$ to the second moment of the same order as that produced

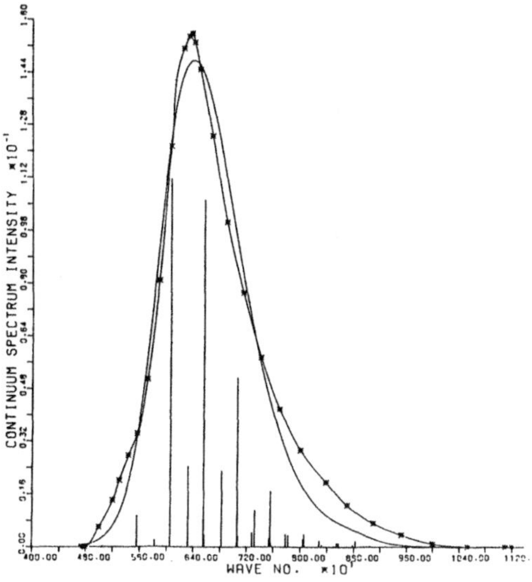

Figure 12. Calculated line and continuum spectra for near infra-red absorption band of Creutz-Taube ion (300K) with Θ=0.55, [cf. fig. 10], J=0.4 eV, S=0.62, ω_o=500 cm^{-1}, and linewidth 1043 cm^{-1}. Asterisked line is experimental envelope.[12]

by frequency dispersion with Θ=.55. In Fig. 13 the calculated temperature dependences[12] of these components are shown as a function of temperature. It is interesting that according to this interpretation the bandwidth is dominated by the distance dispersion term at low temperatures. Also, it has a smaller temperature dependence than $\mu_2(\Gamma)$, so that an approximate estimate of Θ is in fact obtained from the temperature dependence of the bandwidth at high temperatures. It is probable that the very flat ground state potential proposed by Schatz et al. (Fig. 11) results from attempting to accommodate all sources of vibrational band broadening into frequency dispersion, which is, however, a comparatively inefficient promoter of broadening.

It seems reasonable to conclude as follows: in order to analyse the vibrational absorption envelope of the mixed-valence band of a delocalized complex, the simplest model will require five parameters. These are J, χ, ω_0, S, where S is the number of quanta of the symmetric mode excited in the transition, and a linewidth assumed equal for all components. If it is not assumed that the effective excited-state symmetric mode has the same frequency as ω_o, a sixth parameter is required. It follows

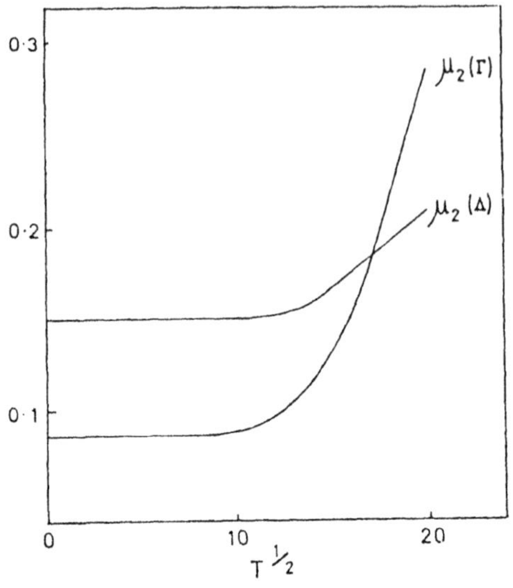

Figure 13. Calculated temperature dependence of contributions to second central moment μ_2 of intervalence absorption envelope of Creutz-Taube ion.[12]
$\mu_2(\Gamma)$: contribution from frequency dispersion of non-totally symmetric mode
$\mu_2(\Delta)$: contributions from equilibrium distance dispersion in coupling to symmetric mode.

from this that the methods of analysis proposed in refs. 27 and 28 cannot be expected to lead to a correct interpretation of spectroscopic data.

8. INFRARED TRANSITIONS IN DELOCALIZED COMPLEXES

The upper and lower potential surfaces on the single-frequency model have quadratic force constants and frequencies are higher and lower respectively than those of the unperturbed localized mode ω_o. For $\Theta<1$ the frequencies are given approximately by $\omega_o(1+\Theta)^{\frac{1}{2}}$ and $\omega_o(1-\Theta)^{\frac{1}{2}}$ respectively (eq. (26)). Thus for a value of Θ of 0.55 for the Creutz-Taube complex, upper and lower frequencies of 340 cm^{-1} and 620 cm^{-1} are expected for ω_o of 500 cm^{-1}. As can be seen from Fig. 10, the lowest adiabatic eigenvalues are 351 and 617 cm^{-1}, respectively, in quite reasonable agreement. Much larger perturbations and departures from harmonicity are obtained with the value Θ=1.215 suggested by Schatz et al.[27,28]. For the lower surface, the two lowest adiabatic eigenvalues are, in that case, 135 and 366 cm^{-1} respectively.

We have seen that there are good reasons for believing that the value of Θ for this complex is nearer to 0.55, and consequently that the frequency shifts will be not nearly so large as in the interpretation of Schatz et al. The mode is assigned by these authors to a Ru-pyrazine stretch: if the model were taken to be literally applicable, the asymmetric Ru-pyrazine stretch in the Creutz-Taube complex would be particularly low, and on the assumption of Θ=1.215 would give rise to absorption at 135, 366 and to corresponding hot bands. The infra-red data obtained by Taylor, Beattie and Hush[25] in fact show only minor shifts** in this frequency between monomer and dimer or on changing oxidation state:

	monomer	dimer	dimer	dimer
ν(Ru-pyrazine)	2+	4+	5+*	6+
(cm^{-1})	316	310	316	320

* Creutz-Taube complex

This brings us to a point of fundamental importance. The localized mode ω_o of the single-frequency model represents an <u>effective</u> frequency. It could only refer to an actual frequency if all electron-vibrational coupling constants g_j to modes j of the localized complex were zero except for one. In practice, this is generally unlikely, particularly where complex organic ligands are involved. In TCNQ radical anion, for example,[33,34] electron-vibrational coupling constants are nonvanishing for all 10 a_{1g} vibrational modes. The 'effective' nature of the frequency in a single-frequency model is clearly recognized in the usual intervalence transfer theory for trapped-valence systems, where both internal and external modes are assumed to be in fact operative. The frequency shifts modelled by the single effective potential will in actuality usually be expected to be not confined to one mode only. We would not therefore <u>in general</u> expect to find characteristic low-frequency infrared bands in delocalized complexes, contrary to the proposal of Piepho, Krausz and Schatz[27].

Numbers quoted in the above discussion refer to results of single-potential model calculations by the adiabatic Born-Oppenheimer approach rather than to the vibronic solution. Although there is little to choose between them computationally in this instance, vibronic calculation is more difficult where more than one mode in involved, as will always be the case, as indicated in the last section, in mixed-valence band analysis. From the theoretical point of view it is also unsatisfactory to proceed beyond the adiabatic step unless it is quite certain that definite normal modes rather than effective averages are involved.

The frequencies in square brackets in Fig. 11 are vibronic[27], and it will be noted that they are essentially identical with the adiabatic values, even for this unusually flat potential.

9. ELECTRON SPIN RESONANCE OF DELOCALIZED COMPLEXES

While interpretation of the electron spin resonance characteristics of a delocalized mixed-valence complex involves no new concepts, particular care is needed in deriving the spin Hamiltonian; in addition, data obtained from powder rather than single crystal spectra must be regarded with caution. In a recent publication, for example,[35] it is claimed that powder ESR measurements on the tosylate salt of the Creutz-Taube complex provide unequivocal evidence for valence trapping. As this conclusion is at variance with the weight of evidence for a delocalized ground state, it will be useful to discuss briefly the theory of the g-tensor for a delocalized complex.

The geometry of the Creutz-Taube ion is known from the X-ray crystallographic measurements of Beattie, Hush, Raston and White[36]. In the orthorhombic chlorotetrabromide, the point group symmetry around the centre of the dimer unit is $C_{2h}(m/2)$, with the c-axis being the 2-fold axis. Four NH_3 ligands lie at 45° to the pyrazine plane around each Ru centre. The shape of the ion is illustrated in Fig. 1. With χ as the molecular Ru-Ru axis, Z the normal to the pyrazine plane and Y in the pyrazine plane, we may define one-electron kets $|a\rangle$ and $|b\rangle$ as:

$$|a\rangle = \frac{1}{\sqrt{2}}(|z^2-y^2\rangle + i|xy\rangle)$$

$$|b\rangle = \frac{1}{\sqrt{2}}(|z^2-y^2\rangle - i|xy\rangle) \quad . \qquad (30)$$

The tetragonal (t) and rhombic (2δ) components of the crystal field are

$$t = E(|xz\rangle) - \tfrac{1}{2}E(|xy\rangle) + E(|z^2-y^2\rangle)\}$$

$$2\delta = E(|xy\rangle) - E(|z^2-y^2\rangle) \quad . \qquad (31)$$

With these definitions, the matrix of spin-orbit coupling and crystal field perturbation is as shown in Table 1.

TABLE 1

SPIN-ORBIT COUPLING AND CRYSTAL-FIELD PERTURBATION MATRIX FOR t_{2g} LEVELS. THE MATRIX ELEMENTS t, δ OF THE CRYSTAL FIELD ARE DEFINED IN EQ. (31).

	$\|a^-\rangle$	$\|xz^+\rangle$	$\|b^-\rangle$
$\|a^-\rangle$	$-\frac{1}{2}\zeta$	0	$-\delta$
$\|xz^+\rangle$	0	t	$\frac{1}{\sqrt{2}}\zeta$
$\|b^-\rangle$	$-\delta$	$\frac{1}{\sqrt{2}}\zeta$	$\frac{1}{2}\zeta$

If we neglect the small rhombic component, the components of the g-tensor can be derived to be, in a notation due to Stevens[37]:

$$g_{\parallel} = 2|(1+k)\cos^2\alpha - \sin^2\alpha|$$

$$g_{\perp} = 2|\sqrt{2}k\cos\alpha\sin\alpha + \sin^2\alpha| \qquad (32)$$

where k is the orbital reduction factor and $\tan 2\alpha = \sqrt{2}\left(\frac{1}{2} - \frac{t}{\zeta}\right)^{-1}$ with $o<2\alpha<\pi$. The free-ion one-electron 4d spin-orbit coupling parameter for Ru (4d) is 1050 cm^{-1} [37]. The parallel component of the tensor lies perpendicular to the plane of the ring. The g-values, as a function of t/ζ, are shown in Fig. 14.

The tetragonal splitting energy t is estimated to be about equal to or somewhat less than $|J|$, so that a value of t/ζ in the range +2 - +3 would be anticipated for $k \approx 1$. From Fig. 14, it can be seen that this corresponds to a predicted $g_{\parallel}$ value in the range ∿1.3-1.6 and a $g_{\perp}$ of ∿2.5-2.6.

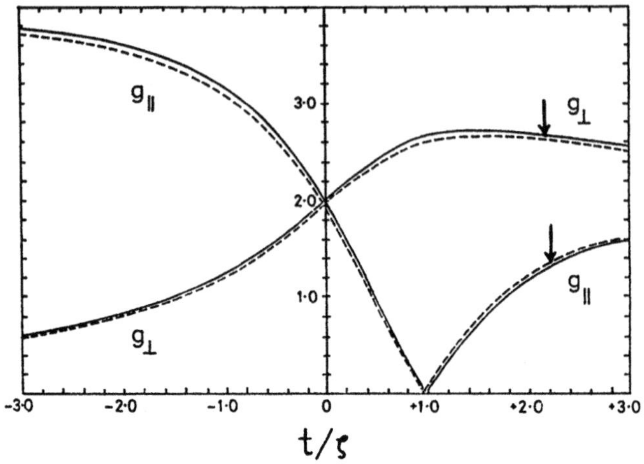

Figure 14. Calculated parallel ($g_{\parallel}$) and perpendicular ($g_{\perp}$) g-factors for Creutz-Taube ion as a function of ratio of possible values of the splitting constant t in the ground state to spin-orbit coupling constant ζ (eq. (32)). [Parallel component is normal to the pyrazine plane and the Ru-Ru axis]. Solid line: orbital reduction factor k=1; dotted line: k=0.96. Arrows indicate single crystal esr determination.[38]

Single-crystal ESR measurements on this salt by Hush, Edgar and Beattie[38] yield (at 113K) $g_{\parallel} = 1.334 \pm 0.010$ and $g_{\perp} = 2.634 \pm 0.005$ respectively, with $g_{\parallel}$ perpendicular to the pyrazine plane. These values are entirely consistent with the assumption of a delocalized ground state. In a trapped-valence complex, it might be supposed that the p_{π}- d_{π} interaction was weak, and that the parallel axis of the g tensor would lie along the axis of distortion, i.e., the Ru-pyrazine axis. This is assumed by Bunker, Drago, Hendrickson, Richman and Kessell, who report a powder spectrum value of $g_{\perp}$ for the tosylate as $g_{\perp} = 2.66$, and conclude that $g_{\parallel}$ is very small (∿0.6) and unobservable with the field strength available. As can be seen from Fig. 14, $g_{\perp}$ is not a reliable index, as a quite wide range of values of t/ζ correspond to closely similar values of $g_{\perp}$ for slightly different k. The failure of these authors to see the signal for the correct $g_{\parallel}$ (∿1.3) is presumably due to the usual poor statistics of $g_{\parallel}$ in powder samples, and their incomplete results in fact provide no evidence in support of a trapped-valence ground state for the Creutz-Taube ion, and their estimate of an activation barrier of 3.4-6.7 kcal/mole for intramolecular electron transfer in the complex cannot be accepted. However, their work does draw attention to the difficulty of obtaining reliable data from any other than single crystal measurements on compounds of this type.

10. RESONANCE SPLITTING AND DYNAMICS OF TUNNELLING TRANSFER

We have been discussing principally symmetrical mixed-valence complexes, with most emphasis on the properties of delocalized systems. We now return to symmetrical trapped-valence systems, in which there is internal electron transfer. In §3, the high-temperature adiabatic electron-transfer rate theories of Marcus and Hush were mentioned. These are each, from different points of view, formulated in terms of transition-state theory. In both approaches, it is assumed that transmission coefficients are generally of order unity and that the transfer process is essentially completely dominated by over-barrier transitions, i.e., barrier tunnelling is not important. These theories were formulated for homogeneous or heterogeneous processes involving ions in fluid media (ionizing solvents), in which the low temperatures at which departures from simple Arrhenius behaviour would not normally be expected to be accessible. However, in mixed-valence materials, it is possible to study rates down to temperatures at which these assumptions can no longer hold. In this case, it is appropriate to approach the calculation of transfer dynamics from a more fundamental point of view, in which the behaviour of individual quantum transitions is examined by time-dependent theory.

The usual approach to this has been to consider systems for which J is sufficiently small (i.e., weak- or very weak interaction systems) that the rate process can be treated by perturbative coupling of nearly localized states. The probability w_{if} of transfer from initial state (i) to final state (f) is then obtained by applying the Fermi Golden Rule:

$$w_{if} = \frac{2\pi}{h}|J|^2\rho_f(E_i) \quad . \tag{33}$$

In this expression, the coupling element J is an average for all final states with energy $E_f = E_i$ and $\rho_f(E_i)$ is the density of final states of energy E_i, which is assumed to form a dense continuum. For a potential containing a term describing linear coupling to phonon modes, a similar rate expression has been derived for a number of different physical processes, including radiationless transitions and small-polaron hopping in ionic solids (for references, see, e.g., ref. 39).

For one-dimensional motion, in which the transferring particle is coupled to a single mode of frequency ω_o, the expression for w_{if} averaged over the initial state Boltzmann distribution has been derived by a number of authors:

$$k = \frac{J^2}{2}\left[\frac{8\sinh(\hbar\omega_o/2k_BT)}{\chi\hbar\omega_o}\right]^{\frac{1}{2}} \times \exp-[-\frac{\chi}{\hbar\omega_o}\cdot\tanh(\frac{\hbar\omega_o}{4k_BT})] \tag{34}$$

This is valid for $T > \hbar\omega_o/4k_B$, but not at very low temperature. The energy

$$E = \frac{\chi}{4}\left[\frac{4k_BT}{\hbar\omega_o}\tanh(\hbar\omega_o/4k_BT)\right] \tag{35}$$

is less than the intersection height $\chi/4$. It represents the energy at which there is the best compromise between over-barrier transitions and transitions achieved by tunnelling through the barrier. In the high temperature limit this reduces to an equation of the usual form of eq. (9).

One feature which is common to these approaches in that the site- to-site sub-barrier tunnelling is assumed to be a non-resonant process. This is evident from the fact that the coupling integral J occurs quadratically rather than linearly. In a symmetrical double-well potential, each below-barrier level ω_i of each well splits into a symmetric and an antisymmetric global wavefunction with an energy splitting ΔE_i. In an isolated system in its lowest state, this leads to oscillation back and forth (pendulation) with a period of $\tau = 2\pi\hbar/\Delta E_i$, as for example, in the inversion vibration of the NH_3 molecule. In the weak-coupling case, this would thus lead to oscillation with a period which is linearly proportional to J. Thus the probability for resonant transfer in the lowest state is approximately the square root of that for non-resonant transfer, and is therefore, for a given potential, a very much more rapid process.

We have referred here to isolated systems. In order to discuss rate processes, we have to introduce coupling of the system to its surroundings (crystal lattice, solvent, etc.) in order to obtain decay rates. The question of the relative importance of resonant and non-resonant processes in symmetrical double well potentials must then be considered in relation to the coupling processes involved. There is rapidly developing interest in this area. In a recent series of papers[40,41], Cribb, Nordholm and Hush have developed the theory of one-dimensional double well transfer in a general way, applicable to molecular tunnelling processes as well as to electron or proton tunnelling. A simple approach termed the Generalized Method of Beats (GMOB) was introduced[40a] to calculate the rates of symmetrical double well transfers on a single surface (corresponding, in the electron-transfer case, to a value of J sufficiently large for the upper surface to play no appreciable part in the kinetics). This develops the above concept of a time period related to energy level splitting by relating the decay rate of a given vibrational level to the energy dispersion of that level. The basic principle is the same as that used in spectroscopy to deduce the lifetime of a level from its linewidth. This has been developed further[40b] into a master equation theory (in which microcanonical equilibrium is not assumed) and also to a density matrix approach; asymmetric as well as symmetric reactions are considered[40a].

The question now arises of the applicability of resonant or nonresonant approaches to a given physical process. Traditionally, for example, tunnelling in symmetrical proton transfer has been assumed to be non-resonant[42] but recent experiments on synchronous proton migration in the inner ring of tetra-phenylporphine[43] appear to be most easily interpreted in terms of resonant tunnelling at low temperature. The motion of Li^+ in KCl:Li, in which the alkali ion can tunnel between eight equivalent sites has also been shown to exhibit phonon resonance absorption at low temperature[44]. On the other hand, hydrogen migration in metals such as niobium and tantalum is thought by Weiner[45] to be best described by non-resonant tunnelling. However, it has been shown by Cribb, Nordholm and Hush[40d] that the experimental data can be equally well (perhaps rather better) accounted for if the tunnelling is assumed to be resonant in nature. (Note, that, for a given rate, non-resonant tunnelling requires a much shorter transfer distance).

This is thus an active area of investigation, and one which may well have interesting consequences for electron transfer dynamics in trapped-valence systems. An example of the kind of issues that are raised is illustrated by the theoretical data in Fig. 15, in which the rate of electron transfer in a symmetrical one-dimensional potential of the type of Fig. 7a is calculated[40a,b] as a function of temperature by the resonant GMOB theory. (Similar calculations have been reported earlier[40a,b]. Fig. 16 shows the comparison of classical and quantum rates for one barrier height. The system is assumed to be coupled to a heat bath maintaining thermal equilibrium of the sub-levels. The barrier height ranges from 0.2112 to 0.8450 eV. The rate constant for each barrier height exhibits the usual high-temperature Arrhenius behaviour, and at lower temperature falls off to a temperature-independent rate as resonant tunnelling from the lowest level becomes the limiting process. The higher the barrier, the lower

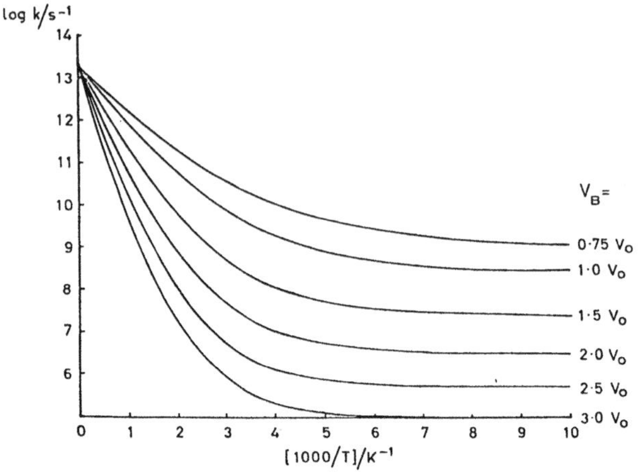

Figure 15. Temperature dependence of high pressure equilibrium form for model symmetric adiabatic electron transfer reaction calculated by the Generalized Method of Beats (GMOB). Potential of form of eq. (16) assumed, with distance between minima 0.15 Å, effective mass 50 a.u. and barrier heights V_B. ^{40e} (V_0=.28166)

the temperature at which departures from Arrhenius behaviour commence, as would be anticipated. A noticeable feature is that since resonant tunnelling processes are relatively rapid, the curvature on the Arrhenius plot commences at relatively high temperatures and for this set of fairly typical parameters is appreciable or even very marked at 300 K. The net result is that apparent activation energies at 300 K are only close to the barrier height for very low barriers: for V_B=0.5633 eV, on the other hand, the apparent activation energy at 300 K is 0.1444 eV, which is only about one-quarter of the barrier height. The numerical data are given in Table 2, where it can be seen that for this particular set of potentials the calculated apparent 300 K activation energy is practically invariant to change in barrier height.

Such behaviour would raise quite new problems of interpretation of rate measurements in mixed-valence systems if calculations of the above type were directly applicable to them. I wish to emphasize that it is not proposed here that this is so. The above calculations are essentially exact for the postulated model, but the applicability of the model to particular types of exchange is still currently under investigation. It is clear, however, that the concept of resonant barrier tunnelling opens up new

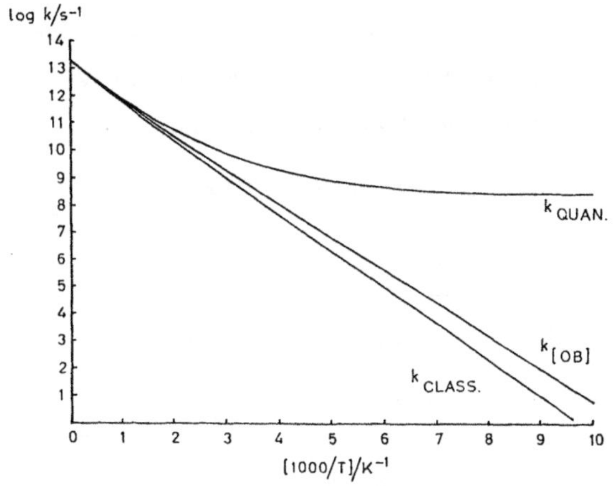

Figure 16. Comparison of temperature variations of electron transfer rate constant k_{QUAN} calculated by GMOB for barrier height 0.28165 eV (cf. fig. 10) with rates of over-barrier only transition.[40e]

$k_{[OB]}$: calculation as for k_{QUAN}, with resonant tunnelling contributions omitted.

k_{CLASS}: classical rate constant calculated as

$$\frac{2kT_B}{h} \sinh(h\nu/2k_BT) \times \exp-(V_o/k_BT).$$

possibilities both of interpretation and of measurement in mixed-valence systems, and indicates important connections with other types of symmetrical-double-well transfers.

11. CONCLUSION

Mixed-valence systems, in addition to their importance in chemistry and their recently-recognized importance in biology, provide unique opportunities for the correlation of structure and reactivity by reasonably simple theoretical approaches, and also for pursuing these in greater depth. It is also clear that it is necessary to pay particularly close attention to their actual properties in applying theoretical ideas. "I am not yet so lost in lexicography" Johnson remarked - and we must concur - "as to forget that words are the daughters of heaven and that things are the sons of earth."[46]

TABLE 2.

APPARENT ACTIVATION ENERGIES E_A [300K] FOR SYMMETRICAL ADIABATIC ELECTRON TRANSFER[a] WITH RESONANT TUNNELLING

Barrier Height/0.28166 eV

	0.75	1.0	1.5	2.0
T^{-1}/K^{-1}		$\ell nk/s^{-1}$		
2×10^{-3}	26.088923	24.900814	22.611024	20.454249
3×10^{-3}	24.427791	22.878801	20.087932	17.651683
4×10^{-3}	23.164061	21.465709	18.591153	16.202018
E_A[300K]/eV	0.11462	0.13053	0.14374	0.14437
$\nu/s^{-1} \times 10^{-13}$ [b]	1.1546	1.29066	1.5230	1.7222
$(V_B-\frac{1}{2}h\nu)$/eV [c]	0.18737	0.25497	0.39099	0.52770

[a] Parameters in Fig. 15

[b] ν is frequency of vibration in one well

[c] Barrier height minus zero pint energy

* A very similar spectrum was obtained for the perchlorate shortly afterwards by Dr. N. Winograd (University of Wisconsin) (private communication), and also for a sample of our pentabromide salt by Hewlett-Packard Photoelectron Division, San José. This is worth noting in view of the possibility of complications due to radiation-induced decomposition.

** We note parenthetically that this mode is therefore not useful in diagnosing whether or not the complex is delocalized owing to its relative insensitivity. The Ru-NH_3 stretch is more important in this respect, as it has the value 438, 449 and 461 cm^{-1} for dimer 4+, 5+ and 6+ respectively: the main point here is the appearance of only <u>one</u> absorption rather than two in the 5+ complex, consistent with a delocalized structure.

REFERENCES

1. Robin, M.B., and Day, P.: 1967, Adv. Inorg. Chem. Radiochem. 10, p. 247.

2. Allen, G.C., and Hush, N.S.: 1967, Prog. Inorg. Chem. 8, p. 357.

3. Hush, N.S.: 1967, Prog. Inorg. Chem. 8, p. 391.

4. Taube, H.: New York Academy of Sciences, Epstein, A.J., and Miller, J.S.; eds., 313, p. 481, "Synthesis and Properties of Low-Dimensional Materials", and references cited therein.

5. Meyer, T.J.: 1978, Accounts of Chemical Research, 11, p. 94 and references cited therein.

6. Creutz, C., and Taube, H.: 1969, J. Am. Chem. Soc. 91, p. 3988.

7. Mott, N.F., and Twose, W.D.: 1961, Adv. Phys. 10, p. 107.

8. Anderson, P.W.: 1958, Phys. Rev. 109, p. 1492.

9. Haas, C.: 1979, "Current Topics of Materials Science", Kaldis, E. ed., North-Holland Publishing Co., Amsterdam.

10. Hush, N.S.: 1968, Electrochim. Acta. 13, p. 1004.

11. General Discussion on Distance in "Tunnelling in Biological Systems", ed. B. Chance et al., Academic Press, New York, 1979, p. 595.

12. Hush, N.S.: to be published.

13. Marcus, R.A.: 1956, J. Chem. Phys. 24, p. 966; 1959, Can. J. Chem. 37, p. 155; 1960, Disc. Far. Soc. 29, p. 21; 1964, Ann. Rev. Phys. Chem. 15, p. 155.

14. Hush, N.S.: 1956, Proc. 4th Moscow Conf. of Electrochemistry, [Eng. trans. Consultants Bureau, New York, 1961], 1, p. 99; 1958, J. Chem. Phys. 28, p. 962; 1957, Z. Elektrochem. 61, p. 734; 1961, Trans. Faraday Soc. 57, p. 557; 1963, Brookhaven National Laboratory Publication No. 6561.

15. Hush, N.S., and Segal, G.A.: 1968, Disc. Faraday Soc. 45, p. 23.

16. Hush, N.S.: 1968, Disc. Faraday Soc. 45, p. 52.

17. Dogonadze, R.R.: 1971, "Reactions of Molecules at Electrodes", Hush, N.S. ed., Wiley/Interscience, London and New York, p. 135.

18. (a) Tom, G.M., Creutz C. and Taube, H.: 1974, J. Am. Chem. Soc. 96, p. 7287.
(b) Callahan, R.W., Meyer, T.J., Powers, M.J. and Salmon, D.J.: 1976, Inorg. Chem. 15, p. 1457.

19. Berkooz, G., Malamud, M. and Shtrikman, S.: 1968, Solid State Commun. 6, p. 185.

20. Bransky, I., Tallan, N.M., and Hed, A.Z.: 1970, J. Applied Phys. 41, p. 1787.

21. Dyke, J.M., 1970: Ph.D. Thesis, Bristol.

22. Wong, K.Y., and Hush, N.S.: to be published.

23. Glauser, R., Hauser, V., Herren, F., Ludi, A., Roder, P., Schmidt, E., Siegenthaler, H., and Wenk, F.: 1973, J. Am. Chem. Soc. 95, p. 8457.

24. Hush, N.S.: 1975, Chem. Phys. 10, p. 361.

25. Taylor, P., Beattie, J.K., and Hush, N.S.: 1976, Inorg. Chem. 15, p. 992.

26. Merrifield, R.E.: 1963, Radiat. Res. 20, p. 154.

27. Piepho, S.B., Krausz, E.R., and Schatz, P.N.: 1978, J. Am. Chem. Soc. 100, p. 2996.

28. Wong, K.Y., Schatz, P.N., and Piepho, S.E.: 1979, J. Am. Chem. Soc. 101, p. 2793.

29. Citrin, P.H.: 1973, J. Am. Chem. Soc. 95, p. 6472.

30. Sherwood, D.E., Jr., and Hall, M.B.: 1978, Inorg. Chem. 17, p. 3397.

31. Clark, D.T., and Müller, J.: 1977, Chem. Phys. 23, p. 429.

32. Domcke, W., and Cederbaum, L.S.: 1977, Chem. Phys. 25, p. 189.

33. Rice, M.J., Duke, C.B., and Lipari, N.O.: 1975, Solid State Commun. 17, p. 1089.

34. Rice, M.J., Pietronero, L., and Bruèsch, P.: 1977, Solid State Commun. 21, p. 757.

35. Bunker, B.C., Drago, R.S., Hendrickson, D.N., Richman, R.M., and Kessell, S.L.: 1978, J. Am. Chem. Soc. 100, p. 3805.

36. Beattie, J.K., Hush, N.S., Raston, C., and White, A.: 1977, J. Chem. Soc. Dalton Trans., p. 1121.

37. Stevens, K.W.H.: 1953, Proc. Roy. Soc. (London) A219, p. 542.

38. Hush, N.S., Edgar, A., and Beattie, J.K.: 1980, Chem. Phys. Lett. 69, p. 128.

39. Mott, N.F.: 1974, "Metal-Insulator Transitions" (Taylor and Francis, London).

40. Cribb, P.C., Nordholm, S., and Hush, N.S.: Chem. Phys. (a) 29, p. 31 (1978); (b) 29, p. 43 (1978); (c) 44 (1979), p. 315 (d) in course of publication; (e) work in progress.

41. Cribb, P.C., Nordholm, S., and Hush, N.S.: 1979, "Tunnelling in Biological Systems", Chance, B., ed. et al., Academic Press, New York, p. 139.

42. Bell, R.P.: (a) 1973, "The Proton in Chemistry" (Cornell Univ. Press, Ithaca; (b) 1933, Proc. Roy. Soc. A139, p. 466.

43. Hennig, J., and Limbach, H.H.: (a) 1979, Chem. Soc. Faraday Trans. II, 75, p. 752; (b) 1979, J. Chem. Phys. 71, p. 3120.

44. Peressini, P.P., Harrison, J.P., and Pohl, R.O.: 1969, Phys. Rev. 180, p. 926.

45. Weiner, J.H.: (a) 1978, J. Chem. Phys. 68, p. 2492; (b) 1978, J. Chem. Phys. 69, p. 4743.

46. Samuel Johnson, Preface to "Dictionary of the English Language", London: J. and P. Knapton and T. Longman, C. Hitch and L. Hawes, A. Millar, and R. and J. Dodsley (1755).

III. EXPERIMENTAL PROBES TO MIXED-VALENCE COMPOUNDS

MAGNETISM OF MIXED-VALENCE COMPOUNDS

William E. Hatfield

Department of Chemistry
University of North Carolina
Chapel Hill, North Carolina 27514

Mixed valence compounds exhibit a wide range of magnetic properties. These properties range from simple diamagnetism which is temperature independent and which may be estimated rather accurately from Pascal's constants, to Curie-Weiss paramagnetism, to exceptionally complicated behavior which may arise from single ion effects, exchange coupling interactions, or both. Studies of the magnetic properties of mixed valence compounds often yield valuable information concerning the chemical nature and electronic structure of the substances. In this chapter, results of representative studies of magnetic properties of mixed valence compounds will be described with the aim of illustrating the phenomena which are exhibited and the manner in which the results are used in the solution of chemical problems. Systems of discrete clusters, chains, sheets, and three dimensional polymeric materials will be discussed here. For the most part, these will be compounds in which metal ions are bridged by ligands such as halides, oxide, pyrazine, and other ions and small molecules, although representative results for compounds containing metal-metal bonds will be included. Because of space limitations and a recent review[1] on the subject, results on mixed valence metalloproteins and enzymes will not be covered.

Since the magnetic properties of mixed valence compounds which contain paramagnetic metal ions that are not exchange coupled, or involved in rapid electron transfer, or delocalized, may be understood in terms of the appropriate theories for single ion molecular species, and since this is a complete topic in itself, class I systems in the Robin and and Day[2] scheme will not be discussed here.

D.B. Brown (ed.), Mixed-Valence Compounds, 191-241.

SURVEY OF THEORETICAL RESULTS

Much of the theory necessary for the analysis of the magnetic properties of mixed valence compounds is the same as that used for compounds in which all paramagnetic species are in the same formal oxidation state. For discrete clusters, the general approach described by Kambe[3] is usually followed, although the technique becomes very tedious when the number of paramagnetic species becomes large and the symmetry of the cluster decreases.[4] Additional complications arise for mixed valence complexes especially if these involve random mixtures of spins. Kudo, Matsubara, and Katsura have discussed the expected magnetic behavior of random mixtures of S = 1/2, S = 0 and S = 1/2, S = 1 ions within a statistical framework,[5] and Wrobleski and Brown have illustrated the problems in their study of mixed valence iron oxalate, squarate, and dihydroxybenzoquinone coordination polymers.[6] Random exchange in linear chains has been the subject of extensive studies and results are discussed in a subsequent section.

Theoretical studies of polymeric materials have provided several exact solutions which are valuable for the analysis of magnetic data. These results include the one-dimensional Ising chain,[7] the two-dimensional Ising layer,[8] the one-dimensional XY chain,[9] and the one-dimensional Heisenberg chain with infinite spin.[10] Results of calculations using various degrees of approximations are also available for several other systems. Bonner and Fisher[11] have carried out exact calculations on short chains of S = 1/2 Heisenberg spins and have extrapolated their results to the infinite one-dimensional limit. Weng,[12] using a similar procedure, has calculated the magnetic properties of linear Heisenberg chains with $S \geq 1$. Wagner and Friedberg[13] have shown that the infinite spin model of Fisher can be scaled to the exact results[14] of the high temperature series expansion for the limiting case of large spins, e.g. S = 5/2. Development of additional models has been presented by Smart,[15] and the application of the results to chemical systems has been critically reviewed by DeJongh and Miedema.[16]

Curie-Weiss Law. We will begin our discussion by considering the Curie-Weiss law. In equation (1), χ is field independent

$$\chi = C/(T-\theta) \tag{1}$$

paramagnetic susceptibility, C is the Curie constant, T is the temperature, and θ is the Weiss constant. Severe deviations of experimental data from the Curie-Weiss law signal electronic state complexities, and in favorable cases it is possible to develop a model which accounts for the deviation, thereby suggesting an explanation for the phenomena that give rise to the deviation.

Let us assume that the deviations arise from exchange interactions, in which case the effective magnetic field H (effective) is the sum of the applied, external magnetic field H (external) plus the internal magnetic field. The internal magnetic field H (internal) is related to the magnetization by the molecular field constant γ. That is, H (internal) = γM, and H (effective) = H (external) + γM. Magnetization is given by

$$\underline{M} = Ng^2\mu_B^2 J(J+1)\underline{H}/3kT \tag{2}$$

where J is the total angular momentum quantum number and the other symbols have their usual meaning. Substitution of H (effective) in equation (2) for H yields M = C[H (external) + γM]/T, from which C, the Curie constant, is given by $Ng^2\mu_B^2J(J+1)/3k$. Since M = χH (external), then $\chi = C/(T-\gamma C) = C/(T-\theta)$. The sign of θ may be positive or negative depending on the nature of the exchange interactions being negative for antiferromagnetic interactions and positive for ferromagnetic interactions, and the magnitude of θ is a gauge of the strength of these interactions.

The Heisenberg-Dirac-Van Vleck Hamiltonian. The Hamiltonian that is used as the starting point in discussion of exchange theory for systems with unpaired electrons localized on the paramagnetic ions is

$$\underline{H} = -2 \sum_{n<m} J_{nm}\hat{\underline{S}}_n \cdot \hat{\underline{S}}_m \tag{3}$$

where the summation is overall paramagnetic ions in the system under consideration, J_{nm} is the exchange integral between ions *n* and *m*, and S_n and S_m spin angular momentum operators. In the Heisenberg-Dirac-Van Vleck model it is assumed that the interacting ions are orbitally nondegenerate, that any orbital contribution will be accommodated in the parametrized g value, and usually nearest neighbor interactions only are considered. Almost always J_{ij} is taken to be isotropic. In a magnetic field the term $-g\mu_B\sum_i(\hat{L}_i + 2\hat{S}_i)\cdot\underline{H}$ must be added to the Hamiltonian.

The development of the theory for two interacting indistinguishable particles, say electrons in orbitals ψ^a and ψ^b, on centers a and b, illustrates the approach to the theoretical description of the problem. The wave equation for two non-interacting electrons takes the form $[H_1 + H_2]\psi^a_{(1)}\psi^b_{(2)} = [E_a + E_b]\psi^a_{(1)}\psi^b_{(2)}$ where H_1 operates only on $\psi(1)$ and H_2 operates only on $\psi(2)$, and the energy eigenvalue is two fold degenerate because the electrons are indistinguishable. The degeneracy is lifted when the Coulomb interactions between the electrons and nuclei are taken into consideration by addition of the term

$H' = e^2/R_{ab} + e^2/r_{12} - e^2/r_{a2} - e^2/r_{b1}$ to the Hamiltonian. Using perturbation theory, the secular determinantal equation

$$\begin{array}{cc} |\psi_1\rangle & |\psi_2\rangle \end{array}$$
$$\begin{vmatrix} H'_{11}-E & H'_{12}-ES \\ H'_{21}-ES & H'_{22}-E \end{vmatrix} = 0$$

must be solved. Here $\psi_1 = \psi^a(1)\psi^b(2)$ and $\psi_2 = \psi^a(2)\psi^b(1)$, $H = H_1 + H_2 + H'$, and $H_{ij} = \int\psi_i^* H\psi_j d\tau$. By definition,

$$J_{ab} = H'_{12} = H'_{21} = \int\psi_a^*(1)\psi_b^*(2)\{H'\}\psi_b(1)\psi_a(2)d\tau_1 d\tau_2.$$

$$K_{ab} = H'_{11} = H'_{22} = \int\psi_a^*(1)\psi_b^*(2)\{H'\}\psi_a(1)\psi_b(2)d\tau_1 d\tau_2.$$

$$= \int\psi_a^*(2)\psi_b^*(1)\{H'\}\psi_a(2)\psi_b(1)d\tau_1 d\tau_2.$$

The eigenvalues are $E_{\pm} = E_a + E_b + (K \pm J)/(1 \pm S)$ where S is the overlap integral, and the eigen functions are the symmetric and antisymmetric combinations

$$\psi_{sym} = [\psi_a(1)\psi_b(2) + \psi_a(2)\psi_b(1)]/\sqrt{2}$$

$$\psi_{anti} = [\psi_a(1)\psi_b(2) - \psi_a(2)\psi_b(1)]/\sqrt{2}$$

If we ignore the overlap integral, then the energy difference between these two eigenstates is 2J. Up to this point, the spin coordinates have been left out to emphasize the fact that exchange is an electrostatic orbital interaction. However, the total wave function must be a product of the orbital part and a spin part.

Now consider the spin properties. Each electron has spin angular momentum $\hbar/2$, and by the vector addition rule the total spin of the system may be either S = 1 (triplet state) or S = 0 (singlet state). For the triplet state, the possible values for the z component of the total spin are 1, 0, -1. If, in the usual manner, $\alpha = +1/2$ and $\beta = -1/2$, then the spin functions $|S,M_S\rangle$ are

$$|1,1\rangle = |\alpha_1\alpha_2\rangle$$

$$|1,0\rangle = [|\alpha_1\beta_2\rangle + |\beta_1\alpha_2\rangle]/\sqrt{2}$$

$$|1,-1\rangle = |\beta_1\beta_2\rangle$$

and

$$|0,0\rangle = [|\alpha_1\beta_2\rangle - |\beta_1\alpha_2\rangle]/\sqrt{2}$$

where the triplet state spin functions are symmetric and the singlet state spin function is antisymmetric.

According to the Pauli principle the total wave function must be antisymmetric with respect to exchange of every pair of electrons. Therefore, the symmetric spin functions (triplet state) must be combined with the ψ_{anti}, and the antisymmetric spin function (singlet state) must be combined with ψ_{sym}. It is in this way that the energy of the system depends on the total spin.

In terms of a vector model, we take $\underline{s}_1$ and $\underline{s}_2$ to be the spin angular momentum operators in units of $\hbar$ with the vector sum $\underline{s}_{12} = \underline{s}_1 + \underline{s}_2$, and

$$(\underline{s}_{12})^2 = (\underline{s}_1 + \underline{s}_2)^2 = (\underline{s}_1)^2 + (s_2)^2 + 2s_1 \cdot s_2$$

The eigenvalues of the operators $\underline{s}_{12}^2$, s_1^2, and s_2^2 are $S(S+1)$, $s_1(s_1+1)$, and $s_2(s_2+1)$. Application of the operator

$\underline{s}_1 \cdot \underline{s}_2 = 1/2[(\underline{s}_{12})^2 - (s_1)^2 - (s_2)^2]$ yields:

$$\text{for } S = 1,\ \underline{S}_1 \cdot \underline{S}_2 = 1/2[1(1+1) - 1/2(1/2+1) - 1/2(1/2+1)] = 1/4$$

$$\text{for } S = 0,\ \underline{S}_1 \cdot \underline{S}_2 = 1/2[0(0+1) - 1/2(1/2+1) - 1/2(1/2+1)] = -\frac{3}{4}$$

This permits us to write the exchange Hamiltonian as $H' = -2J\hat{\underline{S}}_1 \cdot \hat{\underline{S}}_2$. For the multielectron case where for atom a the total spin is $\underline{S}_a = \Sigma \underline{s}_i$, and for atom b, $S_b = \Sigma \underline{s}_j$ (where i and j denote the electrons on a and b, respectively), then we may write $H = -2J_{ab}\hat{\underline{S}}_a \cdot \hat{\underline{S}}_b$. For a cluster of exchange coupled ions we generalize to the Hamiltonian $H = -2 \sum_{n<m} J_{nm}\hat{\underline{S}}_n \cdot \hat{\underline{S}}_m$, which was given at the beginning of this section.

Energy Levels in Exchange Coupled Clusters. Van Vleck has developed a convenient formula for the spin states which result from exchange interactions in clusters with n magnetically equivalent paramagnetic ions.[17] For the assembly of n spins, in the vector model, the resultant spin vector is $\underline{S}' = \sum_n S_i$, with the square of $\underline{S}'$ being

$$[S']^2 = [\sum_{i=1}^{n} \underline{S}_i]^2 = \sum_{i=1}^{n} [\underline{S}_i]^2 + \sum_{i,j} \underline{S}_i \cdot \underline{S}_j$$

In the dot product the indices i and j designate adjacent paramagnetic ions. Combined with the results $[\underline{S}']^2 = S'(S'+1)$ and $[\underline{S}_i]^2 = S_i(S_i+1)$, and the realization that there are n terms in

$\Sigma_n [\underline{S}_i]^2$ and $n(n-1)$ terms in the dot product summation, we may write

$$\underline{S}_i \cdot \underline{S}_j = \frac{S'(S'+1) - nS_i(S_i+1)}{n(n-1)}$$

since all spins are equivalent. There are $n(n-1)/2$ terms in the Hamiltonian $H = -2J \sum_{i \neq j} \underline{S}_i \cdot \underline{S}_j$. The energies of the spin states are given by

$$E(S') = -J[S'(S'+1) - nS_i(S_i+1)] \qquad (4)$$

where S' can take the values allowed by the vector summation rule. For $n = 2$, $S' = 0,1$; for $n = 3$, $S' = 1/2, 3/2$; for $n = 4$, $S' = 0,1,2$.

It is clear that more than one spin state with a given S' may arise. Van Vleck[17] has shown that the number of states W with a given S' is given by

$$W(S') = \Omega(S') - \Omega(S'+1) \qquad (5)$$

where $\Omega(S')$ is the coefficient of $X^{S'}$ in the expansion of

$$(X^S + X^{S-1} + \ldots + X^{-S})^n$$

Consider for example an assembly of three paramagnetic ions with $S_i = 1/2$. The expression to be expanded is $(X^{1/2} + X^{-1/2})^3$, which gives $X^{3/2} + 3X^{1/2} + \ldots$, and the number of states with $S' = 3/2$ is 1 while the number with $S' = 1/2$ is 2. If all interacting ions are magnetically equivalent then all states of a given S' are degenerate.

The Van Vleck Equation. In the magnetic susceptibility experiment, one measures the bulk property as a function of temperature and magnetic field strength, and analyzes the data in terms of the distribution of energy levels which are thermally populated. Usually the energies of these levels are not known, but frequently their relative energies in terms of pertinent parameters, say exchange coupling constants, are known. In such cases appropriate theoretical equations are fit to the experimental data to yield the desired parameters. The most widely used formalism was developed by Van Vleck.

By definition, the magnetic moment $\mu_{n,m}$ of an energy level $E_{n,m}$ is given by $\mu_{n,m} = -\partial E_{n,m}/\partial H$, where H is the magnetic field. The total moment of the system, M is

$$M_i = \frac{N\sum_{n,m} \mu_{n,m} \exp(-E_{n,m}/kT)}{\sum_{n,m} \exp(-E_{n,m}/kT)} \qquad (i = x,y,z)$$

where N is Avogadro's number, and the summation is over all states. Consider one of these states: The energy of the nth state can be expressed in terms of a power series in H as

$$E_{n,m} = E^{\circ}_{n,m} + HE_{n,m}^{(1)} + H^2E_{n,m}^{(2)} + \ldots$$

where $E^{o}_{n,m}$ is the zero field energy and $E_{n,m}^{(1)}$, $E_{n,m}^{(2)}$, ... are the first, second,... Zeeman coefficients. The moment of this state is

$$\mu_{n,m} = -E_{n,m}^{(1)} - 2E_{n,m}^{(2)}H - \ldots$$

It is convenient to expand the exponent

$$\exp(-E_{n,m}/kT) = \exp\left(\frac{-E^{\circ}_{n,m} - HE_{n,m}^{(1)} - H^2E_{n,m}^{(2)}}{kT} \ldots\right)$$

$$\simeq \left[1 - \frac{HE_{n,m}^{(1)}}{kT}\right] \exp\left(\frac{-E^{\circ}_{n,m}}{kT}\right)$$

and to express the total moment as

$$M_i = \frac{N \sum_{n,m}(-E_{n,m}^{(1)} - 2HE_{n,m}^{(2)} - \ldots)(1-HE_{n,m}^{(1)}/kT)\exp(-E^{\circ}_{n,m}/kT)}{\sum_{n,m}(1-HE_{n,m}^{(1)}/kT)\exp(-E^{\circ}_{n,m}/kT)}$$

Paramagnetic substances do not possess magnetic moments in the absence of magnetic fields, so

$$\sum_{n,m} [\{-E_{n,m}^{(1)}\}\exp(-E^{\circ}_{n,m}/kT)] = 0$$

If data are collected at moderate temperatures and magnetic fields then only first order terms in H in the numerator and terms independent of H in the denominator need be retained. By definition $\chi_i = M_i/H_i$, and we have the Van Vleck equation:

$$\chi_i = \frac{N\sum_{n,m}[(E_n^{(1)})^2/kT - 2E_{n,m}^{(2)}]\exp(-E^{\circ}_{n,m}/kT)}{\sum_{n,m} \exp(-E^{\circ}_{n,m}/kT} \qquad (6)$$

This equation will be utilized frequently in the section on Representative Results.

Anisotropic and Antisymmetric Exchange. Frequently, the exchange Hamiltonian $H = -2 \sum_{n<m} J_{nm}\underline{S}_n \cdot \underline{S}_m$ cannot adequately explain

experimental data, and it is necessary to consider additional kinds of interactions. The Hamiltonian as written above implies that the exchange interactions are isotropic and symmetric. If we rewrite the Hamiltonian as

$$H_{nm} = -2\sum_{n<m} J_{nm}\{\hat{S}_{1z}\hat{S}_{2z} + \alpha(\hat{S}_{1x}\hat{S}_{2x} + \hat{S}_{1y}\hat{S}_{2y})\}$$

the multiplier α can be used to account for axial anisotropic symmetric exchange interactions. The Heisenberg case arises when $\alpha = 1$, the Ising case results when $\alpha = 0$, and there are experimental examples for both extreme cases as well as for intermediate values of α, that is, for $0 < \alpha < 1$.

It is instructive to examine the Ising case in detail, and we will use as our example the pair of S = 1/2 exchange coupled ions for which the spin wave functions were given above. The exchange coupling operator for this case is $H = -2J_{12}{}^z\hat{S}_{1z}\hat{S}_{2z}$. Applying the operator to the basis set yields the determinantal equation

$$\begin{matrix} |\alpha\alpha\rangle & |\alpha\beta+\beta\alpha\rangle/\sqrt{2} & |\beta\beta\rangle & |\alpha\beta-\beta\alpha\rangle/\sqrt{2} \end{matrix}$$

$$\begin{vmatrix} -\frac{1}{2}J_{12}^z-E & 0 & 0 & 0 \\ 0 & 1/2J_{12}^z-E & 0 & 0 \\ 0 & 0 & -1/2J_{12}^z-E & 0 \\ 0 & 0 & 0 & 1/2J_{12}^z-E \end{vmatrix} = 0$$

with two roots $E = +J_{12}{}^z$ and two roots $E = -J_{12}{}^z$. The energy separation between the two doublet states is $|J_{12}{}^z|$.

Frequently anisotropic exchange is taken into account by adding the term $\underline{S}_n\cdot\underline{\underline{\Gamma}}_{nm}\cdot\underline{S}_m$ to the Hamiltonian. In this term $\underline{\underline{\Gamma}}_{nm}$ is a symmetric tensor, and is given approximately by $(\Delta g/g)^2J_{nm}$ where $\Delta g = |g-2|$. The effect of anisotropic exchange is to cause a zero-field splitting of the triplet state. In axial cases a singlet and doublet result, while all degeneracy is lifted in rhombic cases.

Antisymmetric exchange interactions may arise between pairs of exchange coupled ions when the single ion g values are different, a condition which implies that the exchange coupled ions are not related by a center of inversion. The term, which is added to the Hamiltonian to account for antisymmetric exchange is $\underline{D}_{nm}\cdot[\underline{S}_n \times \underline{S}_m]$, was theoretically established by Moriya following a phenomenological development by Dzialoshinski. Essentially, this interaction tends to align the spins perpendicular to each other, and in this manner tends to oppose the isotropic exchange interaction which, depending on the sign of J_{nm}, tends to align the spins either parallel or antiparallel to one another.

The magnitude of the antisymmetric coupling constant may be estimated from the expression $|\underline{D}_{nm}| \approx (\Delta g/g) J_{nm}$. Consider a case with $\Delta g = 0.1$ and $J_{nm} = 100\ cm^{-1}$; we calculate $|\underline{D}_{nm}| \approx 5\ cm^{-1}$. This effect can be readily seen in EPR measurements, and in magnetic studies especially if J_{nm} is positive.

Let us examine the antisymmetric exchange interaction for the case of two exchange coupled S = 1/2 ions. The antisymmetric term is

$$\underline{D}_{12} \cdot [\underline{S}_1 \times S_2] = \underline{D}_{12} \cdot \begin{vmatrix} \underline{i} & \underline{j} & \underline{k} \\ S_{1x} & S_{1y} & S_{1z} \\ S_{2x} & S_{2y} & S_{2z} \end{vmatrix}$$

$$= \underline{D}_x(S_{1y}S_{2z} - S_{2y}S_{1z}) + \underline{D}_y(S_{2x}S_{1z} - S_{1x}S_{2z}) + \underline{D}_z(S_{1x}S_{2y} - S_{2x}S_{1y})$$

If we take $\underline{D}_{12}$ along the Z axis, then

$$H_{antisymmetric} = \frac{iDz}{2} [S_{1-}S_{2+} - S_{1+}S_{2-}]$$

The Hamiltonian matrix for symmetric exchange, including an antisymmetric contribution, is

$$\begin{array}{cccc} |\alpha\alpha> & |\alpha\beta+\beta\alpha>/\sqrt{2} & |\beta\beta> & |\alpha\beta-\beta\alpha>/\sqrt{2} \end{array}$$

$$\begin{vmatrix} -1/2J-E & 0 & 0 & 0 \\ 0 & -1/2J-E & 0 & iD_z/2 \\ 0 & 0 & -1/2J-E & 0 \\ 0 & -iD_z/2 & 0 & 3/2J-E \end{vmatrix} = 0$$

This interaction mixes the $|S, M_s>$ states $|1,0>$ and $|0,0>$ and results in a zero-field splitting of the triplet state. The roots of the determinantal equation are

$$E(|1,0>) = [J-J(4-D_z^2/J^2)^{1/2}]/2$$

$$E(|1,0>) = [J-J(4-D_z^2/J^2)^{1/2}]/2$$

$$E(|0,0>) = [J+J(4+D_z^2/J^2)^{1/2}]/2$$

<u>Exchange in Linear Chains</u>. The exchange Hamiltonian for an infinite linear chain may be written as

$$H = -2J \sum_{i=1}^{\infty} \{\alpha \underline{S}_i^z S_{i+1}^z + \gamma(\underline{S}_i^x \underline{S}_{i+1}^x + S_i^y S_{i+1}^y)\} - g\beta H \sum_{i}^{\infty} \underline{H}\cdot\underline{S}_i \quad (7)$$

For α, $\gamma = 1$, the isotropic Heisenberg case arises; while for $\alpha = 1$ and $\gamma = 0$, the anisotropic Ising case obtains; and for $\alpha = 0$, $\gamma = 1$, the relatively rare anisotropic X-Y case results. Fisher[7] has derived exact expressions for both the parallel and perpendicular magnetic susceptibilities for the anisotropic Ising case when $S = 1/2$. The parallel susceptibility is

$$\chi_{||} = \frac{Ng^2\beta^2}{4kT} \exp(2J/kT) \quad (8)$$

while the perpendicular susceptibility is

$$\chi_{\perp} = \frac{Ng^2\beta^2}{8J} \{\tanh(J/kT) + (J/kT)\operatorname{sech}^2(J/kT)\} \quad (9)$$

Bonner and Fisher carried out machine calculations on systems of chains containing from four to twelve spins, and extrapolated their results to the infinite-length chain.[11] The maximum in susceptibility is uniquely defined by the following relations:

$$\frac{kT_{max}}{|J|} \simeq 1.282 \; ; \qquad \frac{\chi_{max}|J|}{Ng^2\beta^2} \simeq 0.07346$$

Bonner and Fisher's numerical results can be reproduced by the expression[18a]

$$\chi_m = \frac{Ng^2\beta^2}{kT} \cdot \frac{A + Bx^{-1} + Cx^{-2}}{1 + Dx^{-1} + Ex^{-2} + Fx^{-3}} \quad (10)$$

where $x = kT/|J|$, $A = 0.25$, $B = 0.14995$, $C = 0.30094$, $D = 1.9862$, $E = 0.68854$, and $F = 6.0626$. An alternate expression which gives comparable results has been given by Jotham.[18b] These expressions have the advantage of permitting the use of computer programs in the analysis of experimental data. This approach will surely be applied to other spin systems as the need arises.

<u>Random Exchange in Linear Chains</u>. The magnetic susceptibilities of mixed valence compounds having one dimensional chain structures frequently reflect antiferromagnetic spin-spin coupling at moderate temperatures, but show an increase in susceptibility at low temperatures. In a regular chain of localized electrons, the susceptibility of a system of antiferromagnetically coupled metal ions with single localized electrons which can be described by the Heisenberg exchange Hamiltonian

(7) has a maximum value at $T_{max} = 1.282|J|/k$, and the susceptibility decreases to a finite value at 0°K. In periodically alternating chains the susceptibility decreases to zero at 0°K. Thus, neither of these models can explain the increase in susceptibility, and this feature is often attributed to the presence of small amounts of paramagnetic impurities which are trapped in the samples during the synthetic procedures. There is now good evidence from careful experimental work on highly purified samples that the low temperature increase in susceptibility is a intrinsic property of some of these compounds and that it arises from random exchange. This can be understood if one divides the entire system of spins into several noninteracting subsystems, of which some will have odd numbers of spins. These noninteracting residual spins will follow the Curie law, resulting in an increase in magnetic susceptibility at low temperatures.

Bulaevskii, et al[19] have derived very useful expressions for the thermodynamic properties of the system described above. Central to their derivation is the expression

$$\rho(\varepsilon) = Ak^{\alpha-1}\varepsilon^{-\alpha}$$

for the density of states, where A and α are parameters to be determined from experiment, and it is assumed that any dependence of ρ on T can be neglected at low temperatures, i.e. 0.1 to 10°K. The results from their derivation which are necessary for our discussion are equation (11) for magnetic susceptibility in the low field limit, i.e. where $g\mu_B H << kT$, and equation (12)

$$\chi(T,H\rightarrow 0) = 2(1-2^{1+\alpha})\zeta(-\alpha)\Gamma(1-\alpha)Ag^2\mu_B^2k^{-1}NT^{-\alpha} \qquad (11)$$

for the magnetization in the high field limit, i.e. where $g\mu_B H >> kT$.

$$M(0,H) = (1-\alpha)^{-1}g\mu_B NA(g\mu_B Hk^{-1})^{1-\alpha} \qquad (12)$$

In equation (11), $\zeta(x)$ is the Riemann zeta function and $\Gamma(x)$ is the gamma function. In practice, the adjustable parameters are determined by fitting these expressions to magnetic susceptibility and magnetization data.

Results which are close to these have been obtained by Theodorou and Cohen,[20] who used a disordered Hubbard model and investigated the effects of random antiferromagnetic exchange between spins. In their derivation the interaction parameter J followed a distribution function of the form $\rho(J) \propto J^{-\alpha}$, where $0 < \alpha < 1$. Theodorou and Cohen suggest that ε and $\rho(\varepsilon)$ of the Bulaevskii model are related to J and $\rho(J)$ of their work, thereby affording an explanation for the phenomenological density of states expression. Application of these ideas to exchange in quinolinium $(TCNQ)_2$ by Azevedo and Clark[21] will be presented in Section Representative Results.

Exchange in Two-Dimensional Sheets. The Hamiltonian appropriate for a two-dimensional magnetic system may be given by

$$H = -2J\sum_{i,j}(\underline{S}_{iz}\underline{S}_{jz} + \xi\underline{S}_{ix}\underline{S}_{jx} + \eta\underline{S}_{iy}\underline{S}_{jy}) \tag{13}$$

$$-2J'\sum_{k,l}\underline{S}_n\cdot\underline{S}_l \ , \quad 0 < \xi , \eta < 1$$

where the S_{iw}(w = x, y, or z) are the components of the spin operator along the Cartesian coordinates, J and J' are the intra- and interlayer exchange constants, and the parameters ξ and η represent the anisotropy in the intralayer exchange. If the z-axis is taken as the easy axis of magnetization, then the anisotropy is Ising-like when $\xi = \eta < 1$ and when $\xi = 1$, $\eta < 1$ it is XY-like. The Heisenberg case arises when there is no anisotropy in the exchange. The various anisotropy parameters can be represented by effective fields by the use of the following molecular field equations for the parameters $D = 1 - \xi$ and $E = 1 - \eta$:[22,23]

$$H_{ex} = 2ZJ/g\mu_\beta$$

$$H_a^{in} - D\ H_{ex}$$

$$H_a^{out} = E\ H_{ex}$$

where H_{ex} is the effective exchange field arising from the intralayer exchange, and Z is the number of near neighbors within the layer. The contribution to the anisotropy arising from the anisotropic exchange interaction which favors parallel alignment within a layer can be estimated by the equations

$$H_{A,E} \approx 2Z\Delta JS/g\mu_\beta$$

$$\Delta J \approx J_{ac} - J_b \approx (\Delta g/\langle g\rangle)^2 J$$

where $\Delta g = g_{ac} - g_b$ and $\langle g\rangle = (2g_{ac} + g_b)/3$. In some cases it has been reported that the experimentally observed H_a^{out} is substantially smaller than the sum of the theoretical components which are expected to be most important.[24,25] This immediately suggests that there must be an additional contribution to the anisotropy arising from exchange interactions either within the layer or between layers. One such additional term which does exist for these systems is antisymmetric (Dzialoshinki - Moriya) exchange which was discussed above.

Exchange in Three Dimensional Polymers. Exchange interactions between paramagnetic ions in three dimensional arrays may lead to ferromagnetism, ferrimagnetism, or antiferromagnetism. In ferromagnetic substances the magnetic dipoles are all oriented in the same direction, while in ferrimagnetic or antiferromagnetic substances there are two or more interpenetrating lattices of magnetic dipoles which generally are oriented in different directions. The description of magnetic properties of three dimensional mixed-valence compounds usually requires two or more sublattices since the paramagnetic ions are nonequivalent. Strictly speaking, all of these compounds should be ferrimagnets, but if all sublattices have their magnetic dipoles oriented in the same direction, then the properties would not be different from those of ferromagnets. For the purposes of this discussion the theory of ferrimagnetism developed by Néel will be required.[26]

Consider a situation in which there are *A* and *B* sites forming two sublattices. The molecular field experienced by an ion on site *A* is $\underline{H} = (2zJ_{ab}/N_b g_a g_b \mu_B^2)\underline{M}_b$, where z is the number of nearest neighbors and J_{ab} is the coupling constant. The multiplier of the magnetic moment of the *B* sublattice is the molecular field coefficient γ_{ab}. Interactions between the paramagnetic ions lead to the following molecular fields: $\underline{H}_{aa} = \gamma_{aa}\underline{M}_a$; $\underline{H}_{ab} = \gamma_{ab}\underline{M}_b$; $\underline{H}_{bb} = \gamma_{bb}\underline{M}_b$; $\underline{H}_{ba} = \gamma_{ba}\underline{M}_a$.

The total fields experienced by *A* and *B* atoms are given by

$$\underline{H}_a = \underline{H}_o + \gamma_{aa}\underline{M}_a \pm \gamma_{ab}\underline{M}_b$$

and

$$\underline{H}_b = \underline{H}_o + \gamma_{bb}\underline{M}_b \pm \delta_{ab}\underline{M}_a .$$

where $\underline{H}_o$ is the applied field. Following Smart,[15] let *N* be the total number of paramagnetic ions, λ be the fraction on *A* sites, μ be the fraction on *B* sites, and name the sites such that $\lambda \leq \mu$. Also, let $\alpha = \gamma_{aa}/\gamma_{ab}$, $\beta = \gamma_{bb}/\gamma_{ab}$, $\sigma_a = \underline{M}_a/\lambda N g \mu_B S$, and $\sigma_b = \underline{M}_b/\mu N g \mu_B S$. The molecular fields become

$$\underline{H}_a = \underline{H}_o + Ng\mu_B S\gamma_{ab}(\lambda\alpha\underline{\sigma}_a \pm \mu\underline{\sigma}_b)$$

$$\underline{H}_b = \underline{H}_o + Ng\mu_B S\gamma_{ab}(\pm \lambda\underline{\sigma}_a + \mu\underline{\sigma}_b)$$

Two cases can be identified, one of these has $J_{ab} > 0$ and the other has $J_{ab} < 0$. For the case with a positive *A*-*B* interaction the high temperature susceptibility becomes

$$\chi = \frac{C[T + \lambda\mu C\gamma_{ab}(2 - \alpha - \beta)]}{T^2 - C\gamma_{ab}(\lambda\alpha + \mu\beta) + \lambda\mu C^2\gamma_{ab}^2(\alpha\beta - 1)} \qquad (14)$$

Néel recognized that the $\chi^{-1} \cdot T$ curve is a hyperbola of the form

$$\chi^{-1} = \frac{T-\theta}{C} - \frac{\zeta}{T-\theta'}, \qquad (15)$$

with $\theta = C\gamma_{ab}\lambda\mu(2 + \lambda\alpha/\mu + \mu\beta/\lambda)$

$$\theta' = -C\gamma_{ab}\lambda\mu(2 - \alpha - \beta)$$

$$\zeta = C\gamma_{ab}{}^{2}\lambda\mu(\lambda\{1 - \alpha\} - \mu\{1 - \beta\})^{2}$$

The hyperbola may be alternately expressed as

$$\chi^{-1} = \frac{(T - T_c)(T - T_c')}{C(T - \theta')} \qquad (16)$$

The hyperbola is sketched in Figure 1, and the constants are designated there. In order to treat the case in which $J_{ab} < 0$, the signs of γ_{ab}, α, and β are taken to be negative, and it is possible to obtain the parameters in the theory by fitting the expression for susceptibility to experimental data. Many examples are given in the recent treatise by Craik.[27]

DATA ANALYSIS

While measurements of magnetic properties at one temperature (say, room temperature) or over a limited temperature range yield data which are more valuable than none at all, it must be emphasized that measurements over wide ranges of temperatures and fields are usually essential for a detailed description of electronic structure and bonding.

The first step in the analysis of the data is to plot the molar magnetic susceptibility, χ_m, which has been corrected for the diamagnetism of the constituents, versus T^{-1}. If a straight line which passes through the origin is obtained, then Curie law behavior, $\chi_m = C/T$, is followed. From the Curie constant, one can calculate μ_{eff} using the expression $\mu_{eff} = 2.828\ (\chi \cdot T)^{1/2}$ and use the Bohr magneton number along with complementary data to solve the problem at hand. If a straight line which intercepts the positive χ-axis at $T^{-1} = 0$ is obtained, it is possible that temperature independent paramagnetism is present, or that the diamagnetic correction has been overestimated. If the straight line intercepts the negative χ-axis at $T^{-1} = 0$, it is likely that the diamagnetic correction has been underestimated.

For mixed-valence, it more likely that a Curie-Weiss law will be required to fit the data, and that this law will hold only over a limited temperature range. The Curie-Weiss law was

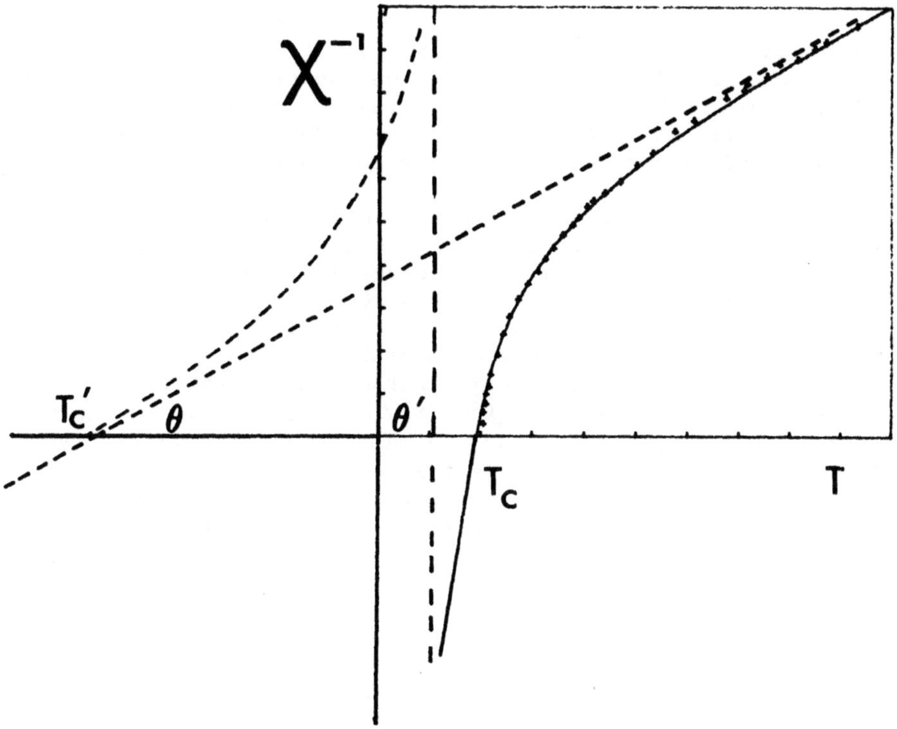

Figure 1. A sketch of the Néel hyperbola with the parameters indicated.

derived above. It is usually instructive to plot $\chi \cdot T$ versus T. Curie law behavior holds if a straight line parallel to the T axis is obtained. For orbitally nondegenerate systems deviations from straight line behavior toward larger $\chi \cdot T$ values indicate ferromagnetic interactions, while deviations toward smaller $\chi \cdot T$ values indicate antiferromagnetic behavior.

A serious complication in the interpretation of magnetic data may arise if ferromagnetic impurities are present even in small quantities. In favorable cases it is possible to account for such impurities by making measurements as a function of magnetic field strength. Generally, for a saturated ferromagnet the force exerted in a Faraday measurement, for example, is proportional to $d\underline{H}/dz$, while that exerted on a paramagnetic substance is proportional to $\underline{H}(d\underline{H}/dz)$. The intrinsic susceptibility of the sample χ_i is given by the measured susceptibility χ_H minus the quantity $C\sigma_s/\underline{H}$ where C is the concentration of the ferromagnetic impurity and σ_s is the saturation magnetization. If a plot of χ_H versus 1/H yields a straight line the correction can be made readily.

Units for Magnetic Properties. There are two systems of units commonly used in publications dealing with magnetic properties. These are the cgs-emu system and the International System of Units (SI). Quickenden and Marshall[28] and Mulay[29] have presented definitions of the quantities that are encountered and have discussed the relationship of these quantities in the two systems. A very convenient table of conversion factors has been compiled by Bennett, Page, and Swartzendruber[30] and is reproduced here in Table I.

REPRESENTATIVE RESULTS

Since a complete survey of the magnetic properties of mixed valence systems is outside the scope of this chapter, results from representative studies only are summarized here. The discussion in this section is intended to illustrate the scope and the complexity of the field and hopefully will serve as a stimulus for additional research; some specific problem areas are identified.

Clusters

Di-μ-oxobis[bis(2,2'-bipyridine)manganese(III,IV)]. This unusual complex was first prepared by Nyholm and Turco,[31] and the structure was determined by Palenik and coworkers.[32] As the bond distances in Figure 2 show, the two manganese ions are distinctly different, and the shorter bond distances from the donor atoms to Mn(2) indicates that Mn(2) is the Mn(IV) ion. In

Table I. *Conversion Factors for Magnetic Quantities.*

Multiply the Number for		By	To Obtain the Number for	
Gaussian Quantity	Unit		SI Quantity	Unit
flux density, B	G	10^{-4}	flux density, B	T(≡Wb/m^2 ≡Vs/m^2)
magnetic field strength, H	Oe	$10^3/4\pi$	magnetic field strength, H	A/m
volume susceptibility, χ	emu/cm^3 (dimensionless)	4π	rationalized volume susceptibility, κ	dimensionless
mass susceptibility, χ_ρ	emu/g(≢cm^3/g)	$4\pi \cdot 10^{-3}$	rationalized mass susceptibility, κ_ρ	m^3/kg
molar susceptibility, χ_{mole}	emu/mol(≡cm^3/mol)	$4\pi \cdot 10^{-6}$	rationalized molar susceptibility, κ_{mole}	m^3/mol
magnetization, M	G or Oe	10^3 $4\pi \cdot 10^{-4}$	magnetization, M magnetic polarization, J	A/m T
magnetization, $4\pi M$	G or Oe	$10^3/4\pi$ 10^{-4}	magnetization, M magnetic polarization, J	A/m T
magnetization, M	μ_B/atom or μ_B/form. unit, etc.	1	magnetization, M	μ_B/atom or μ_B/form, unit, etc.
magnetic moment of a dipole, m	erg/C	10^{-3}	magnetic moment of a dipole, m	J/T(≡Am^2)
demagnetizing factor, N	dimensionless	$1/4\pi$	rationalized demagnetizing factor, N	dimensionless

addition, the lengthening of the quasi axial bonds of Mn(1) is comparable to the lengthening of such bonds in Mn(III) complexes, and it is reasonable to assign the oxidation state +3 to Mn(1).

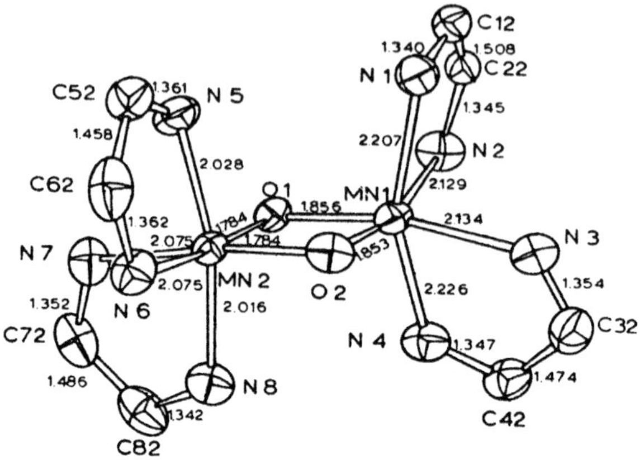

Figure 2. The structure of di-μ-oxobis[bis(2,2'-bipyridine)-manganese(III,IV)]. Reprinted with permission from Reference 32.

Cooper, Calvin, and coworkers[33,34] have recently carried out extensive studies including spectroscopic, magnetic, and electrochemical measurements on the di-μ-oxo-2,2'-bipyridine complex as well as the 1,10-phenanthroline analogue. Their magnetic susceptibility data are shown in Figure 3, where the solid line is the best fit to the data of equation 17 for an exchange coupled S = 2, 3/2 pair using J = $-150\ cm^{-1}$, the EPR

$$\chi_m = \frac{Ng^2\mu_B{}^2}{4kT} \cdot \frac{84\exp(6J/kT) + 35\exp(-J/kT) + 10\exp(-6J/kT}{4\exp(6J/kT) + 3\exp(-J/kT) + 2\exp(-6J/kT)} \cdots$$

$$\cdots \frac{+ \exp(-9J/kT)}{+\exp(-9J/kT)} + N\alpha \qquad (17)$$

<g> value of 2.0, and an assumed value of 120×10^{-6} cgs for the temperature independent paramagnetism, Nα. The 1,10-phenanthroline compound has J = $-134\ cm^{-1}$; Cooper and Calvin[33] attribute the small difference in J values to a more favorable exchange interaction geometry for the more flexible 2,2'-bipyridine ligand. The compounds exhibit moments of 1.75 B.M. and 1.74 B.M. at the low temperature limit of the experimental measurements consistent with the expected spin-only moment of

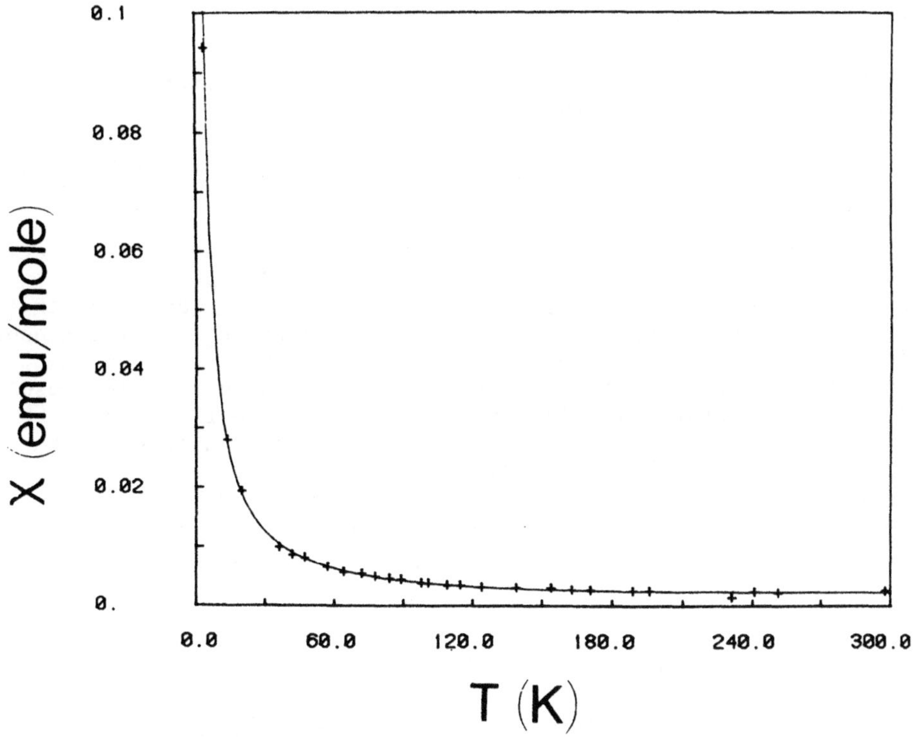

Figure 3. *The magnetic susceptibility data for di-μ-oxobis-[bis(2,2'-bipyridine)manganese (III,IV)]. The experimental data are given by +, and the solid line is the best fit to the expression for an exchange coupled S = 2, 3/2 pair.*

1.73 B.M. for the odd electron of the antiferromagnetically coupled S = 2, 3/2 pair. The features of the EPR spectrum of a frozen solution of the phenanthroline(III,IV) dimer may be accounted for by the spin Hamiltonian

$$H = g_{||}\beta H_z S_z + g_{\perp}\beta(H_x S_x + H_y S_y) + (A_1 I_1 + A_2 I_2)S - 2JS_1 \cdot S_2$$

where $g_{||}$ and $g_{\perp}$ are the principal g-tensor components for the assumed axial symmetry, S_1 and I_1 are the electron and nuclear spin operators, respectively, A_1 is the hyperfine coupling tensor for the Mn(III) ion. The terms with subscript 2 refer to the Mn(IV) ion. The spectrum shows that the exchange coupling results in a doublet state with little anisotropy in the g-tensor. The large difference in the hyperfine couplings constants of A_1 = 167 ± 3G and A_2 = 79 ± 3G indicates that the unpaired electron in the ground spin state is localized on one of the manganese ions, or transferred between them at a rate much slower than $|A_1| - |A_2|$. The slow rate of the electron transfer is a consequence of the electronic structure of the high spin Mn(III) ion; the electron being transferred is in an e_g antibonding orbital, and transfer to the Mn(IV) ion requires considerable changes in the bond distances about the manganese ions.

μ-Oxo-μ hydroxobis[(Schiff base)manganese(III,IV). Magnetic susceptibility studies on a series of bimetallic manganese compounds with the Schiff bases

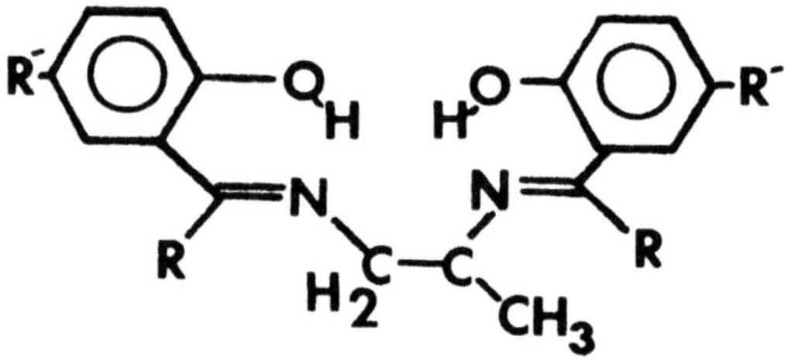

R = H (≡Sol), R = CH_3 (≡Acet)

revealed that the compounds should be formulated as mixed oxidation state dimers of manganese(III) and manganese(IV), i.e. as $[Mn_2(SB)_2(OH)O]\cdot nH_2O$.[35] Magnetic moments of 10% TMS-$CHCl_3$ solutions of the complexes (∿0.002 M) were determined by the NMR method at ambient temperature with a 250-MHz spectrometer. The compound $[Mn_2(BuSalpn)_2(OH)O]\cdot H_2O$ exhibited a magnetic moment of 2.30 BM while $[Mn_2(BuAcetpn)_2(OH)O]\cdot 2H_2O$ has a moment of 2.64 BM. These compounds, which may be examples of Class II, have a new band in their visible absorption spectrum at ∿21-22,000 cm^{-1}. This absorption band, which is responsible for the cherry red color of the compounds, has been assigned as the intervalence band. If this assignment is correct then the degree of interaction must be rather small since the frequency of the transition occurs in the denominator for α.

Biferrocene(II,III) Picrate. Kaufman and Cowan [36,37] recognized that the ligands usually employed in mixed valence chemistry, up to advent of their work, were not readily susceptible to structural variations and conjectured that organometallic compounds would be very good systems for systematic work. They selected metallocenes as the first candidates for their studies for the following reasons: (1) Strong interactions between mixed valence sites are to be expected when the orbitals on these centers have the same symmetry and are close enough for appreciable overlap, and metallocenes have these properties. (2) Delocalization is enhanced when the ligand field strengths of the mixed valence ions is similar. (3) Available crystallographic data on ferrocene and its salts had revealed that the formal oxidation state of the iron atom has little effect on the interatomic distances, and consequently the reorganization energy would not be expected to be very large. (4) Metallocenes exhibit strong interactions between the metal orbitals and the unsaturated π system of the cyclopentadienyl rings, and mixed valence delocalization could possibly be transmitted by metal-ligand orbital overlap. The spectroscopic and electrical conductivity measurements of Kaufman and Cowan on biferrocene(II,III) picrate (19) demonstrated that electron transfer in this system is quite facile.

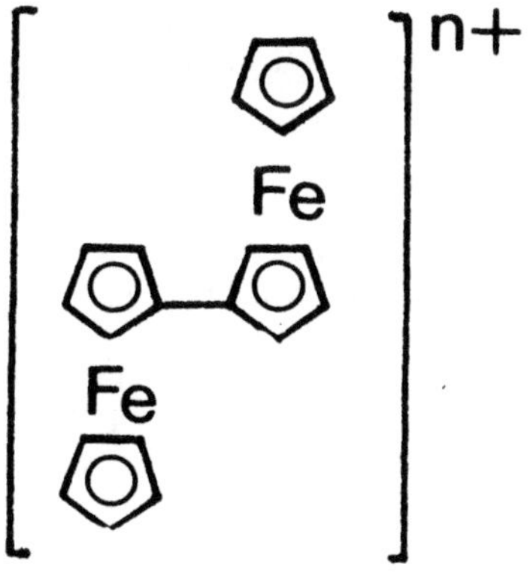

The ultraviolet to near infrared spectrum of a solution of biferrocene(II,III) picrate in acetonitrile exhibited bands at 220 and 295 nm which were assigned to the substituted ferrocene portion of the molecule, a broad band at 550-600 nm which was assigned to ferrocenium portion of the molecule, and a new, broad band centered at 1900 nm (ε 551), which was assigned to the electron-transfer transition. A rate of electron transfer of 1.3 x 10^{10} sec^{-1} was calculated from the energy of this band on the basis of Hush's theory.[38] The interaction parameter α was also calculated using equation (20)

$$\alpha^2 = \frac{4.5 \times 10^{-4}\ \varepsilon(\max)\Delta(1/2)}{(\bar{\nu})(r^2)} \tag{20}$$

where $\varepsilon(\max)$ is the molar absorptivity at the band maximum, $\Delta(1/2)$ is the band half-width in wavenumbers, $\bar{\nu}$ is the energy of the intervalence transfer band in wavenumbers, and r is the distance between donor and acceptor sites in Angstroms. This equation, which is valid only for cases in which there are weak interactions, yields the value 0.095 for α, which signifies a weak interaction between the mixed valence sites of this Class II system.

The magnetic properties of the picrate and tetrafluoroborate salts of biferrocene(II,III) were measured over the range 2-300 K.[39] Expressions for the temperature dependence of the magnetic susceptibility had been derived earlier[40] for the set of two Kramer's doublets of the $^2E_{2g}$ ground state of the ferrocenium ion and these were used for the analysis of the data. The equations are

$$\chi_{||}(E'') = \frac{N\beta^2}{kT}\left[\left(1 + \frac{2k'(1-\zeta^2)}{(1+\zeta^2)}\right)^2 + \frac{16\zeta^2 k'^2 (kT)}{(\zeta^2+\delta^2)^{1/2}(1+\zeta^2)^2}\right] \tag{21a}$$

$$\chi_{\perp}(E'') = \frac{N\beta^2}{kT}\left[\frac{4\zeta^2}{(1+\zeta^2)^2} + \frac{2kT(1-\zeta^2)^2}{(1+\zeta^2)^2(\xi^2+\delta)^{1/2}}\right] \tag{21b}$$

where k' is the orbital reduction parameter, ξ is the spin-orbit coupling constant, δ is the splitting of the energy levels due to low symmetry components, and ζ is the wave function mixing parameter which results from the departure from axial symmetry. These constants may be estimated from EPR results using the g-value expressions derived by Maki and Berry:[41]

$$\begin{aligned} g_{\perp} &= 4\zeta/(1+\zeta^2) \\ g_{||} &= 2 + 4k'(1+\zeta^2)/(1+\zeta^2) \\ |\zeta| &= x/[1 + (1 + x^2)]^{1/2} \\ |x| &= \delta/k'\xi \end{aligned} \tag{22}$$

Table II. *Magnetic Parameters for Ferrocenium Compounds.*

Compound	T, °K	$g_{\|\|}$	$g_{\perp}$	μ_{eff} [a]	μ_{eff} [b]	$\|\zeta\|$	k'	$\|\delta\|$ [c]	$\|\delta\|$ [d]
Biferrocene(3,3)$(BF_4)_2$ [e]	298	3.2	1.91	3.42	3.53	0.74	1.0	660	750
Biferrocene(2,3) picrate	77	<g> =	2.42	2.10	2.21	--	-	985	1000
Biferrocene(2,3) picrate	77	3.52	1.85	2.18	2.23	0.67	1.0	---	--
$(C_6H_5C_5H_4)_2Fe^+$ [e,f]	20	3.62	1.75	--	--	0.58	0.82	580	--
$(C_5H_5)_2Fe^+$ [e,f]	20	4.35	1.26	--	--	0.36	0.76	270	--

[a] Determined from $\mu = [(g_{||}{}^2/3 + 2g_{\perp}{}^2/3)S(S+1)]^{1/2}$.

[b] From direct measurement.

[c] Units of cm^{-1} per ferrocenium unit.

[d] Best fit from susceptibility.

[e] Frozen solution.

[f] Sohn, Y.S.; Hendrickson, D.N.; Gray, H.B. J. Am. Chem. Soc. 1970, 92, 3233.

The magnetic parameters of the mixed valence salts, which resulted from the analysis of the data, are given in Table II and compared with data for related iron compounds.

The very large value of the low symmetry component parameter δ, in comparison to those of other ferrocenium derivatives, reflects the influence of the ferrocene substituent on the ferrocenium portion of the molecule. This perturbation greatly increases the splitting of the two Kramer's doublets of the orbitally degenerate configuration $[\ldots(a_{1g})^2(e_{2g})^3]$, results in an increase in the relaxation time T_1, and permits EPR absorptions to be detectable at higher temperatures. An extensive study of the effects of substituents on the cyclopentadienyl rings on the intensity and energy of the intervalence transfer band has been carried out,[42] but unfortunately complementary magnetic studies are not yet available, so it is not possible to attempt a correlation of the interaction parameter α with structural properties which could be inferred from such data.

Bisfulvalenediiron(II,III) Salts and Related Compounds. The interesting properties of the mixed valence compound biferrocene (II,III) picrate led to additional research on ferrocene-like systems and to preparation of bisfulvalenediiron(II,III) salts (23),[43,44]

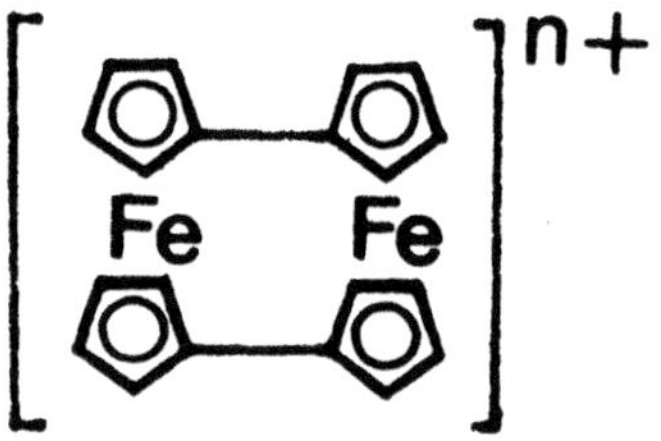

(23)

compounds which shall be referred to as BFD salts. Magnetic susceptibility, optical spectra, EPR measurements, and Mössbauer studies of these compounds revealed that they were of a different nature than the corresponding biferrocene compounds.[45-47] The magnetic susceptibility of BFD(II,III) picrate was measured over the range 2-300 K, and when corrected for diamagnetism, was found to follow the expression $\chi = (0.372/T + 250 \times 10^{-6})$ cm^3 mole^{-1}, yielding a value of 0.375 cm^3 K mole^{-1} for the Curie constant, C. From the equation $\mu_{eff} = 2.828C^{1/2}$, a value of 1.72 μ_B is obtained for the magnetic moment. As given in Table II, biferrocene(II,III) picrate has a moment of 2.2 μ_B, a value which is comparable to those reported for many substituted ferrocenium salts.

The Mössbauer spectrum of BFD(II,III) salts exhibit only two lines, and this observation leads to the conclusion that the two iron atoms are identical on the Mössbauer time scale. There are, of course, two possible explanations for this observation. There could be rapid electron transfer of a class II system or delocalization in a class III system. This latter description is further supported by ESCA and EPR data. BFD(II,III) picrate and tetrafluoroborate have the same g-values in the solid and in glasses, where $g_1 = 1.87$, $g_2 = 2.00$, and $g_3 = 2.27$. The effective magnetic moment calculated from $\mu_{eff} = [1/3(g_1^2 + g_2^2 + g_3^2)S(S+1)]^{1/2}$ is 1.78 μ_B, a value which is in good agreement with the magnetic susceptibility result.

Although there is a broad, intense band in the near infrared at 1550 nm, with a second band at 1140 nm, from which a delocalization of 3% may be calculated, the magnetic properties, EPR, ESCA, and Mössbauer data rule out the assignment of this band to intervalence transfer.

Principally through the work of Davidson, Smart and coworkers,[48,49] additional examples of bisfulvalene-dimetal compounds and mono- and dioxidized salts have been prepared. Compounds of iron (see above), nickel, and cobalt are known. For these, all BFD-M(II,II) compounds are diamagnetic, all BFD-M(II,III) compounds are paramagnetic with $\mu_{eff} \sim 1.7$ B.M. and the moments are temperature independent, and all BFD-M(III,III) compounds are diamagnetic. $[BFD\text{-}Co]^{z+}$ (z = 0, 1) exhibit bands in the electronic absorption spectrum at 980 nm with $\varepsilon = 1150$ for z = 0 and $\varepsilon = 7000$ for z = +1. The electron spectrum of $[BFD\text{-}Ni]^{z+}$ (z = 0, 1, 2) also exhibit similar bands. Smart and Pinsky[49] state that these bands "are not intervalence transfer transitions associated with electron transfer from one localized metal center to another".

The compound 3-vinylbisfulvalenediiron(II,II) has been prepared and homopolymerized in benzene using azobis(isobutyronitrile) initiation. The polymer was oxidized with TCNQ to yield mixed valence polymers with a range of stoichiometries as measured by the ratio $BFD^+/(BFD^+ + BFD)$.[50] The dark conductivity of the sample in which 71% of the bisfulvalenediiron units were monooxidized was 6-9 x 10^{-3} ohm^{-1} cm^{-1}. Magnetic data are not available for these substances.

Oxo-bridged Mixed Valence Diiron(III,IV) Complexes. Wollman and Hendrickson[51] have measured the mixed valence compounds $[Fe_2(TPP)_2O]X$ (TPP = tetraphenylporphinates, X ≡ PF_6^-, BF_4^-) and $[Fe_2(salen)_2O]X$ [salen = N,N'-ethylenebis(salicylideniminato), X^- = PF_6^-, ClO_4^-, BF_4^-, I_3^-], and have analyzed the data in terms

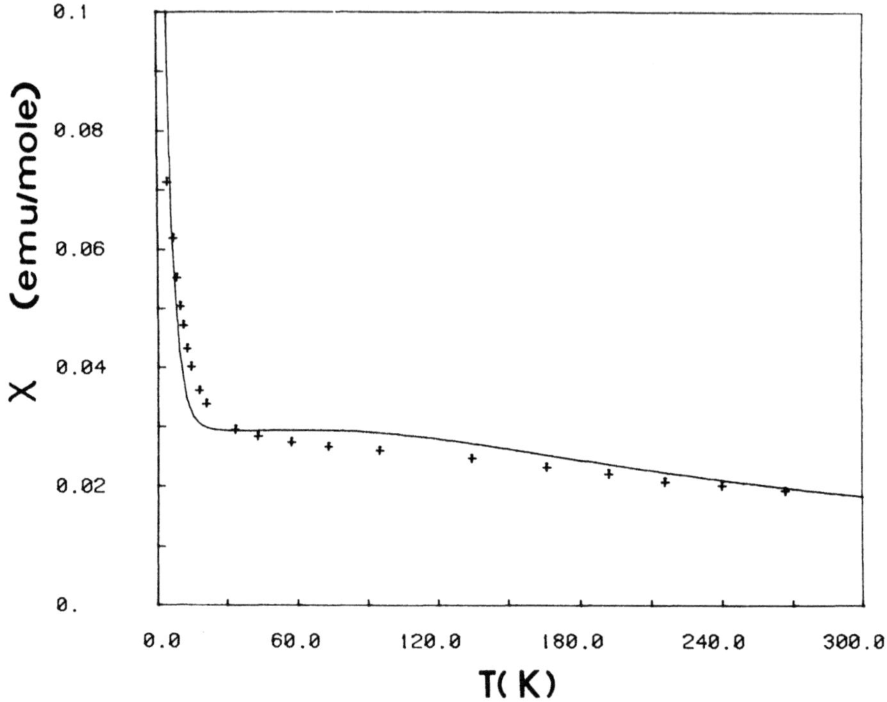

Figure 4. Magnetic susceptibility data for μ-oxo-bis[salen Fe-(III,IV)].

of isotropic exchange of an S = 5/2, 2 pair with a small zero field splitting for the salen compounds and a large zero field splitting for the porphyrin complexes. Unfortunately, as shown in Figure 4 the fit of the Van Vleck equation including a mean field interdimer exchange term to the data for $[Fe_2(salen)_2O](PF_6)$ is not especially good, but the low temperature limiting moment of 1.2 μB is in good agreement with the expectation of one unpaired electron per pair of iron ions. The magnetic parameters for these compounds are collected in Table III, where the range of

Table III. *Magnetic Parameters for Oxo-bridged Iron(III,IV) Dimers*

Compound	J, cm^{-1}	Z'J',K	$\|D\|, cm^{-1}$	$g_{\|\|}$	$g_{\perp}$
$[Fe_2(salen)_2O]PF_6$	-11.6	-0.05			
$[Fe_2(salen)_2O]ClO_4$	-17.6	-1.6			
$[Fe_2(salen)_2O]BF_4$	-8.7	-2.3			
$[Fe_2(salen)_2O]I_3 \cdot CHCl_3$	-7.5	-5.9			
$[Fe_2(TPP)_2O]PF_6$	-119	...	11.7	2.34	5.47
$[Fe_2(TPP)_2O]BF_4$	-82.5	...	19.9	3.11	4.80

exchange coupling constants, mean field intercluster correction terms, zero-field splitting energies, and g-values may be seen to vary over a very wide range.

The Mössbauer spectra of $[(Fe_2salen_2)O]X$ and $[(Fe_2TPP_2)O]X$ were recorded at 77° and 4.2°K, and consisted of a single quadrupole-split doublet in each case. Thus, the mixed valence iron compounds have identical iron ions on the Mössbauer time scale. This means that the rate of electron transfer is faster than $\sim 10^7 sec^{-1}$ in both series of compounds. The EPR spectra of the *salen* compounds exhibit one line whose line-width is slightly temperature dependent, a property to be expected if the rate of electron transfer is comparable to the EPR time scale, about $10^{10} sec^{-1}$. The EPR spectrum of $[Fe_2TPP_2O]PF_6$ was reported to be highly anisotropic, a result which is at variance with earlier observations[52] on $[Fe_2TPP_2O](ClO_4)$. Furthermore, no EPR signal was seen for the tetrafluoroborate salt, and no absorption bands were observed in the near-IR region of the spectra of any of the compounds. It is clear that the oxy-bridged mixed valence iron compounds are not understood very well and that more careful work is needed.

A Nitrido-Bridged Iron Complex. The only example of a nitrido-bridged complex of a first-transition series element was prepared by Summerville and Cohen.[53] The thermal decomposition of [Fe(III)(TPP)N_3], where TPP is tetraphenylporphine, in the absence of oxygen yields [{Fe(TPP)}$_2$N] plus nitrogen gas. The Mössbauer spectra show that the two iron atoms are equivalent, and the Mössbauer parameters are consistent with a higher oxidation state for the iron atoms than those in [{Fe(TPP)}$_2$O].

A magnetic susceptibility study on [{Fe(TPP)}$_2$N] yielded a temperature independent moment of 2.04 μ_B per dimer. Based on a comparison of the Mössbauer quadrupole splitting constant of 1.08 mm sec^{-1} for the nitrido-bridged complex with the usual values of 1.25 to 2.11 mm sec^{-1} for low spin iron(III) hemichromes, Cohen concluded that the iron ions were high-spin and strongly coupled.[1] The data do not permit a classification in the Robin and Day[2] scheme of this fractional oxidation state complex.

[Ni_2(1,8-naphthyridine)$_4X_2$]Y. A series of complexes of this general formula with X = halide, NCS^-, NO_3^- and Y = PF_6^- or $B(C_6H_5)_4^-$ may be prepared by mixing boiling butan-1-ol solutions of the nickel(II) salt, the ligand, and sodium tetraphenylborate.[54a] On further boiling black crystals may be obtained. In the case of the hexafluorophosphate salts, it is necessary to add sodium tetrahydroborate in order to get the desired product. X-ray structural results for [Ni_2(napy)$_4Br_2$]$B(C_6H_5)_4$ reveal that the nickel ions of formal oxidation state +1.5 have identical coordination environments which are square pyramidal. The base of the pyramid is formed by four nitrogen atoms from four different ligand atoms and the apex of the pyramid is occupied by a bromide ion. These two equivalent pyramids are held together by the four bridging naphthyridine ligands with a very short nickel-nickel separation of 2.415(4) Å. The bases of the two tetragonal pyramids are staggered with a twist angle of about 25° along the Br-Ni-Ni-Br axis, which is nearly linear.

The magnetic moments of these complexes are in the range 4.19-4.33 μ_B per formula unit, values which are in agreement with the presence of three unpaired electrons. A variable temperature study of [Ni_2(napy)$_4Br_2$]$B(C_6H_5)_4$ down to 85 K yielded a Weiss constant of −3 K. More extensive magnetic measurements would be of much interest since EPR studies show that the molecule has a quartet ground state as a result of a large ferromagnetic intracluster coupling.[54b] An analogous copper(I)-copper(II) complex has also been characterized.[54c]

Macrocyclic Mixed Valence Copper(I)-Copper(II) Complex. Condensation of 5-methyl-2-hydroxyisophthalaldehyde with 1,3-diaminopropane in the presence of copper(II) perchlorate yields the complex[55]

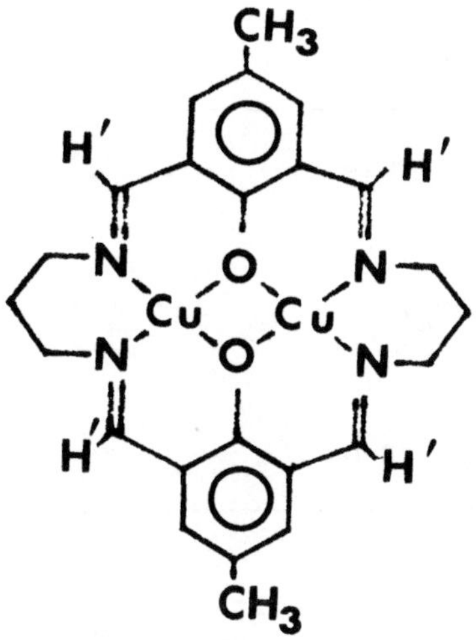

(24)

Electrolysis of an acetonitrile solution at -0.7 V followed by an addition of ether yields a nearly black solid with the formula [Cu(I)Cu(II)L](ClO_4). A new solvent dependent band in the electronic spectrum of the mixed valence complex at about 900-1200 nm was assigned to an intervalence transfer transition. At room temperature, the EPR spectrum exhibits seven lines, while in frozen solutions only four lines are present.[56] These observations suggest that the odd electron interacts with both copper ions by a rapid transfer process at room temperature but is localized on a single copper ion at low temperature. If the primed hydrogen atoms in (24) above are replaced by methyl groups, a four line spectrum is observed even at room temperature.[57] The steric requirements of the methyl groups probably affect the conformation of the macrocycle and inhibit thermal electron transfer on the EPR time scale.

The room temperature magnetic moments of a number of related complexes are given in Table IV. The data show that the copper(II) ions in the complex [Cu(II)Cu(II)L]$(ClO_4)_2 \cdot 2H_2O$ are antiferromagnetically exchange coupled. An explanation of the significant difference between the moments of [Cu(II)Cu(I)L](ClO_4) and the carbonyl adduct will require more extensive magnetic data.

μ-Oxo-Bridged Ruthenium Complexes. Meyer and coworkers[58] have studied a number of mixed valence oxo-bridged ruthenium complexes, and based on chemical, electrochemical, ESCA, magnetic, and spectroscopic data have concluded that complexes typified by the salt $[(bipy)_2ClRuORuCl(bipy)_2](PF_6)_3$ belong to class III. As usual in this research, the approach adopted was a comparison

Table IV. *Magnetic Moments at 25°C*

Compound	μ_{eff}, μ_B
$[Cu^{II}Cu^{II}L](ClO_4)_2 \cdot 2H_2O$	0.60 ± 0.04
$[Cu^{II}Cu^{I}L](ClO_4)$	1.81 ± 0.04
$[Cu^{II}Cu^{I}L(CO)](ClO_4)$	1.94 ± 0.04
$[Cu^{I}Cu^{I}L]$	0.19 ± 0.25
$[Zn^{II}Zn^{II}L](ClO_4)_2 \cdot 2H_2O$	0.00 ± 0.25

of properties of complexes in a range of oxidation states. The structure of the like valence compound $[(bipy)_2(NO_2)Ru\text{-}O\text{-}Ru(NO)_2\text{-}(bipy)_2](ClO_4)_2 \cdot 2H_2O$ has been reported,[59] and the oxo-bridge formulation confirmed.

The (III,III) compounds are paramagnetic at room temperature with moments on the order of 1.8 μ_B per ruthenium ion. The magnetic moments are consistent with the ruthenium(III) ions being in the low-spin $4d^5$ electronic configuration. Variable temperature magnetic measurements were made on $[(bipy)_2(NO_2)Ru\text{-}O\text{-}Ru(NO_2)(bipy)_2](PF_6)_2$ and $[(phen)_2(NO_2)Ru\text{-}O\text{-}Ru(NO_2)(bipy)_2]\text{-}(ClO_4)_2$. The magnetic susceptibility of the 2,2'-bipyridine complex maximizes at about 155 K, while that of the 1,10-phenanthroline salt shows the same property at about 110 K. The magnetic susceptibility data may be fit very accurately with the Van Vleck equation for systems consisting of a singlet ground state and a low lying triplet state. The magnetic parameters which resulted from a least squares fitting of the Van Vleck equation to the data are

	2,2'-bipyridine salt	1,10-phenanthroline salt
$2J$, cm^{-1}	-173	-119
$\langle g \rangle$	2.48	2.29

The magnetic data can be interpreted in two ways. One interpretation would assume strong coupling in a delocalized system, while the second approach would invoke moderate spin-paired interaction between two low-spin ruthenium(III) ions. In view of the short ruthenium-oxygen bonds which should result in strong spin-spin coupling and much larger $|2J|$ values, and the chemical and spectroscopic properties of the complexes it is reasonable to conclude that the system is delocalized. The following qualitative molecular orbital diagram based on the model developed by Dunitz and Orgel[60] should apply:

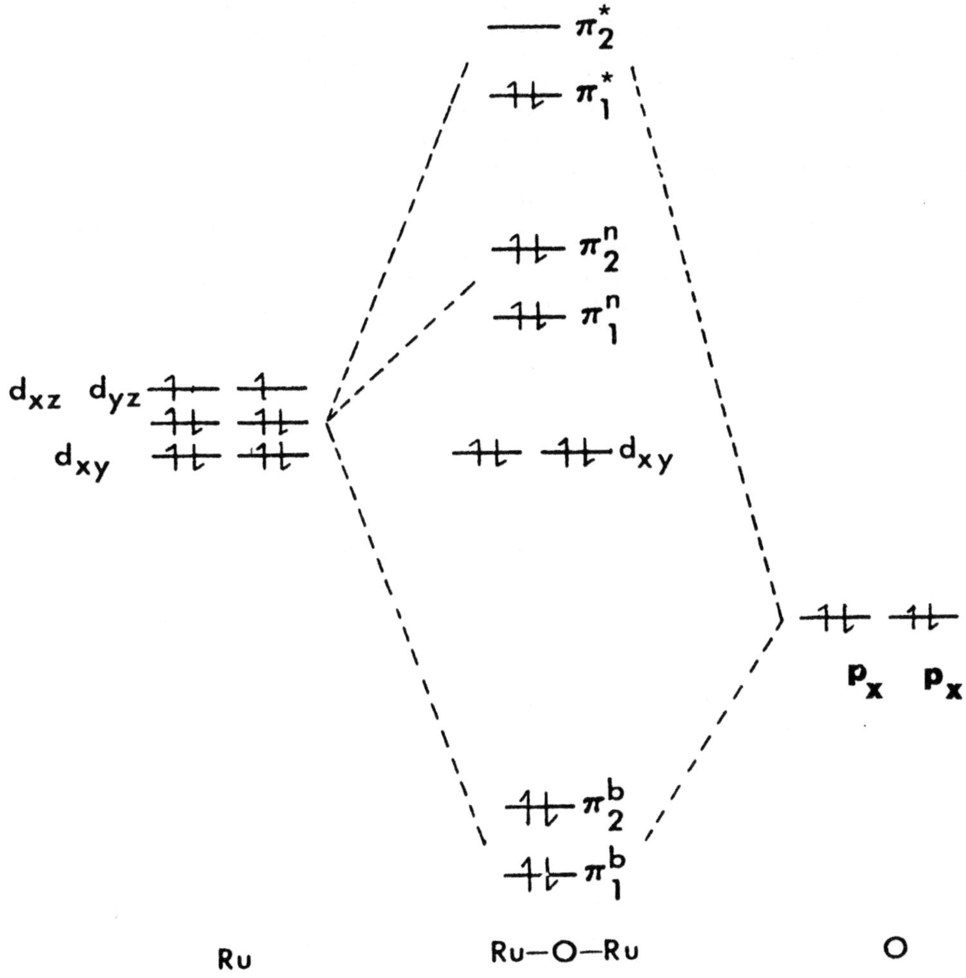

If this scheme applies, then 2J is a measure of the energy difference between a singlet state of the type $^1(\pi_1{}^{*2})$ and the triplet state $^3(\pi_1{}^*\pi_2{}^*)$, with the singlet state $^1(\pi_1{}^*\pi_2{}^*)$ being at higher energy than $^3(\pi_1{}^*\pi_2{}^*)$ because of Hund's rules. The low energy intense bands in the electronic spectra can be assigned to the transitions $\pi^b \rightarrow \pi^*$. Formation of the mixed valence complexes, such as $[(bipy)_2ClRu\text{-}O\text{-}RuCl(bipy)_2]^+$ would involve addition of an electron to $\pi_2{}^*$, and this process should destabilize the Ru-O-Ru linkage. The chemical instability of salts of $[(bipy)_2\text{-}ClRu\text{-}O\text{-}RuCl(bipy)_2]^+$ are consistent with this explanation.

μ-Pyrazine-bis[pentaammineruthenium(II,III)] Ion. The mixed valence ion $[\{Ru(NH_3)_5\}_2pyr]^{5+}$ has been the subject of many discussions since the results of some experiments may be interpreted to indicate complete localization of the odd electron on one of the ruthenium ions, while results of other experiments suggest complete delocalization or very rapid transfer of the electron. Drago, Hendrickson and their coworkers[61] have presented extensive evidence which favors trapped valences in the tosylate salt of $[\{Ru(NH_3)_5\}_2\ pyr]^{5+}$. The variation of the magnetic moment with temperature for the like valence [III,III]-pyrazine dimer is shown in Figure 5, where the moment has been calculated per mole of dimer. The decrease of the moment with temperature parallels the behavior of the monomeric compound $[Ru(NH_3)_6]Cl_3$ down to 15 K. The abrupt decrease in moment at about 15 K for $[\{Ru(NH_3)_5\}_2\text{-}pyr]^{6+}$ was thought to be indicative of a small intramolecular antiferromagnetic interaction or of an intermolecular interaction. Magnetic data below 4.2 K are necessary to resolve this important question. The very small J value was interpreted to mean that there is negligible overlap between the orbitals of the ruthenium ions and the pyrazine bridge orbitals, and that the complex is best described in terms of a localized scheme.

EPR results for the ruthenium(III) monomer is identical with those for the pyrazine bridged [III,III] dimer, and other than a resolution of the $g_\perp$ transition into two components, the spectrum of the pyrazine bridged mixed valence compound yields a comparable $g_\perp$ value of 2.66. Furthermore, the line-widths of the resonances bands of the [III,III] and [II,III] compounds are comparable. These results were said to indicate that the presence of a ruthenium ion at one end of a pyrazine bridge did little or nothing to perturb the ruthenium(III) on the other end of the bridging ligand, and to verify the conclusion that the mixed valence compound belongs to Class II. A line-width study was interpreted in terms of a rate of electron transfer slower than $10^9\ sec^{-1}$. An NMR study showed that the rate was greater than $10^5\ sec^{-1}$. Thus, from the Arrhenius equation, the energy barrier lies between 3.4 kcal $mole^{-1}$ and 6.7 kcal mol^{-1}. Hush disagrees and described a delocalized model and new EPR data in his presentation at the Institute.[111]

A Mixed Valence Cobalt-Copper Complex. Kahn and coworkers[62-64] have investigated a series of heterometallic systems of the general formula NM'L where the ligand is the anion of the N,N'-bis(2-hydroxy-3-carboxybenzylidene)-1,2-diamminoethane:

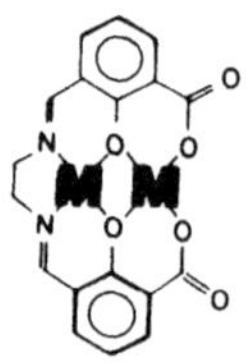

(25)

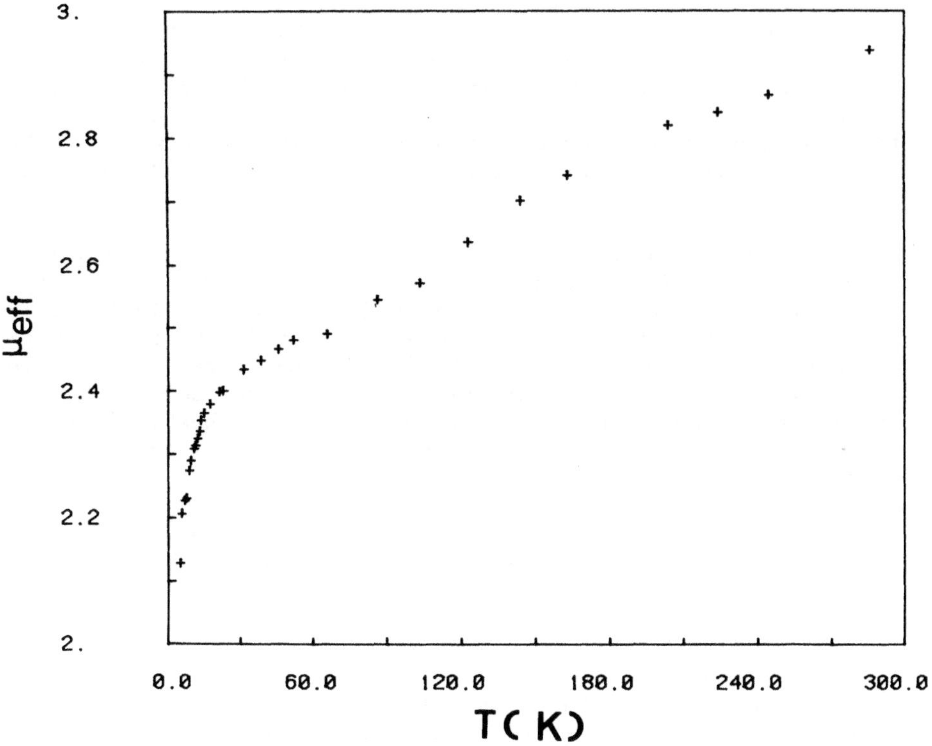

Figure 5. The temperature variation of the magnetic moment of $[\{Ru(NH_3)_5\}_2 \text{ pyrazine}]^{6+}$.

The magnetic susceptibility of samples of the cobalt-copper complex $CoCuL \cdot 3H_2O$ which have the same chemical composition may be described by the Curie-Weiss laws $\chi_m \cdot T = 1.69$ k with $0.4 < k < 1$ in the temperature range 30 to 300 K. All of these samples obey the law with k = 1 after they have been heated in a vacuum. It is thought that the solid phase equilibrium $Co^{II}Cu^{II} \rightleftharpoons Co^{III}Cu^{I}$ is shifted to the left by the heat treatment.

The Curie constant suggests that the triplet spin state, only, is occupied in the temperature range 30-300 K. This may be understood if the single-ion ground state of cobalt(II) is a doublet, and this doublet is interacting ferromagnetically with the copper(II) ion yielding a triplet ground state. The singlet state arising from the interaction must lie much too high to be populated in the temperature interval studied, thus implying a J value of several hundred wave numbers, and the strongest ferromagnetic coupling reported to date.

Below 30 K, $\chi_m \cdot T$ decreases and reaches the value 1.55 cm^3 $mole^{-1}$ at 3.6 K. This low temperature behavior probably arises from an antiferromagnetic intermolecular coupling.

$[Fe(II)Fe(III)_2(CH_3COO)_6O(H_2O)_3] \cdot 2H_2O$. The trimetallic basic acetates have been the subject of many investigations of exchange dating from the classical work of Welo[65] and Kambe.[3] Lupu[66] has prepared and studied iron(II)iron(III) mixed valence basic acetate compounds. From magnetic susceptibility data in the temperature range 94-296 K, he found the exchange coupling constant between the iron(III) ions to be -36 cm^{-1}, a value which is comparable to that found for the iron(III) basic acetates. However, the exchange integral between the iron(II) and iron(III) ions was found to be temperature dependent, being 2.8 cm^{-1} at 100.7 K and -10 cm^{-1} at 288 K. This behavior was attributed to changes in bond distances. As Lupu suggested, more extensive magnetic studies should be carried out on these interesting compounds. Brown and co-workers have made such measurements;[108] they found J[Fe(II)-Fe(III)] to be -50 cm^{-1}, J[Fe(III)-Fe(III)] to be -14.5 cm^{-1}, and both coupling constants to be temperature independent.

Mössbauer studies on $[Fe(II)Fe(III)_2(CH_3COO)O(H_2O)_3] \cdot 2H_2O$ and $[Fe(II)Fe(III)_2(CH_3COO)O(py)_3] \cdot 0.5$ py show that the iron ions are indistinguishable on the Mössbauer time scale at room temperature, but are readily distinguishable below 200 K.[67] These data, along with optical absorption spectra, support the assignment of these compounds to class II.

Triangular Cobalt Clusters. The compounds $Co_3(h^5\text{-}C_5H_5)_3S_2$, $Co_3(h^5\text{-}C_5H_5)_3(CO)S$, and $[Co_3(h^5\text{-}C_5H_5)_3S_2]I$ have structures consisting of three metal ions in a triangular array which are capped above and below by the two sulfur atoms or by one sulfur atom and the carbonyl groups, with the cyclopentadienyl ions bonded to the cobalt atoms.[68] These three compounds exhibit a fascinating array of magnetic properties. $Co_3(h^5\text{-}C_5H_5)_3(CO)S$ is diamagnetic, and the monoxidized iodide salt is a simple paramagnet with a magnetic moment which decreases linearly from 1.95 μ_B at 297 K to 1.77 μ_B at 83 K. At temperatures greater than 195 K the magnetic susceptibility of the disulfide complex obeys the Curie-Weiss law, but at 195 K there is an abrupt discontinuity and the magnetic susceptibility decreases to approximately one-half of its maximum value within a 30° temperature range. Additional evidence for a phase transition at 192.5 K is provided by the calorimetric study by Sorai and coworkers.[69]

A ^1H magnetic resonance study of a CS_2 solution of the disulfide complex over the temperature range 163 to 324 K revealed three distinct regions of paramagnetic shift dependence. For $T > 260$ K the paramagnetic shift varies as $1/T$ and $S = 1$; below 173 K the shift is zero as expected for a diamagnetic substance; while in the region $173 < T < 260$, the shift reflects the spin state equilibrium. However, the shift data can be fit by the expression for the thermal equilibrium between singlet and triplet species only if a temperature dependent energy separation is used.

<u>Cubane Type Cluster Systems Containing Iron</u>. Tetrametallic cubane type cluster systems containing iron have been of considerable interest since it is known that a distorted cubic cluster Fe_4S_4 is present in some non-heme iron-sulfur proteins, and complexes of the type $[(RS)FeS]_4^{n-}$ (where RS is an alkyl or aryl mercaptide) have been shown to be close structural and electronic analogues of the proteins.[70] The Fe_4S_4 cubane structure is also present in dithiolene complexes[71] of the type $[\{(CF_3)_2C_2S_2\}FeS]_4^{n-}$, and in cyclopentadienyl complexes of the type[72] $[(h^5\text{-}C_5H_5)FeS]_4^{n-}$. An analogous tetrametallic complex with CO replacing the bridging sulfur atoms is also known.[73]

In the reduction of $[(SR)FeS]_4^{2-}$ to the trianion, there are significant changes in the Mössbauer isomer shifts, quadrupole interactions, and magnetic hyperfine interactions at the iron sites. These properties have been interpreted in terms of a molecular orbital delocalized over the four metal ions. In the dithiolene complexes $[(R_2C_2S_2)FeS]_4^{2-}$, there are only small changes in the Mössbauer parameters. This observation indicates that the orbital to which the electron is added is predominately ligand based. Oxidation of $[(h^5\text{-}C_5H_5)Fe(CO)]_4$ to the monocation results in a change in the quadrupole interaction and a small magnetic hyperfine interaction, but there is no change in the isomer shift. These results suggest that the electron is removed from an orbital with some metal as well as ligand character.[74]

The electron distribution in the series $[(h^5\text{-}C_5H_5)FeS]_4^{n+}$

(n = 0, 1, 2) is very complex. Mössbauer spectra consist of single quadrupole doublets indicating that all iron atoms are identical on the Mössbauer time scale. The puzzling feature is that the mono-and dication exhibit the same electric quadrupole splittings even though there are significant changes in the iron-sulfur bond distances in going from the neutral molecule to the cations.[75] Reiff and coworkers conclude that the "invariance of the isomer shift upon oxidation from the [neutral molecule] to the [dication] suggests either that electrons are removed from ligand-based orbitals with no change in metal 3d, 4s population or that the electrons are removed from metal-based molecular orbitals in the successive oxidations... and that in the resulting products... there is a redistribution of ligand electron density to the metal atoms so as to result in little overall change of metal orbital electron population."

Magnetic data for the mono-and dications are given in Table V, where it may be seen that both compounds are paramagnetic and that the magnetic moment decreases as the temperature decreases.

Table V. *Magnetic Data for* $[(h^5\text{-}C_5H_5)FeS]_4^{+/2+}$

$[CpFeS]_4^{+}$		$[CpFeS]_4^{2+}$	
T,K	μ_B	T,K	μ_B
303	1.33	298	0.80
276	1.23	288	0.78
255	1.20	252	0.73
232	1.16	216	0.68
190	1.07	180	0.62
131	0.95	144	0.57
99	0.87	108	0.53
52	0.77	78	0.48
25	0.67		
11.2	0.59		
4.2	0.52		
2.14	0.45		
1.50	0.41		

A magnetic moment of 0.87 μ_B/Fe would be expected for one unpaired electron in the monocation cluster, i.e. 1.73 $\mu_B/\sqrt{4}$; the high moments above 100 K indicate that higher cluster spin states are occupied, while the decrease in magnetic moment at low temperatures suggests an antiferromagnetic intercluster interaction. The magnetic behavior of the dication suggests that paramagnetic states are being occupied, but Reiff and coworkers noted that a large TIP contribution could not be ruled out. It is clear that the electronic structures of these clusters are affected by the terminal as well as the bridging ligands. These effects can be rationalized by qualitative molecular orbital calculations,[75b] but more powerful calculations as well as more experimental data are needed before the behavior of these important class III clusters can be thoroughly understood.

Di-μ-acetato-tetrakis[μ_3-methoxo-2,4-pentanedionatocobalt-(II,III)]. A mixed valence compound containing cobalt(II) and cobalt(III) may be prepared by the oxidation with hydrogen peroxide of a mixture of cobalt(II) acetate, 2,4-pentanedione, and potassium hydroxide.[76] The cluster has a cubane type structure with cobalt ions and methoxide oxygen atoms at alternate corners of the cube, and quasi-octahedral coordination of the cobalt ions is completed by a chelated 2,4-pentanedionato group and by an oxygen of one of the acetate groups which bridge the top and bottom faces of the cube. Since the cobalt-oxygen bond distances for one pair of these cobalt ions are 0.16-0.20 Å shorter than those of the second pair, it is likely that the former are the cobalt(III) ions.

The magnetic moment of the cluster is temperature dependent, being 4.98 μ_B per cobalt(II) at room temperature and 4.62 μ_B at 77°K. Although the geometry might suggest a class I system containing a pair of exchange coupled cobalt(II) ions, the dark brown color of the material makes such a classification suspect.

One-Dimensional Chains

Partially Oxidized Metallophthalocyanines. Reactions of Fe, Co, Ni, Cu, Zn, Pt and metal-free phthalocyanines with iodine vapor or solutions in solvents such as chlorobenzene yield darkly colored solids of the general formula M(phthalocyanine)I_x, where x can take on values from less than one to nearly four, depending on reaction conditions and the phthalocyanine.[77a] The mixed valence character of these materials has been established by resonance Raman data which shows that the iodine is present as I_3^-, by iodine-129 Mössbauer studies, and by X-ray structural studies on Ni(phthalocyanine)$(I_3)_{.33}$. The $[Ni(Pc)]^{0.33+}$ molecules stack to form one-dimensional chains, and the I_3^- molecules are present as disordered chains in channels formed by the phthalocyanine units.[77b] The electrical conductivity measured along the stacking direction (the crystallographic c axis) for $[Ni(Pc)](I_3)_{.33}$ is in the range 250-650 ohm^{-1} cm^{-1}, values which are comparable to the electrical conductivities of the most highly conductive molecular materials yet reported.[78] The electrical conductivity increases as the temperature decreases until a maximum value of $\sigma(T)/\sigma_{RT}$ of about four is reached near 90 K at which point there is an abrupt drop in conductivity. Remarkably, there is no sharp break in the magnetic susceptibility, which is weakly paramagnetic and increases with temperature, coincident with the "metal-insulator" transition.

Iridium Carbonyl Chloride. For a number of years it was thought that the compound originally characterized as $Ir(CO)_3Cl$ was nonstoichiometric,[79-82] but recent work by Reiss, Hagley, and Peterson has confirmed the initial formulation.[83] The chlorotricarbonyliridium(III) units of $Ir(CO)_3Cl$ stack in a staggered manner with an Ir-Ir distance of 2.844 (1) Å and an Ir-Ir-Ir angle of 178.53 (2)°. While it is true that data from physical measurements on the stoichiometric material are yet to be had, the data reported by Ginsberg and coworkers[82] on their samples are very interesting. It is likely that these latter workers were carrying out measurements on a mixed valence compound, whose identity is yet to be established. A detailed discussion of the experimental observations including Mössbauer and magnetic susceptibility must await the completion of this task.

Perylene-Metal Dithiolate Complexes. Ion-radical complexes of the general formula $(perylene)_2M(mnt)_2$ (where M = Ni, Cu, Pd, Pt and mnt = maleonitriledithiolate) exhibit relatively high electrical conductivities. The magnetic properties of these compounds have been investigated in attempts to understand the electronic structure of the constituents and to identify the conduction pathways.[84-86] A study of the temperature dependence of the EPR spectrum of $(perylene)_2Pd(mnt)_2$ revealed that the paramagnetism was associated with two spin subsystems, and that each of these is exchange coupled either pairwise or in linear stacks. The magnetic susceptibility data for the copper complex suggested that the $Cu(mnt)_2$ sites were not completely in the monoanionic form, thus implying that the average charge on the perylene sites is less than +1/2. An analysis of the susceptibility data in terms of the contributions from the constituents showed that the magnetic susceptibility of the perylene spin system was largely temperature independent over the range 100-300 K but increased abruptly at 77 K in a manner similar to the temperature dependence of the magnetic susceptibility of other conducting salts such as quinolinium $(TCNQ)_2$. The magnetic moment per formula unit of $Cu(mnt)^{n-}$ was found to be 0.85 μ_B while that of $Ni(mnt)^{n-}$ was 1.99 μ_B. For $Pd(mnt)^{n-}$ it was estimated that there were 0.145 monoanions per formula unit.

$K_2Pt(CN)_4Br_{0.3}\cdot 3H_2O$. The magnetic susceptibility of a single crystal of KCP(Br) has been reported by Menth and Rice.[87] χ is independent of temperature and equal to -160×10^{-6} cm^{3}mole^{-1}, a value which is very close to that expected for the diamagnetism of the constituent atoms in view of the measured diamagnetic susceptibility of -136×10^{-6} cm^{7}mole^{-1} for $K_2Pt(CN)_4\cdot H_2O$.[88]

χ is temperature dependent and in the temperature range 4.2 to 40 K may be described by the expression

$$\chi_{||} = -100 \times 10^{-6} + \frac{1.83}{T} \times 10^{-4}$$

In the temperature range 40 to 250 K the expression for $\chi_{||}$ is

$$\chi_{||} = -160 \times 10^{-6} + \frac{2.50}{T} \times 10^{-3}$$

The need for two Curie-Weiss laws to describe the temperature dependence of the magnetic susceptibility also arises for iridium carbonyl chloride and $(H_3O)_{1.6}[Pt(C_2O_4)_2]\ nH_2O$.

Menth and Rice proposed an interrupted strand model for $\chi_{||}$ of the form

$$\chi_i = 1/2N(\varepsilon_F)\left[\frac{\mu_B g_i}{2}\right]^2 + \frac{1}{2}\left[\frac{Na}{l_o}\right]\left[\frac{1}{k_B T}\right]\left[\frac{\mu_B g_i}{2}\right]^2 \qquad (26)$$

where $N(\varepsilon_F)$ is the thermal density of states at the Fermi energy, N_a is the number of strands per unit area perpendicular to the strand axis, and l_o is the mean length of the 1-D metallic boxes which are formed by the random distribution of lattice defects along the strand. The expression for χ is valid for $k_B T << \Delta$, where $\Delta \simeq 2N_a/(N(\varepsilon_F)l_o)$. Δ is the mean spacing of the energy levels within the box in the vicinity of the Fermi energy.

The first term in the expression represents the Pauli contribution which arises from the boxes which have an even number of carriers and which may be estimated. In a 1-D structure $N(\varepsilon_F) = 2N_a/\pi v_F$, where for KCP(Br) $N_a \sim 10^{14}$ cm^{-2} and $v_F \sim 10^8$cm sec^{-1}. Substitution yields $\chi_p \sim 5 \times 10^{-6}$cm^{3}mole^{-1}.

The second term in expression 26 arises from the boxes which have an odd number of carriers. In principle, it should be possible to deduce l_o, but the result was thought to be unreasonably large. Also, the fraction of unpaired spins per platinum site, f, may be estimated from the expression $\chi = n_s\left[\frac{\mu_B g}{2}\right]^2 f/k_B T$, where n_s is the number of platinum sites per unit volume. For $T < 40$, $f \simeq (2/g_{||})^2 \times 0.6 \times 10^{-3}$, and for $T > 40$, $f \simeq (2/g_{||})^2 \times 7 \times 10^{-3}$. This parameter was thought to be unrealistically small, and it must be concluded that this model is not applicable to KCP(Br).

Kuindersma and Sawatzky have also measured the magnetic susceptibility of KCP(Br) and attributed the Curie behavior at low temperature to the presence of impurity platinum ions in which the d_{xz}, d_{yz} orbitals are the highest occupied orbitals.[89] They argued that the model used by Menth and Rice was inappropriate and that a tight binding model would have been preferable. They suggested that the electrons were strongly paired and that a possible explanation may be found in terms of a semiconductor model, where the band gap was much larger than 300 K. No details on this proposal have been presented.

Phillips[90] has suggested that the abrupt change in slope of the Curie-Weiss plots is a third order phase transition which is intrinsic and is associated with a lateral alignment of paramagnetic domain tips. Along a given metal chain in KCP(Br), or iridium carbonyl chloride, or $(H_3O)_{1.6}[Pt(C_2O_4)_2]nH_2O$ there are quasi periodic microdomains which may have localized spins associated with electrons in tip states. At high temperatures these spins are uncorrelated, but at low temperatures interchain alignment may become possible.

$\underline{(H_3O)_{1.6}[Pt(C_2O_4)_2]nH_2O}$. Magnetic susceptibilities of polycrystalline, partially oxidized samples of $(H_3O)_{1.6}[Pt(C_2O_4)_2]nH_2O$ with two different values of n, and the unoxidized potassium salt have been measured in the temperature range 1.5 - 300 K.[91] The potassium salt is diamagnetic as expected, while the partially oxidized acids are diamagnetic down to 15-20 K at which point they become paramagnetic. Two Curie-Weiss laws are required to fit the temperature dependence of the magnetic susceptibility. The sample with a relatively large water content has a magnetic moment of 0.097 μ_B above 17 K and the moment is 0.079 μ_B below 17 K. The corresponding values for the sample with a relatively small water content are 0.11 μ_B above 22 K and 0.097 μ_B below 22 K. It is thought that unusual magnetic behavior of these compounds are characteristic of linear chain, metal-metal bonded, partially oxidized systems. The paramagnetic behavior may arise from defects or localized singly occupied states, but no explanation has been put forth for the abrupt change in slope of the Curie-Weiss plots.

Random Exchange in Quinolinium $(TCNQ)_2$. Quinolinium $(TCNQ)_2$ is a charge transfer compound consisting of separated stacks of quinolinium ions and TCNQ ions. The very high electrical conductivity of the material has attracted a great deal of attention and a number of elegant studies have been described in the literature. The magnetic properties are of especial interest. At relatively high temperatures the magnetic susceptibility is essentially constant, but at low temperatures there is a Curie-like tail. Azevedo and Clark[21] have been successful in treating the magnetic behavior of quinolinium $(TCNQ)_2$ using the Bulaevskii model for a random exchange Heisenberg chain with a density of states having the form $\rho(\varepsilon) = Ak^{(\alpha-1)}|\varepsilon|^{-\alpha}$, where A and α are parameters selected to fit the data. The best fit of the magnetic susceptibility data to the Bulaevskii model yield A = (1.81 $\pm$ 0.1) x 10^{-3} $K^{\alpha-1}$ and α = 0.82 $\pm$ 0.01. If one takes these values and integrates the density of states equation, a 3% molar concentration of spins with $\varepsilon/k < 10$ K is obtained. The other 97% of the spins must be involved in exchange interactions for which $J/k > 10$ K. A detailed description of the susceptibility follows from the specific form of $\rho(J)$ given above.

Two Dimensional Sheets

Potassium Copper Sulfide, KCu_4S_3. The mixed valence compound KCu_4S_3, which contains three copper ions in formal oxidation state +1 and one copper ion in formal oxidation state +2, has a layered structure of sulfur bridged copper ions where the layers are separated by potassium ions.[92,93] The copper ions are tetrahedrally coordinated by sulfur atoms, and with all copper ions are equivalent. The tetrahedra form the layers by sharing edges and vertices in this Class IIIB compound.

The conductivity of pressed pellets of KCu_4S_3 is 4.1 $\pm$ 1.1 x 10^3 $ohm^{-1}cm^{-1}$ at room temperature and rises to 6.1 x 10^4 ohm^{-1} cm^{-1} at 18 K. Although it has not been possible to obtain crystals for complete anisotropic measurements, the conductivity in the plane has been measured at room temperature and is $\sim$ 8 x 10^3 $ohm^{-1}cm^{-1}$. As expected, this metallic material possesses a very small paramagnetic susceptibility that is nearly temperature independent. Below 15 K there is an increase in susceptibility which follows the Curie Weiss law with C = 0.00477 and θ = 0.16. This behavior suggests the presence of a small concentration of paramagnetic sites, either defects or paramagnetic impurities. If the Curie-Weiss paramagnetism is subtracted from the observed magnetic susceptibility, then a residual molar susceptibility of $\sim$ 7 x 10^{-5} cgs $mole^{-1}$ results. This is a sum of second order single ion TIP and Pauli paramagnetism, and it is difficult to evaluate either of these independently.

Layered Transition Metal Dichalcogenides. Layered transition metal dichalcogenides such as 2H...TaS_2 and 2H...$NbSe_2$ may be readily intercalated with organic molecules, metal atoms, and other σ donors. The intercalated substances frequently have higher superconducting transition temperatures than the precursors, but in some cases the transition temperature is shifted to lower values. Hydrogen may be intercalated in 2H...TaS_2, 2H...$NbSe_2$, 2H...NbS_2, and TiS_2 (but not in 1T...TaS_2, HfS_2, ZrS_2, or MoS_2) by electrolysis in acid solution.[94] In these compounds hydrogen acts as an electron donor.

Magnetic susceptibilities of 2H...TaS_2 and four hydrogen intercalated compounds H_xTaS_2 are given in Figure 6. There is a marked decrease in χ_g for 2H...TaS_2 near 80 K. This decrease in magnetic susceptibility, which is caused by superlattice formation, is suppressed by hydrogen intercalation and is absent when $X \geq 0.11$. Some pertinent magnetic data are summarized in Table VI, where it may be seen that the compound with the highest superconducting transition temperature (X = 0.11) shows the smallest, positive $\Delta\chi$. It is thought that the bonding is due to charge transfer from hydrogen to the *d* band of the tantalum sulfide layer, and that the enhancement of the superconducting transition temperatures is due to a suppression of a charge density wave near 80 K.

Table VI. *Magnetic Properties and T_c of Tantalum Sulfides*

Compound	χ(max) x 10^6 emu/g	$\Delta\chi$ [χ(max)-χ(4.2)]	T_c, K
2H...TaS_2	+.725	+0.134	0.8
$H_{0.06}TaS_2$	+.622	+0.078	
$H_{0.09}TaS_2$	+.590	+0.020	
$H_{0.11}TaS_2$	+.606	+0.002	4.2
$H_{0.15}TaS_2$	+.523	-0.003	
$H_{0.87}TaS_2$	(-0.190)	0	<0.5
$Ta_{1.05}S_2$	+.472	+0.050	∿3.6

None of the H_xTiS_2 compounds were superconducting above 0.5K, and the compound H_xNbSe_2 with x = 0.01 exhibited an enhancement in T_c of 0.2 K (to 7.45 K); while for the compound with x = 0.2, T_c was less than 0.5 K. 2H...$NbSe_2$ also has a distortion at about 35 K, which is suppressed by intercalation.

Three-Dimensional Polymers

Prussian Blue. The mixed valence compound Fe(III)$_4$[Fe(II)-(CN)$_6$]$_3$·14H_2O has been a frequent object of study.[95] In this compound the iron(II) ion, being coordinated by six carbon atoms, occupies a strong-field octahedral site, while the iron(III) ions are in two different kinds of weak-field sites. In one of these high spin iron(III) ions are coordinated by six nitrogen atoms from the cyano ligands coordinated to the iron(II) ions, while in the second site the high spin iron(III) ions are coordinated

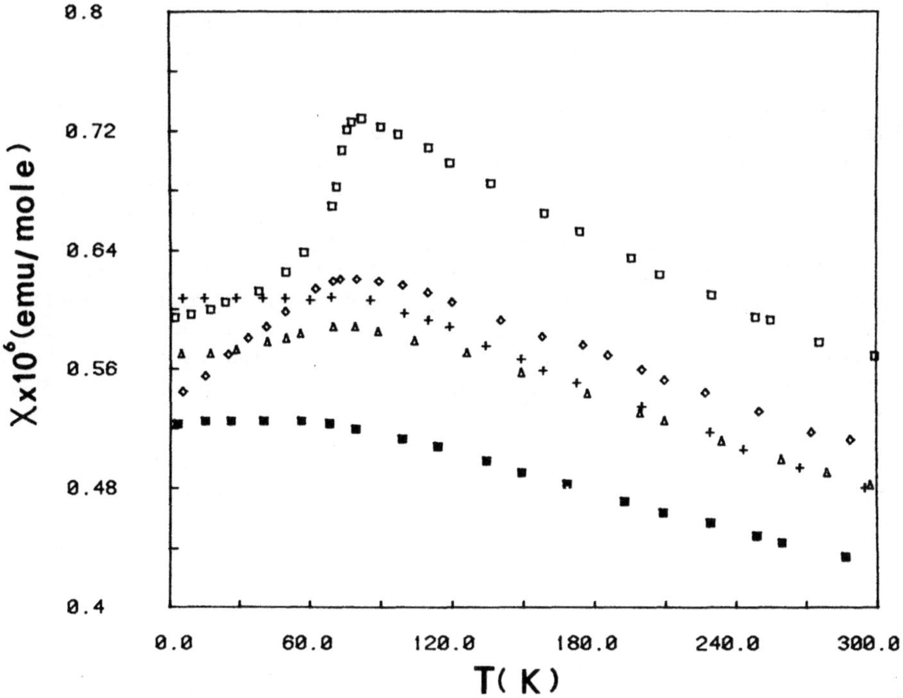

Figure 6. *Magnetic susceptibility data for* $2H\ldots TaS_2$ (□), $H_{0.06}TaS_2$ (◇), $H_{0.09}TaS_2$ (Δ), $H_{0.11}TaS_2$ (+), $H_{0.15}TaS_2$ (■).

by four nitrogen atoms and two water molecules. Although the paramagnetic iron(III) ions are at least 10.16 Å apart, and present in the linear bridge arrangement Fe(III)-NC-Fe(II)-CN-Fe(III), the substance orders ferromagnetically above 5 K. Using molecular orbital theory, Mayoh and Day[96] have calculated the ferromagnetic coupling energy in terms of a "valence delocalization coefficient". This latter parameter was calculated from perturbation theory to be 0.083 and was estimated from the intensity of the mixed valence band to be 0.106. From the relationship $E_{fm} \sim k_BT_c$, they estimated T_c to be 11.2 K from the calculated coefficient and 6.6 K from the experimentally estimated coefficient. Ludi and coworker have recently measured the magnetic properties of deuterated Prussian blue and found T_c to be 5.6 $\pm$0.1 K,[109] a value which is in good agreement with T_c of 5.5 $\pm$ 0.5 K for $Fe_4[Fe(CN)_6]_3 \cdot 14H_2O$.[110]

The close agreement between the experimental and calculated Curie temperatures lends support to a mechanism which involves delocalization of formally t_{2g} electrons of the iron(II) ions onto the iron(III) ions. Since the t_{2g} and e_g orbitals of the iron(III) ions are half occupied, they may accept electrons of only one spin. Since there are six iron(III) ions about a given iron(II) ion, the spins of these six ions are correlated, and the process is repeated throughout the lattice. Mayoh and Day[96] calculated the energy per each iron(II) center when the spins of the six neighboring iron(III) ions were all arranged parallel to one another, and when the six spins were arranged randomly, and attributed that energy to the ferromagnetic coupling energy. The recent prediction[97] of a very low ordering temperature in voltaite[98] demonstrates the power of the computational technique.

Iron Fluoride Dihydrate. Brown, Reiff and coworkers[99] have made Mössbauer and magnetic susceptibility measurements on $Fe_2F_5 \cdot 2H_2O$ which show that the material orders ferrimagnetically at 48 K. Above the critical temperature distinct quadrupole-split Mössbauer absorptions were observed for both the iron(II) and the iron(III) sites. These observations demonstrate the nonequivalence of the oxidation states and establish $FeF_5 \cdot 2H_2O$ as a class II mixed valence compound. Ferrimagnetic ordering was established by the observation of spontaneous magnetization, field-dependent susceptibilities, and a saturation moment corresponding to a formula unit spin of S = 1/2. Below the critical temperature the Mössbauer spectra exhibit the magnetic hyperfine structure expected for ferrimagnetic ordering.

The magnetic susceptibility data above the Neel temperature are plotted as χ^{-1} versus temperature in Figure 7, where the solid line is the best fit to the equation for the Néel hyperbola

$$\chi^{-1} = (T-T_c)(T-T_c')/[C(T-\theta')]$$

with the parameters T_c = 48 K, T_c' = -200.9 K, θ' = 25.8 K, and C = 6.74. At high temperatures the Néel hyperbola asymptotically approaches Curie-Weiss behavior, and at the ferrimagnetic Néel temperature the hyperbola becomes infinite. For $Fe_2F_5 \cdot 2H_2O$, the limiting high-temperature magnetic susceptibility obtained from a fit of the Curie-Weiss law to data in the temperature range 200-250 K is given by $\chi = 6.74/(T+179)$. The effective magnetic moment calculated from $\mu_{eff} = 2.828\ C^{1/2}$ is 7.34 μ_B, a value to be compared to the moment of 7.68 μ_B which is predicted for two noninteracting high spin iron(II) and iron(III) ions with g = 2. The negative θ in the Curie-Weiss law implies antiferromagnetic interactions between Fe(II) and Fe(III).

Below the Néel temperature long range order is expected for ferrimagnetic materials, and, as expected, $Fe_2F_5 \cdot 2H_2O$ exhibits spontaneous magnetization in zero applied field. Qualitatively the behavior agrees with that predicted by a Brillouin function with S = 1/2 and provides additional evidence for ferrimagnetic ordering. Finally, the saturation moment is 1.2 μ_B. For a negative coupling constant between Fe(II) and Fe(III), a saturation moment of 1.0 μ_B would have been expected, while a positive coupling between these ions would yield S = 9/2 and a saturation moment of 9 μ_B.

Iron Fluoride Heptahydrate, $Fe_2F_5 \cdot 7H_2O$. The magnetic susceptibility of a powdered sample of $Fe_2F_5 \cdot 7H_2O$ has been measured as a function of temperature using a vibrating sample magnetometer operating at 10 kOe.[100] A transition to an antiferromagnetic state was observed at 3.0 $\pm$ 0.2 K. Using data collected in the range 4.2 - 60 K, a magnetic moment of 7.68 μ_B and a Weiss constant of 6.9 K may be calculated. This is precisely the magnetic moment expected from the spin-only values of the two iron ions. Although ferrimagnetism may have been expected, no spontaneous magnetization was observed below T_N, so the authors proposed a four-sublattice model to account for the observed behavior. Additional measurements are needed in order to sort out the interactions in this mixed valence compound since the Mössbauer and spectroscopic results of Brown and coworkers[101] lead to the conclusion that $Fe_2F_5 \cdot 7H_2O$ is a class I material.

Chevrel Phase Compounds. Compounds of the general formula $M_xMo_6S_8$ with M = Cu, Pb, Sn, Ag, rare earths, etc., which were first reported by Chevrel et al.[102] have attracted considerable attention since they undergo a transition to a superconducting state.[103] The structure of these compounds consist of Mo_6S_8 units with the second metal atom occupying positions between the clusters. The identity and concentration of the second metal atom has a dramatic effect on T_c.[104,105] The rare earth molybdenum sulfides $(RE)_xMo_6S_8$ ($x \sim 1$ for the light rare earths and $x \sim 1.2$ for the heavier ones) are superconducting even in the

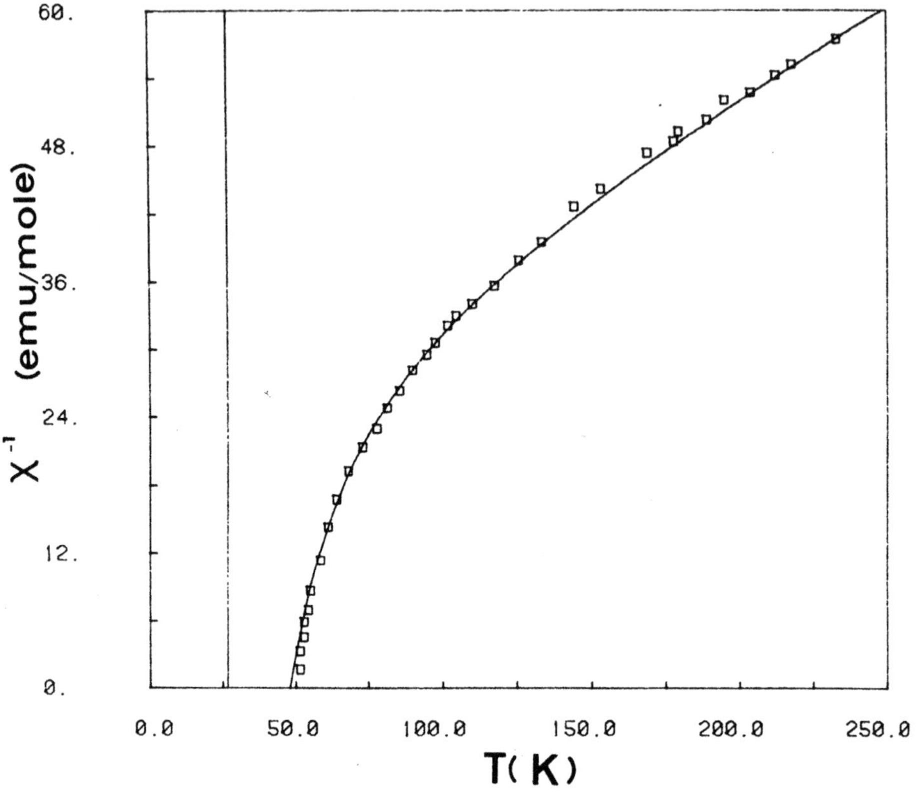

Figure 7. Magnetic susceptibility data for $Fe_2F_5 \cdot 2H_2O$. The solid line is the best fit of Néel hyperbola to the experimental data.

presence of a high concentration of magnetic ions. Extensive magnetization measurements have been reported for the series of compounds $(RE)_xMo_6Se_8$ ($x \sim 1$; RE = La, Ce, Pr, Eu, Gd, Yb, and Lu).[106] The susceptibilities of $CeMo_6Se_8$, $PrMo_6Se_8$, and $YbMo_6Se_8$ deviate marked from Curie-Weiss behavior above 10 K because of crystal field effects. There is a small exchange interaction between gadolinium ions in superconducting $Gd_{1.2}Mo_6Se_8$ which leads to long range order at 0.8 K but $CeMo_6Se_8$, which apparently becomes ferromagnetic, does not superconduct above 0.05 K. The compound $Er_xMo_6Se_8$ undergoes a transition to a superconducting state at T_c = 5.70 - 6.17 K, and the Er^{3+} ions order antiferromagnetically at 1.07 K. The ordering does not result in the loss of superconductivity.[107] These compounds provide very good examples of substances which exhibit coexisting superconductivity and long range order.

ACKNOWLEDGEMENTS

This work was supported in part by the Office of Naval Research and by the North Atlantic Treaty Organization.

REFERENCES

1. Cohen, I. Structure and Bonding, in press.
2. Robin, M.B.; Day, P. Adv. Inorg. Chem. And Radiochem. 1967, 10, 247.
3. Kambe, K. J. Phys. Soc. Japan 1950, 5, 48.
4. Hatfield, W.E. In "Theory and Applications of Molecular Paramagnetism", Boudreaux, E.A. and Mulay, L.N., Ed.; Wiley: New York, 1976; Chapter 7, pp. 349ff.
5. Kudo, T.; Matsubara, F.; Katsura, S. Physica 1978, 93A, 255.
6. Wrobleski, J.T.; Brown, D.B. Inorganic Chemistry, in press.
7. Fisher, M.E. J. Math. Phys. 1963, 4, 123.
8. Onsager, L. Phys. Rev. 1944, 65, 90.
9. Katsura, S. Phys. Rev. 1963, 129, 2835.
10. Fisher, M.E. Am. J. Phys. 1964, 32, 343.
11. Bonner, J.C.; Fisher, M.E. Phys. Rev. 1964, 135A, 640.
12. Weng, C.-Y Ph. D. Dissertation, Carnegie-Mellon University, 1968.
13. Wagner, G.R.; Friedberg, S.A. Phys. Lett. 1964, 9, 11.
14. Rushbrooke, G.S.; Wood, P.J. Molec. Phys. 1958, 1, 257.
15. Smart, J.S. "Effective Field Theories of Magnetism", Saunders: Philadelphia, 1966.
16. DeJongh, L.J.; Miedema, A.R. Adv. Phys. 1974, 23, 1.
17. Van Vleck, J.H. "The Theory of Electric and Magnetic Susceptibilities"; Oxford University Press: London, 1932.
18a. Hall, J.W. Ph.D. Dissertation, University of North Carolina, Chapel Hill, 1977; b. Jotham, R.W. J. Chem. Soc. Dalton Trans. 1977, 266.
19. Bulaevskii, L.N.; Zvarykina, A.V.; Karimov, Yu. S.; Lyubovskii, R.B.; Shchegolev, I.F. Sov. Phys.-JETP 1972, 35, 384.
20. Theodorou, G.; Cohen, M.H. Phys. Rev. Lett. 1976, 37, 1014.
21. Azevedo, L.J.; Clark. W.G. Phys. Rev. B 1977, 16, 3252.
22. Dupas, A.; LeDang, K.; Renard, J.P.; Veillet, P.; Daoud, A.; Perret. R. J. Chem. Phys. 1976, 65, 4099.
23. Oguchi, T. J. Phys. Soc. Japan, 1971, 31, 1021, and references therein.
24. Yamada, I. J. Phys. Soc. Japan, 1972, 33, 979.
25. Erdos, P. J. Phys. Chem. Solids, 1966, 27, 1705 and references therein.
26. Néel, L. Ann. Phys. (Paris) 1948, 3, 137.
27. Craik, D.J. "Magnetic Oxides"; Wiley: New York, 1975.
28. Quickenden, T.I.; Marshall, R.C. J. Chem. Educ. 1972, 49, 114.
29. Mulay, L.N. in "Theory and Applications of Molecular Paramagnetism", Boudreaux, E.A. and Mulay, L.N., Ed.; Wiley: New York, 1976; Chapter 9.
30. Bennett, L.H.; Page, C.H.; Swartzendruber J. Research, National Bureau of Standards, 1978, 83, 9.
31. Nyholm, R.S.; Turco, A. Chem. Ind. (London) 1960, 74.
32. Plaskin, P.M.; Stoufer, R.C.; Mathew, M.; Palenik, G.J. J. Am. Chem. Soc. 1972, 94, 2122.

33. Cooper, S.R.; Calvin, M. J. Am. Chem. Soc. 1977, 99, 6623.
34. Cooper, S.R.; Dismukes, G.C.; Klein, M.P.; Calvin, M. J. Am. Chem. Soc. 1978, 100, 7248.
35. Boucher, L.J.; Coe, C.G. Inorg. Chem. 1976, 15, 1334 and references therein.
36. Cowan, D.O.; Kaufman, F. J. Am. Chem. Soc. 1970, 92, 219.
37. Kaufman, F.; Cowan, D.O. J. Am. Chem. Soc. 1970, 92, 6198.
38. Hush, N.S. Prog. Inorg. Chem. 1967, 8, 391.
39. Cowan, D.O.; Candela, G.A.; Kaufman, F. J. Am. Chem. Soc. 1971, 93, 3889.
40. Sohn, Y.S.; Hendrickson, D.N.; Gray, H.B. J. Am. Chem. Soc. 1970, 92, 3233.
41. Maki, A.H.; Berry, T.E. J. Am. Chem. Soc. 1965, 87, 4437.
42. LeVanda, C.; Bechgaard, K.; Cowan, D.O.; Rausch, M.D. J. Am. Chem. Soc. 1977, 99, 2964.
43. Cowan, D.O.; LeVanda, C. J. Am. Chem. Soc. 1972, 94, 9271.
44. Mueller-Westerhoff, U.T.; Eilbracht, P. J. Am. Chem. Soc. 1972, 94, 9272.
45. Cowan, D.O.; LeVanda, C.; Collins, R.L.; Candela, G.A.; Mueller-Westerhoff, U.T.; Eilbracht, P. J.C.S. Chem. Comm. 1973, 329.
46. Mueller-Westerhoff, U.T.; Eilbracht, P. Tetrahedron Letters 1973, No. 21, 1855.
47. LeVanda, C.; Bechgaard, K.; Cowan, D.O.; Mueller-Westerhoff, U.T.; Eilbracht, P.; Candela, G.A.; Collins, R.L. J. Am. Chem. Soc. 1976, 98, 3181.
48. Davison, A.; Smart, J.C. J. Organomet. Chem. 1973, 49, C43.
49. Smart, J.C.; Pinsky, B.L. J. Am. Chem. Soc. 1977, 99, 956.
50. Pittman, Jr., C.U.; Surynarayanan, J. Am. Chem. Soc. 1974, 96, 7916.
51. Wollman, R.G.; Hendrickson, D.N. Inorg. Chem. 1977, 16, 723.
52. Felton, R.H.; Owen, G.S.; Dolphin, D.; Forman, A.; Borg, D.C.; Fajer, J. Ann. N.Y. Acad. Sci. 1973, 206, 504.
53. Summerville, D.A.; Cohen, I.A. J. Am. Chem. Soc. 1976, 98, 1747.
54a. Sacconi, L.; Mealli, C.; Gatteschi, D. Inorg. Chem. 1974, 13, 1985; b. Bencini, A.; Gatteschi, D.; Sacconi, L. Inorg. Chem. 1978, 17, 2670; c. Gatteschi, D.; Mealli, C.; Sacconi, L. Inorg. Chem. 1976, 15, 2774.
55. Gagne, R.R.; Koval, C.A.; Smith, T.J. J. Am. Chem. Soc. 1977, 99, 8367.
56. Gagné, R.R.; Koval, C.A.; Smith, T.J.; Cimolino, M.C. J. Am. Chem. Soc. 1979, 101, 4571.
57. Addison, A.W. Inorg. Nucl. Chem. Lett. 1976, 12, 899.
58. Weaver, T.R.; Meyer, T.J.; Adeyemi, S.A.; Brown, G.A.; Eckberg, R.P.; Hatfield, W.E.; Johnson, E.C.; Murray, R.W.; Untereker, D. J. Am. Chem. Soc. 1975, 97, 3039.
59. Phelps, D.W.; Kahn, E.M.; Hodgson, D.J. Inorg. Chem. 1975, 14, 2486.
60. Dunitz, J.D.; Orgel, L.E. J. Chem. Soc. 1953, 2594.
61. Bunker, B.C.; Drago, R.S.; Hendrickson, D.N.; Richman, R.M.; Kessell, S.L. J. Am. Chem. Soc. 1978, 100, 3805.

62. Kahn, O.; Tola, P.; Coudanne, H. Inorg. Chim. Acta 1978, 31, L405.
63. Kahn, O.; Tola, P.; Galy, J.; Coudanne, H. J. Am. Chem. Soc. 1978, 100, 3931.
64. Kahn, O.; Claude, R.; Coudanne, H. J. Chem. Soc. Chem. Comm. 1978, 1012.
65. Welo, L.A. Phil. Mag. 1928, [7], 6, 481.
66. Lupu, D. Rev. Roum. Chem. 1970, 15, 417.
67. Lupu, D.; Barb, D.; Filoti, G.; Morariu, M.; Tarina, D. J. Inorg. Nucl. Chem. 1972, 34, 2803.
68. Frisch, P.D.; Dahl, L.F. J. Am. Chem. Soc. 1972, 94, 5082.
69. Sorai, M.; Kosaki, A.; Suga, H.; Seki, S. Yoshida, T.; Otsuka, S. Bull. Chem. Soc. Japan 1971, 44, 2364.
70. Holm, R.H. Endeavour 1975, 34, 38.
71. Frankel, R.B.; Reiff, W.M.; Bernal, I.; Good, M.L. Inorg. Chem. 1974, 13, 493.
72. Wong, H.; Sedney, D.; Reiff, W.M.; Frankel, R.B.; Meyer, T.J.; Salmon, D. Inorg. Chem. 1978, 17, 194.
73. Neumann, M.A.; Trinh-Toan; Dahl, L.F. J. Am. Chem. Soc. 1972 94, 3383.
74. Frankel, R.B.; Reiff, W.M.; Meyer, T.J.; Cramer, J.L. Inorg. Chem. 1974, 13, 2515.
75a. Trinh-Toan; Teo, B.K.; Fergusson, J.A.; Meyer, T.J.; Dahl, L.F. J. Am. Chem. Soc. 1977, 99, 408; See, for example, Gall, R.S.; Chu, C.T.W.; Dahl, L.F. J. Am. Chem. Soc. 1974, 96, 4019.
76. Bertrand, J.A.; Hightower, T.C. Inorg. Chem. 1973, 12, 206.
77a. Petersen, J.L.; Schramm, C.S.; Stojakovic, D.R.; Hoffman, B.M.; Marks, T.J. J. Am. Chem. Soc. 1977, 99, 286; b. Schramm, C.J.; Stojakovic, D.R.; Hoffman, B.M.; Marks, T.J. Science 1978, 200, 47.
78. Hoffman, B.M.; Ibers, J.A.; Marks, T.J.; Scaringe, R.P.; Schramm, C.J.; Stojakovic, D.R. Symposium on Properties of Low Dimensional Solids, American Chemical Society National Meeting, Honolulu, Hawaii, April 2-5, 1979.
79. Hieber, H.; Lagally, H. Z. Anorg. Allg. Chem. 1940, 245, 321.
80. Fischer, E.O.; Brenner, K.S. Z. Naturforsch. B. 1962, 17, 774.
81. Krogmann, K.; Binder, W.; Hausen, H.D. Angew. Chem., Int. Ed. Engl. 1968, 7, 812; Angew. Chem. 1968, 80, 844.
82. Ginsberg, A.P.; Cohen, R.L.; DiSalvo, F.J.; West, K.W. J. Chem. Phys. 1974, 60, 2657.
83. Reis, Jr., A.H.; Hagley, V.S.; Peterson, S.W. J. Am. Chem. Soc. 1977, 99, 4184.
84. Alcácer, L.; Maki, A.H. J. Phys. Chem. 1974, 78, 215.
85. Alcácer, L.; Maki, A.H. J. Phys. Chem. 1976, 80, 1912.
86. Alcácer, L.; Novais, H.; Pedroso, F. In "Molecular Metals"; Hatfield, W.E., Ed.; Plenum: New York, 1979; p. 415.
87. Menth, A.; Rice, M.J. Solid State Comm. 1972, 11, 1025.
88. Krogmann, K.; Hausen, H.D. Z. Anorg. Allgem. Chem. 1968, 358, 67.

89. Kiundersma, P.I.; Sawatzky, G.A. Solid State Comm. 1973, 13, 39.
90. Phillips, J.C. Phys. Stat. Sol. 1977, 79B, 111.
91. Heitkamp, D.; Rade, H.S.; Keller, H.J.; Rupp, H.H. J. Solid State Chem. 1975, 15, 292.
92. Rudorff, W.; Schwarz, H.G.; Walter, M. Z. Anorg. Allg. Chem. 1952, 269, 141.
93. Brown, D.B.; Zubieta, J.A.; Valla, P.A.; Wrobleski, J.T.; Watt, T.; Hatfield, W.E.; Day, P. to be published.
94. Murphy, D.W.; DiSalvo, F.J.; Hull, Jr., G.W.; Waszczak, J.V.; Mayer, S.F.; Stewart, G.R.; Early, S.; Acrivos, J.V.; Geballe, T.H. J. Chem. Phys. 1975, 62, 967.
95. For a list of references, see Mayoh, B.; Day, P. J. Chem. Soc. Dalton Trans. 1976, 1483.
96. Mayoh, B.; Day, P. J. Chem. Soc. Dalton Trans. 1974, 846.
97. Beveridge, D.; Day, P. J. Chem. Soc. Dalton Trans. 1979, 648.
98. Herman, E.; Hanned, R.; Simkin, D.; Brand, D.E.; Muir, W.B. Canad. J. Phys. 1976, 54, 1149.
99. Walton, E.G.; Brown, D.B.; Wong, H.; Reiff, W.M. Inorg. Chem. 1977, 16, 2425.
100. Jones, Jr., E.R.; Hendricks, M.E.; Anel, T.; Amma, E.L. J. Chem. Phys. 1977, 66, 3252.
101. Walton, E.G.; Corvan, P.J.; Brown, D.B.; Day, P. Inorg. Chem. 1976, 15, 1737.
102. Chevrel, R.; Sergent, M.; Prigent, J. J. Solid State Chem. 1971, 3, 515.
103. Matthais, B.T. Science 1972, 175, 1465.
104. Fischer, Ø.; Odermatt, R.; Bongi, G.; Jones, H.; Chevrel, R.; Sergent, M. Phys. Lett. 1973, 45A, 87.
105. Fischer, Ø.; Treyvaud, A.; Chevrel, R.; Sergent, M. Solid State Comm. 1975, 17, 721.
106. Johnston, D.C.; Shelton, R.N. J. Low Temp. Phys. 1977, 26, 561.
107. McCallum, R.W.; Johnston, D.C.; Shelton, R.N.; Fertig, W.A.; Maple, M.B. Solid State Comm. 1977, 24, 501.
108. Dziobkowski, C.T.; Wrobleski, J.T.; Brown, D.B. to be published.
109. Herren, F.; Fischer, P.; Ludi, A.; Hälg, W. to be published.
110. Ito, A.; Suenaga, M.; Ono, K. J. Chem. Phys. 1968, 48, 3597.
111. Hush, N.S. Presentation at the NATO-ASI on "Mixed Valence Compounds", Oxford, England: September 15, 1979.

APPLICATIONS OF THE MÖSSBAUER EFFECT TO THE STUDY OF MIXED-VALENCE COMPOUNDS

David B. Brown and James T. Wrobleski*

Department of Chemistry, University of Vermont, Burlington, Vermont 05405

Abstract

The principal utility of Mössbauer Effect spectroscopy in the study of mixed-valence materials is its ability (through measurement of the chemical isomer shift) to distinguish between different oxidation states of the same element in similar chemical environments. By using Mössbauer Effect spectroscopy it is possible to distinguish between trapped and delocalized valence states in mixed-valence compounds. In addition, the rather long lifetime of the Mössbauer experiment, as compared to other forms of spectroscopy, permits evaluation of the rate of intervalence electron exchange in mixed-valence materials.

INTRODUCTION

In recent years the Mössbauer Effect (ME), a form of nuclear γ-ray resonance spectroscopy, has been applied to a broad range of research problems involving mixed-valence materials. As a result, our understanding of the role of mixed valency in certain areas of catalysis, corrosion science, biological redox reactions, mineralogy, and the study of impurity structures in semiconductors and metals has progressed substantially.

ME spectroscopy is a powerful tool for the study of mixed-valence materials because it can distinguish different oxidation states of the same element in similar chemical environments. Through application of the ME one may obtain information about

*Present address: Monsanto Co., St. Louis, Missouri.

D. B. Brown (ed.), Mixed-Valence Compounds, 243-270.

both structural and magnetic properties of mixed-valence compounds. In certain cases it is also possible to probe dynamic features of the mixed-valence state such as kinetics of intervalence electron transfer in semiconductors and isolated metal ion clusters.

The purpose of this article is to illustrate the types of research problem involving mixed-valence compounds which have been examined with ME spectroscopy. Experimental results which demonstrate the applicability of ME spectroscopy to mixed-valence materials are preceded by a brief, operational introduction to the ME. Those readers who wish more detailed discussion of the ME are encouraged to consult works listed in the general references list at the end of this paper.

DESCRIPTION OF THE MÖSSBAUER EFFECT

The fundamental nuclear event which describes ME spectroscopy is the recoilless (zero-phonon) emission or absorption of a γ ray. In the usual ME experiment (Figure 1a) a radioactive parent nuclide decays to a daughter (the source) with excited state $|\Psi_e^S>$ and nuclear half life τ_n. A fraction $\underline{f}$ of the excited state nuclei decay to the ground state $|\Psi_g^S>$ with emission of a γ ray γ_M which has not suffered energy loss due to recoil effects. This recoilless fraction $\underline{f}$ is a function of both the energy of γ_M and the lattice dynamics of the source material. When the source is brought in the vicinity of a material (the absorber) whose nuclear energy levels exactly match its own, γ_M may be resonantly absorbed

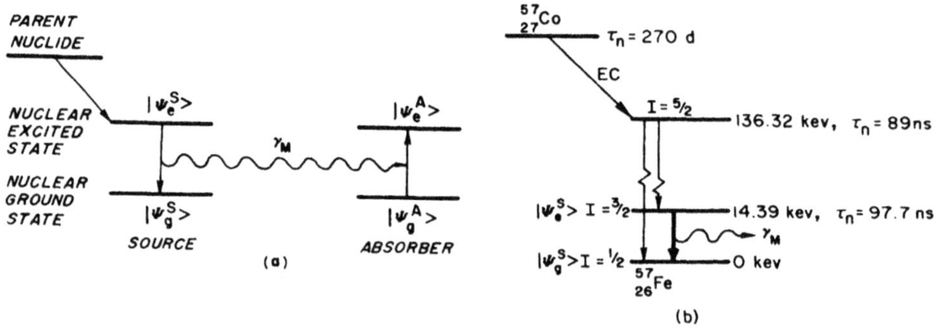

Figure 1. Schematic illustrations of nuclear processes important in ME spectroscopy. (a) Recoilless emission + absorption of ME gamma ray, γ_M. (b) Specific decay scheme for $^{57}_{27}Co$ decay to $^{57}_{26}Fe$ with production of 14.4 keV γ_M.

to populate a nuclear excited state $|\Psi_e^A\rangle$ in the absorber. Figure 1b illustrates the decay scheme of ^{57}Co which is used to populate the nuclear excited state of ^{57}Fe, the most widely-studied ME nuclide.

Two types of source + absorber geometries are commonly employed in ME spectroscopy. These geometries are illustrated in Figure 2. In transmission geometry (Figure 2a) the absorber is positioned between source and detector and the γ ray energy distribution transmitted through the absorber is recorded. A sketch in the lower portion of Figure 2a represents a spectrum which is obtained in this manner. In the backscatter experiment (Figure 2b) the detector and source are shielded from one another on the same side of the absorber (scatterer) and the detector records the energy distribution of resonant fluorescence emitted from $|\Psi_e^A\rangle$.

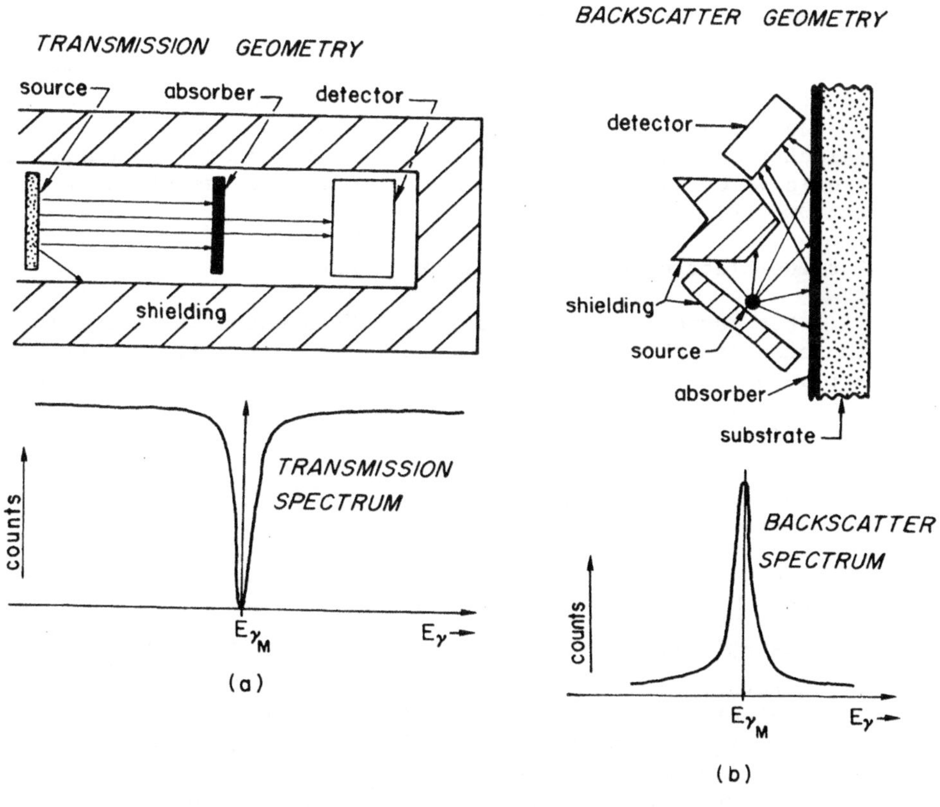

Figure 2. The two types of source + absorber ME experimental geometries. (a) Transmission geometry. (b) Backscatter geometry.

A spectrum obtained in the backsactter geometry is illustrated in the lower portion of Figure 2b.

In the experiments described above it is apparent that if the absorber nuclear ground-excited state splitting does not exactly match the energy of γ_M the system is off resonance: that is, no γ ray absorption will occur. In order to bring the system into resonance it is necessary to move the source or absorber. This relative motion modulates the energy of γ_M by an amount $E'(v)$ dictated by the Doppler effect $[E'(v) = \pm(v/c)E_{\gamma_M}]$, where v is the relative velocity of source with respect to absorber, c is the speed of light, and E_{γ_M} is the energy of the unmodulated Mössbauer γ ray. By convention the positive sign in this Doppler equation refers to closing velocities. Because the nuclear energy shifts of importance in the ME are of the order 1×10^{-3} cm^{-1}, it is seldom necessary for $|v|$ to exceed 10 or 20 mm/sec. However, it is necessary to define a zero of velocity and have available a precisely-known velocity scale. The zero-velocity calibrant of choice for a given Mössbauer nuclide is the centroid of the ME spectrum of a conveniently-prepared standard absorber. A variety of techniques exist for determining the functional form of the velocity scale. Included among these techniques are purely electronic velocity calibrants which involve the use of Moiré interferrometry.

Nucleus-Electron Hyperfine Interactions

For purposes of analyzing most ME spectra it is sufficient to recognize three hyperfine effects which result from the interaction of electric and magnetic fields with the nuclear energy levels. These interactions are normally expressed in terms of a hyperfine hamiltonian $\hat{H}_{hf}$, eq. 1. In this equation $\hat{E}_o$ represents

$$\hat{H}_{hf} = \hat{E}_o + \hat{M}_1 + \hat{E}_2 \tag{1}$$

the Coulomb potential which exists between the electrons and nucleus, $\hat{M}_1$ refers to coupling of the nuclear magnetic moment with any magnetic fields in the nuclear environment, and $\hat{E}_2$ describes the interaction between the nuclear quadrupole moment and the local electric field gradient (efg). Although these three hyperfine interaction terms are normally sufficient to describe ME spectra, one must be aware of the time dependence of electron-nuclear hyperfine effects. For example, one should anticipate complicated spectra when local efg's and magnetic fields are fluctuating at a rate comparable to the Mössbauer excited state relaxation rate $k_n (\equiv 1/\tau_n)$. In these instances it is necessary to compute the time dependent part of $\hat{\underline{H}}_{hf}$ by assuming a suitable relaxation model. In the area of mixed-valence materials these complicating features are encountered when, for instance, the intervalence electron transfer rate $\underline{k}_e$ is of order $\underline{k}_n$.

Coulomb Interactions (Isomer Shift)

In most chemical applications it is convenient and permissible to a high degree of accuracy to treat electronic and nuclear properties as separable (the Born-Oppenheimer approximation). This assumption implies that nuclear energy levels are insignificantly perturbed by their electronic environment. The ME, however, is capable of detecting the minute energy shifts which arise from nucleus-electron coupling. For example, the Coulomb interaction between the nuclear energy levels and their electronic environment depends upon the penetration of electron density into the finite nuclear volume. Because $\underline{s}$ electrons have a finite probability of existing at the nucleus, the Coulomb interaction depends on the $\underline{s}$ electron probability density at the nucleus $|\Psi_s(o)|^2$. In addition, because the rms nuclear radius changes by an amount $\delta\underline{R}$ upon emission or absorption of a γ ray, the Coulomb interaction between the nucleus and its environment changes by an amount Δw which is given by eq. 2.

$$\Delta w = \text{constant} \times \delta\underline{R} \times |\Psi_s(o)|^2 \quad (2)$$

In a ME source + absorber experiment one measures the difference between Δw for the source and Δw for the absorber. This difference is called the isomer shift δ and is given by eq. 3.

$$\delta = \text{constant} \times \delta\underline{R}\{|\Psi_s^A(o)|^2 - |\Psi_s^S(o)|^2\} \quad (3)$$

It is thus evident that measurement of the Mössbauer isomer shift will provide information about the electronic (chemical) environment in which the nucleus resides. The importance of δ for study of mixed-valence compounds is nicely illustrated by considering the range of δ values given in Table I for some commonly-encountered oxidation states of Mössbauer nuclei. For elements in which $\delta\underline{R}$ is positive, a decrease in the number of $\underline{p}$ or $\underline{d}$ electrons increases δ because of decreased shielding of the $\underline{s}$ electrons from the nucleus. Conversely, for ^{57}Fe, in which $\delta\underline{R}$ is negative, a decrease in 3$\underline{d}$ electron density results in a decrease in the isomer shift.

Quadrupole Interactions (Quadrupole Splitting)

Nuclear states with spin quantum number $\underline{I}$ greater than 1/2 possess non-zero quadrupole moments $\underline{Q}$ and therefore may be split

Table I

Average Isomer Shift Values of Selected Mössbauer Nuclides

Mössbauer Nuclide	Formal Oxidation State	Formal Electron Configuration	Average δ, mm/s
$^{57}Fe^{a}$	Fe^{2+}	$[Ar]4s^0 3d^6$	1.0
$(-\delta R)$	Fe^{3+}	$[Ar]4s^0 3d^5$	0.4
	Fe^{4+}	$[Ar]4s^0 3d^4$	0.0
	Fe^{6+}	$[Ar]4s^0 3d^2$	- 0.8
$^{99}Ru^{b}$	Ru^{2+}	$[Kr]5s^0 4d^6$	- 0.7
$(+\delta R)$	Ru^{3+}	$[Kr]5s^0 4d^5$	- 0.5
	Ru^{4+}	$[Kr]5s^0 4d^4$	- 0.3
	Ru^{6+}	$[Kr]5s^0 4d^2$	0.4
	Ru^{7+}	$[Kr]5s^0 4d^1$	0.8
	Ru^{8+}	$[Kr]5s^0 4d^0$	1.0
$^{119}Sn^{c}$	Sn^{2+}	$[Kr]4d^{10} 5s^2 5p^0$	> 2.9
$(+\delta R)$	Sn^{4+}	$[Kr]4d^{10} 5s^0 5p^0$	< 2.0
$^{151}Eu^{d}$	Eu^{2+}	$[Xe]6s^0 4f^7$	-13.0
$(+\delta R)$?	Eu^{3+}	$[Xe]6s^0 4f^6$	0.0

δ relative to: [a] Fe metal, [b] Ru metal, [c] SnO_2, and [d] Eu_2O_3.

by interaction with a finite efg. In any principal axis system the efg tensor is diagonal and traceless. If the three diagonal elements of the efg tensor are V_{xx}, V_{yy}, and V_{zz} and if $|V_{zz}| = eq$ is the largest element then $\hat{E}_2$ in eq. 1 is given by eq. 4,

$$\hat{E}_2 = e^2qQ[3\hat{I}_z^2 - I(I + 1) + \eta(\hat{I}_x^2 - \hat{I}_y^2)]/4I(2I - 1) \qquad (4)$$

where $\hat{I}_x$, $\hat{I}_y$, and $\hat{I}_z$ are spin operators and $\eta = (V_{xx} - V_{yy})/V_{zz}$, the asymmetry parameter, indicates the degree of asymmetry of the efg tensor. Thus $\eta = 0$ implies an axially-symmetric efg while $0 < \eta \leq 1$ is indicative of an efg with less than axial symmetry.

In materials containing, for example, ^{57}Fe, ^{119}Sn, ^{193}Ir, or ^{195}Pt ME transitions may be observed between nuclear spin levels 1/2 and 3/2 (written as 1/2↔3/2). For ^{99}Ru 3/2↔5/2 transitions are observed and for ^{151}Eu 5/2↔7/2 transitions are possible. The effect of quadrupole coupling with each of these level systems is shown in Figure 3 for a positive V_{zz}. If V_{zz} is negative both excited and ground state splittings are inverted. Numbers given in the circles which accompany each transition indicate the relative transition probability for randomally-oriented polycrystalline samples. In the case of 1/2↔3/2 quadrupole-split spectra the separation between individual transitions of the quadrupole doublet is referred to as the quadrupole splitting and is numerically equal to $1/2e^2qQ(1 + \eta^2/3)^{1/2}[\equiv 1/2eQV_{zz}(1 + \eta^2/3)^{1/2}]$. The quadrupole splitting is given the symbol Δ (some authors prefer ΔE_Q). Examination of the quadrupole spectra for the 3/2↔5/2 and 5/2↔7/2 systems allows a determination of both the sign and magnitude of V_{zz}. However, because the 1/2↔3/2 quadrupole splitting is symmetric, quadrupole-split ME spectra of these systems provide only the absolute value of Δ.

The quadrupole-split spectrum for a compound with a 1/2↔3/2 nucleus, $^{57}Fe(C_2O_4)(H_2O)_2$, is compared in Figure 4 with a more complicated quadrupole spectrum for $^{129}I_2$ which is a 5/2↔7/2 spin system.

Although it is very difficult to calculate Δ from principles of bonding theory, a knowledge of the magnitude, sign, and temperature dependence of Δ is important in any qualitative picture of the electronic structure and chemical bonding in the material of interest. This is particularly true for ^{57}Fe-containing materials for which an enormous number of ME spectra have been obtained

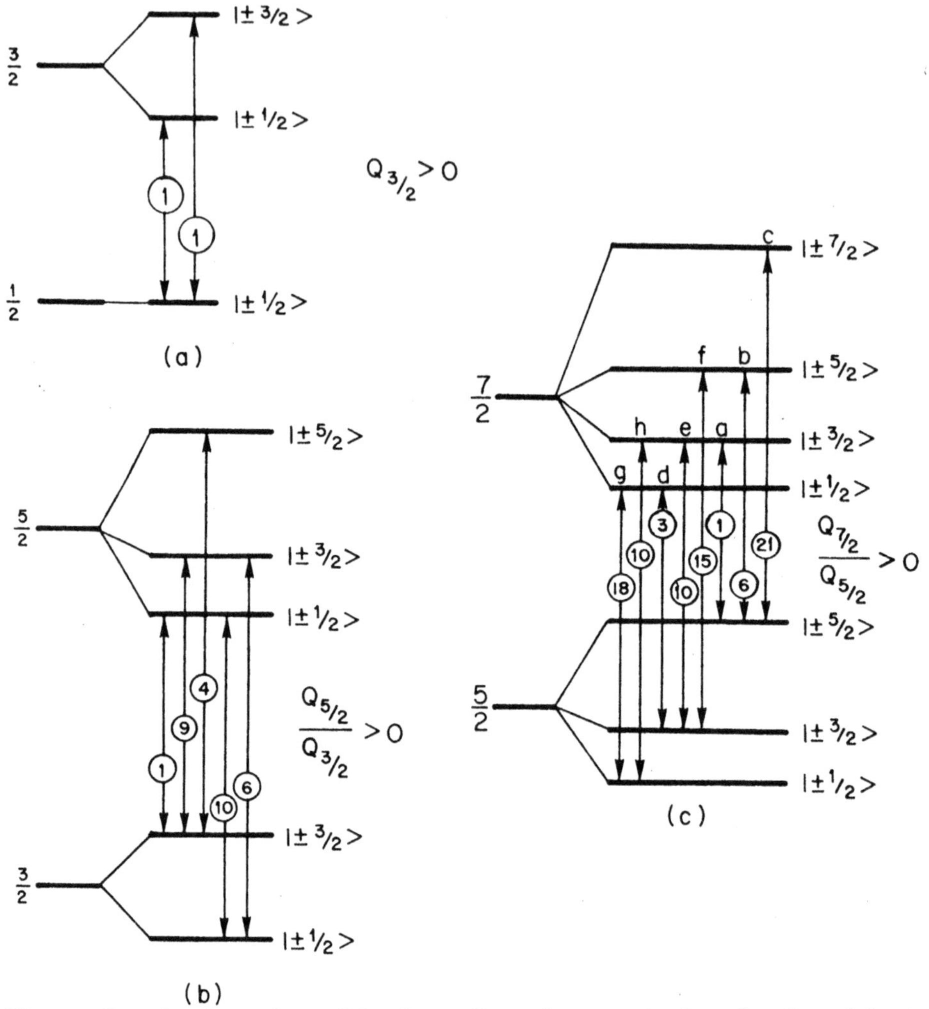

Figure 3. Quadrupole splitting of nuclear spin levels for (a) 1/2↔3/2, (b) 3/2↔5/2, and (c) 5/2↔7/2 transitions.

under a wide array of experimental conditions. For example, in the normally-encountered high-spin electronic configurations ($\underline{O}_h$ microsymmetry) of $Fe^{2+}(\underline{t}_{2g}^4\underline{e}_g^2)$ and $Fe^{3+}(\underline{t}_{2g}^3\underline{e}_g^2)$ it is always observed that the quadrupole splitting for compounds which contain the former ion is approximately 2-3 mm/s whereas the latter ion gives Δ near zero. A viable interpretation of this difference assumes that the additional $\underline{t}_{2g}$ electron in Fe^{2+} adds a very substantial non-spherical contribution to the efg as compared to the spherical

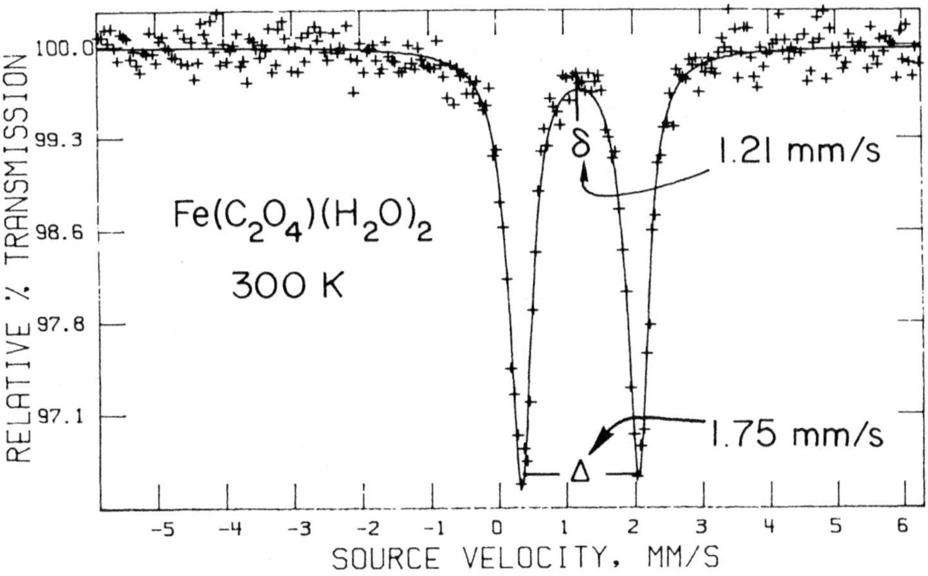

(a)

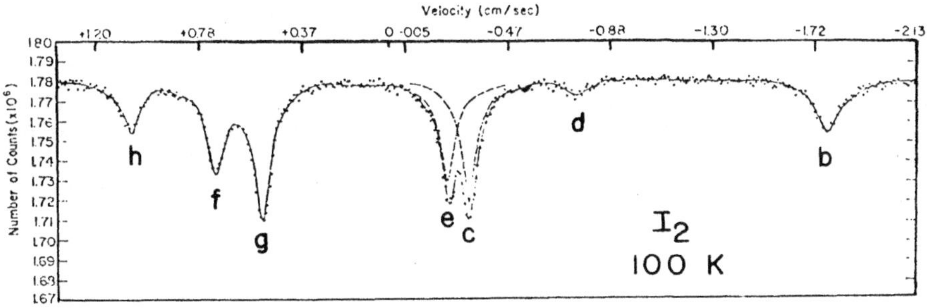

(b)

Figure 4. Quadrupole-split ME spectra for (a) $Fe(C_2O_4)(H_2O)_2$ relative to Fe metal [taken from Wrobleski, J. T., and Brown, D. B., 1979, Inorg. Chem., 18,] and (b) I_2 relative to $^{66}Zn^{129}Te$ source [taken from Pasternak, M., Simopoulos, A., and Hazony, Y., 1965, Phys. Rev., 140, p. A1892].

(half-filled) ferric configuration. By contrast, low-spin ferrous complexes ($\underline{t}_{2g}^{6}\underline{e}_{g}^{o}$) have $\Delta \sim 0.0$ mm/s, whereas low-spin ferric complexes ($\underline{t}_{2g}^{5}\underline{e}_{g}^{o}$) have $\Delta \sim 0.5$ mm/s. These arguments may be extended to a large number of other electronic configurations, not only for ^{57}Fe, but also for other elements.

Magnetic Interactions (Magnetic Hyperfine Splitting)

Magnetic perturbations of the nuclear energy levels in a Mössbauer nucleus may arise from magnetic fields applied externally to the sample, H_{ext}, and/or internal magnetic fields, H_{int}, which are present as a result of preferential electron spin alignment in the solid (magnetic ordering). Figure 5 depicts two spectra which illustrate the splitting caused by an internal magnetic field. The right side of this figure shows the spin level splitting which is obtained in the absence of quadrupole splitting effects. The result of this magnetic hyperfine splitting is the symmetrical ^{57}Fe-metal spectrum shown at the bottom of the figure. In the case of combined magnetic and quadrupole splittings, the resulting ME spectrum is no longer symmetrical. This asymmetry is inferred from the splitting diagram on the left side of Figure 5. The 4.2 K spectrum of FeF_2, illustrated at the top of the figure, shows the effect of combined magnetic and quadrupole hyperfine perturbations on the nuclear spin levels. It should be noted that above the Curie temperature of iron metal its ME spectrum consists of a single absorption which indicates $\Delta \sim 0$. Above the Neél temperature of FeF_2 $\Delta \sim +2.8$ mm/s.

The effect of H_{ext} on a quadrupole split $1/2 \leftrightarrow 3/2$ ME spectrum is illustrated in Figure 6. Application of a large external magnetic field provides a very useful method for determining the sign of Δ. Comparison of the right with the left side of the figure indicates that if Δ is positive (that is $|\pm 3/2\rangle$ lie above $|\pm 1/2\rangle$) then a spectrum of the type illustrated for $[Fe^{II}(H_2O)_6]SO_4$ is observed. If, however, Δ is negative then an essentially mirror image ME spectrum results, as illustrated for $[Fe^{II}(H_2O)_6]SiF_6$. One must remember that if $\eta \neq 0$ the magnetically-perturbed spectra are more complicated. The effect of $\eta > 0$ on these spectra is discussed in several of the general references.

APPLICATION OF ME SPECTROSCOPY TO MIXED-VALENCE COMPOUNDS

The type of ME spectra observed for mixed-valence compounds depends on the relative rate of Mössbauer nuclear excited state decay $\underline{k}_n$ compared to the rate of intervalence electron transfer $\underline{k}_e$. Three spectral regimes are experimentally encountered. In

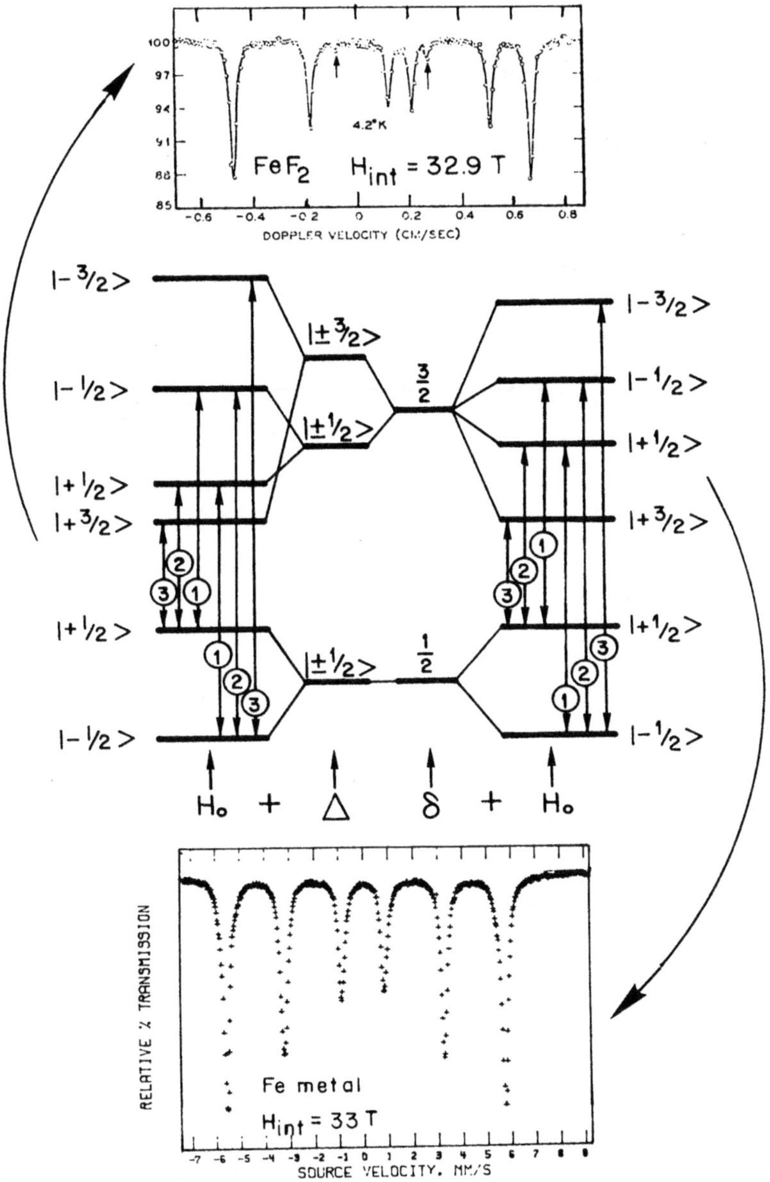

Figure 5. The effect of magnetic and magnetic + quadrupole hyperfine splitting on the 1/2 and 3/2 nuclear spin levels. The arrows on the spectrum of FeF_2 refer to two weak absorptions which arise from |-1/2> ↔ |+3/2> and |+1/2> ↔ |-3/2> transitions which are formally forbidden by the M1 selection rules [taken from Wertheim, G. K., and Buchanan, D. N. E., 1967, Phys. Rev., 161, p. 478].

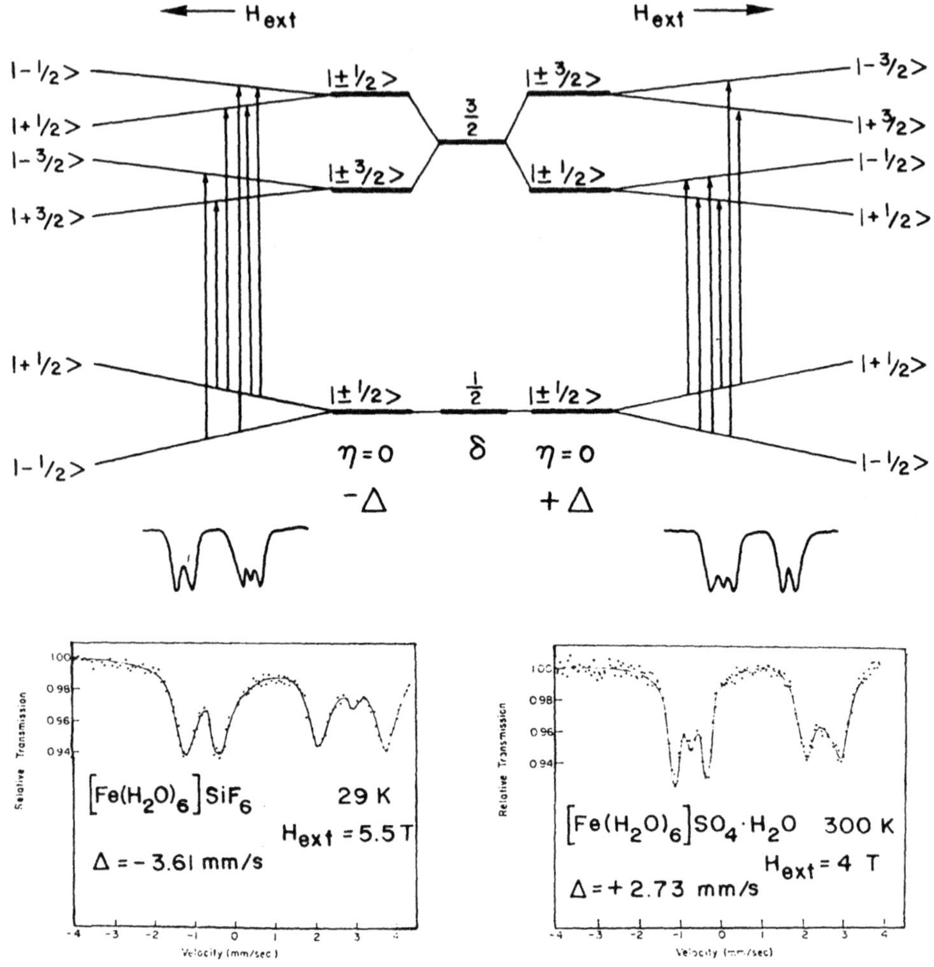

Figure 6. The effect of an applied magnetic field on the nuclear energy levels and ME spectra for a 1/2↔3/2 system with Δ positive and negative [taken from Grant, R. W., Wiedersich, H., Muir, A. N., Jr., Gonser, U., and Delgass, W. N., 1966, J. Chem. Phys., 45, p. 1015].

the first regime $\underline{k}_{\underline{e}} \ll \underline{k}_{\underline{n}}$, that is, the intervalence electron transfer rate is much slower than the nuclear excited state relaxation rate. Under this condition, discrete (deeply-trapped) oxidation states are observed on the ME time scale. In the second regime $\underline{k}_{\underline{e}} \gg \underline{k}_{\underline{n}}$, which implies that the rate of intervalence electron transfer greatly exceeds the nuclear excited state decay rate. If this criterion is met, the ME records a spectrum which is the weighted average of the individual oxidation state spectra. In the third regime $\underline{k}_e \simeq \underline{k}_n$ and ME spectra are observed which are intermediate between those of the first and second regimes. In the third regime ME spectra are very temperature sensitive in response to the exponential temperature dependence of $\underline{k}_e$.

It is obvious that information about the dynamic aspects of mixed valency is accessible from ME experiments provided a suitable range of temperature (or perhaps pressure) is found such that $\underline{k}_e \simeq \underline{k}_n$. Unfortunately many mixed-valence materials cannot withstand these experimental conditions or else the recoilless fraction of the particular nuclide under investigation is so small as to render the ME ineffective. Because of these factors only a relatively few mixed-valence compounds have provided the type of information about electron transfer rates in solids which is theoretically possible with the ME. Several of these systems, and compounds which show limiting behavior, are discussed in the following sections.

Compounds with $\underline{k}_e << \underline{k}_n$

Although compounds with $\underline{k}_e << \underline{k}_n$ fail to show mixed-valence interactions on the Mössbauer time scale, the ME does provide useful information about their structural and electronic properties. ME spectra of these materials are normally analyzed in terms of a simple superposition of the ME spectra of the individual oxidation site fragments. An example of this analysis is illustrated in Figure 7. Here the ME spectrum of mixed-valence $Fe_2F_5 \cdot 7H_2O$ (d) is compared with spectra of $[Fe(II)(H_2O)_6]SO_4 \cdot H_2O$ (a), $K_2[Fe(III)F_5(H_2O)]$ (b), and a 1:1 mixture of $[Fe(II)(H_2O)_6]SO_4 \cdot H_2O$ and $K_2[Fe(III)F_5(H_2O)]$ (c). The similarity of (c) and (d) strongly suggests that $Fe_2F_5 \cdot 7H_2O$ be formulated as $[Fe(II)(H_2O)_6][Fe(III)F_5(H_2O)]$ (1).

In the spectrum of $Fe_2F_5 \cdot 7H_2O$ (Figure 7d) the area ratio of the Fe(III) to the Fe(II) quadrupole doublet, $\underline{A}(Fe^{3+})/\underline{A}(Fe^{2+})$, is 1.0. This ratio is consistent with the ratio of ferric to ferrous ion determined by chemical analysis. Because the recoilless fraction may vary from site to site in a mixed-valence material, the areas of various ME absorptions in these compounds are not necessarily a direct measure of the relative amounts of the different oxidation states. Nonetheless, as in $Fe_2F_5 \cdot 7H_2O$, it is frequently observed that sites with different oxidation states in similar chemical environments do have comparable recoilless fractions, and therefore site populations may be at least semi-quantitatively related to ME area ratios.

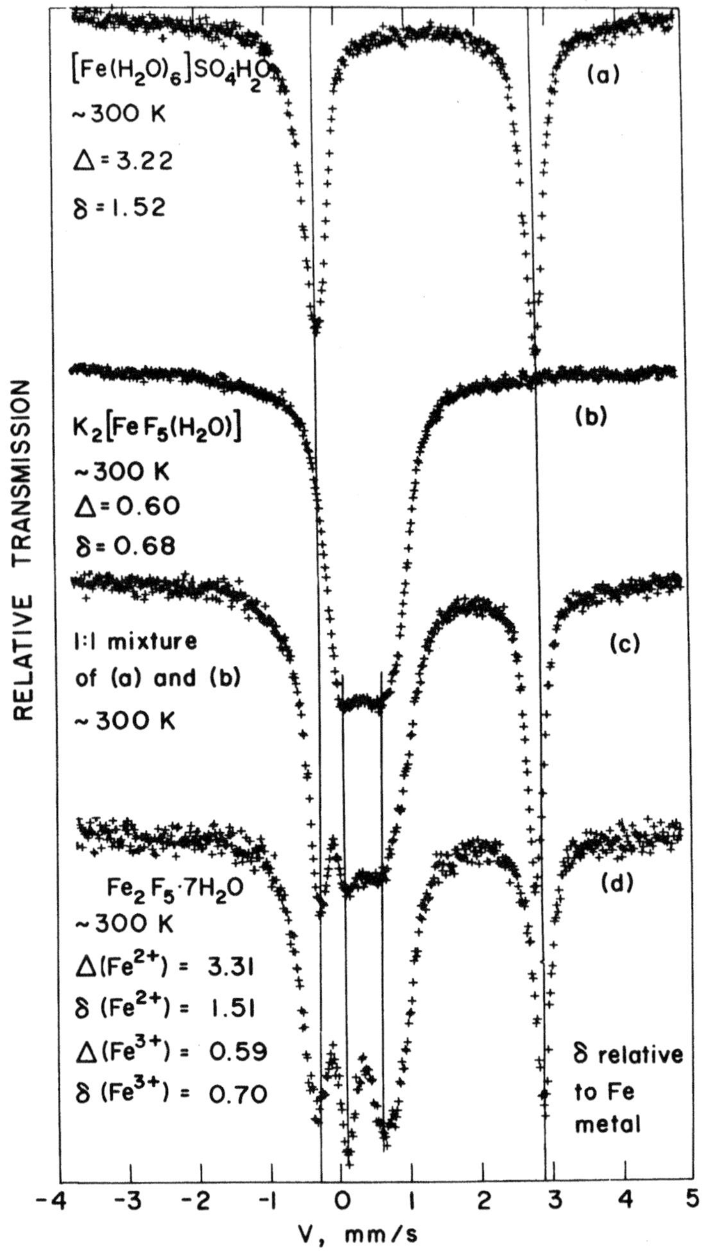

Figure 7. Room temperature ^{57}Fe Mössbauer spectra of (a) $[Fe(H_2O)_6]SO_4 \cdot H_2O$, (b) $K_2[FeF_5(H_2O)]$, (c) 1:1 mixture of (a) and (b), and (d) $Fe_2F_5 \cdot 7H_2O$.

A somewhat similar analysis has been extremely useful in clarifying the nature of the Prussian Blues (2,3). By using the ME, it has been possible to show that Prussian Blue and Turnbull's Blue (prepared by eqs. 5 and 6, respectively) are in fact identical, and that both contain low-spin Fe(II) (C_6 coordination) and

$$Fe_2(SO_4)_3 + K_4Fe(CN)_6 \rightarrow Fe(III)_4[Fe(II)(CN)_6]_3 \quad (5)$$

$$FeCl_2 + K_3Fe(CN)_6 \rightarrow Fe(III)_4[Fe(II)(CN)_6]_3 \quad (6)$$

high spin Fe(III) (N_6 or N_4O_2 coordination). The ME spectral overlap due to the presence of both types of iron site makes interpretation difficult, but the problem is simplified by selective isotopic substitution. Naturally-occurring iron contains only 2.17% ^{57}Fe, the ME active nuclide. By labelling one of the reactants in eqs. 5 or 6 with either ^{57}Fe or ^{56}Fe, materials can be prepared whose ME spectra show either selective enhancement or disappearance of absorption due to a particular site. This substitution allows not only identification of spectral components, but in the case of eq. 6 demonstrates that the product forms as the result of electron transfer, rather than ligand transfer. Of course, the use of isotopic substitution in this fashion is dependent on the iron sites maintaining their uniqueness throughout the reaction.

Several Ru-containing mixed-valence compounds have been studied with ^{99}Ru ME spectroscopy. Because of the low recoilless fraction of ^{99}Ru (γ_M = 89.36 keV) above 4.2 K, ME spectra are difficult to obtain at temperatures where thermally-activated electron-transfer processes are likely to occur. Rather, low-temperature ME spectra have been used to discriminate between the limiting cases of Robin and Day Class II and Class III behavior (4) (or of trapped versus delocalized electrons). ME spectra of representative Ru mixed-valence compounds are illustrated in Figure 8. From the earlier discussion (see Figure 3) the ^{99}Ru ME spectrum of a single type of Ru is expected to consist of five absorptions because ^{99}Ru has ground- and excited-state nuclear spins of 5/2 and 3/2, respectively. In fact, two line quadrupole split spectra are normally observed for ^{99}Ru compounds. This two line splitting pattern arises because the nuclear quadrupole moment ratio $\underline{Q}(3/2)/\underline{Q}(5/2)$ is very large for ^{99}Ru. Therefore only transitions from the nearly degenerate $\underline{m}_I$ = ±5/2, ±3/2, and ±1/2 components of the $\underline{I}$ = 5/2 ground state to the strongly quadrupole-split $\underline{m}_I$ = ±3/2 and ±1/2 components of the $\underline{I}$ = 3/2 excited state are observed.

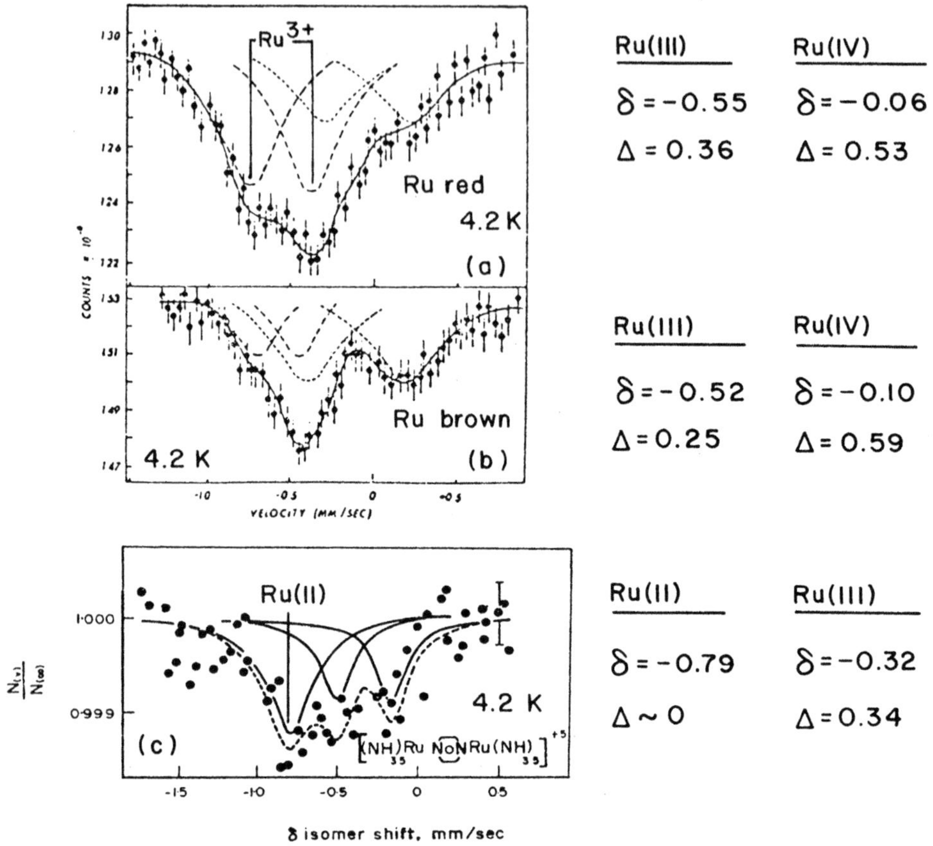

Figure 8. 4.2 K ^{99}Ru Mössbauer spectra of (a) Ruthenium Red, (b) Ruthenium Brown, and (c) Creutz and Taube ion, all of which are trapped-valence, Class II compounds.

Isomer shifts for the three illustrated oxidation states increase in the order Ru(IV) > Ru(III) > Ru(II) as anticipated on the basis of increased s-electron shielding by valence d electrons in this series (Table I). In the case of Ruthenium Red, $[(NH_3)_5Ru(III)ORu(IV)(NH_3)_4ORu(III)(NH_3)_5]^{6+}$ [Figure 8, spectrum (a)], and its one-electron oxidation product Ruthenium Brown, $[(NH_3)_5Ru(IV)ORu(III)(NH_3)_4ORu(IV)(NH_3)_5]^{7+}$ [spectrum (b)], the Ru(III)/Ru(IV) area ratios are 2/1 and 1/2, respectively (5). Figure 8 (c) shows the ^{99}Ru ME spectrum of the Creutz and Taube ion (6),

This spectrum is clearly consistent with the presence of trapped Ru(II) and Ru(III) valences.

This example points out the clear utility of ME spectroscopy as a probe for the extent of electron delocalization. For a totally-delocalized (Class III) material, all valence sites must be equivalent under any conditions of measurement. By contrast, valence sites which are trapped (Class II) but which may be inter-changed by electron transfer may appear to be equivalent or inequivalent depending upon the relative rates of the electron transfer and the measuring technique. The literature contains a number of conflicting claims concerning the Creutz and Taube ion, but the ME evidence is unambiguous in demonstrating that this ion is a Class II material. Because of the exponential temperature dependence of the (phonon-assisted) electron-transfer process, at temperatures much greater than 4.2 K even very fast measurement techniques may fail to detect trapped valence states.

Several other compounds in which $k_e << k_n$ have been investigated by ME spectroscopy. Among these are materials of the type A_2SbX_6 in which A is a monovalent cation and X is a halide. These compounds were previously thought to be rare examples of Sb(IV) compounds. Original examination of ^{121}Sb ME spectrum of Cs_2SbCl_6 indicated that discrete Sb(III) (δ = -19.0 mm/s rel. to $Ba^{121}SnO_3$) and Sb(V) (δ = -3.0 mm/s) sites were present at 4.2 K. At higher temperatures additional absorptions with intermediate δ values were observed (7). These additional absorptions have not been found in more recent work (8), and on the basis of ME and infrared results these compounds are presently thought to have the composition $A_4\{[Sb(III)X_6][Sb(V)X_6]\}$.

A mixed-valence linear chain polymer, $Fe(II,III)(C_2O_4)(H_2O)_{1.4}Br_{0.6}$, has recently been studied by zero-field ^{57}Fe ME spectroscopy (9). In the temperature range 30-400 K the spectrum consists of superimposed Fe(II) and Fe(III) quadrupole doublets. At 30 K the resonant absorptions begin to broaden and below ca. 25 K a large number of lines appear in the spectrum. This complicated low-temperature spectrum results from the superposition of magnetic hyperfine absorptions for the Fe(II) and Fe(III) sites in this magnetically-ordered material.

mixed-valence Sn compound, octakis-μ-(o-nitrobenzoato)-di-μ$_3$-[illegible]-bis(tetrahydrofuran)ditin(II)ditin(IV), has been studied in the temperature range 77–295 K by using ^{119}Sn ME spectroscopy (10). The spectrum at 295 K consists of three resonant absorptions, two of which are derived from quadrupole interaction with the Sn(II) site (δ = 3.906 and Δ = 2.133 mm/s) and the other from the Sn(IV) site (δ = -0.021 mm/s).

Temperature-dependent ME spectra of mixed-valence meso-tetraferrocenylporphyrin, whose structure is shown in Figure 9(a), have recently been investigated (11). Spectra taken at 295, 90, and 4.2 K are illustrated in Figure 9(b). Although these spectra are not characteristic of oxidation state averaging, there is a very strong temperature dependence to the Fe(II)/Fe(III) line intensity

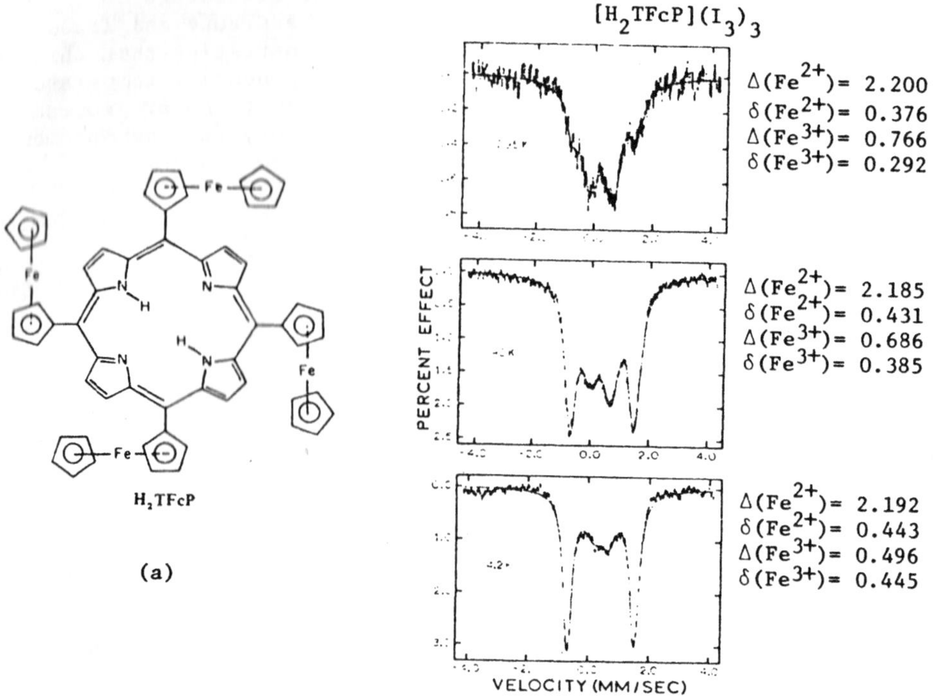

(a) (b)

Figure 9. (a) Structure of Fe(III)$_3$Fe(II)-tetraferrocenylporphyrin. (b) ^{57}Fe ME spectra of this cation at 295, 90, and 4.2 K.

ratio. The authors (11) attribute this temperature dependence to a temperature-dependent intramolecular redox process in which electrons are transferred from ferricenium-like groups to the porphyrin ring. An alternative explanation for these observations was discussed by the authors, but ultimately rejected by them. As discussed above, the areas of ME absorptions are not necessarily a direct reflection of individual valence populations. Because the recoilless fraction may have a different temperature dependence for each valence site in a mixed-valence material, the area variations of these spectra may simply reflect such differences.

Many naturally-occurring substances, including a large number of minerals, are mixed-valence iron compounds in which intervalence electron exchange is slow. Because of the wide distribution of these materials, the ME has become a powerful tool in geology, the analysis of art objects, and archaeology. As only one of many examples, the ME has been used to study two different forms of pottery shards, one red and the other gray (12). Both of these types of shards were found in Iran and dated at approximately 3000 B.C. ME spectra of these shards were compared with ME spectra of local clays fired at various temperatures and the relative amounts of Fe^{3+} and Fe^{2+} in the samples were determined. These investigations led to the conclusion that the red and gray pottery shards both had a local origin but differed in their firing temperature and in the use of oxidizing or reducing atmospheres in the firing process.

Compounds with $\underline{k}_e \gg \underline{k}_n$

Mixed-valence materials in which the electron transfer rate is much larger than the Mössbauer nuclear excited state relaxation rate are expected to possess intermediate ME isomer shifts and quadrupole splittings which are weighted averages of δ for the isolated oxidation states. (This averaging process is only an approximation, however, because of the undetermined effect of covalency on the isomer shift of the mixed-valence state. Thus one would anticipate that the degree of covalency in the mixed-valence state might be different than the degree of covalency in the isolated oxidation states.) The observation in the ME spectrum of a single type of absorption for a material which is formally mixed valent cannot by itself establish the material as belonging to Class III although it can set a lower limit on the rate of an electron transfer process. As discussed in the following section, Class II materials in which electron exchange is rapid on the Mössbauer time scale exhibit averaged spectra of this type, but at some temperature the distinct oxidation states become distinguishable. If the barrier to electron transfer is zero ($\underline{k}_e = \infty$, Robin and Day Class III behavior) then there is no temperature at which oxidation state ordering is observed. Although there are very few examples of Class III mixed-valence compounds which contain Mössbauer-active nuclei, this situation apparently obtains for the mixed-valence material SmB_6 in which the ^{149}Sm isomer shift (-0.38 mm/s at 1 K) is the weighted average of the isomer shift of Sm^{2+} and Sm^{3+} compounds [-0.92 (SmF_2) and -0.1 (SmF_3) mm/s, respectively] (13).

Another example of a material in which $\underline{k}_e$ >> $\underline{k}_n$ is the cubic phase of Fe_3O_4. This inverse spinel Fe(II,III) oxide phase which exists above 120 K gives rise to a twelve line magnetically-ordered ME spectrum (14). One six line hyperfine pattern arises from trapped Fe^{3+} ions in the tetrahedral sites while the other six line pattern is due to oxidation state averaged iron ions in octahedral sites. Because Fe_3O_4 undergoes a structural phase transition at 120 K it is not possible to determine if the Class III behavior of the cubic phase persists to very low temperatures.

Perhaps the most striking examples of non-integral oxidation state materials, and certainly the most extensively studied, are compounds related to biferrocene, I. Although isomer shifts for

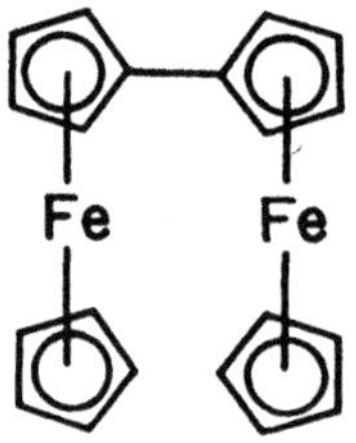

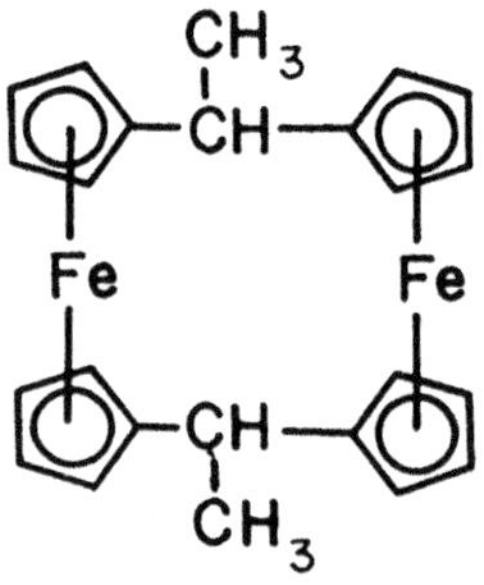

species similar to ferrocene and ferricenium ion are virtually identical, the large difference in quadrupole splittings (ferrocene Δ = 2.4 mm/s and ferricenium ion $\Delta \cong$ 0-0.8 mm/s) permits facile identification of Fe^{2+} and Fe^{3+} sites in similar compounds. Thus, monooxidized salts of I are shown by ME spectroscopy to contain trapped valence states with $\Delta(Fe^{2+})$ = 2.14 mm/s and $\Delta(Fe^{3+})$ = 0.288 mm/s (15). Trapped valences are also observed for [1.1] ferrocenophanes such as II (16). However, the monooxidized salts of both biferrocenylene, III, and [2.2]ferrocenophane-1,13-diyne, IV, exhibit intermediate (averaged) oxidation state ME spectra (16, 17). Thus, at 78 K salts of the monocations of III and IV give ME

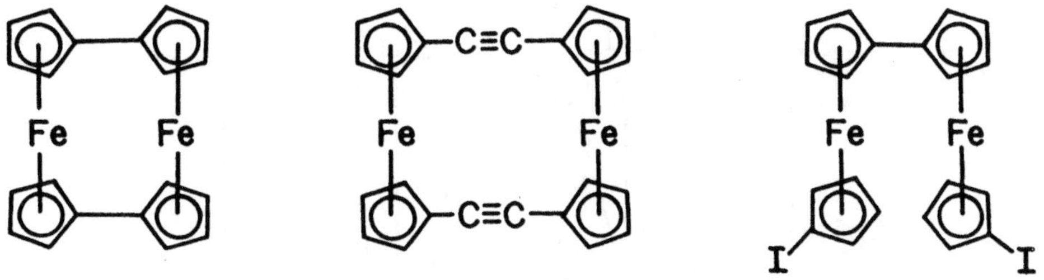

spectra which consist of single quadrupole doublets with Δ = 1.78 and 1.61 mm/s, respectively. These Δ values are approximately the averages of those obtained for the individual Fe^{2+} and Fe^{3+} complexes. This averaging of Δ values indicates equivalence of all Fe sites on the Mössbauer time scale. Because in each case the sites remain averaged at 4.2 K, the monocations of III and IV are clearly Class III compounds. Surprisingly, the monooxidized I_3^- salt of 1',6'-diiodobiferrocene, V, which lacks the rigid, conjugated structural features of III and IV, also exhibits valence state delocalization. The coupling mechanisms by which ion sites in III-V are rendered equivalent are not at present understood.

Compounds with $\underline{k}_e \sim \underline{k}_n$

Compounds in which the rate of nuclear excited state decay and the rate of intervalence electron transfer are comparable are probably the most interesting cases for study by ME techniques. In principle it is possible to extract rates of electron transfer, and consequently energy barriers to electron transfer, from a study of the temperature dependence of ME spectra of these materials. In these cases the ME spectra vary in a predictable fashion with the rate of electronic relaxation because $\underline{k}_n$ is essentially independent of chemical environment and temperature.

Possibly the most thoroughly understood mixed-valence material in which $\underline{k}_e \sim \underline{k}_n$ is Eu_3S_4, $Eu(III)_2Eu(II)S_4$. In this compound all Eu ions occupy crystallographically-equivalent sites at room temperature. Temperature-dependent $^{151}Eu(\tau_n = 8.8$ ns) ME spectra of Eu_3S_4 are illustrated in Figure 10 (18). At 85 K the spectrum consists of two absorptions for which $\delta \cong -12.0$ mm/s (Eu^{2+} site) and $\delta \cong 0.0$ mm/s (Eu^{3+} site) in the approximate ratio 1:2, respectively. The spectrum of Eu_3S_4 at 4.2 K (19) is essentially unchanged. (No quadrupole splitting of either site is observed because the quadrupole interaction is considerably less than the

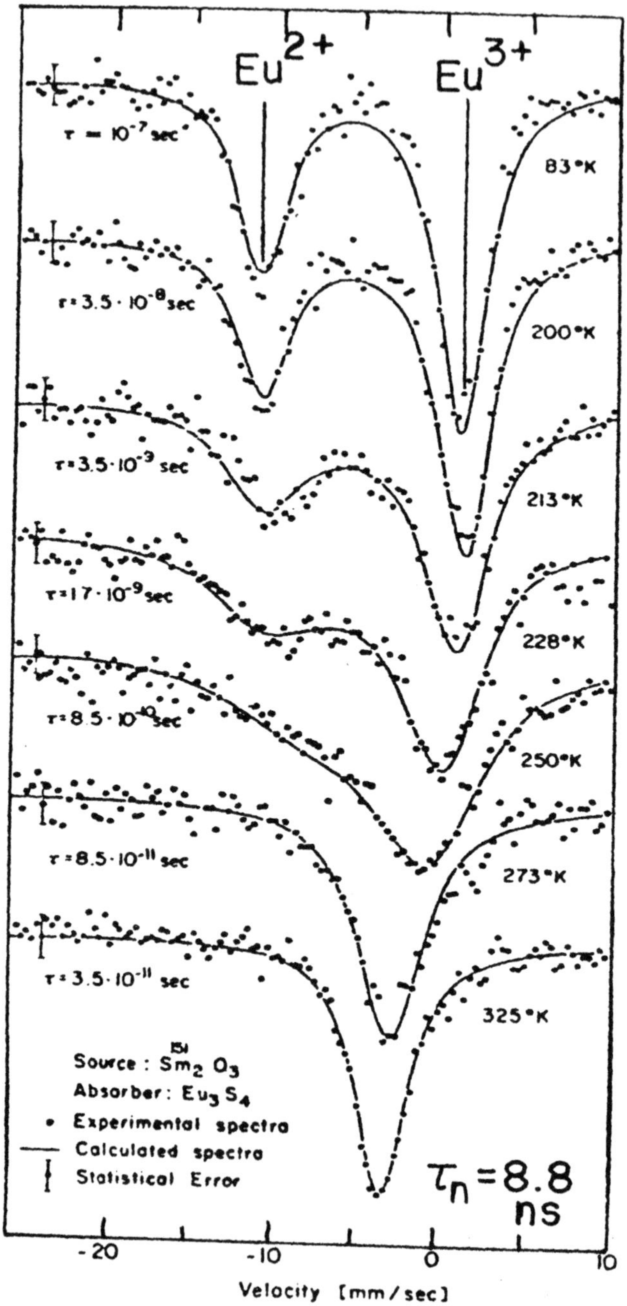

Figure 10. Temperature variation of the ^{151}Eu ME spectrum of Eu_3S_4 and fits for the relaxation model.

width of γ_M.) As the temperature is increased the resonant absorptions broaden and approach one another. At 325 K the spectrum consists of an unbroadened resonant absorption at $\delta \cong -4$ mm/s. This high-temperature spectrum arises from the oxidation state averaged Eu sites. The high temperature isomer shift is exactly equal to the weighted average (eq. 7) of the Eu^{2+} and Eu^{3+} isomer shifts.

$$\delta_{average} = [2\delta(Eu^{3+}) + \delta(Eu^{2+})]/3 \qquad (7)$$

In order to determine the thermal barrier to electron transfer in Eu_3S_4 from ME data, the electron relaxation time, $\tau_e(\equiv 1/\underline{k}_e)$, was obtained by fitting experimental spectra to the semi-quantum mechanical rate equations of Wickman (20). Resulting relaxation times are given in Figure 10. The electron relaxation times may be fit to an Arrhenius equation (eq. 8) where $\underline{E}$ is the intervalence

$$\tau_e(T) = \tau_o \exp(\underline{E}/\underline{k}_B\underline{T}) \qquad (8)$$

electron transfer energy barrier and $\underline{k}_B$ is Boltzmann's constant. A plot of log τ_e versus $1/\underline{T}$ for Eu_3S_4 gives $\underline{E} = 0.24 \pm 0.03$ eV and $\tau_o = 25 \times 10^{-14}$ s. This value of $\underline{E}$ is identical to the band gap energy for Eu_3S_4 determined by conductivity measurements. This congruence of experimental results demonstrates that the electronic relaxation phenomenon observed with ME spectroscopy is a consequence of an intervalence electron transfer process.

Electron hopping in the mixed-valence compound $[Fe(II)Fe(III)_2O(CH_3COO)_6(H_2O)_3]$ has recently been investigated by ME spectroscopy (21). The structure of this compound is shown below and several representative ME spectra in the range 17-300 K are illustrated in Figure 11. At 17 K the spectrum consists of quadrupole

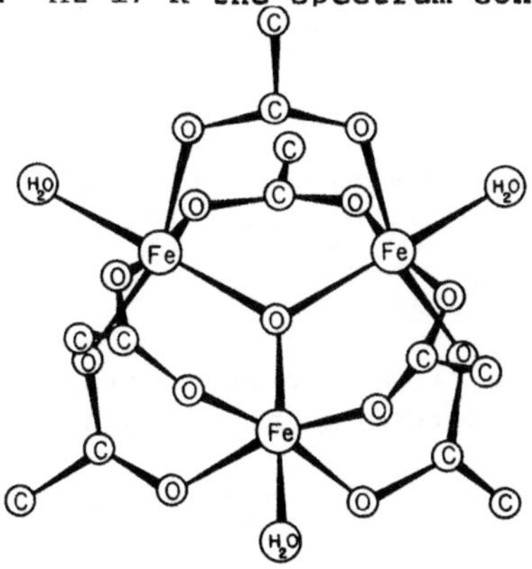

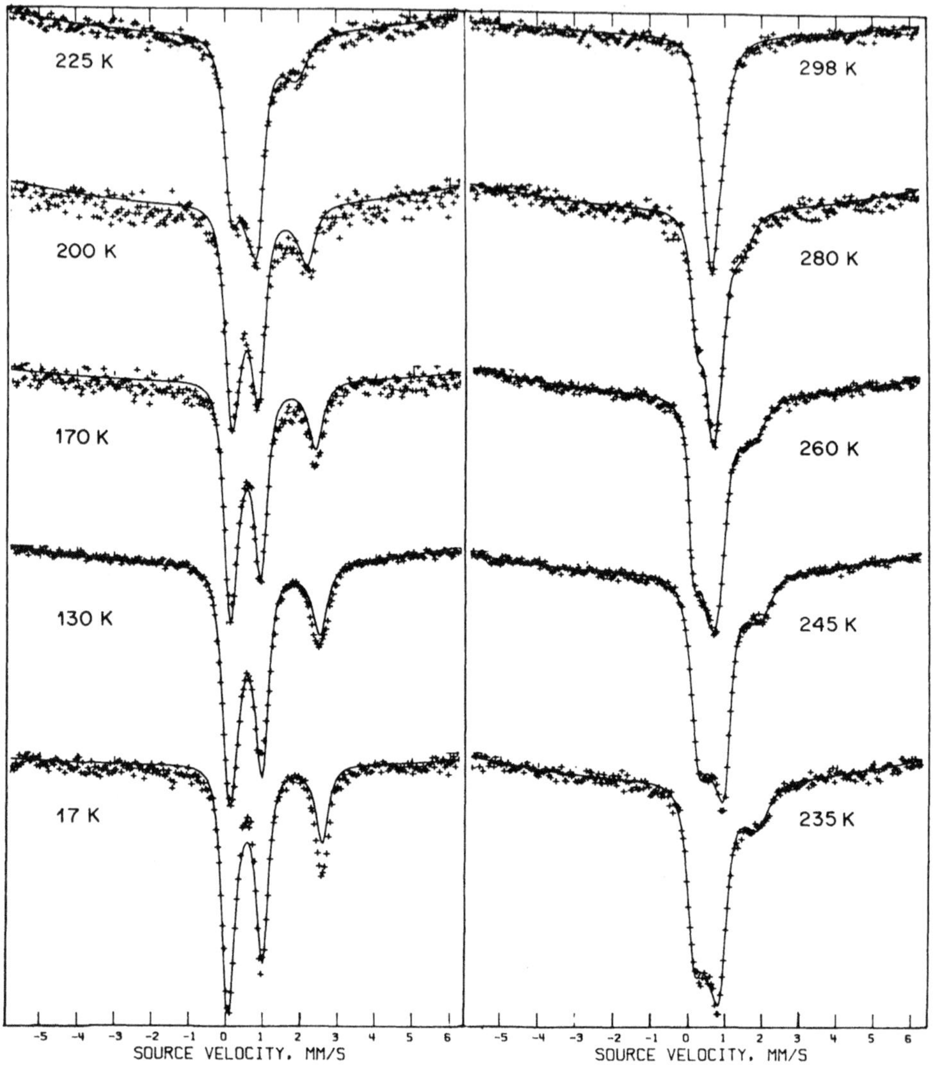

Figure 11. Temperature variation of the ^{57}Fe ME spectrum of $[Fe(III)_2Fe(II)O(CH_3COO)_6(H_2O)_3]$. The solid lines represent fits to a relaxation model (21).

doublets which arise from discrete Fe(II) and Fe(III) sites. As the temperature is increased the sites become indistinguishable on the ^{57}Fe ME time scale (τ_n = 98 ns) until finally at 295 K a single resonance with δ = 0.75 mm/s is obtained. It is interesting to note that the quadrupole splitting "averages" to zero in the delocalized spectrum of this complex. This observation is consistent with a negative $\Delta Fe(II)/\Delta Fe(III)$ ratio (21). If a modified relaxation model based on the equations of Wickman (20) is applied to these data (21), the calculated spectra shown in Figure 11 are obtained. Relaxation times range from >2400 ns at 17 K to 65 ns at 225 K to 5 ns at 298 K. A plot of log τ_e vs. $1/\underline{T}$ yields an electron hopping barrier of 470 cm^{-1}. This value of $\underline{E}$ is consistent with a phonon-assisted electron transfer process in that infrared spectral studies (21) indicate that the asymmetric Fe_3O stretching frequency of this compound is near 520 cm^{-1}. Although the symmetric Fe_3O stretching absorption has not been observed, it is predicted to lie in the infrared at a frequency near $\underline{E}$ determined from ME studies. It is most likely this low-frequency Fe_3O stretching mode which is responsible for intervalence electron transfer in this trimer.

In order to properly interpret electron hopping rates determined from ME measurements it is necessary to know the crystallographic distribution of oxidation sites. Thus in both Eu_3S_4 and $[Fe_3O(CH_3COO)_6(H_2O)_3]$ all the metal sites are equivalent both on the ME and X-ray crystallography time scales. In some instances not all of the metal centers are oxidation-state averaged on the ME time scale. The problem of deciding among which crystallographic sites electron hopping is occurring is particularly severe in the case of mixed-valence minerals as was recently shown for hematite-ilmenite (22). In the case of the silicate mineral ilvaite, $CaFe_2^{2+}Fe^{3+}Si_2O_8(OH)$, it is apparently necessary to include five quadrupole doublets in an analysis of the temperature dependent ME spectra of this material (23). In addition, the details of these spectra depend to a large extent on the geological history of the mineral specimens. Several other mixed-valence, iron-containing minerals, which have not been investigated in detail, appear to undergo thermal intervalence electron hopping with rates on the order of the ME time scale. These minerals include pentlandite $[(Ni,Fe)_9S_8]$ (24) and wustite $(Fe_{\sim 0.9}O)$ (25).

Although the ME has the potential for probing directly the dynamics of intervalence electron transfer processes in materials with $\underline{k}_e \sim \underline{k}_n$, practical considerations limit its use. In particular, the ability to perform ME measurements over a wide temperature range, and hence a range of electron transfer rates, is limited to a few nuclei including iron and europium. Thus ruthenium complexes, such as the Creutz and Taube ion, can be examined by ME only near 4 K, and therefore the rapid intervalence electron transfer which is known to occur at higher temepratures cannot be observed in ME experiments. Because of the exponential temperature dependence of the electron transfer process, many iron compounds might be expected to fulfill the condition $k_e \sim k_n$ at elevated temperatures. Although ^{57}Fe ME spectra can be obtained at elevated temperatures (>300 K), the thermal stability of the materials often is the limiting factor under these conditions.

CONCLUSIONS

ME spectroscopy has proven to be an extremely useful tool for the study of both static and dynamic features of the mixed-valence state. Although a wide range of mixed-valence materials have been studied by ME spectroscopy, a large number of interesting systems remain uninvestigated. In addition, a good deal of fundamental theoretical work is required to fully understand the effects on ME spectra of electron transfer in mixed-valence solids. The unique ability of the ME to directly probe oxidation states insure its continuing use in the study of mixed-valence materials.

ACKNOWLEDGEMENT

Our work in the area of mixed-valence chemistry has been supported in part by the Office of Naval Research.

GENERAL REFERENCES ON THE MÖSSBAUER EFFECT

A. N. N. Greenwood and T. C. Gibb, "Mössbauer Spectrocopy," Chapman and Hall, London (1971).

B. T. C. Gibb, "Principles of Mössbauer Spectroscopy," Chapman and Hall, London (1976).

C. V. I. Goldanskii and E. F. Makarov, "Fundamentals of Gamma-Resonance Spectroscopy," Chapter 1 in "Chemical Applications of Mössbauer Spectroscopy," V. I. Goldanskii and R. H. Herber, Eds.; Academic Press, New York (1968).

D. R. L. Cohen, "Elements of Mössbauer Spectroscopy," Chapter 1 in "Applications of Mössbauer Spectroscopy," Vol. 1, R. L. Cohen, Ed.; Academic Press, New York (1976).

E. G. M. Bancroft, "Mössbauer Spectroscopy An Introduction for Inorganic Chemists and Geochemists," John Wiley and Sons, New York (1973).

SPECIAL REFERENCES

1. Walton, E. G., Corvan, P. J., Brown, D. B., and Day, P.: 1976, Inorg. Chem. 15, p. 1737.

2. Maer, K., Jr., Beasley, M. L., Collins, R. L., and Milligan, W. O.: 1968, J. Am. Chem. Soc. 90, p. 3201.

3. Bonnette, A. K., Jr., and Allen, J. F.: 1971, Inorg. Chem. 10, p. 1613.

4. Robin, M. B., and Day, P.: 1967, Adv. Inorg. Chem. Radiochem. 10, p. 247.

5. Clausen, C. A., III, Prados, R. A., and Good, M. L.: 1971, Inorg. Nucl. Chem. Lett. 7, p. 485.

6. Creutz, C., Good, M. L., and Chandra, S.: 1973, Inorg. Nucl. Chem. Lett. 9, p. 171.

7. Aleksandrov, A. Y., Ionov, S. P., Baltrunas, D. A., and Makarov, E. F.: 1972, Soviet Phys. JETP Lett. 16, p. 147.

8. Longworth, G., and Day, P.: 1976, Inorg. Nucl. Chem. Lett. 12, p. 451.

9. Wrobleski, J. T., and Brown, D. B.: 1979, Inorg. Chem. 18, p. 0000.

10. Ewings, P. F. R., Harrison, P. G., Morris, A., and King, T. J.: 1976, J. Chem. Soc. Dalton Trans., p. 1602.

11. Wollmann, R. G., and Hendrickson, D. N.: 1977, Inorg. Chem. 16, p. 3079.

12. Bouchez, R., Coey, J. M. D., Coussement, R., Schmidt, K. P., Van Rossum, M., Aprahamian, J., and Deshayes, J.: 1974, Proc. Int. Conf. Appl. Mössbauer Effect, Bendor, 1974, J. Phys. Colloq. 35, C6, p. 541.

13. Varma, C. M.: 1978, Rev. Mod. Phys. 48, p. 219.

14. Banerjee, S. K., O'Reilly, W., and Johnson, C. E.: 1967, J. Appl. Phys. 38, p. 1289.

15. Cowan, D. O., LeVanda, C., Park, J., and Kaufman, F.: 1973, Acc. Chem. Res. 6, p. 1.

16. Morrison, W. H., Jr., and Hendrickson, D. N.: 1975, Inorg. Chem. 14, p. 2331.

17. Motoyama, I., Watanabe, M., and Sano, H.: 1978, Chem. Lett., p. 513.

18. Berkooz, O., Malamud, M., and Shtrikman, S.: 1968, Sol. State Comm. 6, p. 185.

19. Görlich, E., Hyrnkiewicz, H. U., Kmiec, R., Latka, K., and Tomola, K.: 1974, Phys. Stat. Sol. (b) 64, p. K147.

20. Wickman, H. H.: 1966, Mössbauer Effect Methodology 2, p. 39. Wickman, H. H., Klein, M. P., and Shirley, D. A.: 1966, Phys. Rev. 152, p. 345.

21. Dziobkowski, C. T., Wrobleski, J. T., and Brown, D. B.: Inorg. Chem., submitted for publication.

22. Warner, B. N., Shive, P. N., Allen, A. L., and Terry, C. J.: 1972, Geomag. Geoelectr. 24, p. 353.

23. Nolet, D. A.: 1978, Sol. State Comm. 28, p. 719. A three quadrupole doublet analysis of ilvaite has been presented: Heilmann, I. U., Olsen, N. B., and Olsen, J. S.: 1977, Phys. Scripta 15, p. 285.

24. Vaughan, D. J., and Ridout, M. S.: 1971, J. Inorg. Nucl. Chem. 33, p. 741.

25. Elias, D. J., and Linnett, J. W.: 1969, Trans. Farad. Soc. 65, p. 2673.

ELECTRONIC, RAMAN, AND RESONANCE-RAMAN SPECTROSCOPIC STUDIES OF MIXED-VALENCE COMPLEXES

Robin J.H. Clark

Christopher Ingold Laboratories
University College London
20 Gordon St., London WC1H 0AJ, U.K.

Resonance Raman spectroscopy is shown to be a strikingly effective technique with which to obtain information on the vibrational and electronic properties of class II and class III mixed-valence complexes and of the nature of the structural changes consequent upon the change by a molecule from the ground to the mixed-valence excited state.

INTRODUCTION

A Raman spectrum recorded using exciting radiation whose wavenumber corresponds with that of an electronic transition of a molecule is known as a resonance Raman (RR) spectrum. Such a spectrum differs from a normal Raman spectrum in that the intensities of certain bands are enhanced in the resonance case (sometimes by several orders of magnitude), and overtone and combination tone progressions may also be observed. Clearly, therefore, there is a vast amount of vibrational information made available by this technique, information which would (in the absence of the band enhancement feature of the technique) remain inaccessible. Thus the technique is of particular importance for the study of the vibrational properties of species in low concentrations e.g. haemoglobin in blood, pollutant gases in the atmosphere, radical species (both stable as well as short-lived). However, the technique is of real value on a wider front than is implicit from these considerations alone, in that it is found that those vibrations which are 'active' in the electronic transition, i.e. provide intensity in an absorption band, are frequently also the ones which give rise to the enhanced bands and progressions in a RR spectrum. Therefore,

D.B. Brown (ed.), Mixed-Valence Compounds, 271-292.

since the widths of Raman bands are normally very much less than those of electronic absorption bands, it is possible, by monitoring the changes in the Raman spectrum as the wavenumber of the exciting radiation is changed, to deduce the modes which are active in the resonant electronic transition. It is also possible to deduce the polarization of the resonant electronic transition of a scattering molecule from measurements of the depolarization ratios of Raman bands at resonance. These additional features of the technique render it of major importance for the study of excited states of molecules in general and, in the context of mixed-valence complexes, of probing the nature of the mixed-valence excited states of such molecules.

A variety of reviews has appeared in the last few years covering both experimental and theoretical aspects of the technique e.g. see refs. (1-7). The essence of the theory is as follows. The intensity of a Raman transition, $I(\nu_R)$, averaged over all molecular orientations, can be expressed in terms of the Raman scattering tensor, $\alpha_{\rho\sigma}$, by

$$I(\nu_R) = \frac{2^7\pi^5}{3^2 c^4} \nu_R^4 I(\nu_o) \sum_{\rho} \sum_{\sigma} |\alpha_{\rho\sigma}|^2$$

where $I(\nu_0)$ is the intensity of the exciting line, frequency ν_0, ν_R is the frequency of the Raman scattered radiation, and c is the velocity of light. It is possible to express $\alpha_{\rho\sigma}$ so that it incorporates both Herzberg-Teller as well as Born-Oppenheimer non-adiabatic coupling. Thus, according to the equation

$$\alpha_{\rho\sigma} = A + B$$

the Raman tensor is the sum of terms which describe scattering arising from Franck-Condon allowed transitions involving a single excited electronic state (A terms) together with those which derive intensity through vibronic coupling between two or more excited electronic states (B terms). For a transition from vibrational level i to j in the ground electronic state (g), these terms can be expressed as follows:

$$A = \frac{1}{h} \sum \left[\frac{(g^0|\mu_\rho|e^0)(e^0|\mu_\sigma|g^0)\langle gj|ev\rangle\langle ev|gi\rangle}{\nu_{ev,gi} - \nu_0 + i\Gamma_{ev}} + \frac{[\rho \leftrightarrow \sigma]}{\nu_{ev,gi} + \nu_0 + i\Gamma_{ev}} \right]$$

$$B = -\frac{1}{h^2} \sum_{e,v} \sum_{s,u}' \left[\frac{(g^0|\mu_\rho|s^0)(s^0|(\delta H/\delta Q)_0|e^0)(e^0|\mu_\sigma|g^0)\langle gj|su\rangle\langle su|Q|ev\rangle\langle ev|gi\rangle}{(\nu_{su,gi} - \nu_0 + i\Gamma_{su})(\nu_{ev,gi} - \nu_0 + i\Gamma_{ev})} + \frac{[\rho \leftrightarrow \sigma]}{(\nu_{su,gi} + \nu_0 + i\Gamma_{su})(\nu_{ev,gj} + \nu_0 + i\Gamma_{ev})} \right]$$

In these expressions, $|e^0)$ represents the adiabatic electronic wave function of excited electronic state e, evaluated at the equilibrium nuclear coordinates of the ground electronic state, μ is the electric dipole transition operator, and $|ev\rangle$ is the wave function of vibrational level v of the electronic state e. The terms $\nu_{ev,gi}$ are the frequency separations between vibronic levels, and the Γ_{ev} represent damping terms describing the radiative widths of the vibronic states, ev. Excited electronic states other than e are denoted by s, the prime on the summation implies exclusion of terms where s = e, and $(\delta H/\delta Q)_0$ represents the vibronic coupling operator which describes how the electronic Hamiltonian varies with displacement along normal coordinate Q. This operator is evaluated at the ground-state nuclear coordinates and transforms in the same way as Q. The second term in each equation is non-resonant. From the equations it is evident that, for A terms, totally symmetric modes only are expected to be resonance enhanced, whereas for B terms, only those vibrational modes are expected to be enhanced which vibronically couple states e and s. Those normal coordinates which transform the molecule from the ground state nuclear configuration to a significantly displaced nuclear configuration in an excited electronic state should, other factors being equal, experience the greatest resonance enhancement.

The present article relates specifically to results on mixed-valence complexes, using the scheme for their classification proposed by Robin and Day (8). On this basis, class II mixed-valence compounds comprise those in which the two different valences of a given metal atom are trapped with distinguishable sites but in which there is sufficient overlap between orbitals on adjacent metal atoms, sometimes via the intermediacy of the s and p orbitals of the bridging halogen atoms, to permit electron transfer between the two sites. Class III mixed-valence complexes are those in which all metal atom sites are equivalent but in which the oxidation state of the metal atoms is non-integral. Both classes of mixed-valence complex possess intense mixed-valence electronic transitions in the visible region, and it is the purpose of this article to survey the nature of the RR spectra obtained on irradiating within the contours of such bands.

CLASS II MIXED-VALENCE COMPLEXES

Linear Chain Complexes

Of the many mixed-valence complexes known, the halogen-bridged derivatives of platinum have been the most intensively studied. These are of the general types (9):

$[Pt^{II}L_4][Pt^{IV}L_4X_2]X_4.yH_2O$, where L = ethylamine or propylamine, X = Cl, Br or I, and y = 4 or 0.

$[Pt^{II}L_2X_2][Pt^{IV}L_2X_4]$, where L = NH_3, X = Cl or Br.

$[Pt^{II}(L-L)_2][Pt^{IV}(L-L)_2X_2]Y_4$, where (L-L) = 1,2-diaminoethane (en), 1,2-diaminopropane (pn) or 1,3-diaminopropane (tn), X = Cl, Br or I, Y = ClO_4^-, BF_4^- or, in some cases, discrete ions such as $CuBr_4^{3-}$ or chain ions such as $(Cu_3Br_5)_n^{2n-}$.

In these complexes (typified by Wolffram's red, L = ethylamine Figure 1) the two metal-atom sites are structurally distinguishable but may be interconverted by a concerted movement of the axial halogen atoms in phase away from the platinum(IV) atoms towards the platinum(II) ones. Such complexes are high dichroic, intense absorption associated with the inter-valence charge-transfer transition occurring where the electric vector of the incident beam is parallel to the long (chain) axis of the needle-like crystals, but little or no absorption occurring where the electric vector of the incident beam is perpendicular thereto. A number of other properties of such complexes are highly anisotropic viz. their electrical conductivity which, although small, is typically about 300 times greater in the chain direction than in the perpendicular directions. The complexes are thus, at best, one dimensional semi-conductors at atmospheric pressure. A hopping process has been proposed for the conductivity of these complexes at ambient pressures, although at high pressure this

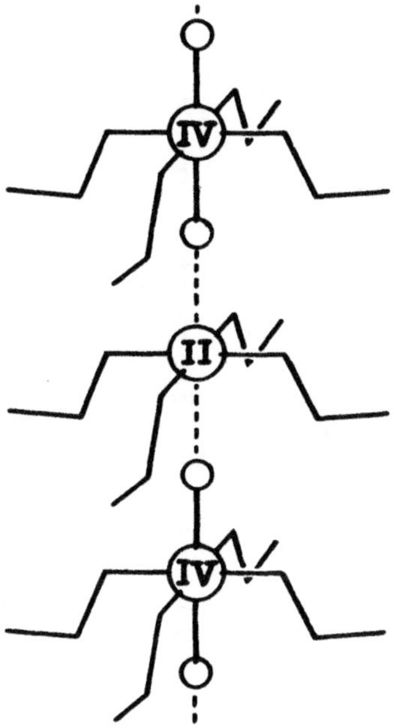

Figure 1

localised process may give way to a band-type conduction process because the metal d_{z^2} and ligand p_z orbital interactions are known to increase under pressure. Excitation of this type of mixed-valence complex within the contour of the mixed-valence bands leads to an enormous enhancement to the Raman band associated with the $\nu_1(X\text{-}Pt^{IV}\text{-}X)$ fundamental, and the the development of a long associated overtone progression. This progression is typically so intense that it completely dominates the Raman spectrum, although in some cases other weaker progressions are also observed; in these cases, however, the progression-forming mode is still the ν_1 mode, while the enabling mode is another Raman-active mode of the complex (10, 11).

The case of $[Pt(tn)_2][Pt(tn)_2Cl_2](ClO_4)_4$ is typical of those for mixed-valence complexes under discussion (12). Excitation within the contour of the broad mixed-valence band centred around 23,000 cm^{-1}, and polarised parallel to the chain axis, leads to the enhancement of the strong ν_1 band of the complex, $\nu(Cl\text{-}Pt^{IV}\text{-}Cl)$ at 315 cm^{-1}, together with its associated overtone progression (which in this case, at ca. 80 K, reaches as far as 13 members, (Figure 2), but may in other cases reach as far as 18 members). The ν_1 band and its associated overtone bands are all polarised parallel to the chain axis (Figure 3), consistent with their assignment to a mode in which the halogen atoms move in the chain direction.

Typical results on a large number of linear-chain complexes of platinum are summarised in Table 1. Included in the Table are crystal colours, band maxima of the mixed-valence bands, maxima of the excitation profiles of the ν_1 bands (i.e. in the plot of the intensity of the ν_1 band versus excitation wavenumber), the nature of the disc into which the complex has been pressed, the harmonic wavenumber (ω_1) and an harmonicity constant (x_{11}) of the ν_1 mode, the number of members of the main progression in ν_1, and the maximum observed value of the intensity of the first overtone of ν_1 to that of the fundamental. Several features of these results deserve comment. These are: (a) The value of ω_1 is almost independent, not only of the equatorial ligands attached to the platinum atom, but also of the interchain counter-ion. Thus the following ranges cover all the data listed: 312±4 (chlorides), 176±7 (bromides), 123±8 cm^{-1} (iodides). Clearly the electrical and mechanical coupling between the axial and equatorial modes

TABLE 1

Resonance Raman Data on Mixed-Valence Linear-Chain Complexes of Platinum at ca. 80 K

Complex	Crystal Colour	Mixed Valence band max./ cm^{-1}	E.P. max./ cm^{-1}	Sample form	ω_1/ cm^{-1}	x_{11}/ cm^{-1}	Progn.	$I(2\nu_1)/I(\nu_1)$
Chlorides								
$[Pt(etn)_4][Pt(etn)_4Cl_2]Cl_4.4H_2O$	red	21,000	17,450	KCl disc	312.3	-1.29	16 ν_1	1.1
$[Pt(en)Cl_2][Pt(en)Cl_4]$	red	23,500	18,200	KCl disc	307.8	-0.98	14 ν_1	0.96
$[Pt(en)_2][Pt(en)_2Cl_2](ClO_4)_4$	red	20,000	17,000	KCl disc	313.3	-1.09	14 ν_1	0.61
$[Pt(pn)_2][Pt(pn)_2Cl_2](ClO_4)_4$	yellow	25,000	~20,000	NaCl disc	310	-1.0	8 ν_1	0.47
$[Pt(tn)_2][Pt(tn)_2Cl_2](ClO_4)_4$	red	22,900	18,600	$KClO_4$ disc	315.0	-0.71	13 ν_1	0.64
Bromides								
$[Pt(etn)_4][Pt(etn)_4Br_2]Br_4.4H_2O$	green	18,250	15,950	KBr disc	177.0	-0.3	12 ν_1	~0.6
$[Pt(etn)_4][Pt(etn)_4Br_2]Br_4$	orange	23,600	20,000	KBr disc	182.5	-0.65	12 ν_1	~0.6
$[Pt(en)Br_2][Pt(en)Br_4]$	gold-green	19,000	~14,000	KBr disc	171.7	-0.25	10 ν_1	0.4
$[Pt(en)_2][Pt(en)_2Br_2](ClO_4)_4$	bronze-green	18,400	17,700	$KClO_4$ disc	173.2	-0.38	10 ν_1	0.75
$[Pt(en)_2][Pt(en)_2Br_2](CuBr_2)_4$	gold	15,800	13,300	KBr disc	168.6	-0.52	12 ν_1	0.51
$[Pt(pn)_2][Pt(pn)_2Br_2](Cu_3Br_5)_2$	gold-green	19,000	15,000	KBr disc	176.6	-0.27	17 ν_1	0.46
$[Pt(tn)_2][Pt(tn)_2Br_2](ClO_4)_4$	gold-green	20,000	18,600	KBr disc	183.6	-0.12	18 ν_1	0.54

TABLE 1 (cont.)

Complex	Crystal Colour	Mixed Valence band max./ cm^{-1}	E.P. max./ cm^{-1}	Sample Form	ω_1/ cm^{-1}	x_{11}/ cm^{-1}	Progn.	$I(2\nu_1)/I(\nu_1)$
Iodides								
$[Pt(etn)_4][Pt(etn)_4I_2]I_4$	green-black	20,600	17,600	KI disc	120.3	−0.33	8 ν_1	~0.6
$[Pt(en)I_2][Pt(en)I_4]$	bronze	15,000	≤13,000	CsI disc	121.2	−0.3	6 ν_1	0.45
$[Pt(en)_2][Pt(en)_2I_2](ClO_4)_4$	gold	14,500	14,100	$KClO_4$ disc	125.3	−1.3	9 ν_1	0.60
$[Pt(pn)_2][Pt(pn)_2I_2](ClO_4)_4$	gold	12,500	<12,000	powder (R.T.)	121.0	~0.0	4 ν_1	0.35
$[Pt(tn)_2][Pt(tn)_2I_2](ClO_4)_4$	gold	13,500	≤12,500	$KClO_4$ disc	131.4	0.1	4 ν_1	0.48

Abbreviations: etn = ethylamine, en = 1,2-diaminoethane, pn = 1,2-diaminopropane, tn = 1,3-diaminopropane, R.T. = room temperature

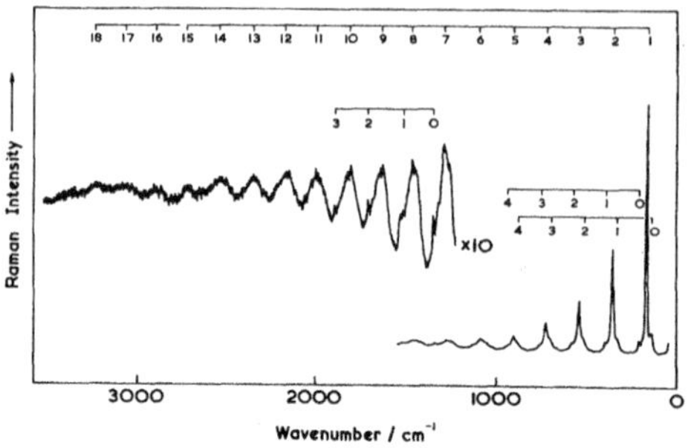

Figure 2

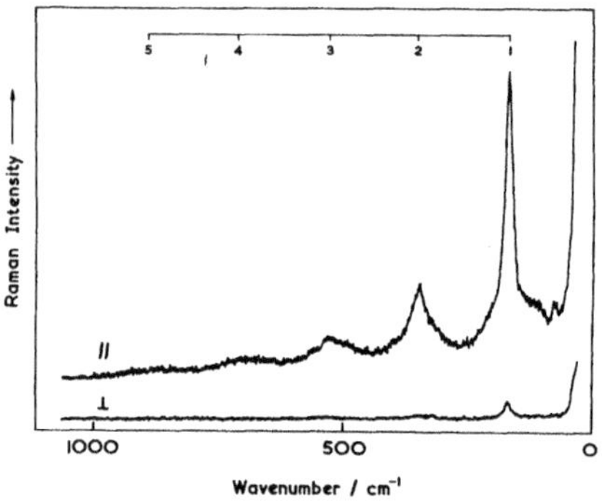

Figure 3

of the complexes is only slight.

(b) The anharmonicities of the ω_1 values, although bigger (relative to the value of ω_1) than found for discrete complexes (7), are nevertheless small, and imply little deviation from simple harmonic behaviour up to $v_1 = 18$.

(c) The colours of the complexes as finely ground powders frequently differ markedly from those of the crystals, owing to the differing relative importance of specular and diffuse reflectance as a function of particle size.

(d) Although the wavenumber of the mixed-valence band is somewhat dependent on the counter-ion (ClO_4^- or BF_4^- etc.), the trend for a related series of salts is clearly Cl>Br>I, implying that the halogen is to some extent involved in the charge transfer process. Moreover (Table 2), the ratio $r(Pt^{IV}-X)/r(Pt^{II}-X)$ approaches unity in the order Cl<Br<I, implying the greatest equivalence of positive charge on the geometrically distinct platinum atoms in the case of the iodide, i.e. this salt is nearest to containing a hypothetical Pt^{III} chain. The results of X-ray photoelectron spectral measurements confirm this conclusion in that the $4f_{7/2}$ and $4f_{5/2}$ binding energies of both the Pt(II) and Pt(IV) constituents move towards the average values in the order Cl→Br→I.

(e) The value of $I(2\nu_1)/I(\nu_1)$ can reach as much as unity in some cases, a feature which is only possible under resonance conditions.

(f) The excitation profiles of the ν_1 bands usually maximise at, or within 4000 cm^{-1} on the low wavenumber side, of the maximum of the mixed-valence band in the disc transmission spectrum of the powdered solids. Recent studies of the specular reflectance from single crystals of these complexes suggest that

TABLE 2

1,3-Diaminopropane (tn) Complexes[a]

Complex	$\tilde{\nu}_{max}$ /10^3 cm^{-1}	$r(Pt^{IV}-X)$ /Å	$r(Pt^{II}-X)$ /Å	$\frac{r(Pt^{IV}-X)}{r(Pt^{II}-X)}$
$[Pt(tn)_2][Pt(tn)_2Cl_2](ClO_4)_4$	22.9	-	-	-
$[Pt(tn)_2][Pt(tn)_2Cl_2](BF_4)_4$	20.6	2.299	3.096	0.74
$[Pt(tn)_2][Pt(tn)_2Br_2](ClO_4)_4$	18.4	2.546	2.955	0.86
$[Pt(tn)_2][Pt(tn)_2Br_2](BF_4)_4$	14.0	2.541	2.921	0.87
$[Pt(tn)_2][Pt(tn)_2I_2](ClO_4)_4$	13.5	-	-	0.91

[a] Crystallographic data from Matsumoto, N., Yamashita, M., and Kida, S.: 1978, Bull.Chem.Soc.Japan, 51, pp. 3514-3518.

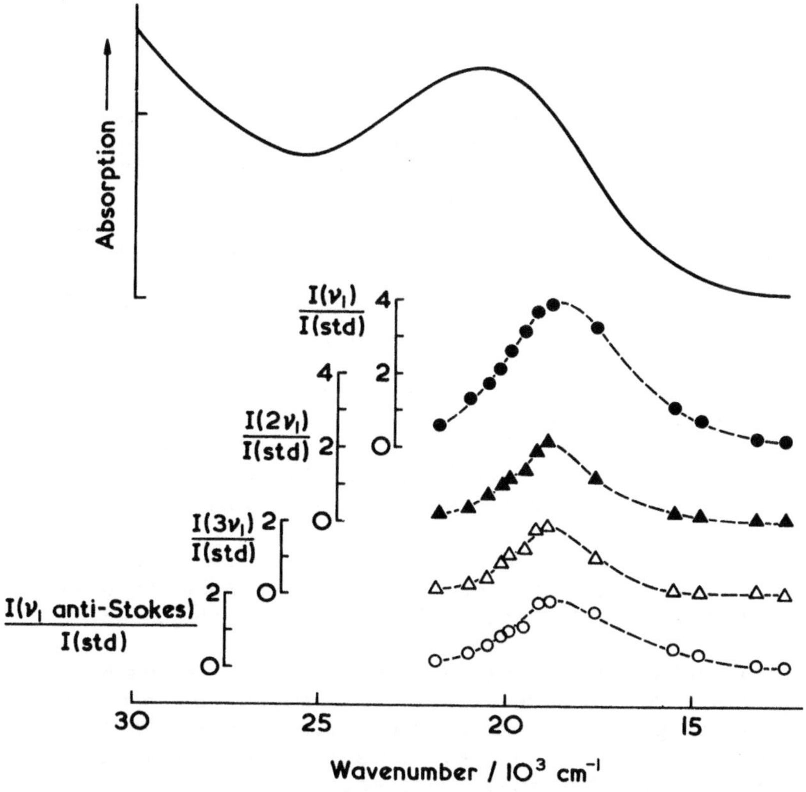

Figure 4

the excitation profile of the ν_1 band actually maximises most nearly to the maximum in the I_{11} specular reflectance spectrum of the crystals (Figure 4).

The wider interest in these results lies in the fact that the ν_1 normal coordinate is related to the halogen movements involved in the hopping process proposed for the conductivity of these complexes. The chain halogen atoms would need to move, on average, 0.54, 0.38 and 0.22 Å for chlorides, bromides and iodides, respectively, in order to reach the point mid-way between the platinum(II) and platinum(IV) atoms i.e. to a situation of a platinum(III) chain. These are not unrealistic distances for a halogen atom to move in a normal coordinate since the root-mean-square amplitudes of vibration of ν_1 in the v_1 = 16 states are 0.22 Å (X = Cl) and 0.20 Å (X = Br) (9). Such distance changes are related to the shift in the equilibrium Pt^{IV}-X bond lengths on excitation of the complex from the ground to the mixed-valence (resonant) excited state. In principle, it is possible to calculate these distance changes from the excitation profiles of the ν_1, $2\nu_1$, $3\nu_1$ etc. bands, but only if the resonant band is structured and the band origin well defined. Such is not in general the case for mixed-valence bands (13).

Single valence species such as Magnus Green, $[Pt^{II}(NH_3)_4]$ $[Pt^{II}Cl_4]$, although stacked, do not display a RR spectrum, presumably since there is no axial coordinate which is likely to change its equilibrium bond length in an excited state of the complex. However the partially oxidised versions of the salt with the reported overall composition $Pt_6(NH_3)_{10}X_{10}(HSO_4)_4$ (X = Cl or Br) and oxidation state +2.33 do display a RR spectrum, a progression in the fundamental (ν(PtCl) = 313 cm^{-1} to $8\nu_1$; ν(PtBr) = 178 cm^{-1} to $7\nu_1$) being observed on irradiation within the contour of the lowest mixed-valence band in each case. (465 nm for the chloride, 540 nm for the bromide) (14). This indicates that the oxidised form of the salt is a chlorine-bridged polymer, directly analogous to Wolffram's red etc. This deduction is seemingly incompatible with the X-ray result (15) that the c-axis of the oxidised chloride salt is exactly three times that of Magnus green whereas the a-axis of the one is identical with that of the other, a result which has been interpreted to imply that the Pt-Pt distances in the two salts are identical. Further work on these partially oxidised salts will be required in order that their structures are known.

The nature of the vibrations of these systems has recently been considered in some detail. The normal modes may be regarded as consisting of (a) lattice modes i.e. collective (phonon) modes and (b) molecular modes i.e. almost independent local modes associated with the constituent molecules (16-18). The former consist of axial (chain) molecular vibrations which propagate along the chain as longitudinal collective modes. The latter consist of vibrations of the molecules perpendicular to the chain, and these ones have frequencies almost the same as the corresponding ones of the isolated constituent species, since they are not coupled along the chain and the inter-chain interactions are weak. New collective modes appear when the chains are formed (e.g. the vibrations of the octahedral units against the square planar units), and these cannot be associated with any modes of the constituent species themselves. By considering the complexes in question as infinite linear chains, involving four masses, two different (ML_4 and X = Cl, Br or I), and two different axial force constants, the longitudinal phonon spectrum is shown to consist of one acoustic and three optical modes (ν_1, ν_2 and ν_3). ν_1 and ν_2 correlate with the symmetric and antisymmetric X-Pt-X stretching modes of the isolated Pt^{IV} entities, whereas ν_3 is the longitudinal mode in which $X-Pt^{IV}L_4-X$ vibrates against $Pt^{II}L_4$. The results lead to Pt^{IV}-X and Pt^{II}-X stretching force constants of ca. 2.0 and ca. 0.2 mdyn $Å^{-1}$ and to ν_3 values of ca. 50 cm^{-1}. A band attributable to ν_3 is often observed either directly or indirectly (as part of a combination tone progression) in the RR spectrum of these complexes in this region. It is worth noting that the ν_1 band of these complexes does not display the chlorine isotopic splitting pattern expected for a two chlorine atom vibrator. Clearly, this implies that several $PtCl_2$ units must be involved in the normal mode, consistent with the collective mode description of the vibrations as indicated above. The above discussion has been confined to linear chain complexes of platinum simply because these form the most readily. Many structurally similar complexes of palladium also occur, and display essentially similar features to those found for platinum (12, 19). In addition, in rare cases halogen-bridged chain species of other metals are known, and have been studied by RR spectroscopy. Thus the gold complex [AuCl SBz_2][$AuCl_3$ SBz_2], which is orange, displays a seven-membered progression in the symmetric Au-Cl stretching mode of the chain under resonance conditions (12).

Discrete Linear Species

The simplest and best known of the class II discrete species are the Br_3^- and I_3^- ions; the triodide may be given the $I^0-I^{-1}-I^0(p^5,p^6,p^5)$ description. A simple m.o. scheme for linear I_3^- ($D_{\infty h}$ symmetry), based on the d_σ and d_π orbitals only, leads to the assignment σ_u^* $(a_{2u})\leftarrow\sigma_g(a_{1g})$ for the lowest allowed z-polarised transition, which occurs at 350 nm (20-22). Irradiation within the contour of this band leads, for a CsI disc at 80 K, to the development of a 17-membered progression in the symmetric stretching vibration of the ion (7), and indicates a substantial bond length increase in the excited state (Table 3). Similar results have been obtained for the Br_3^- ion, although they

TABLE 3

Resonance Raman Data on Radical Cations and Anions[a]

Species	Oxid. State	λ_{max} /nm	λ_L /nm	ωe /cm^{-1}	ωexe /cm^{-1}	Main Progn.	$I(2\nu_1)/I(\nu_1)$
			Diatomic Species				
Br_2^+	$+\frac{1}{2}$	510	514.5	361.5	0.9	6ν	1.0
I_2^+	$+\frac{1}{2}$	640	632.8	238	0	4ν	0.70
Xe_2^+	$+\frac{1}{2}$	710	676.4	123	0	4ν	0.5
S_2^-	$-\frac{1}{2}$	431[b]	454.5	599.8	2.6	10ν	0.83
		400[c]	457.9	594.3	2.5	5ν	–
SSe^-	$-\frac{1}{2}$	–	488.0	464	2.0	3ν	–
Se_2^-	$-\frac{1}{2}$	–	488.0	325	0.75	3ν	–
$^{35}Cl_2^-$	$-\frac{1}{2}$	450	488.0	249.1	1.61	9ν	~0.5
Br_2^-	$-\frac{1}{2}$	660	647.1	149	–	2ν?	–
I_2^-	$-\frac{1}{2}$	800	647.1	114.2	0.50	6ν	0.28
			Triatomic Species	ω_1/cm^{-1}	x_{11}/cm^{-1}		
$^{16}O_3^-$	$-\frac{1}{3}$ av.	440	488.0	1021.3	-5.33	$3\nu_1$	~1.4
S_3^-	$-\frac{1}{3}$ av.	610[d]	488.0	532	-1	$6\nu_1$	0.46
		610[c]	647.1	550.3	-0.75	$6\nu_1$	0.51
		600[e]	647.1	537.5	-1.3	$6\nu_1$	0.69
Br_3^-	$-\frac{1}{3}$ av.	260?	457.9	157	~0.0	$6\nu_1$	0.28
I_3^-	$-\frac{1}{3}$ av.	350[f]	363.8	111	0.6	$8\nu_1$	0.59
		350[g]	457.9	111	~0	$17\nu_1$	≥1
			Tetratomic Species				
Se_4^{2+}	$+\frac{1}{2}$	410	457.9	321	~0	$5\nu_1$	
Te_4^{2+}	$+\frac{1}{2}$	520	514.5	214.1	-0.2	$6\nu_1$	

[a] Abstracted partly from Clark, R.J.H., and Stewart, B.: Structure and Bonding, 36, pp. 1-80.
[b] Doped in NaI, 80 K.
[c] As found in ultramarine blue.
[d] Doped in NaCl.
[e] Dimethylformamide solution.
[f] K^+ Salt in MeOH solution.
[g] CsI, 80 K.

are less spectacular owing to the present lack of suitable excitation lines in the region of the resonant transition (260 nm) (7). Polyiodide ions have also been identified by RR spectroscopy in the partially oxidised one-dimensional mixed-valence bis-benzoquinonedioximates and bis-diphenylglyoximates of nickel and palladium, although they are not considered responsible for the charge transport. The species, studied in great detail, include $[Ni(bqd)_2]_6^+ I_3^-$ and $[Ni(dpg)_2]_5^+ I_5^-$ (23, 24).

RR studies of $(SNBr_{0.4})_x$ and $(SNBr_{0.27})_x$ have led to the conclusion that bromine enters not only the interfibrillar regions of $(SN)_x$ as Br_3^- but also the $(SN)_x$ lattice as Br_2. The RR spectrum of these materials with 457.9 nm excitation is dominated by the six-membered progression in ν_1 of Br_3^- (25, 26).

Similar kinds of results have been obtained for the ozonide ion (27), and for its sulphur analogue, S_3^-. The latter has been detected in a wide variety of different media, including NaCl matrices, both natural (lapis lazuli) and synthetic samples of ultramarine blue, green and violet, as well as solutions of sulphur-containing species in dimethylformamide, hexamethylphosphoramide etc. In all cases, irradiation within the contour of the lowest allowed transition of the ion ($^2A_2 \leftarrow {}^2B_1$) has led to the observation of a progression in the symmetric stretching frequency (28, 29) from which, under favourable circumstances, the change in the M-M bond length on going from the ground to the resonant excited state can be calculated. The RR results on such ions do not differ in any significant way from those obtainable from discrete species involving only single-valence elements. Thus, although the RR spectra of mixed valence ions are very spectacular, they simply indicate that their allowed transitions are similar in nature to those of single valence species.

Very intense RR spectra are also obtained from studies on longer, more complicated ions involving more than one oxidation state of the metal ions. Examples of such species, recently studied in detail, include ruthenium red and ruthenium brown.

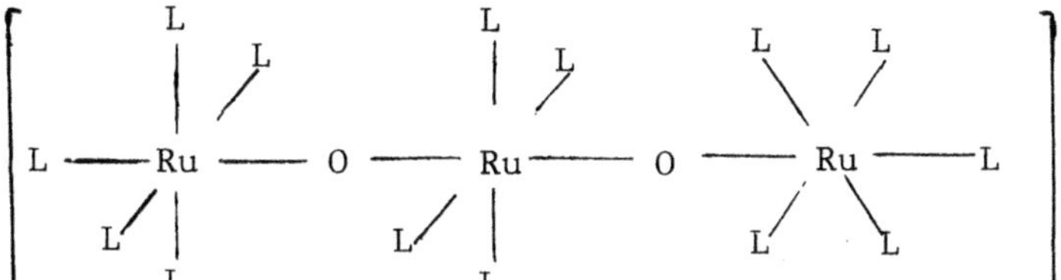

6+ or 7+

($L = NH_3$)

and the analogous 1,2-diaminoethane-substituted species

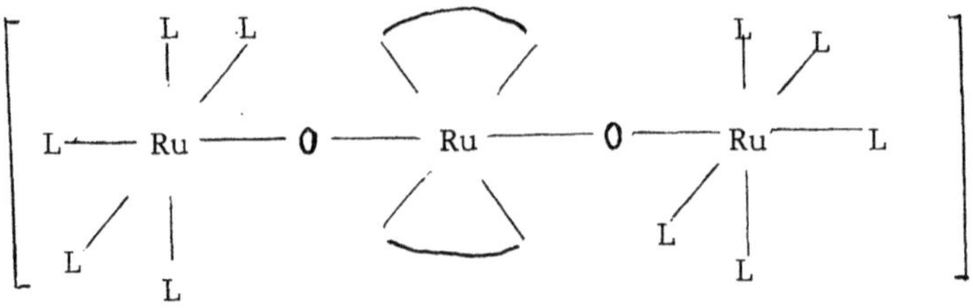

All of these ions can be regarded as involving Ru^{III}-Ru^{IV}-Ru^{III} or Ru^{III}-Ru^{IV}-Ru^{IV} units, the distribution of the 22 valence π-electrons of the former in the ground state being $Ru^{III}(d^5)O(p^4)Ru^{IV}(d^4)O(p^4)Ru^{III}(d^5)$. Studies of both the ^1H, ^2H, ^{16}O and ^{18}O species have led to a consistent set of vibrational assignments for their i.r.- and Raman-active bands (30). Moreover, studies of the excitation profiles of the a_{1g} bands of the ions have led to the following assignments for the very intense electronic transitions responsible for the dye-like quality of their colours, viz. ruthenium red: $e_g(\pi^*) \leftarrow e_u(\pi^*)$, $^1A_{2u} \leftarrow {}^1A_{1g}$ (at 18,900 cm^{-1}), ruthenium brown: $e_g(\pi^*) \leftarrow e_u(\pi^*)$, $^2E_g \leftarrow {}^2E_u$ (at 21,800 cm^{-1}), and to related assignments for the 1,2-diaminoethane-substituted species. These assignments are, of course, based on the idealised D_{4h} symmetry for the cations, a situation which does not necessarily hold in the solid state. Measurements of the depolarization ratios of the a_{1g} bands at resonance leads to the conclusion, consistent with the assignments above, that the resonant electronic transitions are z (i.e. axially) polarized.

Similar studies on the tripalladium complexes $[Pd_3(CNCH_3)_8](PF_6)_2$ and $[Pd_3(CNCH_3)_6\{P(C_6H_5)_3\}_2](PF_6)_2$ have been carried out in the regions (411 and 473 nm) of their lowest allowed electronic transitions. These ions may be regarded as involving the Pd^I-Pd^0-Pd^I (d^9, d^{10}, d^9) unit, and they have the structure where L = $CNCH_3$ and X = $CNCH_3$ or $P(C_6H_5)_3$. Very large intensification to the 114, 262, 312 and 362 cm^{-1} skeletal a_{1g} stretching modes ($\nu(PdPd)$, $\nu(PdC_{ax})$, $\nu(PdC_{eq})$ and $\nu(PdC_{cent})$) takes place under resonance conditions, but overtones of these a_{1g} modes do not appear. Discussion of these matters has appeared elsewhere (31).

A further type of linear mixed valence species which yields very intense RR spectra is that typified by the dinitrogen bridged species

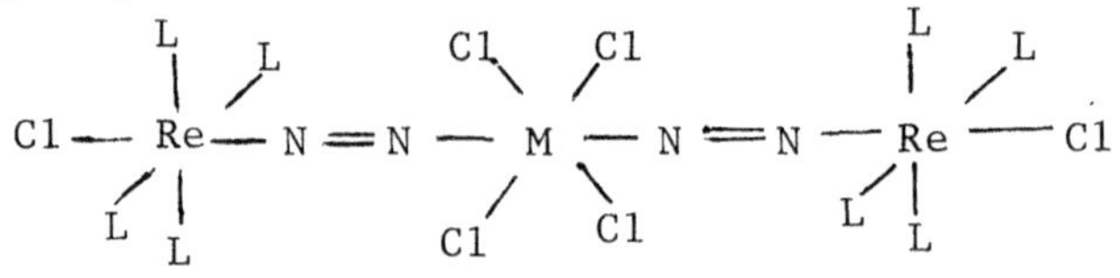

where L = PMe_2Ph and M = Mo or W. In the case of the molybdenum complex, the Raman bands most resonance enhanced in the vicinity of the intense electronic band maximising at 23,300 cm^{-1} are the a_{1g} NN stretching mode (1818 cm^{-1}) and the a_{1g} mode in which the N_2 units vibrate between the rhenium and molybdenum atoms (689 cm^{-1}). Such studies lead to an understanding of the nature of the resonant transition which, among other conclusions, is deduced from measurements of the depolarization ratios of the Raman bands, to be axially polarized.

Three Dimensional Species

The first mixed-valence complex to have its Raman spectrum recorded under resonance conditions was the dark blue solid Cs_2SbCl_6, which contains antimony in oxidation states III and V in the forms of the octahedral ions $SbCl_6^{3-}$ and $SbCl_6^-$ (33). Irradiation of this complex within the contour of the broad charge-transfer transition at 17,900 cm^{-1} (5s←5s) led to the observation of a RR spectrum dominated by bands attributable to the $SbCl_6^-$ ion, with only one band (ν_1) due to the $SbCl_6^{3-}$ ion participating. The main progression-forming mode is ν_1 (a_{1g}) of the $SbCl_6^-$ ion, and the enabling modes include ν_1 (a_{1g}) of the $SbCl_6^{3-}$ ion, ν_5 (t_{2g}) of the $SbCl_6^-$ ion as well as a lattice mode at 60 cm^{-1}. The results thus confirm the mixed-valence nature of the resonant electronic transition in that the RR effects involve fundamentals of both anions. Similar studies have been carried out on mixed valence complexes of bismuth, tin and lead, e.g. on $Cs_4[SbBiCl_{12}]$, $[Co(NH_3)_6][PbCl_6]$ and $[Co(NH_3)_6][PbSnCl_{12}]$ (34). The RR spectra observed for these types of mixed-valence complex, interesting as they are, are less spectacular than those of most single valence complexes, or mixed-valence linear or linear-chain complexes. This suggests that the bond length changes on excitation of such complexes from the ground to the mixed-valence state are less than for the other types of complex mentioned above, presumably as these changes can, for three-dimensional species, by distributed over far more bonds.

A similar RR study of the doped system, $Cs_2[Sb_x^{III}Sb_x^{V}Sn_{1-2x}]Cl_6$ shows resonance enhancement of the host (Cs_2SnCl_6) modes as well as of those of the Sb^{III}–Sb^{V} chromophore. This indicates a strong dynamic involvement of the host lattice during Sb^{III}–Sb^{V} dopant electron transfer, and is regarded as providing the first direct experimental observation of the dynamics associated with outer-sphere chemical activation for electron transfer reactions (35).

Preliminary studies of other class II three-dimensional systems have been disappointing. Species such as Prussian blue (36), $Fe^{III}{}_4[Fe^{II}(CN)_6]_3.xH_2O$, x = 14-16, have yielded only rather weak spectra under resonance conditions. Either all the

Raman radiation scattered by the chromophore is absorbed by the crystal or the bond length changes on going to the mixed-valence state are each only slight owing to the large number of bonds involved. At the present time it seems certain that the most intense RR spectra from mixed-valence complexes are to be expected from one-dimensional species.

Similar difficulties were experienced with $CsAuBr_3$, whose chloro analogue (Wells' salt) contains half of the gold atoms coordinated in a square planar fashion to four chlorine atoms while the other half are coordinated in a linear fashion to two chlorine atoms (37). Since square planar coordination is typical of the low-spin d^8 configuration, and linear coordination of d^{10}, the structure may be regarded, to a first approximation, as containing $Au^{III}Cl_4^-$ and $Au^{I}Cl_2^-$ units. However, the structure is essentially a three-dimensional one, and one which can be considered to be derived from the cubic perovskite structure by displacing the chlorine atoms in a systematic fashion from the centres of the cell edges. Presumably it is this three-dimensional nature of the structure which restricts the quality of the RR spectra, as discussed above, since although Raman spectra can be obtained from the (black) $CsAuBr_3$ (38), they are of poor quality and no overtones of the AuBr stretching modes have been detected under resonance conditions. Further studies of this type of class II complex are needed, since they display several intriguing properties, viz. the electrical conductance of $CsAuCl_3$ increases continuously by 10^9 on application of a pressure of 60 kbar (39), although neutron diffraction data at 28 kbar indicate that at this pressure the two gold sites remain crystallographically inequivalent (40), and X-ray results indicate that this is still so at 52 kbar (41).

CLASS III MIXED-VALENCE COMPLEXES

Discrete Species (Class IIIA)

The class III species which have so far been most effectively studied by RR spectroscopy are the discrete ones (class IIIA) and in particular the diatomic cations (Br_2^+, I_2^+, Xe_2^+) and diatomic anions (Se_2^-, SSe^-, Se_2^-, Cl_2^-, Br_2^-, and I_2^-). In all these cases, irradiation within the contour of the lowest allowed electronic transition of the ion leads to a long progression in the molecular fundamental reaching in favourable cases (e.g. matrices at low temperatures) as far as 10ν. A summary of the known data on these ions is given in Table 3. The results differ in no significant respect from those known or obtainable from single valence species. Clearly, however, this is a growth area for the subject, since attention is now being focused on establishing the properties of an increasingly large number of radical cations and anions, both stable and short-lived. In this context, the adaptation of rapid-flow techniques from electron-spin-resonance spectroscopy is likely to have wide consequences (42).

Another class IIIA radical cation of considerable interest is the red, square planar Te_4^{2+} ion, which formally contains tellurium in the $+\frac{1}{2}$ oxidation state. The ion, which has a 6π-electron delocalised system, was first prepared and studied by Gillespie and co-workers (43) who showed that, in concentrated H_2SO_4 solution, a short RR progression in the ν_1 (a_{1g}) fundamental (at 214 cm^{-1}) of the ion could be obtained on irradiating within the contour of its π^* $(b_{2u})\leftarrow\pi(e_g)$ transition. More recent studies of this ion, as it occurs in $Te_4(Al_2Cl_7)_2$ and while held at ca. 100 K, have led to the observation of much better RR spectra, the $v_1\nu_1$ progression reaching as far as $6\nu_1$ (44). Similar studies of the related, square planar Se_4^{2+} ion, on irradiation within the contour of its analogous 410 nm band, have led to the observation of a comparable $v_1\nu_1$ progression reaching as far as $4\nu_1$ (44). There seems little doubt that RR studies of other radical cations and anions, e.g. S_4^{2+}, will prove to be effective probes of the nature of the excited states of such species.

A further interesting class of complex ion is that of the sort $[M_2OX_{10}]^{n-}$, where M = Ru, Re, Os or W, X = Cl, Br, or I, and n is usually four. These ions each contain a linear M-O-M bonding system, and thus possess D_{4h} symmetry. Irradiation within the contour of the axially polarized $\pi^*\leftarrow\pi$ system of each ion, which is usually near 20,000 cm^{-1}, leads to the observations of up to eight progressions, all of which involve ν_1 (a_{1g}), the symmetric M-O-M stretching mode, as the progression-forming mode. Although most of these ions involve single metal-atom oxidation states, the $[Re_2OCl_{10}]^{3-}$ ion involves $Re^{+4.5}$. It provides one of the most spectacular RR spectra known, the main progression $(v_1\nu_1)$ reaching as far as v_1 = 14. RR and magnetic studies have led to the suggestion that the resonant electronic transition has the assignment $e_u^*\leftarrow e_g$, $^4E_g\leftarrow{}^4E_u$ (45).

The nature of the oxidation state of ruthenium in the Creutz-Taube complex ion, $[(NH_3)_5Ru\ N\langle\bigcirc\rangle N\ Ru(NH_3)_5]^{5+}$, is controversial. Undoubtedly, the rate of electron transfer in the ion is rapid compared to solvent orientation (implying valence delocalisation), since the wavelength of the intervalence charge-transfer transition (1580 nm in D_2O) is virtually independent of the nature of the solvent. Some lines of evidence (electronic, XPS (46) and RR (47)) have been interpreted to imply the presence of two different types of ruthenium atom (i.e. valence trapped, low spin $Ru^{II}(d^6)$ and $Ru^{III}(d^5)$), but other lines (including a reinterpretation of the RR and XPS data (48) as well as further vibrational (49) and electronic (50) data) imply fairly convincingly the presence of only one type, the valence delocalised, class III $Ru^{+2,5}$. The answer to the question may depend simply on the time-scale of the physical measurements under discussion. In this context it is worth noting that the usual criterion for valence delocalisation is that the lifetime of a particular mixed-valence state should be shorter than the vibrational timescale (~10^{-13} s).

A recently developed vibronic coupling model for the calculation of mixed-valence absorption profiles, as applied to the Creutz-Taube complex, indicates that the lower potential surface has a small (~57 cm^{-1}) barrier (implying valence trapping) but that the lowest vibronic state is 30 cm^{-1} above the top of the barrier (implying no effective trapping). Moreover, the calculated probability distribution in configuration space shows complete valence delocalisation (no trapping) at low temperature, although there is a slight hint of trapping at room temperature (13).

Three Dimensional Species

Class IIIB species such as the anion-deficient one-dimensional conductors KCP, $K_2[Pt(CN)_4]X_{0.30}.3H_2O$, X = Cl or Br (51) or the comparable cation-deficient species $M_{1.75}[Pt(CN)_4].1.5H_2O$, M = K, Rb or Cs (52), although the subject of detailed Raman studies, have not been shown to display any resonance effects. As in the case of Magnus green, it seems that the reason for this is the lack of any suitable longitudinal mode whose Raman band intensity could be enhanced under resonance conditions. However, the metallic reflection of such complexes (which appear reddish-bronze) also militates against the obtaining of good Raman spectra under resonance conditions and so the reason may be simply experimental. Few investigations of class IIIB complexes have yet been undertaken, and it may yet prove to be feasible to obtain good vibrational data from such complexes with which to complement the optical data obtainable by specular reflectance techniques.

The future in this area has much in store, and one looks forward to many developments in the technique of RR spectroscopy by which to probe further the nature of the mixed-valence state.

ACKNOWLEDGEMENT

The author is much indebted to his past and present co-workers in the field of mixed-valence complexes, viz. J.R. Campbell, D.G. Cobbold, L.J. Ferris, M.L. Franks, M. Kurmoo, C. Sourisseau, B. Stewart, W.R. Trumble and P.C. Turtle for their expert experimental work, which has been partly funded by the Science Research Council, the University of London Inter-Collegiate Research Service, and Johnson-Matthey Ltd.

REFERENCES

1. Behringer, J., in Chemical Society Specialist Periodical Reports on Molecular Spectroscopy, ed. Barrow, R.F., Long, D.A., and Millen, D.J.: The Chemical Society, London, Vol. 2, 1974, pp. 100-172; Vol. 3, 1975, pp. 163-280.

2. Spiro, T.G. and Loehr, T.M., in Advances in Infrared and Raman Spectroscopy, ed. Clark, R.J.H. and Hester, R.E.: Heyden, London, Vol. 1, 1975, pp. 98-142.

3. Clark, R.J.H., in Advances in Infrared and Raman Spectroscopy, ed. Clark, R.J.H. and Hester, R.E.: Heyden, London, Vol. 1, 1975, pp. 143-172.

4. Kiefer, W., in Advances in Infrared and Raman Spectroscopy, ed. Clark, R.J.H. and Hester, R.E.: Heyden, London, Vol. 3, 1977, pp. 1-42.

5. Rousseau, D.L. and Williams, P.F.: 1976, J.Chem.Phys. 64, pp. 3519-3537.

6. Warshel, A.: 1977, Ann.Rev.Biophys.Bioeng. 6, pp. 273-300.

7. Clark, R.J.H. and Stewart, B.: 1979, Structure and Bonding, 36, pp. 1-80.

8. Robin, M.B. and Day, P.: 1967, Adv.Inorg.Chem.Radiochem. 10, pp. 247-422.

9. Clark, R.J.H.: 1978, Ann.N.Y.Acad.Sci. 313, pp. 672-685.

10. Clark, R.J.H. and Turtle, P.C., 1978, Inorg.Chem. 17, pp. 2526-2531.

11. Campbell, J.R., Clark, R.J.H. and Turtle, P.C., 1978, Inorg.Chem. 17, pp. 3622-3628.

12. Clark, R.J.H., Kurmoo, M., unpublished results.

13. Piepho, S.B., Krausz, E.R., and Schatz, P.: 1978, J.Amer. Chem.Soc. 100, pp. 2996-3005.

14. Papavassiliou, G.C. and Theophanides, T.: 1979, Z.Naturforsch. 34b, pp. 986-988.

15. Tsujikawa, I., Kubota, R., Enoki, T., Miyajima, S. and Kobayashi, H.: 1977, J.Phys.Soc.Japan, 43, pp. 1459-1460.

16. Clark, R.J.H. and Stewart, B., unpublished work.

17. Paraskevaidis, C. and Papatriantafillou, C., 1978, Chem. Phys.Letters, 58, pp. 301-303.

18. Paraskevaidis, C., Papatriantafillou, C. and Papavassiliou, G.C., 1979, Chem.Phys. 37, pp. 389-399.

19. Matsumoto, N., Yamashita, M., and Kida, S.: 1978, Bull. Chem.Soc.Japan, 51, pp. 2334-2337.

20. Hach, R.J. and Rundle, R.E.: 1951, J.Chem.Phys. 73, pp. 4321-4324.

21. Pimental, G.C.: 1951, J.Chem.Phys. 19, pp. 446-449.

22. Robin, M.B.: 1964, J.Chem.Phys. 40, pp. 3369-3377.

23. Brown, L.D., Kalina, D.W., McClure, D.S., Schultz, S., Ruby, S.L., Ibers, J.A., Kannewurf, C.R., Marks, T.J.: 1979, J.Amer.Chem.Soc. 101, pp. 2937-2947.

24. Cowie, M., Gleizes, A., Grynkewich, G.W., Kalina, D.W., McClure, M.S., Scaringe, R.P., Teitelbaum, R.C., Ruby, S.L., Ibers, J.A., Kannewurf, C.R., Marks, T.J.: 1979, J.Amer. Chem.Soc. 101, pp. 2921-2936.

25. Temkin, H. and Street, G.B.: 1978, Solid State Comm. 25, pp. 455-460.

26. Iqbal, Z., Baughman, R.H., Kleppinger, J., and MacDiarmid, A.G.: 1978, Solid State Comm. 25, pp. 409-413.

27. Andrews, L., in Advances in Infrared and Raman Spectroscopy, ed. Clark, R.J.H. and Hester, R.E.: Heyden, London, Vol. 7, 1980, in press.

28. Holzer, W., Murphy, W.F., and Bernstein, H.J.: 1970, Chem.Phys.Letters, 4, pp. 641-642.

29. Clark, R.J.H. and Cobbold, D.G.: 1979, Inorg.Chem., 17, pp. 3169-3174.

30. Campbell, J.R., Clark, R.J.H., Griffith, W.P. and Hall, J.: 1979, J.Chem.Soc.(Dalton), in press.

31. Clark, R.J.H. and Sourisseau, C.: 1979, Nouveau.J.Chim., in press.

32. Campbell, J.R., Clark, R.J.H. and Dilworth, J.: 1979, unpublished work.

33. Clark, R.J.H. and Trumble, W.R.: 1975. J.Chem.Soc.Chem.Comm., pp. 318-319.

34. Clark, R.J.H. and Trumble, W.R.: 1976, J.Chem.Soc.(Dalton), pp. 1145-1149.

35. Clark, H.W. and Swanson, B.I.: 1979, J.Amer.Chem.Soc. 101, pp. 1604-1606.

36. Buser, H.J., Schwarzenbach, D., Petter, W. and Ludi, A.: 1977, Inorg.Chem. 16, pp. 2704-2710.

37. Tindemanns, J.C.M. and Verschoor, S.G.C.: 1974, Mater.Res. Bull. 9, pp. 1664-1667.

38. Bosworth, Y.M. and Clark, R.J.H.: 1975, J.Chem.Soc.(Dalton), pp. 381-385.

39. Keller, R., Fenner, J. and Holzapfel, W.B.: 1974, Mater.Res. Bull., 9, pp. 1363-1370.

40. Day, P., Vettier, C. and Parisot, G.: 1978, Inorg.Chem. 17, pp. 2319-2320.

41. Denner, W., Schultz, H., and d'Amour, H.: 1979, Acta Cryst. A35, pp. 360-365.

42. Hester, R.E., in Advances in Infrared and Raman Spectroscopy ed. Clark, R.J.H. and Hester, R.E.: Heyden, London, Vol. 4, 1978, pp. 1-45.

43. Booth, M., Gillespie, R.J., and Morton, M.J.: 1973, Adv. Raman Spectrosc., pp. 364-369.

44. Clark, R.J.H. and Ferris, L.T.H.: unpublished work.

45. Campbell, J.R. and Clark, R.J.H.: 1978, Mol.Phys. 36, pp. 1133-1148.

46. Citrin, P.H.: 1973, J.Amer.Chem.Soc. 95, pp. 6472-6473.

47. Strekas, T.C. and Spiro, T.G.: 1976, Inorg.Chem. 15, pp. 974-976.

48. Hush, N.S.: 1975,Chem.Phys. 10, p. 361.

49. Beattie, J.K., Hush, N.S. and Taylor, P.R.: 1976, Inorg.Chem. 15, pp. 992-993.

50. Taube, H.: 1978, Ann.N.Y.Acad.Sci. 313, pp. 481-495.

51. Ferraro, J.R., Basile, L.J. and Williams, J.M.: 1976, J.Chem.Phys. 64, pp. 732-735.

52. Ferraro, J.R. Basile, L.J. and Williams, J.M.: 1977, J.Chem.Phys. 67, pp. 742-745.

IV. CLASSES OF MIXED-VALENCE MATERIALS

MIXED-VALENCE MINERALS OF IRON AND TITANIUM: CORRELATIONS OF STRUCTURAL, MÖSSBAUER AND ELECTRONIC SPECTRAL DATA

Roger G. Burns and
D.A. Nolet, K.M. Parkin, C.A. McCammon, K.B. Schwartz

Department of Earth and Planetary Sciences,
Massachusetts Institute of Technology,
Cambridge, Massachusetts 02139

"Since ferrous iron usually colors minerals green, and ferric iron yellow or brown, it may seem rather remarkable that the presence of both together should give rise to a blue color, as in the case of vivianite. It may be pointed out, however, that this is by no means a unique instance of such an effect. Even apart from the artificial substances, "Prussian" and "Turnbull's" blues, which are complex cyanides containing both ferric and ferrous iron, there are several blue minerals in which the color seems explainable only on this basis. The most noteworthy of these are: crocidolite and related amphiboles; iolite; and the blue tourmaline or indicolite. Other instances may perhaps be discovered, should this subject ever be investigated as it deserves to be."

E.T. Wherry, Amer. Mineral, v.3, p.161 (1918)

ABSTRACT

A large variety of mixed-valence silicate and oxide minerals occur in rocks from the Earth, Moon and meteorites. They frequently contain Fe^{2+}-Fe^{3+}, Ti^{3+}-Ti^{4+}, and Fe^{2+}-Ti^{4+} assemblages and often display anomalous color or opacity, high thermal and electrical conductivities, and unusual spectral variations with temperature or pressure. Such properties are not attributable to transitions in single cations and indicate the presence of intervalence transitions. The intervalence transitions are observed either as charge transfer (CT) bands in the visible-near infrared region and frequently cause blue colors in Fe^{2+}-Fe^{3+} and Fe^{2+}-Ti^{4+} bearing minerals (e.g. vivianite, sapphire, cordierite, kyanite,

D.B. Brown (ed.), Mixed-Valence Compounds, 295-336.

glaucophane, aquamarine), or as electron delocalization (ED) often causing opacity and semiconductor behavior (e.g. magnetite, ilvaite, deerite, riebeckite). Many of the physical and spectral properties of mixed-valence silicate and oxide minerals may be correlated with crystal structure features and site occupancies of the metals. Structural variables which affect properties of mixed-valence oxide and silicate minerals include: symmetries of coordination polyhedra (e.g. regular or distorted octahedra, tetrahedra, dodecahedra), type of polyhedral linkage (e.g. face-shared, edge-shared, corner-shared); metal-oxygen (M-O) distances, metal-metal (M-M) distances, and M-O-M bond angles; and orientation and extent of linked polyhedra (e.g. isolated pairs, chains, bands, sheets; blocking by Mg^{2+} and Al^{3+} in solid-solution series). These features may be used to interpret the temperature and pressure dependencies of CT bands in minerals with isolated Fe^{2+}-Fe^{3+} and Fe^{2+}-Ti^{4+} clusters (e.g. vivianite, sapphire, glaucophane, kyanite, cordierite, omphacite, pyroxenes, etc.), and the occurrence of Fe in intermediate electronic environments in minerals with infinite chains of edge-shared Fe^{2+}-Fe^{3+} polyhedra (e.g. magnetite, ilvaite, deerite,schorlomite garnets, riebeckite, and potentially aenigmatite, howieite, babingtonite).

INTRODUCTION

Mixed-valence minerals have attracted attention since antiquity because many gems, including sapphire, aquamarine and iolite, owe their colors to coexisting Fe^{2+}, Fe^{3+} and Ti^{4+} ions in their crystal structures (1). More recently intervalence charge transfer in minerals has been used to explain anomalous color, spectra, electrical and magnetic properties, and unusual temperature variations of these properties in terrestrial, lunar and meteoritic materials (2-7).

Although many minerals contain various transition elements as both major and minor constituents, the relatively high crustal and cosmic abundances of iron and titanium compared to other heavy metals renders these two elements ubiquitous constituents of most common rock-forming minerals found in and on planetary bodies including the Earth, Moon, Mars, Mercury, certain asteroids and meteorites. Redox conditions during the evolutionary histories and mineral parageneses of the Earth, Moon, etc., restrict the oxidation states of iron in iron-bearing minerals to Fe^{0}, Fe^{II} and Fe^{III}, while titanium can occur as both Ti^{IV} and Ti^{III}. As a result, many mixed-valence minerals are known containing coexisting Fe^{2+}, Fe^{3+}, Ti^{4+}, or Ti^{3+}. Rarer assemblages of Mn^{2+}-Mn^{3+}-Mn^{4+}, Cu^{+}-Cu^{2+}, Ni^{2+}-Ni^{3+}, Co^{2+}-Co^{3+}, and perhaps Cr^{2+}-Cr^{3+} occur in oxide, silicate, sulphide, phosphide, and arsenide minerals. The focus of this review, however, is on mixed-valence minerals containing Fe^{2+}-Fe^{3+},

Ti^{3+}-Ti^{4+}, and Fe^{2+}-Ti^{4+} assemblages, in which oxygen is the ligand coordinated to the transition metal ions.

Intervalence transitions in minerals are observed both as charge transfer (CT) in optical spectra, frequently causing blue colors in Fe^{2+}-Fe^{3+} or Fe^{2+}-Ti^{4+} bearing minerals, and as electron delocalization (ED), a thermally activated process often causing opacity and semi-conductor behavior. Many of these electronic properties can be correlated with crystal structure features, including metal-oxygen (M-O) bond strength as evidenced by the M-O interatomic distance; metal-oxygen-metal overlap, indicated by metal-metal (M-M) distance; symmetries of coordination polyhedra, e.g., regular or distorted octahedra, tetrahedra, dodecahedra; type of polyhedral linkage, e.g., face-shared, edge-shared, corner-shared; and orientation and extent of linked polyhedra, e.g., isolated pairs, chains, bands, sheets. Such structural correlations are highlighted in this review.

Minerals exhibit a number of structural properties not generally found in synthetic and biological materials.
(i) Atomic substitution is prevalent, whereby ions of comparable ionic radius replace one another. This is particularly important in common Mg and Al silicates in which Mg^{2+} (octahedral ionic radius 72pm) and Al^{3+} (53pm) may be replaced by Fe^{2+} (77pm), Fe^{3+} (65pm), Ti^{3+} (68pm), Ti^{4+} (61pm), Mn^{2+} (82pm), Mn^{3+}(65pm), Mn^{4+} (54pm), Cr^{2+} (82pm), Cr^{3+} (62pm), etc. Conversely, host Mg^{2+} and Al^{3+} ions may act as substitutional blockers of extended $Fe^{2+} \rightarrow Fe^{3+}$, $Fe^{2+} \rightarrow Ti^{4+}$, etc., interactions when the latter are present as minor constituents.
(ii) Many mineral structures contain two or more distinct crystallographic positions, so that cation ordering (preference for, and relative enrichment in, a specific site) is common. Cation distributions are influenced by temperatures and pressures of crystallization, so that minerals with identical bulk compositions may have different site populations.
(iii) Cation sites with the same coordination number show a variety of distortions resulting from structural constraints of other coordination polyhedra (e.g., linkages of $[SiO_4]$ tetrahedra in silicates; shared edges between adjacent polyhedra; differences of oxygen bond-type, such as O^{2-}, OH^-, H_2O, $Si-O^-$, and Si-O-Si ligands). As a result, a range of regular and distorted octahedral sites may exist in the same mineral structure. Such structural constraints impose severe limitations when attempts are made to apply classification schemes for mixed-valence compounds to mineral systems.

In this review of mixed-valence minerals of Fe and Ti, structural features which influence intervalence transitions are discussed. The approach is to focus on a few representative structure-types which exhibit different $Fe^{2+} \rightarrow Fe^{3+}$ interactions, detectable by electronic absorption and Mössbauer spectroscopy. The coverage is then extended to mineral systems containing Fe^{2+}-Fe^{3+} clusters in unusual coordination symmetries and to Ti^{3+}-Ti^{4+} and Fe^{2+}-Ti^{4+} assemblages.

VIVIANITE: THE STRUCTURE-TYPE WITH ISOLATED CLUSTERS OF (Fe^{2+}, Fe^{3+}) POLYHEDRA

Vivianite, $Fe_3(PO_4)_2 \cdot 8H_2O$, is the best known mineral in which a $Fe^{2+} \rightarrow Fe^{3+}$ intervalence transition occurs across isolated pairs of edge-shared octahedra (2, 3, 8-10). The crystal structure of vivianite illustrated in Figure 1 contains two distinct cation sites (11). In the isolated Fe_A octahedra, Fe^{2+} ions are coordinated to four H_2O ligands in the rhombic plane and to two <u>trans</u>-oxygens of $[PO_4{}^{3-}]$ tetrahedra. In the Fe_B octahedra, two H_2O ligands are coordinated to Fe^{2+} ions in a <u>cis</u> arrangement, and four oxygens of $[PO_4]$ tetrahedra supply the remaining ligands of the octahedron. Two Fe_B octahedra share a common O-O edge across which the Fe^{2+} ions are separated by only 285pm along the <u>b</u> axis. Hydrogen bonding between the H_2O ligands holds together sheets consisting of linked Fe_A octahedra, $[PO_4]$ tetrahedra and double Fe_B octahedra, and accounts for the perfect (010) cleavage of vivianite.

Freshly mined vivianite crystals or newly precipitated $Fe_3(PO_4)_2$ are pale green, but slowly turn blue when exposed to air. Mössbauer spectra of vivianites mechanically ground in air, such as the profiles illustrated in Figure 2, show that Fe^{2+} ions in Fe_A and Fe_B octahedra are oxidized to Fe^{3+} (12,13). However, Fe_B^{2+}/Fe_A^{2+} ratios increase with rising Fe^{3+} concentration, suggesting that there is a greater probability of oxidizing remaining Fe_A^{2+} ions than the second Fe^{2+} of a Fe_B^{2+}-Fe_B^{3+} pair.

The occurrence of paired Fe_B octahedra and the short Fe_B-Fe_B distance provide a simple explanation for the intense blue-pale green pleochroism of vivianite and the polarization dependence of its optical spectra illustrated in Figure 3. The absorption band centered at 15,200 cm^{-1} is responsible for the intense blue color and occurs when light is polarized along the Fe_B-Fe_B vector. The color and polarization-dependence of this band originate from $Fe^{2+} \rightarrow Fe^{3+}$ CT between adjacent Fe^{2+} and Fe^{3+} ions in edge-shared Fe_B octahedra. The intensity of this charge transfer band increases significantly with decreasing temperature (7) and rising pressure (14), supporting its assignment to an intervalence transition between Fe^{2+} and Fe^{3+}.

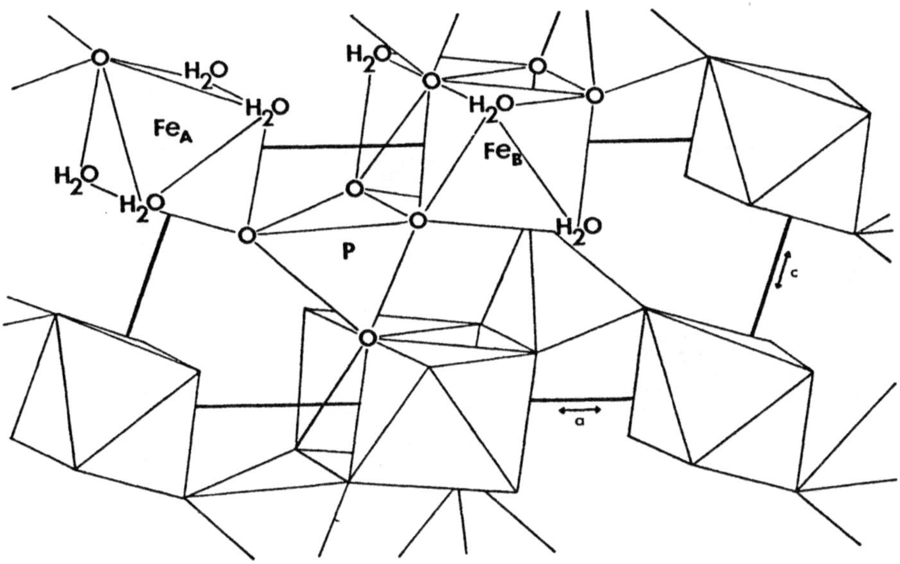

Figure 1 The vivianite crystal structure viewed along the b axis (from refs. 11-13).

The mechanism of oxidation of Fe^{2+} ions in vivianite involves H_2O ligands which are converted to OH^- ions:

$$2[Fe^{2+}-H_2O] + \frac{1}{2} O_2 \rightarrow 2[Fe^{3+}-OH] + H_2O$$

As a result the vivianite structure collapses due to the progressive elimination of hydrogen bonds, facilitating diffusion of oxygen into the structure. Thus, extensively oxidized vivianites not only become deeper blue colored, but they also disintegrate into friable powders.

In summary, vivianite has the potential of being, a Class IIIA mixed-valence compound (10) with respect to the Fe_B octahedra. However, although $Fe^{2+} \rightarrow Fe^{3+}$ CT transitions are observed when light is polarized along the short Fe_B-Fe_B axis, resolution of individual Fe^{2+} and Fe^{3+} ions by Mössbauer spectroscopy indicates that the iron cations have trapped valencies in the vivianite structure.

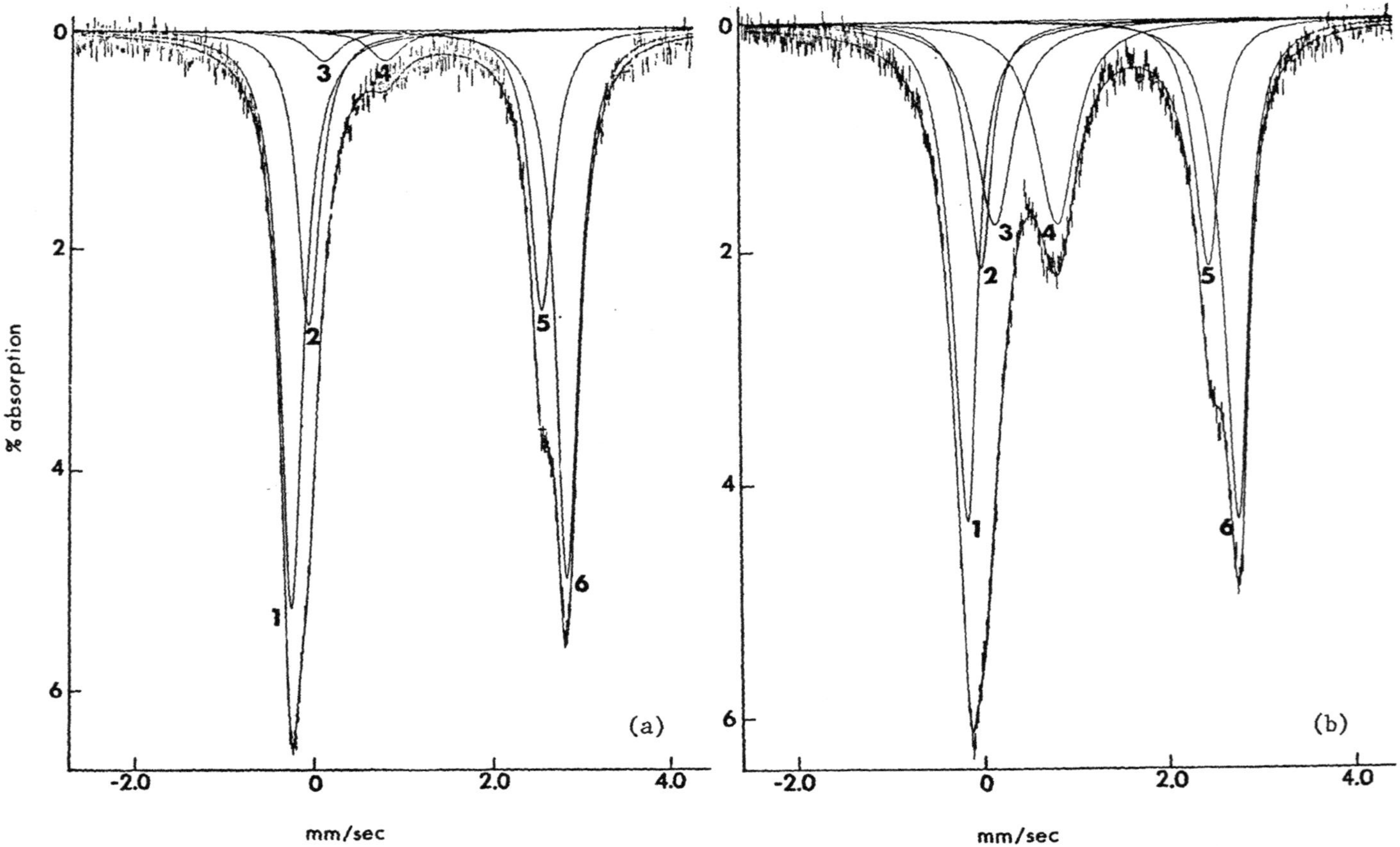

Figure 2 Mössbauer spectra of vivianite (a) before and (b) after oxidation by mechanical grinding in air (from refs. 12 and 13). Peaks 1 and 6: Fe^{2+} in the Fe_B sites; peaks 2 and 5: Fe^{2+} in the Fe_A sites; peaks 3 and 4: octahedral Fe^{3+}.

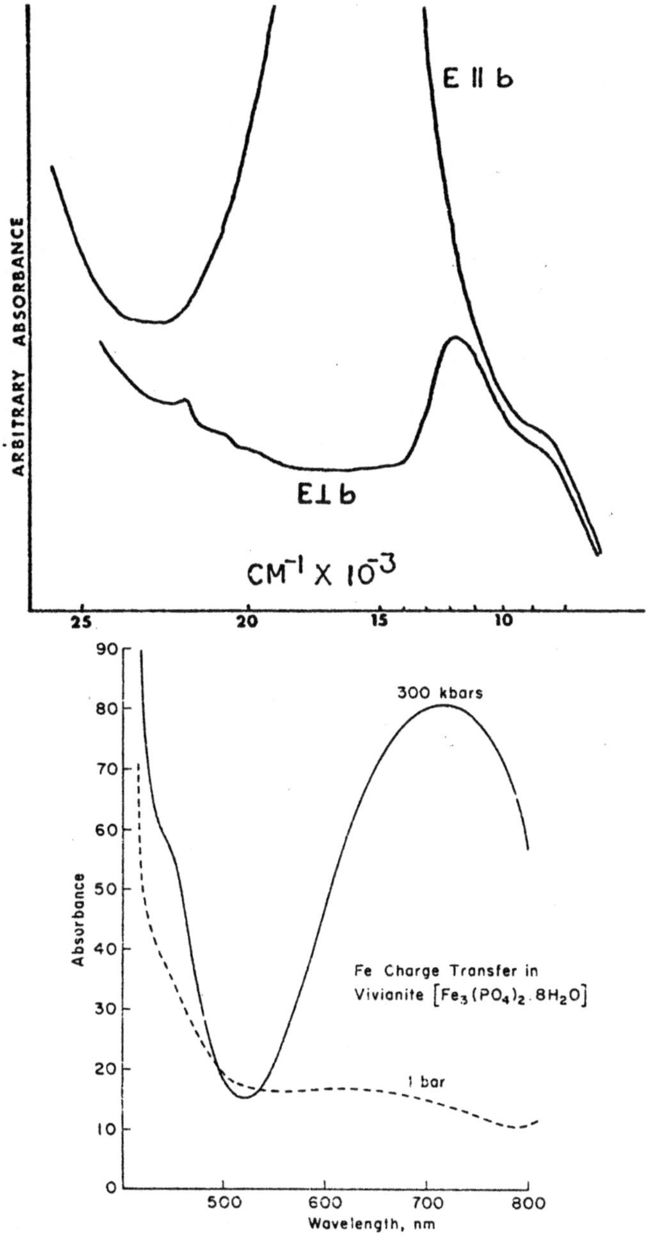

(a)

(b)

Figure 3 Absorption spectra of vivianite (a) single crystal in polarized light at 1 atmos. (from ref. 3); (b) polycrystalline sample in unpolarized light at 1 atmos. and 300 kbar (from ref. 14).

MAGNETITE: THE CLASSIC BUT AMBIGUOUS STRUCTURE-TYPE EXHIBITING ELECTRON DELOCALIZATION ALONG INFINITE CHAINS OF Fe^{2+}-Fe^{3+} OCTAHEDRA

Just as vivianite is regarded as a simple example of a mineral with isolated clusters of Fe^{2+}-Fe^{3+} octahedra showing CT transitions, so too was magnetite, Fe_3O_4, considered to be the classic model of a structure-type with infinite chains of Fe^{2+}-Fe^{3+} octahedra exhibiting electron delocalization. However, a phase change in magnetite at 119K complicates interpretations of ED behavior in this structure-type.

The magnetite structure, an inverse spinel, contains Fe^{3+} ions in tetrahedral A sites, and Fe^{2+} and Fe^{3+} ions separated by 297pm in three-dimensional infinite chains of edge-shared octahedral B sites. Above 119K, ED involving the B-site cations occurs in magnetite, resulting in opacity, and high electrical and thermal conductivities. The Mössbauer spectra show two cation species: tetrahedral Fe^{3+} and mixed-valence octahedral iron (15-17). At 119K, a phase change occurs (18), which doubles the unit cell, leads to magnetic ordering of Fe^{2+} and Fe^{3+}, and produces distinguishable Fe^{2+} and Fe^{3+} octahedral sites with attendant drop in electrical conductivity and changes in the Mössbauer spectra (15-17, 19-21). The transition in magnetite, which is called the Verwey transition, is interpreted as an ionic order-disorder transition (19), but controversy has arisen recently concerning the number of disordering parameters necessary (22) and the validity of the order-disorder model versus a band model or polaron model (20,21). Thus, although magnetite is classified as a Class IIIB mixed-valence compound, application of a Verwey-type model of quenched electron delocalization to other systems not showing a phase change should be made with caution.

ILVAITE: A STRUCTURE SHOWING ELECTRON DELOCALIZATION ALONG INFINITE CHAINS OF Fe^{2+}-Fe^{3+} OCTAHEDRA

The mixed-valence iron silicate mineral ilvaite,

$$CaFe_2^{2+}Fe^{3+}Si_2O_8(OH),$$

possesses many of the attributes necessary for electron delocalization behavior. The ilvaite structure (23, 24) illustrated in Figure 4 (25) consists of a framework of infinite double chains of edge-shared Fe(A) octahedra linked by double tetrahedral Si_2O_7 groups. Six-coordinated Fe(B) sites and seven-coordinated Ca sites are bonded to oxygens of this framework. Neutron diffraction measurements demonstrate that Fe^{2+} and Fe^{3+} ions are located in the Fe(B) sites while Fe(A)

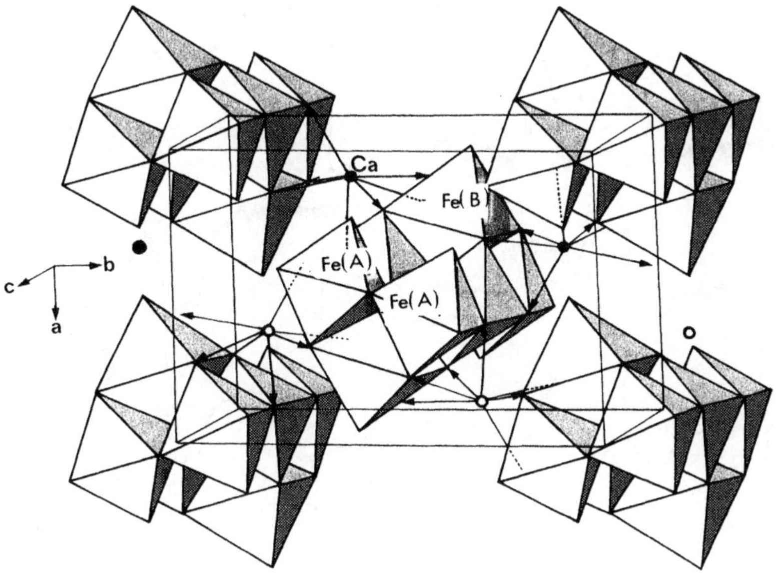

Figure 4 The ilvaite crystal structure, showing the arrangement of infinite double chains of edge-shared Fe(A) octahedra extending along the c axis (from ref. 25).

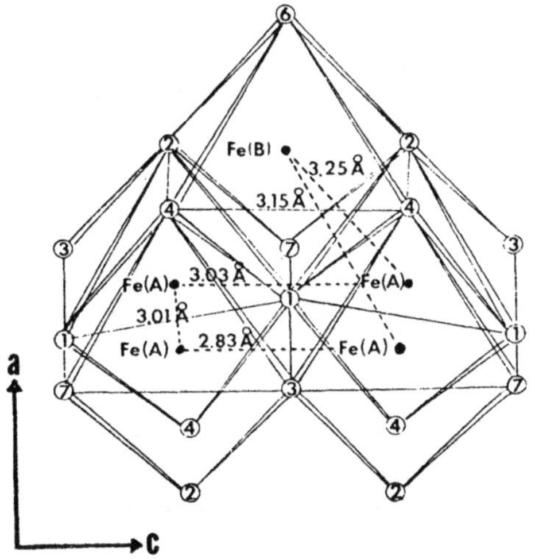

Figure 5 Projection of one unit of the Fe(A) double chain showing possible Fe(A)-Fe(A) and Fe(A)-Fe(B) interactions across shared edges and appropriate metal-metal distances (from refs. 30-33).

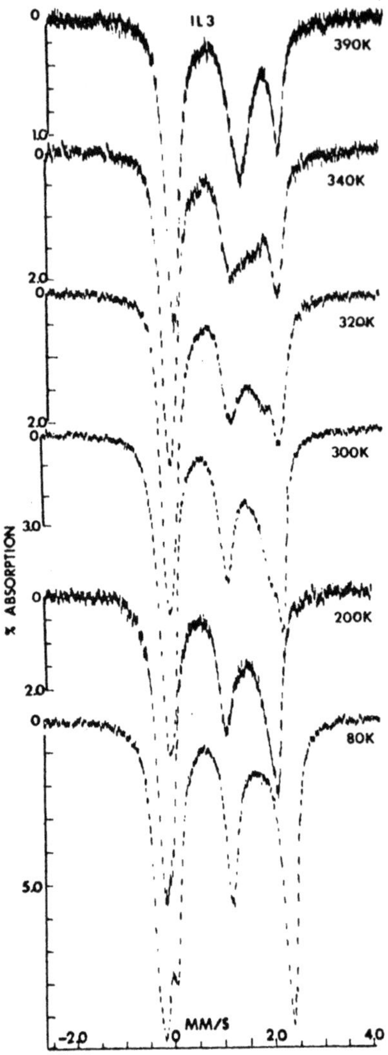

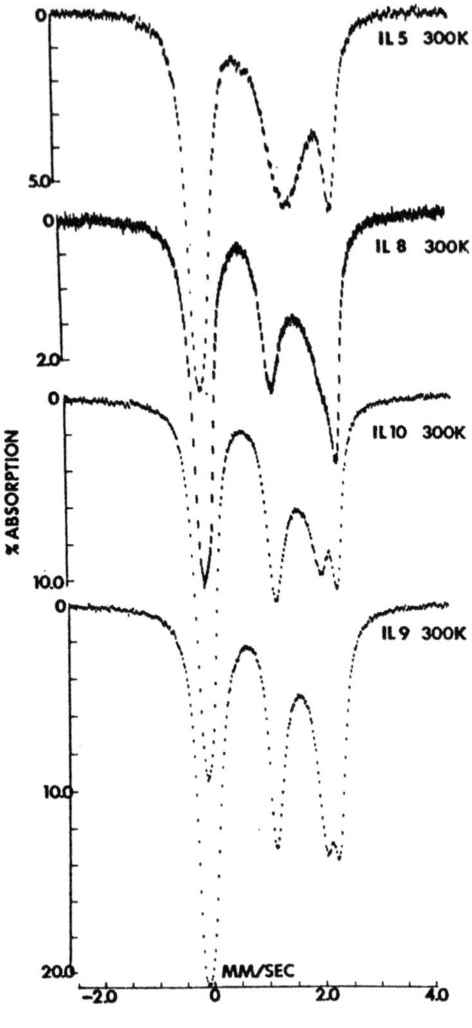

Figure 6 (left) Temperature variations of the Mössbauer spectra of an ilvaite specimen (from refs. 30-33). Note the variations in the 1.5 mm/sec region.

Figure 7 (right) Mössbauer spectra of several ilvaite specimens showing the variability of spectrum profiles (from refs. 30-33). Compare the variations with the temperature variations shown in Figure 6.

sites accomodate Fe^{2+} ions (24). The structural possibilities for Fe^{2+}-Fe^{3+} interactions across a shared edge are many, as seen in Figure 5. Thus, Fe(A)-Fe(A) couples occur along the chain unit in the c direction (Fe-Fe distances of 283 and 303pm), and across the chain in the a-b plane (Fe-Fe = 301pm). Four Fe(B)-Fe(A) interactions occur per Fe(B) site with Fe-Fe distances ranging from 315 to 325pm.

Electron delocalization behavior has been deduced from the temperature variation of the Mössbauer spectra of ilvaite (26-30) illustrated in Figure 6 (31-33). Gerard and Grandjean (26, 27) ascribed the variations to one ferrous iron undergoing a change (i.e., rapid electron transfer) which gives it a ferric character. They fitted their Mössbauer spectra to three doublets assigned to one Fe^{3+} and two Fe^{2+} species. One high velocity Fe^{2+} peak was interpreted to move to lower velocity with increasing temperature so as to overlap the ferric high velocity component. The authors compared this transition at about 320K to the magnetite Verwey transition at 119K, and suggested that an ionic order-disorder transition occurred in ilvaite at about 320K.

The spectral profiles of several ilvaite specimens of similar bulk composition illustrated in Figure 7 show considerable variation in the region 1.2-2.0 mm/sec. Increasing absorption around 1.5 mm/sec of the four samples can be seen from bottom to top of Figure 7 with absorption in this region masking any 1.2 mm/sec absorption in the IL5 spectrum. Also varying is the 1.9 mm/sec peak resolved in IL9 and IL10 but appearing only as a shoulder in IL8 and not at all in IL5. All spectra are characterized by strong overlap of the low velocity components of all doublets centered near -0.1 mm/sec.

The Mössbauer spectra of ilvaites have been successfully fitted to five component doublets (30-33), the assignments of which are given in Figure 8. In addition to one Fe^{3+} doublet and two Fe^{2+} doublets representing cations in the Fe(A) and Fe(B) sites, two doublets representing delocalized $Fe^{2+} \rightarrow Fe^{3+}$ species, one assigned to $Fe^{2+}(A) \rightarrow Fe^{3+}(A)$ parallel, and the other to $Fe^{2+}(A) \rightarrow Fe^{3+}(A)$ perpendicular, to the chains of edge-shared Fe(A) octahedra (Figure 5). The latter two assignments explain the opacity and black color of ilvaite in all directions.

It is immediately apparent from the fitted spectra that the cause of the variations in the ilvaite Mössbauer spectra is not anomalous movement of the peaks with temperature or changing character of the iron species, but intensity changes indicating changing populations of electronic levels. Thus, the reported electronic transition in ilvaite (26, 27) is not

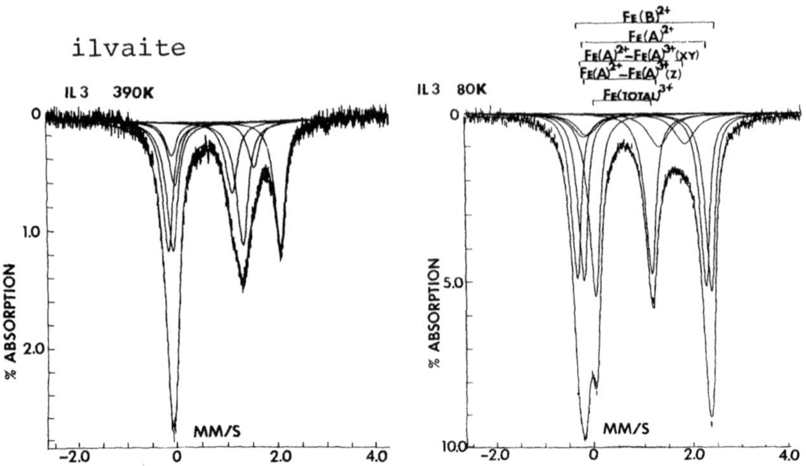

Figure 8 Fitted Mössbauer spectra for ilvaite specimen IL3 (from ref. 33). (a) the 390K spectrum fitted to eight peaks or four doublets, Fe^{2+}(A) is below resolution of spectrometer; (b) the 80K spectrum showing assignments of component doublets.

a localized-delocalized order-disorder transition but is merely a point at which peaks due to $Fe^{2+}(A) \rightarrow Fe^{3+}(A)$ species were resolvable within the experimental limits of earlier work. The more recent study (30-33) indicates rather that certain electronic levels associated with trapped valencies become depopulated with increasing temperature contemporaneous with population of new delocalized levels associated with "intermediate" valencies.

In summary, ilvaite with its infinite chains of edge-shared Fe^{2+}-Fe^{3+} octahedra is a typical example of a Class IIIB mixed-valence compound (10) with respect to the Fe(A) octahedra. The short Fe(A) - Fe(A) distances, 283-303pm, are particularly conducive to ED. The Fe(A) - Fe(B) octahedral clusters, however, probably constitute a Class II compound, the somewhat larger Fe(A) - Fe(B) distances, 315-325pm, being less conducive to ED. Ilvaite is representative of several end-member Fe^{2+}-Fe^{3+} silicate minerals in which ED is predicted or observed along chains of linked octahedra. Further examples are discussed later.

GLAUCOPHANE: A MINERAL WITH EXTENDED CLUSTERS OF Fe^{2+}-Fe^{3+} OCTAHEDRA LIMITED BY SUBSTITUTIONAL BLOCKING

Magnetite and ilvaite represent end-member Fe^{2+}-Fe^{3+} mixed-valence compounds whose crystal structures contain infinite chains of edge-shared [FeO_6] octahedra permitting electron delocalization. Numerous silicate minerals have crystal structures in which infinite chains of edge-shared octahedra exist, but atomic substitution of non-transition metal ions such as Mg^{2+} and Al^{3+} with ionic radii comparable to those of Fe^{2+} and Fe^{3+} limit the extent of $Fe^{2+} \rightarrow Fe^{3+}$ interactions by substitutional blocking. Minerals of this type are dominated by CT optical absorptions instead of ED. The intensities of the CT bands generally increase with the concentration product [Fe^{2+}] [Fe^{3+}] and with the extent of ordering of Fe^{2+} and Fe^{3+} ions in adjacent sites.

Glaucophane, an alkali amphibole with the general formula:

$$Na_2(Mg,Fe^{2+})_3(Al,Fe^{3+})_2Si_8O_{22}(OH)_2,$$

is the model mineral for substitutional blocking. The Fe^{2+}-Fe^{3+} end-member is riebeckite

$$Na_2Fe_3^{2+}Fe_2^{3+}Si_8O_{22}(OH)_2,$$

the fibrous or asbestiform variety of which is called crocidolite. The glaucophane structure (34) illustrated in Figure 9 consists of a band of octahedral sites, designated M1, M2 and M3, extending along the c axis. The larger M1 and M3 are occupied preferentially by Mg^{2+} and Fe^{2+} ions, while the smaller Al^{3+} and Fe^{3+} ions are enriched in the M2 sites. Metal-metal distances across the edge-shared octahedra are in the range 309-331pm (Figure 10). Intervalence $Fe^{2+} \rightarrow Fe^{3+}$ CT transitions, primarily Fe^{2+}(M1) $\rightarrow$ Fe^{3+}(M2) and Fe^{2+}(M3) $\rightarrow$ Fe^{3+}(M2), account for the blue-violet pleochroism and absorption bands in the region 16,130-18,520 cm^{-1} when light is polarized in the bc plane of linked octahedra (Figure 11). Glaucophane is colorless when light is polarized perpendicular to the bc plane because the electric vector does not coincide with Fe^{2+}-Fe^{3+} vectors. Strong charge transfer bands have been measured in the optical spectra of several alkali amphiboles (7, 35-38), and shown to have intensities proportional to the product of donor Fe^{2+} and acceptor Fe^{3+} concentrations (7,35). A similar concentration dependence was found for the electrical conductivity of crocidolite (35) which is significantly high along the fibre axis (the direction of infinite chains of edge-shared [FeO_6] octahedra). However, although the Mössbauer spectra of several compositions along the glaucophane-riebeckite solid solution series have been measured, only doublets attributable

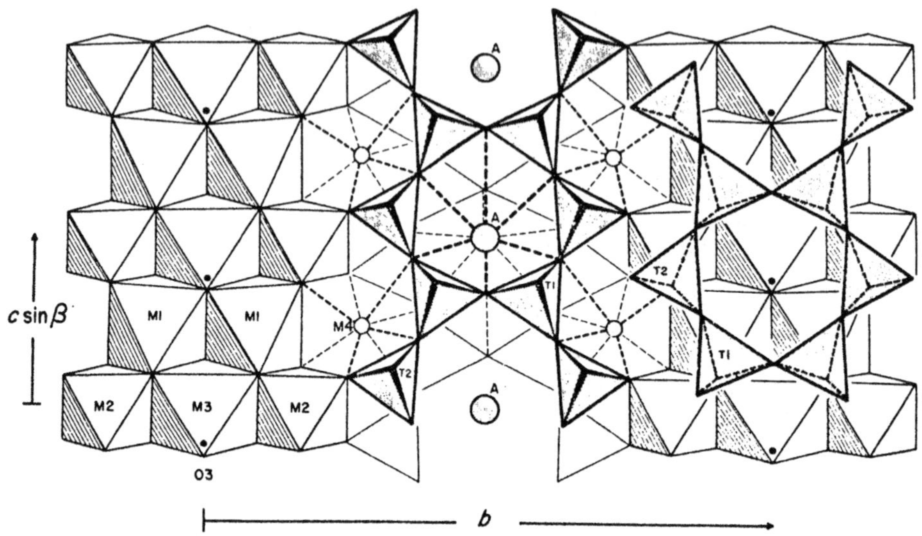

Figure 9 The glaucophane crystal structure, showing the infinite bands of edge-shared octahedra (from ref. 34).

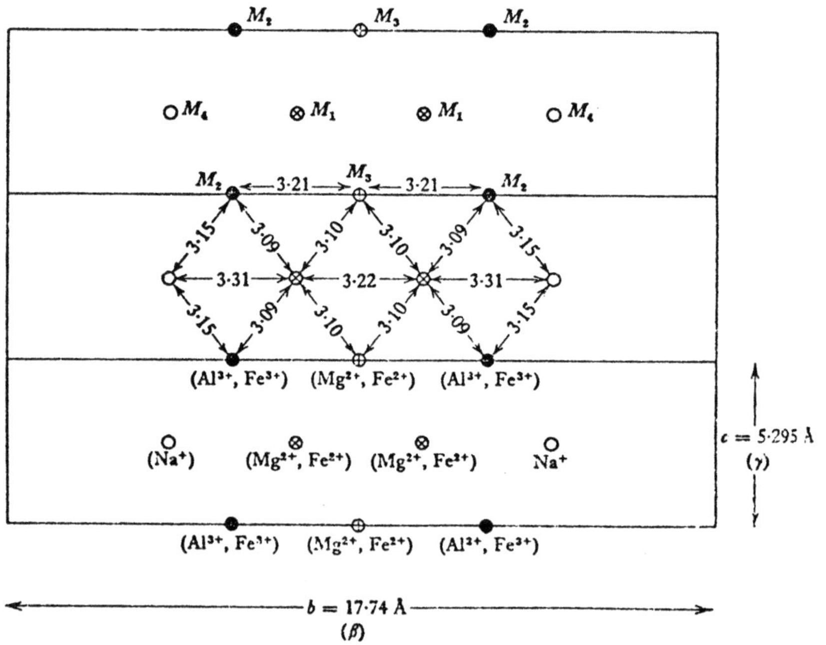

Figure 10 Projection of the octahedral sites on (100) showing metal-metal distances and site occupancies in glaucophane (from refs. 4 and 36).

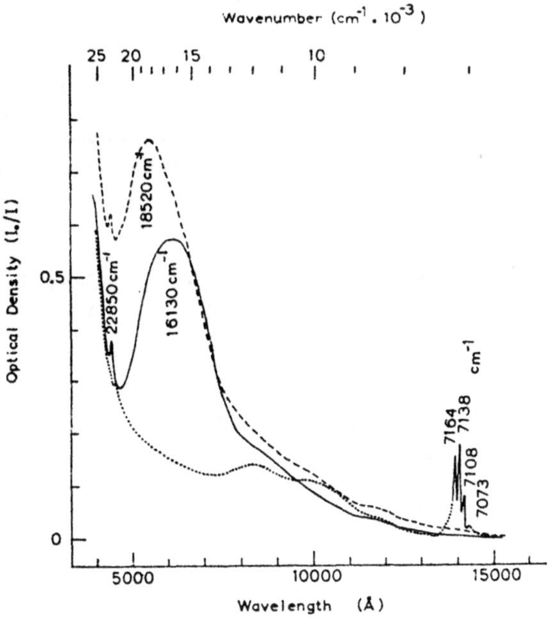

Figure 11 Polarized absorption spectra of glaucophane, showing the intense Fe^{2+}-Fe^{3+} CT band at 16,130-18,520 cm^{-1} when light is polarized in the bc plane of linked octahedra (see Figures 9 and 10) (from refs. 4 and 36).

to trapped valencies Fe^{2+}(M1), Fe^{2+}(M3) and Fe^{3+} (octahedral) have been resolved (36, 39), suggesting that ED is limited or undetectable by Mössbauer spectroscopy. Negligible ED in glaucophane is certainly attributable to Mg^{2+} and Al^{3+} blocking $Fe^{2+} \rightarrow Fe^{3+}$ interactions along the chains of edge-shared octahedra. In the riebeckite end-member composition, however, ED may also be limited by the somewhat larger Fe-Fe distances (compared to ilvaite and magnetite) and the subtle dissimilarities between the M1, M2 and M3 octahedra which render alkali amphiboles Class II mixed-valence compounds. Pairs of edge-shared M1 octahedra are parallel to the b axis (Figures 9 and 10), so that riebeckite is a potential Class IIIA compound. It could become a Class IIIB compound if cation disordering could be achieved leading to a random distribution of Fe^{2+} and Fe^{3+} ions over the M1, M2 and M3 sites. Such disordering may have been induced in earlier conductivity measurements of heated crocidolites (35), while delocalized Fe^{2+}-Fe^{3+} species may contribute to the Mössbauer spectra of riebeckites formed in high temperature mineral parageneses such as granite pegmatites (36).

Table 1 Structural and spectral data for mixed-valence

Mineral (Formula)	Sites (Coord. No.)	M-M Distances (pm)	Polyhedral Linkages (type of sharing; extent of M-M cluster)
vivianite $Fe_3(PO_4)_2 \cdot 8H_2O$	Fe_A (6) Fe_B (6)	Fe_B-Fe_B=285	edge: isolated clusters;
sapphire Al_2O_3/Fe,Ti	Al (6)	⊥c = 279 ∥c = 265	edge: finite clusters; face: isolated clusters
kyanite Al_2SiO_5/Fe,Ti	Al_1 (6) Al_2 (6) Al_3 (6) Al_4 (6)	275-278	edge: infinite Al_1-Al_2 chains; isolated Al_1-Al_2, Al_2-Al_3, Al_1-Al_4, Al_2-Al_4 clusters; substitutional blocking by Al^{3+}
aquamarine (beryl) $Be_3Al_2Si_6O_{18}$	Al (6)	∥c = 460	isolated octahedra; cations possibly in channels
cordierite (iolite) $Al_3(Mg,Fe^{2+})_2$ $Al,Fe^{3+})Si_5O_{18}$	Mg (6) T_1 (4)	Mg-T_1 = 274	edge: isolated octahedral-tetrahedral clusters; cations possibly in channels
tourmaline $Na(Mg,Fe,Mn,Li,Al)_3$ $(Al,Fe)_6(Si_6O_{18})$ $(BO_3)_3(OH,F)_4$	Mg (6) Al (6)	Mg-Mg = 304 Mg-Al = 297 Al-Al = 280	edge: trigonal planar clusters of Mg-Mg and Mg-Al ⊥c; infinite spiral clusters of Al-Al ∥c; substitutional blocking by Mg^{2+} and Al^{3+}
yoderite $(Al,Mg,Fe,Mn)_8$ $Si_4O_{18}(OH)_2$	Al_1 (6) Al_2 (5) Al_3 (5)	Al_1-Al_1=290 Al_1-Al_2=350 Al_1-Al_3=328	edge: infinite Al_1-Al_1 chains; isolated Al_1-Al_2 and Al_1-Al_3 clusters; substitutional blocking by Mg^{2+} & Al^{3+}
orthopyroxenes $(Mg,Fe)_2Si_2O_6$	M1 (6) M2 (6)	M1-M1 = 315 M1-M2 = 308 and 327	infinite M1-M1 and M1-M2 chains ∥c; substitutional blocking by Mg^{2+}

minerals of iron and titanium

CT Energy (cm^{-1}) and assignment	Comments (Moss. Spect.; Elect. Cond.; T variations, etc.)	References
15,200: $Fe^{2+} \rightarrow Fe^{3+}$	M/S spectra:trapped valence inverse T depend. of CT; press. intensifies	8-14
17,000 (E⊥C): $Fe^{2+} \rightarrow Ti^{4+}$ 11,150 (E⊥C): $Fe^{2+} \rightarrow Ti^{4+}$ 12,900 (E∥ C): $Fe^{2+} \rightarrow Fe^{3+}$ 9,700 (E∥ C): $Fe^{2+} \rightarrow Fe^{3+}$	No M/S; inverse T depend. of CT; debate over assignment	7,54,69-76 53
16,500: $Fe^{2+} \rightarrow Ti^{4+}$ 11,500: $Fe^{2+} \rightarrow Fe^{3+}$	M/S confirm Fe^{2+}; inverse T depend. of CT; debate over assignment	7,42,75, 77-80
16,100 (E ∥ C): $Fe^{2+} \rightarrow Fe^{3+}$	M/S confirm Fe^{2+}; debate over assignment	1,42,81-83
17,500 (E⊥C): $Fe^{2+} \rightarrow Fe^{3+}$	M/S detect Fe^{3+} at 77K only; inverse T depend. of CT	2,3,7,40-42,84,85
18,000 (E⊥C >> E∥ c): $Fe^{2+}(Mg) \rightarrow Fe^{3+}(Mg,Al)$	M/S detects Fe^{2+} in Mg and Al: trapped valencies; inverse T depend. of CT; considerable debate over assignment	2,3,7,44, 54, 86-93 53
13,800: $Fe^{2+}(Al_1) \rightarrow Fe^{3+}(Al_1)$ 16,500: $Mn^{2+}(Al_1) \rightarrow Mn^{3+}(Al_1)$ 21,000: $Mn^{2+}(Al_1) \rightarrow Mn^{3+}(Al_2)$	M/S detects Fe^{2+} and Fe^{3+}: trapped valencies	94,95
14,500 (E ∥ C): $Fe^{2+}(Mg,M1) \rightarrow Fe^{3+}(M1)$	M/S detects Fe^{2+} and Fe^{3+}; Fe^{2+} strongly ordered in distorted M2 site	4,96

Table 1 (continued) Structural and spectral data for

Mineral (Formula)	Sites (Coord. No.)	M-M Distances (pm)	Polyhedral Linkages (type of sharing; extent of M-M cluster)
augite $Ca(Mg,Fe,Al)(Si,Al)_2O_6$	M1 (6) M2 (8)	M1-M1 = 311 M1-M2 = 321	same as orthopyroxene
omphacite $(Na,Ca)(Al,Fe,Ti)Si_2O_6$	M1 (6)	M1-M1 = 290	similar to ortho-pyroxene
glaucophane $Na_2(Mg,Fe^{2+})_3(Al,Fe^{3+})_2Si_8O_{22}(OH)_2$	M1 (6) M2 (6) M3 (6) M4 (8)	M1-M1 = 322 M1-M2 = 309 M1-M3 = 310 M2-M3 = 331 M2-M4 = 315 M1-M4 = 331	infinte bands of edge-shared octahedra ‖ c; substitutional blocking by Mg^{2+} and Al^{3+}
riebeckite, crocidolite $Na_2Fe_3^{2+}Fe_2^{3+}Si_8O_{22}(OH)_2$	as for glauco-phane	comparable to glauco-phane	infinite bands of edge-shared $[FeO_6]$ octahedra;
actinolite-hornblende $Ca_2(Mg,Fe,Al)_5(Si,Al)_8O_{22}(OH)_2$	as for glauco-phane	M1-M1 = 323 M1-M2 = 311 M1-M3 = 310 M2-M3 = 320 M2-M4 = 320 M1-M4 = 342	similar to glaucophane
osumilite $K(Mg,Fe^{2+})_2(Al,Fe^{3+})_3(Si,Al)_{12}O_{30}$	similar to cordier-ite	comparable to cordierite	similar to cordierite
biotite (annite-phlogopite) $K(Mg,Fe^{2+},Fe^{3+})_{2-3}(Si_3AlO_{10})(OH)_2$	M1 (6) M2 (6)	M1-M1 and M1-M2 ≈ 310-320pm	infinite 2-D sheets of edge-shared octahedra
babingtonite $Ca_2Fe^{2+}Fe^{3+}Si_5O_{14}(OH)$	Fe_1 (6) Fe_2 (6)	Fe_1-Fe_1=337pm Fe_1-Fe_2=330pm	chains of edge-shared $[FeO_6]$ octahedra

mixed-valence minerals of iron and titanium

CT Energy (cm^{-1}) and assignment	Comments (Moss. Spect.; Elect. Cond.; T variations, etc.)	References
13,000: $Fe^{2+} \rightarrow Fe^{3+}$	M/S detects Fe^{2+} and Fe^{3+}; trapped valencies	4,97,98
15,040: $Fe^{2+} \rightarrow Fe^{3+}$ or $Fe^{2+} \rightarrow Ti^{4+}$	Blue Ti variety; M/S detects Fe^{2+} and Fe^{3+} trapped valencies; press. variations	99-101, 136
18,520 (E ∥ b): Fe^{2+}(M1,M3) $\rightarrow Fe^{3+}$(M2) 16,130 (E ∥ c): Fe^{2+}(M1) $\rightarrow Fe^{3+}$(M2)	M/S detects Fe^{2+} and Fe^{3+}; trapped ; Fe^{2+} ordered in M1 and M3; inverse T depend. of CT	4,7,34, 36,39
15,000-18,000: $Fe^{2+} \rightarrow Fe^{3+}$	High elect. cond.; M/S detects Fe^{2+} and Fe^{3+}: trapped valencies	4,7,9, 35-38
14,200 (E∥ b) and 13,700 (E⊥c): $Fe^{2+} \rightarrow Fe^{3+}$	M/S detects Fe^{2+} in M1, M2, M3: trapped valencies	4, 102-106
15,480cm^{-1} (E ⊥ c): Fe^{2+}(oct) $\rightarrow Fe^{3+}$(tet); or Fe^{2+}(oct) $\rightarrow Fe^{3+}$ (channel)	M/S detect Fe^{2+} in oct. and channel sites; Fe^{3+} in tet.; trapped valencies	107-109
13,650 and 16,400cm^{-1} $Fe^{2+} \rightarrow Fe^{3+}$ (M1 → M2 and M2 → M2)	M/S resolved into four doublets; Fe^{2+} and Fe^{3+} in M1 and M2; trapped valencies	7,53,93, 104,110-112,54
opaque-very dark green	M/S detects Fe^{2+} and Fe^{3+}, each in only one oct. site	113-116

Table 1 (continued) Structural and spectral data for

Mineral (Formula)	Sites (Coord. No.)	M-M Distances (pm)	Polyhedral Linkages (type of sharing; extent of M-M cluster)
ilvaite $CaFe_2^{2+}Fe^{3+}Si_2O_7(OH)$	Fe_A (6) Fe_B (6)	Fe_A-Fe_A=283 and 303pm (∥ c) and 301pm (⊥ c), Fe_A-Fe_B=315 and 325pm	infinite double chains of edge-shared Fe_A octahedra ∥ c
deerite $Fe_6^{2+}Fe_3^{3+}O_3(Si_6O_{17})(OH)_5$	cations in nine octahedral sites	range of M-M dist. 311-331pm	infinite bands, six octahedra wide, of edge-shared $[FeO_6]$ octahedra;
howieite $NaFe_{10}^{2+}Fe_2^{3+}(Si_{12}O_{34})(OH)_{10}$	cations in six octahedral sites	range of M-M dist. 316-329pm	infinite bands, four octahedra wide of edge-shared $[FeO_6]$ octahedra
sapphirine $(Mg,Fe^{2+},Al,Fe^{3+})_7(Si,Al)_6O_{20}$	cations in seven octhedral sites	range of M-M dist. 318-335pm	infinite bands four and three octahedra wide of edge-shared octahedra
aenigmatite $Na_2Fe_5TiSi_6O_{20}$	similar to sapphirine	comparable to sapphirine	similar to sapphirine
magnetite Fe_3O_4	Fe_A (4) Fe_B (6)	Fe_B-Fe_B=297pm Fe_A-Fe_B=348pm	infinite single chains of edge-shared $[FeO_6]$ running in 3-D.

mixed-valence minerals of iron and titanium

CT Energy (cm^{-1}) and assignment	Comments (Moss. Spect.; Elect. Cond.; T variations, etc.)	References
opaque	M/S detect trapped valencies Fe^{2+} in Fe_A and Fe_B, and Fe^{3+} in Fe_A; two ED species detected assigned to $Fe_A^{2+} \rightarrow Fe_A^{3+}$ $\parallel$ c and $\perp$ c.	23-33, 117
opaque	low mag. susc.; M/S detect trapped valencies Fe^{2+} and Fe^{3+} each in two or more oct. sites; at least two ED species detected	118-123
dark-green; no electronic spectra	M/S detect trapped valencies Fe^{2+} and Fe^{3+} ions; suggestion of some ED species	120-124
blue-green;no electronic spectra	M/S detect trapped valencies Fe^{2+} and Fe^{3+}	120,125-127
dark blue; no electronic spectra	M/S detect trapped valencies Fe^{2+} and Fe^{3+}	128
opaque	high elect. cond. above 119K; M/S resolves Fe_A^{3+} (tet) and ED species in Fe_B	15-22

OTHER MINERALS WITH $Fe^{2+} \rightarrow Fe^{3+}$ INTERACTIONS ACROSS EDGE-SHARED OCTAHEDRA

Substitutional blocking typified by glaucophane is an important factor controlling the electronic properties of minerals. Several iron-bearing silicates contain infinite chains, bands, or sheets of edge-shared octahedra, but ED is prevented by atomic substitution of Mg^{2+} and Al^{3+} in solid solution series which block extended $Fe^{2+} \rightarrow Fe^{3+}$ interactions. However, intense CT bands are frequently observed in polarized spectra of such minerals measured in the visible region. Examples of minerals showing trapped valencies as a result of atomic substitution, to which $Fe^{2+} \rightarrow Fe^{3+}$ CT transitions across edge-shared octahedra have been assigned, are listed in Table 1, and include micas of the phlogopite-biotite series, amphiboles of the actinolite-hornblende series, many pyroxenes, omphacite, certain tourmalines, chlorite, kyanite, and potentially sapphirine.

On the other hand, pure end-member Fe^{2+}-Fe^{3+} silicates exhibiting ED between octahedrally coordinated iron cations include deerite in addition to ilvaite and riebeckite (Table 1). Other minerals with infinite linkages of edge-shared $[FeO_6]$ octahedra which might be predicted to display ED include howieite, aenigmatite, babingtonite, annite, and stilpnomelane.

MINERALS WITH $Fe^{2+} \rightarrow Fe^{3+}$ INTERACTIONS IN OTHER COORDINATION SYMMETRIES

The majority of mixed-valence iron-bearing minerals displaying $Fe^{2+} \rightarrow Fe^{3+}$ interactions contain cations in edge-shared octahedra. However, intervalence transitions have been documented between Fe cations in other coordination symmetries.

Cordierite

The best known example of Fe^{2+}(octahedral) $\rightarrow Fe^{3+}$(tetrahedral) interactions is the mineral cordierite,

$$Al_3(Mg,Fe^{2+})_2[(Al,Fe^{3+})Si_5O_{18}]$$

The crystal structure of cordierite (40) contains M octahedra which share edges with two T_1 and one T_2 tetrahedra. Fe^{3+} ions are believed to be concentrated in the T_1 tetrahedra, while Fe^{2+} ions occupy M octahedra. The M-T_1 distance across the isolated M octahedron-T_1 tetrahedron clusters is only 274pm. The absorption spectra of cordierite (2, 3, 7, 41) show a broad intense band at about 17,000 cm^{-1} polarized in the plane of the edge-shared M and T_1 polyhedra, the intensity of which increases at low temperatures (7). The Mössbauer spectra of such cordierites, however, show no evidence of Fe^{3+} at 293K,

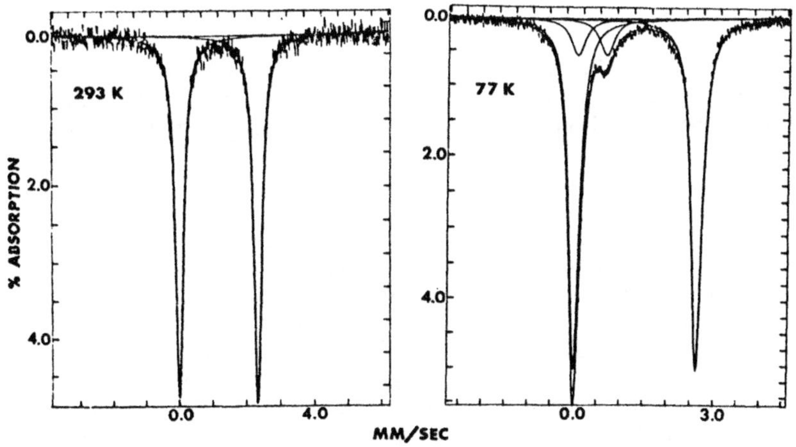

Figure 12 Mössbauer spectra of a cordierite, the chemical analysis of which is reported to contain about 22% of the iron as Fe(III). Same specimen as that used for polarized spectral measurements in ref. 3. Note that the 293K spectrum shows only an Fe^{2+} doublet, at 77K an Fe^{3+} doublet emerges, but accounts for only 13% of the computed area (from ref. 42).

(42), even in specimens reported to contain as much as 22% of the iron in the ferric state (Figure 12). It is only at 77K that ferric iron is detected in the Mössbauer spectra, but then accounting for smaller proportions of the chemically analysed Fe^{3+} ions. The failure to detect Fe^{3+} ions in the 293K spectra might be explained if the electron transfer between adjacent Fe^{2+} and Fe^{3+} ions is faster than the Mössbauer transition (10^{-7} sec). Ferric ions could appear in the 77K spectra because electron hopping is partially quenched. The phenomenon would be comparable to ED observed in magnetite between octahedral Fe^{2+} and Fe^{3+} ions which are indistinguishable above the Verwey temperature (119K). In the Mössbauer spectra of magnetite above 119K, the parameters are intermediate between those expected for octahedral Fe^{2+} and Fe^{3+} (15, 17). Such averaging is not observed, however, in the 293K Mössbauer spectra of cordierites, even in specimens containing Fe^{3+}/total Fe ratios as high as 0.25. Another possible explanation (41), is that the $Fe^{2+} \rightarrow Fe^{3+}$ CT transition involves Fe^{3+} ions located in the channels enclosed by the hexagonal rings of corner-shared $[SiO_4]$ tetrahedra in the cordierite structure. Iron in a loosely bound channel-site would have an anomalously low recoil-free fraction and not contribute to the room-temperature Mössbauer spectra of cordierite. The problem remains unresolved at this time.

Fe-Ti Garnets

Garnets of the andradite-melanite-schorlomite series represent another mineral group containing Fe^{2+}-Fe^{3+} clusters of different coordination symmetries. Many andradites,

$$\{Ca\}_3[Fe^{3+}]_2(SiO_4)_3$$

contain high titanium contents, leading to subsilicic compositions and unusual crystal chemistries of Fe and Ti. In the garnet structure illustrated in Figure 13, {X} cations are coordinated by eight oxygens at the vertices of a distorted cube described as an 8-fold triangular dodecahedron (43). The [Y] cations are octahedrally coordinated and (Z) cations are in tetrahedral coordination, both by oxygens. The structure consists of alternating (ZO_4) tetrahedra and $[YO_6]$ octahedra sharing corners to form a continuous three dimensional framework. The structure is in the orthosilicate class as (ZO_4) tetrahedra share no corners with each other. The $[YO_6]$ octahedra also do not share corners with one another, nor do they share edges as in most silicate minerals. The only edge-sharing interaction involves $\{XO_8\}$ dodecahedra. In particular, three dimensional chains of edge-shared $\{XO_8\}$ dodecahedra and (ZO_4) tetrahedra are present in the garnet structure, and in andradite the X-Z interatomic distance is only 301pm.

Mössbauer measurements of several titanium-bearing andradites (i.e., melanites and schorlomites) have revealed the presence of coexisting Fe^{2+} and Fe^{3+} ions, and have demonstrated that Fe^{3+} ions, and not Ti^{4+}, make up the deficiencies in the tetrahedral sites of these sub-silicic garnets (44-49). Up to five component doublets have been fitted to the spectra illustrated in Figure 14, four of which may be unambiguously assigned to Fe^{3+} ions in octahedral [Y] and tetrahedral (Z) sites and to Fe^{2+} ions in 8-fold dodecahedral {X} and octahedral [Y] sites (45-49). The fifth doublet showing unusual temperature variations (48) may be resolved in the Mössbauer spectra of high Ti garnets, and its assignment has aroused considerable debate (47-49). Its parameters are suggestive of tetrahedral Fe^{2+} ions (47, 48), but the substitution of large Fe^{2+} ions in the small Si^{4+} sites is unlikely (49), particularly when octahedral Fe^{3+} ions predominate. An alternative assignment of the fifth doublet to $Fe^{2+}\{X\} \rightarrow Fe^{3+}(Z)$ ED has been proposed (49). This assignment not only argues against tetrahedral Fe^{2+} replacing Si^{4+} in the garnet structure, but also correlates with the short X-Z separation of 301pm between edge-shared $\{XO_8\}$ and (ZO_4) tetrahedra, and with electronic spectral data. The near infrared spectra of Ti garnets show an absorption band at 5,280 cm^{-1} (44, 47, 50), which was originally assigned to the crystal field $^5E \rightarrow {^5T_2}$ transition in tetrahedral Fe^{2+}

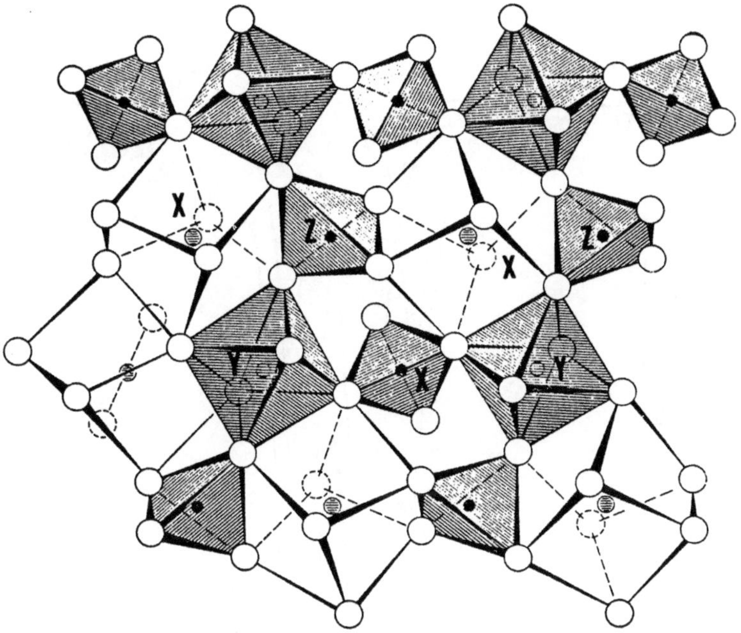

Figure 13 Portion of the garnet structure projected down the c axis (from ref. 43). Note the chains of alternating edge-shared XO_8 dodecahedra and ZO_4 tetrahedra; in andradite the X-Z distance is only 301pm.

ions (47, 51, 52). However, the temperature dependence of the 5,280 cm^{-1} band, which decreases in intensity with rising temperature (50), is not consistent with a crystal field transition. Instead, it conforms with the temperature variations found for CT bands in other minerals (7, 53, 54). Furthermore, electrical conductivity measurements on garnets show that pure andradite is an insulator, but becomes a semiconductor as the Ti content increases (50). The activation energy for conduction was calculated to be about 3,800 cm^{-1} (50), in agreement with the energy of the onset of the CT band centered at 5,280 cm^{-1}. These results support the assignment of a $Fe^{2+}\{X\} \rightarrow Fe^{3+}(Z)$ ED doublet in the Mössbauer spectra of Fe-Ti garnets.

Although the garnet structure contains infinite chains of edge-shared $\{XO_8\}$ and (ZO_4) tetrahedra (Figure 13), the extent of $Fe^{2+}\{X\} - Fe^{3+}(Z)$ interactions is limited because Ca^{2+} and Si^{4+} are the major constituents of the X and Z sites, respectively, of andradite. It has been suggested (49) that the $Fe^{2+}\{X\} - Fe^{3+}(Z)$ cluster overlap is facilitated by the similar relative energy levels of t_{2g} (t_2) and e_g (e) molecular orbitals in dodecahedral and tetrahedral coordinations.

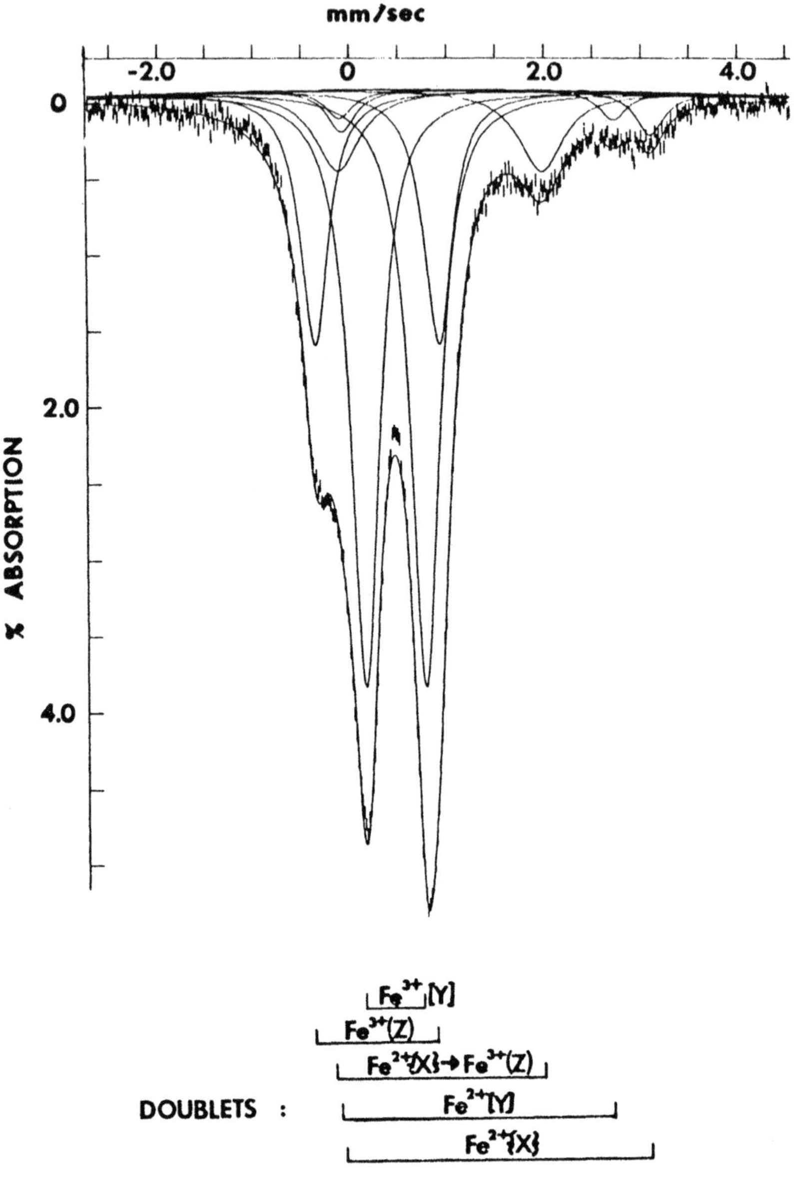

Figure 14 Mössbauer spectrum of a titanian andradite fitted to five component doublets (from ref. 49). Note that one doublet is assigned to Fe^{2+}[X]-Fe^{3+}(Z) electron delocalization.

In summary, the $\{XO_8\}$-(ZO_4) coordination clusters in the garnet structure suggest that this mineral should be a Class I mixed-valence compound (10). However, the ED properties of $Fe^{2+}\{X\}$ -$Fe^{3+}(Z)$ clusters in Fe-Ti garnets are more indicative of a Class III compound.

MINERALS WITH Ti^{3+} Ti^{4+} OCTAHEDRAL CLUSTERS

The Ti^{III} oxidation state is rare in terrestrial minerals due to the comparatively high redox conditions on Earth. However, Ti^{3+} ions may exist in some high pressure phases in the Mantle (55). Trivalent titanium does occur in extraterrestrial materials, however, including titaniferous pyroxenes and glasses from the Moon and certain meteorites (56-59). One such pyroxene occurs in the meteorite that fell at Pueblo de Allende, Mexico, in 1969. Chemical analysis of the Allende pyroxene (59) reveal that it is iron-free, and lead to the formula:

$$Ca_{1.01}Mg_{0.38}Ti^{3+}_{0.34}Ti^{4+}_{0.14}Al_{0.87}Si_{1.26}O_6$$

The crystal structure, like other calcic clinopyroxenes (97), contains chains of edge-shared M1 octahedral sites extending along the c axis, and M1-M1 distances are 315pm. Absorption bands in the polarized spectra illustrated in Figure 15a occur around 20,000 cm^{-1}, and 16,000 cm^{-1}, and were originally assigned (59) to Ti^{3+} crystal field (20,000 cm^{-1}) and $Ti^{3+} \rightarrow Ti^{4+}$ CT (16,000 cm^{-1}) transitions, respectively. An alternative assignment (60), which proposed that both absorption bands originated from crystal field transitions in Ti^{3+} ions in the distorted M1 octahedral sites, was debated (60, 61). Subsequently, measurements of the polarized spectra of the Allende pyroxene at elevated pressures (62) resolved the conflict. The spectra illustrated in Figures 15b and c show that, in addition to the bands at 16,000 and 20,000 cm^{-1}, another occurs around 15,000 cm^{-1} and intensifies at high pressures. Furthermore, the position of the 15,000 cm^{-1} band is insensitive to pressure, whereas the 16,000 cm^{-1} and 20,000 cm^{-1} bands shift to higher energies with rising pressure, which is consistent with their assignment to Ti^{3+} crystal field transitions. Such pressure measurements demonstrate conclusively that the $Ti^{3+} \rightarrow Ti^{4+}$ CT transition in pyroxenes occurs at 15,000 cm^{-1}.

The $Ti^{3+} \rightarrow Ti^{4+}$ CT transition has been assigned to the spectra of other synthetic phases and natural minerals (5), including andalusite at 20,800 cm^{-1} (63) and tourmaline at 22,000 cm^{-1} (64). Pressure-dependent studies of the absorption spectra are required to confirm these assignments.

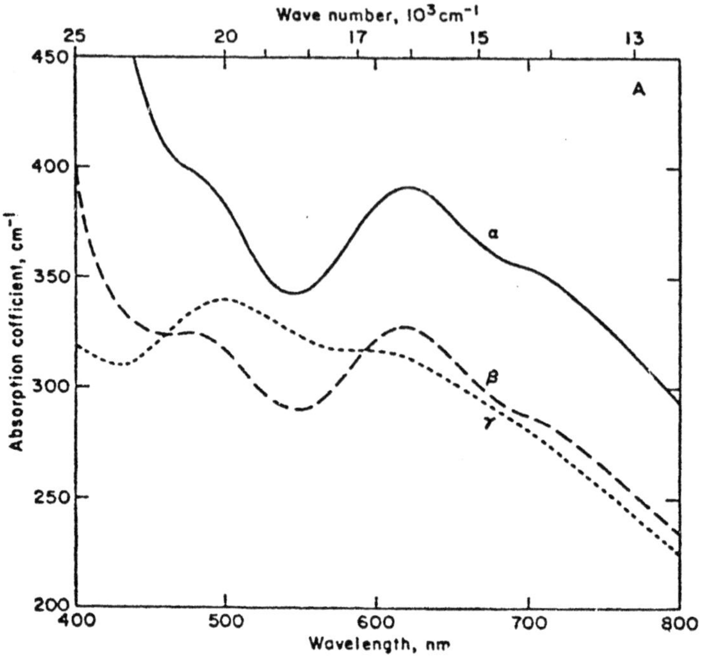
Wave number, 10^3 cm^{-1}
25
20
17
15
13
A
450
400
350
300
250
200
Absorption coefficient, cm^{-1}
a
β
γ
400
500
600
700
800
Wavelength, nm

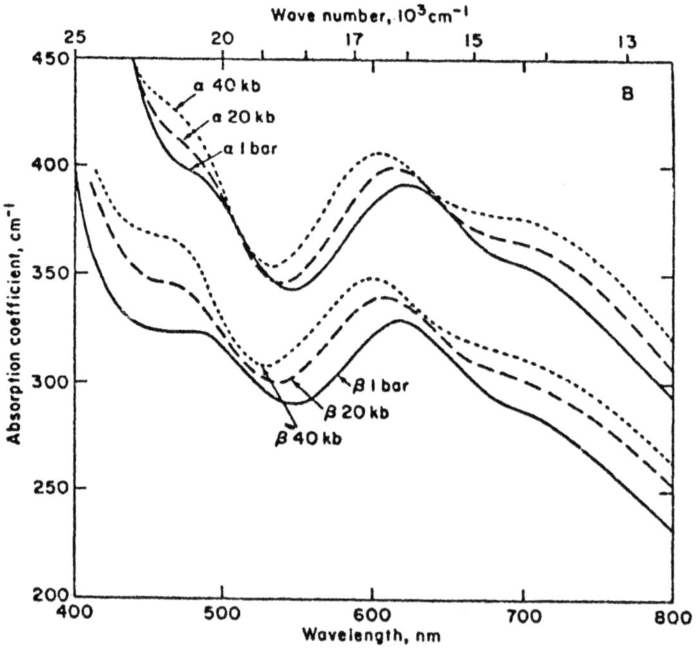
Wave number, 10^3 cm^{-1}
25
20
17
15
13
B
450
400
350
300
250
200
Absorption coefficient, cm^{-1}
α 40 kb
α 20 kb
α 1 bar
β 1 bar
β 20 kb
β 40 kb
400
500
600
700
800
Wavelength, nm

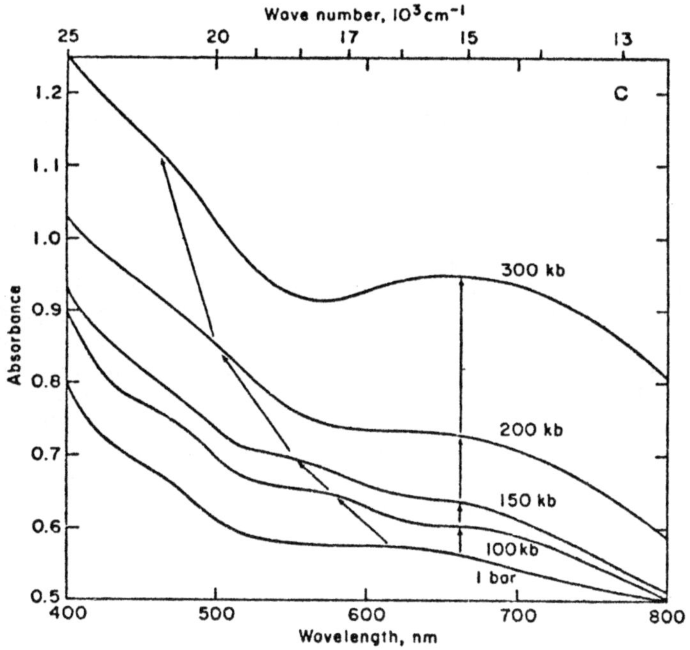

Figure 15 Spectra of the Ti^{3+}-Ti^{4+} pyroxene from the Allende meteorite (from ref. 62). (a) polarized spectra of the Allende pyroxene at 1 atmos; (b) polarized spectra at 1 bar, 20 kbar, and 40 kbar. Note that the doublet at 650 nm splits with increasing pressure, the higher energy peak shifting to higher energy and the lower energy peak remaining stationary with rising pressure; (c) spectra of polycrystalline pyroxene at very high pressures. The CF band shifts to higher energy and the CT band gains intensity but remains stationary with increasing pressure.

MINERALS EXHIBITING $Fe^{2+} \rightarrow Ti^{4+}$ INTERACTIONS

The Ti^{IV} oxidation state is far more common in terrestrial minerals than Ti^{III}, and the existence of Fe^{2+}-Ti^{4+} assemblages is potentially very common. In fact, Fe^{2+}-Fe^{3+} and Fe^{2+}-Ti^{4+} clusters may coexist in the same structure, and give rise to homonuclear $Fe^{2+} \rightarrow Fe^{3+}$ CT and heteronuclear $Fe^{2+} \rightarrow Ti^{4+}$ CT transitions in electronic spectra, thereby complicating assignments of absorption bands. As a result, some of the bands originally assigned to $Fe^{2+} \rightarrow Fe^{3+}$ CT transitions, have been re-assigned to $Fe^{2+} \rightarrow Ti^{4+}$ CT. The role of Fe^{2+}-Ti^{4+} clusters in mixed-valence iron minerals is illustrated by the following examples.

Titanian Pyroxenes.

One mineral in which Fe^{2+}-Ti^{4+} clusters occur and Fe^{3+} ions are absent is the pyroxene from the meteorite that fell at Angra dos Reis, Brazil. The composition of the Angra dos Reis pyroxene,

$$Ca_{0.97}Fe_{0.22}Mg_{0.58}Ti_{0.06}Al_{0.43}Si_{1.79}O_6$$

together with measurements of its crystal structure and Mössbauer spectrum (65-67), indicate that iron and titanium are predominantly in the M1 octahedral sites and that no ferric iron is detectable. Light polarized in the plane of the M1 cations gives rise to a broad intense absorption band centered around 20,600 cm^{-1} (66,67), which may be assigned to $Fe^{2+} \rightarrow Ti^{4+}$ CT. Pressure not only intensifies this intervalence transition, but it also results in a systematic shift of the absorption band to about 19,200 cm^{-1} at 52 kbar (68). Such a pressure-induced shift for the $Fe^{2+} \rightarrow Ti^{4+}$ CT band differs from that observed (62) for the $Ti^{3+} \rightarrow Ti^{4+}$ CT transition measured in the Allende pyroxene (see Figures 15b and c).

The $Fe^{2+} \rightarrow Ti$ CT transition has also been identified in the spectra of titanian pyroxenes from the Moon and in crustal rocks (98). The Mössbauer spectra of terrestrial titanaugites not only show the presence of Fe^{2+} ions in the M1 and M2 sites of the pyroxene structure, but also reveal that Fe^{3+} ions occur in both octahedral and tetrahedral coordinations. As a result the polarized spectra of titanaugites illustrated in Figure 16 are particularly complicated because they contain spin-allowed and spin-forbidden crystal field transitions in multiple-site Fe^{2+} and Fe^{3+} ions in addition to the intervalence $Fe^{2+} \rightarrow Fe^{3+}$ and $Fe^{2+} \rightarrow Ti^{4+}$ CT transitions. The latter are superimposed on Fe^{3+} crystal field bands intensified somewhat due to Fe^{3+} ions in non-centrosymmetric tetrahedral sites.

Kyanite

The major feature in the visible-region spectra of blue kyanites is an intense polarization-dependent band at 16,500 cm^{-1} with a prominent shoulder in the region 11,500 - 12,500 cm^{-1} (78, 79). The 16,500 cm^{-1} band was assigned to $Fe^{2+} \rightarrow Fe^{3+}$ CT (75, 78) between Fe^{2+} and Fe^{3+} ions located in chains of edge-shared $[AlO_6]$ octahedra in the kyanite structure in which Al-Al distances are 276-288 pm (77). The shoulder at 12,500 cm^{-1} was attributed to a spin-allowed crystal field transition in octahedral Fe^{2+} ions (78). Mössbauer spectroscopy subsequently confirmed that octahedral Fe^{2+} and Fe^{3+} ions are both present in blue kyanites (42). However, reported correlations of the intensity of the blue color with Ti contents of kyanites (79, 80) led to the suggestion that the 16,500 cm^{-1} band is due, instead,

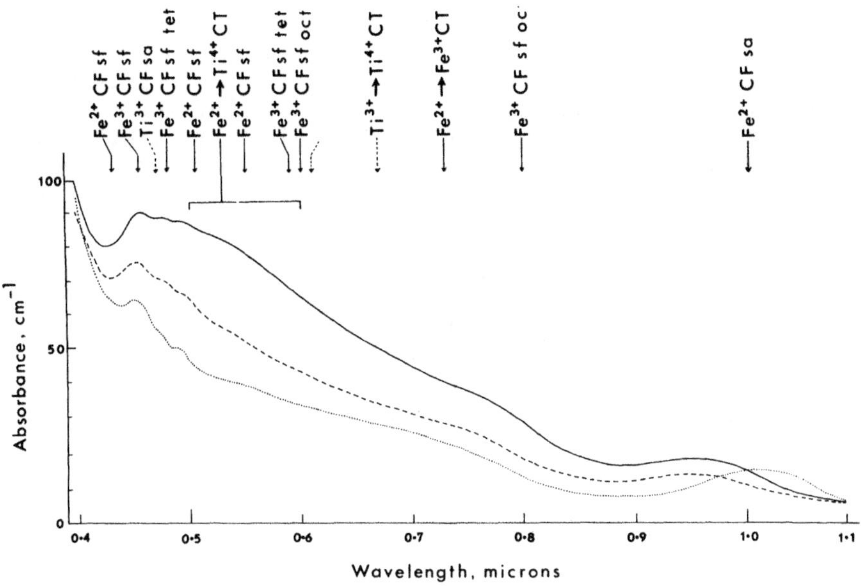

Figure 16 Assignment of peaks and inflexions in the absorption spectra of titanian pyroxenes. CF sa: crystal field, spin-allowed; CF sf: crystal field, spin-forbidden; CT: charge transfer; oct: octahedral; tet: tetrahedral

to the $Fe^{2+} \rightarrow Ti^{4+}$ CT transition (7). A coupled substitution of Fe^{2+} and Ti^{4+} for two Al^{3+} was assumed, since this maintains local charge balance within the structure. It also allows for a high probability of Fe^{2+}-Ti^{4+} clusters existing in adjacent sites, which is necessary to explain the intensity of absorption with rather low concentrations of Ti. The shoulder at 11,500 - 12,500 cm^{-1}, like the 16,500 cm^{-1} band, intensifies at low temperatures (7), leading it to be re-assigned to the $Fe^{2+} \rightarrow Fe^{3+}$ CT transition (7).

Sapphire

A similar assignment to that suggested for the kyanite spectra has been proposed for absorption bands in sapphire (7). The corundum structure consists of hexagonal close-packed oxygen ions in which [AlO_6] octahedra share faces parallel to the $\underline{c}$ axis and share edges perpendicular to $\underline{c}$ (69). Interatomic distances are 265 pm $\parallel \underline{c}$ and 279 pm $\perp \underline{c}$. Thus, two types of cation → cation interactions are possible in the corundum structure. Crystal growth studies have established that minor amounts of both Fe and Ti must be added to Al_2O_3 to generate the blue coloration of sapphire. In polarized spectra of synthetic sapphires (72, 73), bands at 17,000 cm^{-1} (E $\perp$ c) and 12,900 (E $\parallel$ c) occur only when Fe and Ti are present, while a band at about 11,150 cm^{-1} (E $\perp$ c) occurs when Fe alone is present. These bands intensify at low temperatures, indicating that they are CT transitions (7, 54). A fourth band has been identified at 9,700 cm^{-1} (7, 71). As a result, the following assignments have been proposed for the sapphire spectra (7, 54, 70, 72, 73):

17,000 cm^{-1}: $Fe^{2+} \rightarrow Ti^{4+}$ CT $\perp$ c
12,900 cm^{-1}: $Fe^{2+} \rightarrow Ti^{4T}$ CT $\parallel$ c
11,150 cm^{-1}: $Fe^{2+} \rightarrow Fe^{3+}$ CT $\perp$ c
9,700 cm^{-1}: $Fe^{2+} \rightarrow Fe^{3+}$ CT $\parallel$ c

Clearly, Fe-Ti interactions in the corundum structure are important factors determining the color and spectra of sapphire.

Hematite-Ilmenite Solid Solutions

The Fe-Ti clusters manifesting color in sapphire are also responsible for ED in the hematite-ilmenite solid solution series,

$$(1-x)Fe_2O_3 \cdot xFeTiO_3$$

which is isostructural with corundum. The Fe^{3+} ions in hematite (131) occupy face-shared and edge-shared $[FeO_6]$ octahedra as in corundum, with Fe-Fe distances 289pm ($\parallel$ c) and 297pm ($\perp$ c). In ilmenite (132), Fe^{2+} and Ti^{4+} ions are ordered so as to be coupled in the face-shared octahedra (Fe-Ti =294pm $\parallel$ c), so that planes of edge-shared $[FeO_6]$ and $[TiO_6]$ octahedra alternate along the c axis (Fe^{2+}-Fe^{2+} = 300pm $\perp$ c; Ti^{4+}-Ti^{4+} = 299pm $\perp$ c). A Mössbauer study of the hematite-ilmenite series (133) demonstrated that all Fe^{2+} ions in hematite-rich samples participate in electron transfer with an equal number of Fe^{3+} ions. Electron delocalization continues for values of x as great as 0.60, but no transfer is observed in the composition range $0.75<x<1.00$. The cations are believed to be completely disordered in the range $0<x<0.6$, so that $Fe^{2+} \rightarrow Fe^{3+}$ ED occurs both between face-shared ($\parallel$ c) and edge-shared ($\perp$ c) octahedra. Cation ordering increases in the range $0.60<x<0.75$, with the result that $Fe^{2+} \rightarrow Fe^{3+}$ ED between face-shared octahedra predominates. Between $0.75<x<1.00$, Ti effectively blocks ED between Fe^{2+} and Fe^{3+} ions.

Electron delocalization also occurs in the magnetite-ülvospinel solid solution series (134), in which Fe^{2+} and Fe^{3+} ions can occur in edge-shared octahedra throughout the composition range Fe_3O_4-Fe_2TiO_4. Other minerals to which Fe^{2+}-Ti^{4+} CT transitions have been assigned include andalusite (53,63), micas (2), tourmalines (53), vesuvianite (135) and blue omphacite (136).

CONCLUSIONS

Silicate and oxide minerals exhibit a variety of features which render them unique mixed-valence compounds

(1) They are conducive to extensive atomic substitution resulting from similarities of ionic radii. This enables mixed-valence Fe^{2+}-Fe^{3+} assemblages to coexist in many mineral structures. However, the extent of $Fe^{2+} \rightarrow Fe^{3+}$ interactions may be blocked by other substituent cations (Mg^{2+}, Al^{3+}, Ti^{4+}).

(2) Most minerals, particularly silicates, contain two or more distinct crystallographic sites, the oxygen coordination polyhedra about which may show subtle differences of distortion (manifested in variations of O-M-O bond angles, and M-M, O-O and M-O interatomic distances) and oxygen bond-type. Consequently, constituent cations may have different site occupancies in minerals of the same bulk composition which formed at different temperatures, pressures, or oxygen fugacities. This can lead to variations of physical properties and spectral profiles.

(3) Edge-shared octahedra predominate in oxide and silicate minerals. Important exceptions are cordierite (octahedron-tetrahedron shared edge), garnet (distorted cubes-tetrahedra share edges), and corundum (face-shared and edge-shared octahedra).

(4) Infinite chains of edge-shared octahedra occur in many mineral structures. These may consist of single chains one octahedron wide, double chains two edge-shared octahedra wide, or bands two or more octahedra wide extending along one crystallographic axis (e.g., pyroxenes, ilvaite, amphiboles, deerite, aenigmatite); two-dimensional sheets of edge-shared octahedra (e.g., mica, chlorite); or three-dimensional single chains (e.g., magnetite).

(5) Electron delocalization occurs along chains of edge-shared $[FeO_6]$ octahedra, or is predicted to occur, in Fe^{2+}-Fe^{3+} end-member mineral compositions (e.g., magnetite, ilvaite, deerite, riebeckite). However, trapped valencies for Fe^{2+} and Fe^{3+} are found when Mg^{2+} and Al^{3+} ions occur in atomic substitution along the octahedral chains.

(6) Since most silicate and oxide minerals have relatively high thermal stabilities, their spectra may be measured at elevated temperatures. Rising temperature generally decreases the intensity of $Fe^{2+} \rightarrow Fe^{3+}$ (and $Fe^{2+} \rightarrow Ti^{4+}$) CT bands (e.g., biotite, cordierite, kyanite, sapphire).

(7) Rising pressure, on the other hand, generally increases the intensity of CT bands, often with only minor changes of energy (e.g., vivianite, blue omphacite, titanian pyroxenes).

(8) Mössbauer spectroscopy may reveal temperature dependencies of electron delocalization processes (e.g., magnetite, ilvaite, deerite, riebeckite).

ACKNOWLEDGMENTS

Helpful discussions were held with Dr. G. Amthauer, Prof. G.M. Bancroft, Prof. B.J. Evans, Ms. Margery Osborne, Prof. G.R. Rossman, Dr. G. Smith, and Dr. R.G.J. Strens, and we thank these people for providing manuscripts of their work in advance of publication. We also thank Dr. P.M. Bell and Dr. H.K. Mao for permission to reproduce their spectra. We gratefully acknowledge the assistance of Francis Doherty for preparation of the manuscript. We thank Ms. Virginia Mee Burns for much of the bibliographic research. The research is supported by grants from the National Aeronautics and Space Administration (NGR 22-009-551 and NSG 7604).

*Present addresses of coauthors:

D.A. Nolet, Material Control and Applied Research, Manufacturing and Engineering, Corning Glass Works, Corning, New York 14830,

K.M. Parkin, Metal Oxidation and Corrosion, Division of Chemistry, National Research Council of Canada, Ottawa K1A 0R6, Canada

C.A. McCammon, Research School of Earth Sciences, Australian National University, Canberra, ACT 2600, Australia

K.M. Schwartz, Department of Earth and Space Sciences, State University of New York, Stony Brook, New York 11794

REFERENCES

1. Loeffler, B.M., and Burns, R.G.: 1976, Amer. Sci., 64, pp.636-647.

2. Faye, G.H.: 1968, Canad. Min., 9, pp.403-425.

3. Faye, G.H., Manning, P.G., and Nickel, E.H.: 1968, Amer. Min., 53, pp.1174-1201.

4. Burns, R.G.: 1970, "Mineralogical Applications of Crystal Field Theory." Cambridge Univ. Press.

5. Burns, R.G., and Vaughan, D.J.: 1975, In: "Infrared and Raman Spectroscopy of Lunar and Terrestrial Minerals." (C. Karr Jr., ed.; Academic Press), Ch. 2, pp.39-72.

6. Loeffler, B.M., Burns, R.G., and Tossell, J.A.: 1975, Proc. Lunar Sci. Conf., 6th, pp.2663-2676.

7. Smith, G., and Strens, R.G.J.: 1976, In: "The Physics and Chemistry of Minerals and Rocks." (R.G.J. Strens, ed.; J. Wiley & Sons), pp.583-612.

8. Allen, G.C., and Hush, N.S.: 1967, In: "Progress in Inorganic Chemistry." (F.A. Cotton, ed.; Interscience), 8, pp.357-389.

9. Hush, N.S.: 1967, idem ibid., pp.391-444.

10. Robin, M.B., and Day, P.: 1967, Adv. Inorg. Chem. Radiochem., 10, pp.247-423.

11. Mori, H., and Ito, T.: 1950, Acta Cryst., 3, pp.1-6.

12. McCammon, C.A.: 1979, B.Sc. (Hons) Thesis, M.I.T.

13. McCammon, C.A., and Burns, R.G.: m/s, submitted to Amer. Min.

14. Mao, H.K.: 1976, In: "The Physics and Chemistry of Minerals and Rocks." (R.G.J. Strens, ed.; J. Wiley & Sons), pp.573-581.

15. Sawatsky, G.A., Coey, J.M.D., and Morrish, A.N.: 1969, J. Appl. Phys., 40, pp. 1402-1403.

16. Lotgering, F.K., and Diepen, A.M. van.: 1977, J. Phys. Chem. Solids, 38, pp.565-572.

17. Kundig, W., and Hargrove, R.S.: 1969, Solid State Comm., 7, pp.223-227.

18. Hamilton, W.C.: 1958, Phys. Rev., 110, pp.1050-1057.

19. Verwey, E.J., and Haayman, P.W.: 1941, Physica, 8, pp.979-987.

20. Evans, B.J.: 1975, In: "AIP Conf. Proc. on Magnetism and Magnetic Materials." (H.C. Wolfe, ed.), no 24, pp.73-78.

21. Verbel, J.L.: 1974, Phys. Rev., (B), 9, pp.5236-5248.

22. Cullen, J.R., and Callen, E.: 1974, Phys. Rev. Lett., 26, pp. 236-238.

23. Beran, A., and Bittner, H.: 1974, Tschermaks. Min. Petrog. Mitt., 24, pp.11-29.

24. Haga, N., and Takéuchi, Y.: 1976, Zeit. Krist., 144, pp. 161-174.

25. Yamanaka, T., and Takéuchi, Y.: 1979, Phys. Chem. Min., 4, pp.149-159.

26. Gerard, A., and Grandjean, F.: 1971, Solid State Comm., 9, pp.1845-1849.

27. Grandjean, F., and Gerard, A.: 1975, Solid State Comm., 16, pp.553-556.

28. Heilmann, I.U., Olsen, N.B., and Olsen, J.S.: 1977, Phys. Scripta, 15, pp.285-288.

29. Paques-Ledent, M.T., Grandjean, F., and Gerard, A.: 1977, Bull. Soc. R. Sci. Liege, 46, pp.337-342.

30. Nolet, D.A.: 1978, M.Sc. Thesis, M.I.T.

31. Nolet, D.A.: 1978, Solid State Comm., 28, pp.719-722.

32. Nolet, D.A.: 1978, Geophys. Res. Lett., 5, pp.821-824.

33. Nolet, D.A., and Burns, R.G.: 1979, Phys. Chem. Min., 4, pp.227-239.

34. Papike, J.J. and Clark, J.: 1968, Amer. Min., 53, pp.1156-1173.

35. Littler, J.G.E., and Williams, R.J.P.: 1965, J. Chem. Soc., pp.6368-6371.

36. Bancroft, G.M., and Burns, R.G.: 1969, Min. Soc. Amer., Spec. Publ., 2, pp.137-148.

37. Manning, P.G., and Nickel, E.H.: 1969, Canad. Min., 10, pp.71-83.

38. Faye, G.H., and Nickel, E.H.: 1970, Canad. Min., 10, pp.616-635.

39. Ernst, W.G., and Wai, C.M.: 1970, Amer. Min., 55, pp.1226-1258.

40. Gibbs, G.V.: 1966, Amer. Min., 51, pp. 1068-1087.

41. Goldman, D.S., and Rossman, G.R.: 1978, Phys. Chem. Min.

42. Parkin, K.M., Loeffler, B.M., and Burns, R.G.: 1977, Phys. Chem. Min., 1, pp.301-311.

43. Novak, G.A., and Gibbs, G.V.: 1971, Amer. Min., 56, pp.791-825.

44. Burns, R.G.: 1972, Canad. Spectr., 17, pp.51-59.

45. Schwartz, K.B.: 1977, M.Sc. Thesis, M.I.T.

46. Huggins, F.E., Virgo, D., and Huckenholz, H.C.: 1977, Amer. Min., 62, pp.475-490.

47. Huggins, F.E., Virgo, D., and Huckenholz, H.C.: 1977, Amer. Min., 62, pp.646-665.

48. Amthauer, G., Annersten, H., and Hafner, S.S.: 1976, Phys. Chem. Min., 1, pp.399-413.

49. Schwartz, K.B., Nolet, D.A., and Burns, R.G.: 1979, Amer. Min., in press.

50. Moore, R.K., and White, W.B.: 1971, Amer. Min., 56, pp.826-840.

51. Manning, P.G.: 1967, Canad. J. Earth Sci., 4, pp.1039-1047.

52. Manning, P.G., and Harris, D.C.: 1970, Canad. Min., 10, pp.260-271.

53. Smith, G.: 1977, Canad. Min., 15, pp.500-507.

54. Smith, G.: 1979, Phys. Chem. Min., 3, pp. 375-383.

55. Burns, R.G.: 1976, In: "The Physics and Chemistry of Minerals and Rocks." (R.G.J. Strens, ed.; J. Wiley & Sons), pp. 555-572.

56. Burns, R.G., Abu-Eid, R.M., and Huggins, F.E.: 1972, Proc. 3rd Lunar Sci. Conf., 1, pp.533-543.

57. Sung, C-M, Abu-Eid, R.M., and Burns, R.G.: 1974, Proc. 5th Lunar Sci. Conf., 1, pp.717-726.

58. Bell, P.M., Mao, H.K., and Weeks, R.A.: 1976, Proc. 7th Lunar Sci. Conf., 3, pp. 2543-2559.

59. Dowty, E., and Clark, J.R.: 1973, Amer. Min., 58, pp.230-242.

60. Burns, R.G., and Huggins, F.E.: 1973, Amer. Min., 58, pp.955-961.

61. Dowty, E., and Clark, J.R.: 1973, Amer. Min., 58, pp.962-963.

62. Mao, H.K., and Bell, P.M.: 1974, Ann. Rept. Geophys. Lab., Carnegie Inst., Yearbook 73, pp.488-492.

63. Faye, G.H., and Harris, D.C.: 1969, Canad. Min., 10, pp.47-56.

64. Manning, P.G.: 1969, Canad. Min., 10, pp.678-690.

65. Hazen, R.M., and Finger, L.W.: 1977, Earth Planet. Sci. Lett., 35, pp.357-362.

66. Bell, P.M., and Mao, H.K.: 1976, Ann. Rept. Geophys. Lab., Carnegie Inst., Yearbook 75, pp.701-705.

67. Mao, H.K., Bell, P.M., and Virgo, D.: 1977, Earth Planet. Sci. Lett., 35, pp.352-356.

68. Hazen, R.M., Bell, P.M., and Mao, H.K.: 1977, Ann. Rept. Geophys. Lab., Carnegie Inst., Yearbook 76, pp.515-516.

69. Newnham, R.E., and de Haan, Y.E.: 1962, Zeit. Krist., 117, pp.235-237.

70. Townsend, M.G.: 1968, Solid State Comm., 6, pp.81-83.

71. Lehmann, G., and Harder, H.: 1970, Amer. Min., 55, pp.98-105.

72. Ferguson, J., and Fielding, P.E.: 1971, Chem. Phys. Lett., 10, pp.262-265.

73. Ferguson, J., and Fielding, P.E.: 1972, Aust. J. Chem., 25, pp.1371-1385.

74. Eigenmann, K., Kurtz, K., and Gunthard, H.H.: 1972, Helv. Phys. Acta, 45, pp. 452-480.

75. Faye, G.H.: 1971, Amer. Min., 56, pp. 344-350.

76. Nikolskaya, L.V., Terekhova, V.M., and Samoilovich, M.I.: 1978, Phys. Chem. Min., 3, pp.213-224.

77. Burnham, C.W.: 1963, Zeit. Krist., 118, pp.337-360.

78. Faye, G.H., and Nickel, E.H.: 1969, Canad. Min., 10, pp.36-46.

79. White, E.W., and White, W.B.: 1967, Science, 158, pp.915-917.

80. Rost, F., and Simon, E.: 1972, Neues Jahrb. Mineral Monat., 9, pp.383-395.

81. Gibbs, G.V., Breck, D.W., and Meagher, E.P.: 1968, Lithos, 1, pp. 275-285.

82. Wood, D.L., and Nassau, K.: 1968, Amer. Min., 53, pp.777-800.

83. Goldman, D.S., Rossman, G.R., and Parkin, K.M.: 1978, Phys. Chem. Min., 3, pp. 225-235.

84. Pollak, H.: 1976, Phys. Stat. Sol., 74, pp. K31-34.

85. Goldman, D.S., Rossman, G.R., and Dollase, W.A.: 1977, Amer. Min., 62, pp.1144-1157.

86. Buerger, M.J., Burnham, C.W., and Peacor, D.R.: 1962, Acta Cryst., 15, pp.583-590.

87. Wilkins, R.W.T., Farrell, E.F., and Naimen, C.S.: 1969, J. Phys. Chem. Solids, 30, pp.43-56.

88. Townsend, M.G.: 1970, J. Phys. Chem. Solids, 31, pp.2481-2488.

89. Simon, H.F.: 1973, M.Sc. Thesis, M.I.T.

90. Hermon, E., Simkin, D.J., Donnay, G., and Muir, W.B.: 1973, Tschermaks. Min. Petrog. Mitt., 19, pp.124-132.

91. Faye, G.H., Manning, P.G., Gosselin, J.R., and Tremblay, R.J.: 1974, Canad. Min., 12, pp.370-380.

92. Scorzelli, R.B., Baggio-Saitovich, E., and Danon, J.: 1976, J. de Phys., Colloq. C6, 37, pp.801-805.

93. Smith, G.: 1978, Phys. Chem. Min., 3, pp.343-373.

94. Fleet, S.G. and Megaw, H.D.: 1962, Acta Cryst., 15, pp.721-728.

95. Abu-Eid, R.M., Langer, K., and Siefert, F.: 1978, Phys. Chem. Min., 3, pp. 271-289.

96. Burnham, C.W., Ohashi, Y., Hafner, S.S., and Virgo, D.: 1971, Amer. Min., 56, pp.850-876.

97. Clark, J.R., Appleman, D.E., and Papike, J.J.: 1969, Min. Soc. Amer., Spec. Pap. 2, pp. 31-50.

98. Burns, R.G., Parkin, K.M., Loeffler, B.M., Abu-Eid, R.M., and Leung, I.S.: 1976, Proc. 7th Lunar Sci. Conf., 3, pp.2561-2578.

99. Curtis, L., Gittins, J., Kocman, V., Rucklidge, J.C., Hawthorne, F.C., and Ferguson, R.B.: 1975, Canad. Min., 13, pp.62-67.

100. Abu-Eid, R.M.: 1976, In: "The Physics and Chemistry of Minerals and Rocks" (R.G.J. Strens, ed.; J. Wiley & Sons), pp.641-675.

101. Aldridge, L.P., Bancroft, G.M., Fleet, M.E., and Herzberg, C.T.: 1978, Amer. Min., 63, pp. 1107-1115.

102. Mitchell, J.T., Bloss, F.D., and Gibbs, G.V.: 1970, Zeit. Krist., 133, pp.273-300.

103. Burns, R.G., and Greaves, C.: 1971, Amer. Min. 56, pp.2010-2033.

104. Bancroft, G.M., and Brown, J.: 1975, Amer. Min., 60, pp.265-272.

105. Goldman, D., and Rossman, G.R.: 1977, Amer. Min., 62, pp.205-216.

106. Goldman, D.: 1979, Amer. Min. 64, pp.109-118.

107. Brown, G.E. and Gibbs, G.V.: 1969, Amer. Min., 54, pp.101-116.

108. Faye, G.H.: Canad. Min., 11, pp.473-487.

109. Goldman, D.S. and Rossman, G.R.: 1978, Amer. Min., 63, pp.490-498.

110. Hazen, R.M. and Burnham, C.W.: 1973, Amer. Min. 58, pp.889-900.

111. Annersten, H.: 1974, Amer. Miner., 59, pp.143-151.

112. Robbins, D.W. and Strens, R.G.J.: 1972, Min. Mag., 38, pp.551-563.

113. Araki, T., and Zoltai, T.: 1972, Zeit. Krist., 135, pp.355-375.

114. Kosoi, A.L.: 1976, Sov. Phys. Cryst., 20, pp.446-451.

115. Amthauer, G.: 1979, Amer. Min., in press.

116. Nolet, D.A. and Burns, R.G.: unpubl. data.

117. Amthauer, G. and Evans, B.J.: 1979, J. Phys. Chem. Sol., in press.

118. Fleet, M.E.: 1977, Amer. Min., 62, pp.990-998.

119. Carmichael, I.S.E., Fyfe, W.S., and Machin, D.J.: 1966, Nature, 211, pp.1389.

120. Bancroft, G.M., Maddock, A.G., and Burns, R.G.: 1968, Geochim. Cosmochim. Acta, 32, pp.547-559.

121. Frank, E. and Banbury, D.St.P.: 1974, J. Inorg. Nucl. Chem., 36, pp.1725-1730.

122. Bancroft, G.M.: 1979, J. de Phys., 40, pp.C2: 464-471.

123. Amthauer, G., Langer, K., and Schliestedt, M.: 1979, Phys. Chem. Min., in press.

124. Wenk, H.R.: 1974, Amer. Min., 59, pp.86-97.

125. Moore, P.B.: 1969, Amer. Min., 54, pp.31-49.

126. Merlino, S.: 1973, Contrib. Min. Pet., 41, pp.23-29.

127. Burns, R.G.: unpubl. data.

128. Cannillo, E., Mazzi, F., Fang, J.H., Robinson, P.D., and Ohya, Y.: 1971, Amer. Min., 56, pp.427-446.

129. Merlino, S.: 1970, Chem. Comm., 20, pp.1288-1289.

130. Osborne, M.D. and Burns, R.G.: unpubl. data.

131. Blake, R.L., Zoltai, T., Hessevick, R.E., and Finger, L.W.: 1970, U.S. Bur. Mines, rept. no. 7384, pp.1-20.

132. Wenk, H.R., and Raymond, K.N.: 1973, Zeit. Krist., 137, pp.86-105.

133. Warner, B.N., Shive, P.N., Allen, J.L., and Terry, C.: 1972, J. Geomag. Geoelectr., 24, pp.353-367.

134. Banerjee, S.K., O'Reilly, W., Gibb, T.C., and Greenwood, N.: 1967, J. Phys. Chem. Sol., 28, pp.1323-1335.

135. Manning, P.G.: 1977, Canad. Min., 15, pp.508-511.

136. Strens, R.G.J., Mao, H.K., and Bell, P.M.: m/s in prep.

MIXED-VALENCY IN DISCRETE AND CONDENSED TRANSITION METAL CLUSTER SPECIES WITH CLASSICAL LIGANDS

Robert E. McCarley

Ames Laboratory-USDOE and Department of Chemistry, Iowa State University, Ames, Iowa 50011

Abstract

This paper summarizes and illustrates structural and bonding features of mixed-valence cluster compounds with emphasis on most recent developments. All cluster types from dimers through units with twelve metal atoms are surveyed among compounds with discrete cluster units. Some principles also are discussed relating extended, metal-metal bonded systems to the condensation of cluster units. Condensation of octohedral cluster units to form infinite chains, sheets and three dimensional metal-metal bonded systems is viewed from the perspective of the sharing of metal atoms on corners, edges and faces between adjacent units in the condensed systems.

INTRODUCTION

The compounds to be included in this discussion are all characterized by some degree of metal-metal bonding between transition metal atoms. Because of the broad range of compound types included in this area the discussion will be restricted to compounds having "classical" metal-ligand bonds, thus excluding the abundant organometallic derivatives, here defined as any compound having metal-carbon bonds. The compound types are conveniently categorized as dimers, discrete cluster species containing three or more metal atoms, and condensed phases having extended chain, sheet or three dimensional metal-metal bonded (condensed cluster) systems. Within these categories the metal-metal bonding is observed to include a range of formal bond order[1] from

D.B. Brown (ed.), Mixed-Valence Compounds, 337-364.

about 0.25 to 4.0. Thus the bonds range from the weakest to the very strongest known among chemical compounds.

Mixed-valency occurs often in metal cluster compounds and examples with this feature will be emphasized throughout the discussion. We may distinguish three predominant modes of mixed-valency:

1. Cluster compounds or condensed cluster systems where the average oxidation state of the metal atoms is nonintegral but where all metal atoms are chemically and structurally equivalent, e.g. $(Nb_6Cl_{12})Cl_6^{2-}$ (1), Mo_6Se_8 (2), $W_2Br_9^{2-}$ (3), etc.

2. Systems where metal atoms within individual clusters or condensed clusters occupy nonequivalent sites and the average oxidation state is generally nonintegral. Examples here include $CsNb_4Cl_{11}$ (4), $Mo_4I_{11}^{2-}$ (5), Gd_2Cl_3 (6), $Mo_5Cl_{13}^{2-}$ (7) Ti_5Te_4 (8), etc.

3. Compounds containing metal atoms in both cluster and noncluster sites. Generally the metal atoms in noncluster sites exhibit higher oxidation states then those bound within the cluster species, e.g. $ScCl_2^+$ $Sc_4Cl_6^-$ (9), $Sc^{3+}(Sc_6Cl_{12}^{3-})$ (10), $2Ta_6Br_{15}\cdot 3TaBr_4$ (11), $(C_9H_7NH)_2ReBr_6\cdot Re_3Br_9$ (12) and $TiCl_4\cdot 2Ti_3Cl_6$ (13).

In the first two cases the bonding interactions are delocalized making any real difference in effective oxidation state of the metal atoms difficult to quantify. In the third case there appears to be little interaction between the two types of metal atoms and a definite integral oxidation state usually can be assigned to the metal atom in the noncluster site based on bond length data, magnetic or spectroscopic properties.

The following discussion is organized around cluster nuclearity and structural type for the discrete cluster units and increasing degree of condensation of cluster units within the extended systems.

COMPOUNDS WITH DIMERIC UNITS

Mixed-valency within homonuclear dimeric units can result either from one-electron oxidation or reduction of species with equivalent metal atoms and integral metal-metal bond order, or from construction of species with different ligands on the two metal atoms, each set of ligands demanding a different net oxidation state for the attached metal atom. Examples of both these cases are included in Table I.

Dimers with equivalent metal atoms (and net metal-metal bond order) are: $W_2Br_9^{2-}$(2.5) $Ru_2(O_2CR)_4^+$, $Mo_2(SO_4)_4^{3-}$(3.5), and $Tc_2Cl_8^{3-}$(3.5). Among derivatives with non-equivalent metal atoms are: $Re_2Cl_5(CH_3SC_2H_4SCH_3)_2$(3.5) and $Re_2Cl_5(Ph_2PCH_2PPh_2)_2$(3.5). The metal-metal interactions in these examples are well understood compared to those in the other entries given in Table I, viz. $Re_2OCl_3(O_2CC_2H_5)_2(PPh_3)_2$, $La_3Os_2O_{10}$ and $La_4Re_6O_{19}$(better formulated as $La_4(Re_2)_3O_{19}$). In each of the latter cases the formal M-M bond order is thought to be 2.0, i.e. $\sigma^2\pi^2$, with the remaining metal-based electrons residing in M-Oπ^* or nonbonding orbitals. The structure involves M_2O_{10} units with the two metal atoms bridged by oxygen atoms in an edge-shared bioctohedral arrangement. Especially in these dioxo-bridged systems the M-M and M-Oπ-interactions are extensively mixed (17), thus the M-M distance cannot be expected to correlate well with a bond order derived from a simple scheme where the metal-based electrons enter only into M-M bonding. However Sleight, et.al. (17) have shown that in these systems good σ and π M-M orbital overlaps are obtained, and for the cases where 4 or more electrons are available the M-M distance supports the assigned bond order of 2.0. Because of bond angle strain and nonbonded repulsions within the bridged systems the metal-metal bond distances do not correlate well with values for the same bond order derived from compounds where the bond is unsupported by bridging ligands, or unstrained by bridging ligands such as the carboxylates whose "bite" is commensurate with the M-M bond distance.

Among the species listed in Table I with bond order 3.5 the structure of $Re_2Cl_5(dth)_2$ is most unusual. In this molecule both dth ligands are bound as chelates to the same metal atom along with one Cl atom positioned along the Re-Re bond axis. The other Re atom is bound to four Cl's which are in staggered conformation with relation to the S atoms bound to the partner Re atom. The compound then should be considered as a Re(I)-Re(IV) mixed-valence derivative. The compound $MoW(O_2CCMe_3)_4I\cdot CH_3CN$ also is unusual because it is an example of the rare heteronuclear dimer. Along the Mo-W bond axis the acetonitrile is coordinated to Mo and the I atom to W, which then suggests the mixed-valence Mo(II)-W(III) combination.

TRIMETALLIC CLUSTER COMPOUNDS

All of the examples to be discussed here exhibit a triangular arrangement. The basic structure types I, II and III designated in Table II, differ from one another by virtue of the number and location of bridging ligands as shown in Fig. 1. Type

Table I

Some Mixed-Valence Dimers with Strong Metal-Metal Bonding

Compound	Ref.	d(M-M), A	n[a]	Structure Type	Bond Configuration
$(Pr_4N)_2W_2Br_9$	3	2.601(2)	2.5	$W_2Cl_9^{3-}$	$\sigma^2\pi^3$
$Ru_2(O_2CC_3H_7)_4Cl$	14	2.281(4)	2.5	$Re_2(O_2CR)_4Cl_2$	$\sigma^2\pi^4\delta^2\delta^{*1}\pi^{*2}$
$Re_2OCl_3(O_2CC_2H_5)_2(PPh_3)_2$	15	2.514(1)	2.0 (?)	esbo[b]	$\sigma^2\pi^2$
$La_3Os_2O_{10}$	16	2.462(1)	2.0 (?)	esbo[b]	$\sigma^2\pi^2$
$La_4Re_6O_{19}$	17	2.42	2.0 (?)	esbo[b]	$\sigma^2\pi^2$
$K_3Mo_2(SO_4)_4\cdot 3.5H_2O$	18	2.165(1)	3.5	$Mo_2(O_2CR)_4$	$\sigma^2\pi^4\delta^1$
$K_3Tc_2Cl_8\cdot nH_2O$	19	2.117(2)	3.5	$Mo_2Cl_8^{4-}$	$\sigma^2\pi^4\delta^2\delta^{*1}$
$Re_2Cl_5(dth)_2$[c]	20	2.293(2)	3.5	staggered	$\sigma^2\pi^4\delta^2\delta^{*1}$
$Re_2Cl_5(dppm)_2$[d]	21	2.263(1)	3.5	eclipsed	$\sigma^2\pi^4\delta^2\delta^{*1}$
$MoW(O_2CCMe_3)_4I\cdot CH_3CN$	22	2.194(2)	3.5	eclipsed	$\sigma^2\pi^4\delta^1$

[a] Formal metal-metal bond order

[b] Edge-shared bioctahedron

[c] dth = 2,5-dithiahexane

[d] dppm = bis(diphenylphosphino)methane

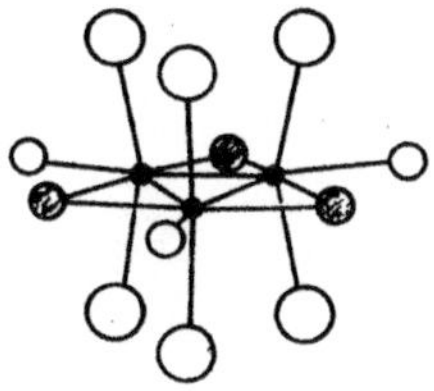

I

Re_3Cl_9

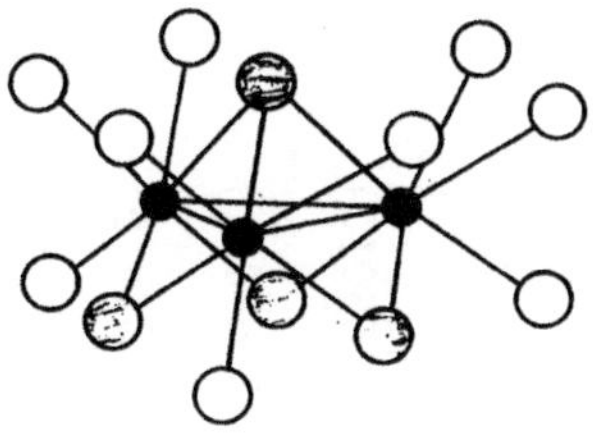

II

$Zn_2Mo_3O_8$

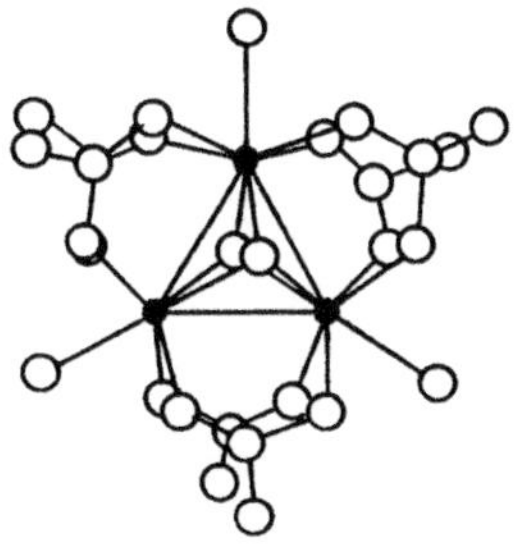

III

$Mo_3(OEt)_2(O_2CMe)_6(H_2O)_3{}^{2+}$

Figure 1. Structure of trinuclear clusters

I is exemplified by the $Re_3Cl_{12}{}^{3-}$ cluster (23), type II by the cluster in $Zn_2Mo_3O_8$ (24) and type III by the recently reported $[Mo_3(OEt)_2(O_2CCH_3)_6(H_2O)_3]^{2+}$ cation (25). These structure types are distinguished also by the number of metal-metal bonding orbitals which can be formed. In type I each metal is bound to a maximum of 5 ligand atoms. This permits use of four d-orbitals from each metal in the metal-metal bonding. It has been shown by Cotton and Haas (26) that the resulting twelve M.O.'s divide into sets of six bonding and six antibonding. If the bonding orbitals are filled the net M-M bond order is 2.0. The Re-Re distances known in the numerous Re_3X_9 derivatives generally fall in the range 2.47±0.03 A which may be compared with 2.566 and 2.385 A, respectively, calculated for bonds of order 1.0 and 2.0 from Pauling's equation D(n)=D(1)-0.6 log n and a value of R_1=1.283 A for the metallic radius of Re (28).

Table II. Trinuclear Mixed-Valence and Related Clusters

Compound	Structure Type	MCE[a]	d(M-M), A	Ref.
$(C_9H_7NH)_2ReBr_6 \cdot Re_3Br_9$	I	12	2.465(8) (Av.)	12
Nb_3X_8 (X=Cl, Br, I)	II	7	2.81(1) (X=Cl)	30
			2.88(3) (X=Br)	33
Ti_7X_{16} (X=Cl, Br)	II	6	2.954(2)	13
$Zn_2Mo_3O_8$	II	6	2.524(2)	24
$(NH_4)_5W_3O_4F_9 \cdot NH_4F \cdot H_2O$	II	6	2.515(1)	34
$Cs_2[Mo_3O_4(C_2O_4)_3(H_2O)_3] \cdot 4H_2O \cdot \frac{1}{2}H_2C_2O_4$	II	6	2.486(1)	32
$[Mo_3OCl_3(O_2CCH_3)_3(H_2O)_3](ClO_4)Cl$	II	8	2.550(2)	31
$W_3(OCH_2CMe_3)O_3Cr_3(O_2CCMe_3)_{12}$	II	8	2.610(1)	35
$[Mo_3(OEt)_2(O_2CCH_3)_6(H_2O)_3](CF_3SO_3)_2$	III	8	2.854(2)	25

[a] Number of metal-based cluster electrons

In compounds with structure type II the metal atoms are coordinated to 6 ligand atoms in distorted octahedral arrangement. Thus each metal atom has three valence orbitals for construction of the M-M MO's. Cotton (29) has derived a rough MO energy level scheme for C_{3v} clusters of this type, with the result that three MO's are strongly bonding ($a_1^{(1)}+e^{(1)}$), five are strongly antibonding ($e^{(2)}+e^{(3)}+a_2$), and one ($a_1^{(2)}$) with either bonding or antibonding character depending upon the structural and electronic parameters of the particular compound. Recently compounds have been structurally characterized having 6, 7 and 8 metal-based cluster electrons (MCE) available for metal-metal bonding; those with 7 and 8 MCE represent mixed-valence clusters. Examples are also known where reversible oxidation-reduction reactions can be effected. These examples have so far failed to provide definitive evidence about the role of the bonding or antibonding character of the $a_1^{(2)}$ orbital. It appears that the relative energy of this orbital is quite sensitive to the bonding of the edge-bridging ligand atoms. The metal valence orbitals from which $a_1^{(2)}$ is constructed may participate in π-bonding with these ligands, i.e. accepting π-electron density from the edge-bridging atoms. This in turn raises the energy and imparts antibonding character to the $a_1^{(2)}$ MO. The $a_1^{(2)}$ orbital may be occupied in complexes such as Nb_3Cl_8(7MCE)(30) and $[Mo_3OCl_3(O_2CCH_3)_3(H_2O)_3]^{2+}$ (8MCE)(31) where π-interactions with the edge-bridging Cl atoms are weak. In $Zn_2Mo_3O_8$(24) and $[Mo_3O_4(C_2O_4)_3(H_2O)_3]^{2-}$ (32), each with 6 MCE, the Mo-O distances (1.92 A) to the three edge-bridging O atoms are the shortest in the cluster, the π-interactions are strong, and the $a_1^{(2)}$ orbital is sufficiently antibonding in character that it remains unoccupied. Exactly the same consideration applies to the tungsten cluster $W_3O_4F_9^{5-}$ (34). However in $W_3(OCH_2R)O_3Cr_3(O_2CR)_{12}$ (35), with 8 MCE, the edge-bridging O atoms are each trigonally hybridized with σ-bonds to two W and one Cr atom, the π-interaction is reduced, and the $a_1^{(2)}$ orbital is filled. The three Cr-atoms are bonded only to O-atoms at the ligands in roughly octahedral coordination, and are present as Cr(III), each with three unpaired electrons.

There are several features of the type III structure, as represented by $[Mo_3(OEt)_2(O_2CCH_3)_6(H_2O)_3]^{2+}$, which are unusual. First we note the metal triangle is capped both above and below by triply-bridging alkoxo ligands. Secondly the edge-bridging acetate ligands are coordinated from above and below the triangle in such a way that each metal atom becomes enclosed in a trigonal prism of O atoms. The M-M and M-OH_2 bonds then are formed in the plane of the triangle through the square faces of the trigonal prisms. Finally, in this arrangement each metal attains 9-coordination with seven M-O and two M-M bonds. This leaves only two valence orbitals available from each metal for M-M interactions.

These combine to form a set of three bonding ($a_1'+e'$) and three antibonding ($e'+a_2'$) MO's. Since there are 8 MCE's, at least two electrons must occupy antibonding orbitals and the net M-M bond order is reduced to 2/3. The increased coordination number and reduced M-M bond order evidently result in the increased Mo-Mo bond distance (2.854A) compared to the other Mo-clusters included in Table II.

TETRANUCLEAR CLUSTERS

Relatively few examples of mixed-valence tetranuclear clusters are known. These may be classified as tetrahedral(I), butterfly(or opened tetrahedral)(II), and planar butterfly(III) types, as shown in Fig. 2.

Types of Tetranuclear Clusters

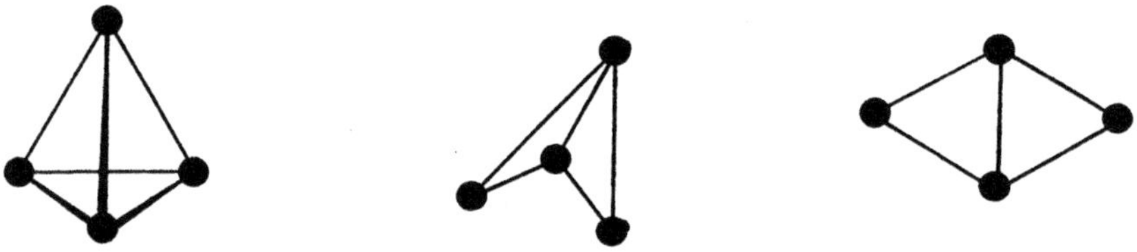

I(T_d) II(C_{2v}) III(C_{2h})

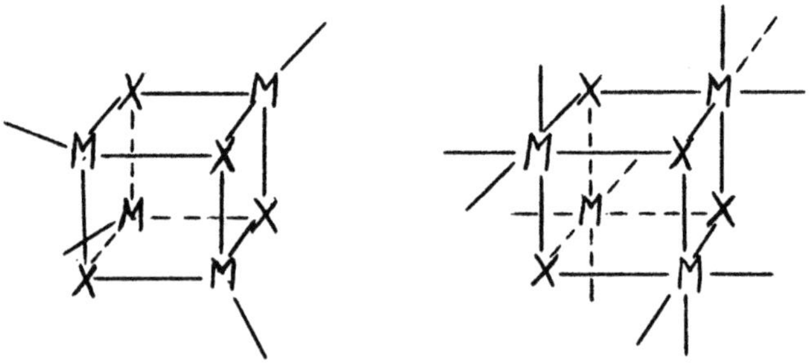

Ia Ib

Figure 2. Structure of tetranuclear clusters

A listing of compounds with each of these cluster types is given in Table III. Essentially all of the known examples of type I have the cubane core with triply bridging ligand atoms situated over each of the four faces of the tetrahedron. Depending upon the ligands or MCE the resulting cluster may have either tetrahedral(T_d) or distorted tetrahedral symmetry (D_{2d},C_{2v}). The metal atoms within the cubane structure may be bound to either 4 or 6 ligand atoms as in Fig. Ia or Ib. In Ia each metal atom has 4 ligands in approximate tetrahedral array, while in Ib each metal has 6 ligands in roughly octahedral array. The most important examples of Ia are the iron-sulfur clusters $Fe_4S_4(SR)_4^{n-}$ which have been extensively investigated by Holm and coworkers as model compounds for the 4-iron proteins (36). A great variety of organometallic clusters of this type also are known (37). Examples of type Ib are $Mo_4S_4Br_{12/3}$ (38) and $GaMo_4S_4S_{12/3}$ (39). An important distinction between types Ia and Ib is the number of valence orbitals which may enter into the M-M interactions. In type Ia each metal atom may contribute five orbitals to the M-M cluster interactions. Dahl et.al. (40) have shown that the 20 derived MO's divide into a strongly bonding set of six (a_1+e+t_2), an intermediate weakly bonding or nonbonding set (e+t_1+t_2), and a strongly antibonding set (t_1+t_2). Because of the nearly nonbonding character of the intermediate set, the occupancy of these orbitals is variable and such clusters show interesting redox or electron-transfer chemistry. In clusters of type Ib only 12 valence orbitals are available for M-M interactions, viz. the t_{2g} set from each octahedrally coordinated metal atom. These combine to form the strongly bonding set (a_1+e+t_2) and the strongly antibonding set (t_1+t_2). Different occupation of these MO's is observed for different compounds but the consequence for a redox process involving any one species is likely to be a severe structural change. Thus reversible electron transfer involving clusters of type Ib is not very likely.

While many examples of organometallic clusters with the bent or planar butterfly structures (types II or III) are known (37), very few clusters of these types are known with only classical ligands. The anion $Mo_4I_{11}^{2-}$ (5) is an example of type II and the compounds MNb_4X_{11}(M=Rb,Cs; X=Cl,Br)(4) examples of type III. As shown in Fig. 3 the structure of the $Mo_4I_7^{2+}$ cluster unit of $Mo_4I_{11}^{2-}$ can also be viewed as a fragment formed by removing two adjacent Mo atoms and one I atom from the Mo_6I_8 cubo-octahedral cluster unit. A particularly interesting feature of the $Mo_4I_{11}^{2-}$ cluster is that it possesses 15 electrons for M-M interactions. The short distances for four of the Mo-Mo bonds, 2.545(5)A average, are indicative of multiple bond character in these bonds. Just how many of the 15 electrons occupy bonding orbitals is not known, although a rough MO analysis indicates as many as 14 electrons may contribute to the bonding. Attempts to obtain oxidized or reduced derivatives in order to derive further information about the bonding have not been successful. In contrast to

Table III. Some Mixed Valence Tetranuclear Clusters

Compound	Structure Type	MCE[a]	Av.d(M-M),A	n[b]	Ref.
$(Et_4N)_3[Fe_4S_4(SCH_2Ph)_4]$	Ia	23	2.759(2)	?	36
$(Et_4N)_2[Fe_4S_4(SCH_2Ph)_4]$	Ia	22	2.746(3)	?	41
$Mo_4S_4Br_{12/3}$[c]	Ib	12	2.798[d]	1.0	38
$GaMo_4S_4S_{12/3}$	Ib	11	2.823[d]	0.917	39
$(Bu_4N)_2Mo_4I_{11}$	II	15	2.570(5)	>1.0	5
$CsNb_4Cl_{11}$	III	10	2.86[d]	1.0	4
$CsNb_4Br_{11}$	III	10	2.98[d]	1.0	4

[a] Total number of metal-centered electrons involved in cluster interactions

[b] Formal M-M bond order

[c] Not mixed valence but prototype for structure type Ib

[d] Estimated standard deviations not given

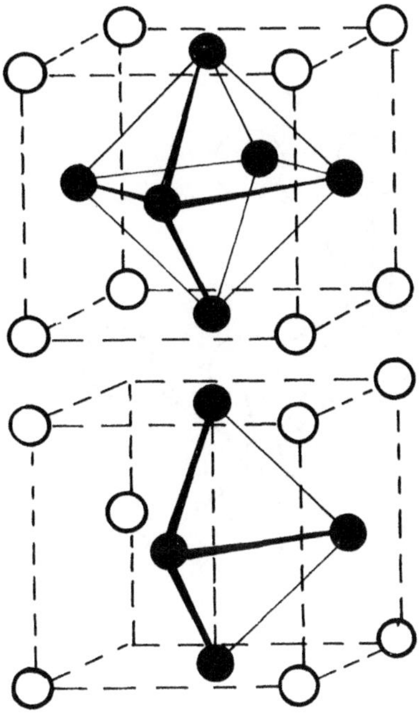

Figure 3. Structural relation between the core $Mo_4I_7^{2+}$ of $Mo_4I_{11}^{2-}$ and the core $Mo_6I_8^{4+}$ of $(Mo_6I_8)I_6^{2-}$.

$Mo_4I_{11}^{2-}$ the planar arrangement of metal atoms in $CsNb_4Cl_{11}$, Fig. 4, provides a more readily understood bonding picture. In this case each metal atom contributes three valence orbitals for M-M bonding. These combine in C_{2h} symmetry to provide a set of five bonding ($2a_g+b_g+a_u+b_u$), two approximately nonbonding (a_g+b_u), and five antibonding MO's ($a_g+b_g+2a_u+b_u$). Thus the 10 M-M electrons fill the five bonding levels and the bond order per M-M bond is 1.0. The two M-M nonbonding orbitals are localized on the Nb atoms most distant from each other, and if not imparted strong antibonding character by the ligands might accommodate additional electrons, e.g. as in the hypothetical $CsNb_2Mo_2Cl_{11}$ or $BaNb_4Cl_{11}$. There is good prospect that other compounds with this structure should be found in future research. So far no clusters with this particular structure have been found outside the solid state and thus available for solution studies.

PENTANUCLEAR CLUSTERS

Only one compound containing a non-organometallic pentanuclear cluster has been described, viz. $(Bu_4N)_2Mo_5Cl_{13}$ (7). The structure of the $Mo_5Cl_{13}^{2-}$ anion is shown in Fig. 5; we note immediately that it may be viewed as a fragment of the $Mo_6Cl_{14}^{2-}$

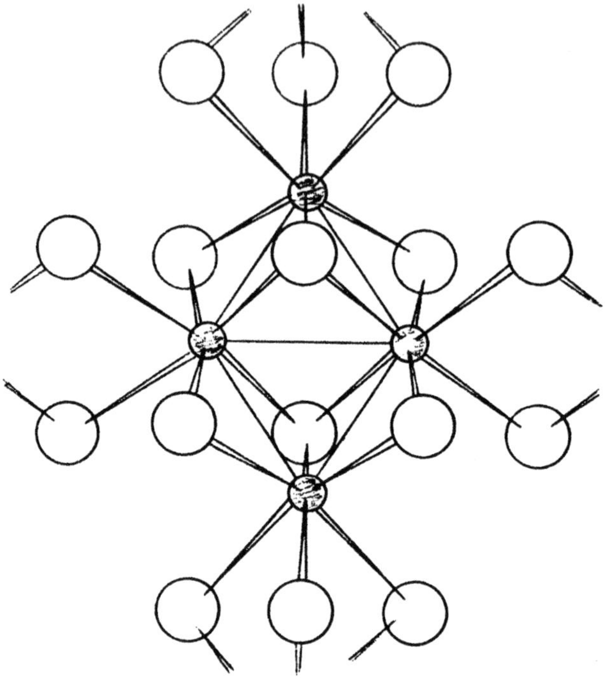

Figure 4. Structural arrangement about the cluster in $CsNb_4X_{11}$(X=Cl or Br).

cubo-octahedron with one Mo and the associated terminal Cl removed. The $Mo_5Cl_{13}^{2-}$ cluster is then left with a total of 19 electrons to form the M-M bonds, where formally only 16 electrons are required if each edge of the square pyramidal cluster has bond order unity. Like the case of $Mo_4I_{11}^{2-}$, the bonding in $Mo_5Cl_{13}^{2-}$ is unclear, particularly in regard to the number of electrons in bonding vs. nonbonding MO's. An average bond order near unity for the M-M bonds is indicated by the average Mo-Mo distance of 2.582(3)A, which may be compared with d(Mo-Mo)=2.61(1)A in Mo_6Cl_{12} (42). Even a cursory examination of the bonding in $Mo_5Cl_{13}^{2-}$ however makes it clear that the HOMO's are centered on the basal Mo atoms. Consequently the cluster should show reactivity at these centers which may be exploited in subsequent studies. The addition of other metal atoms to form $(MMo_5Cl_8)Cl_6^{n-}$ cluster ions is a particularly attractive goal. Finally, further synthetic work is needed to uncover examples of other pentanuclear clusters of this type.

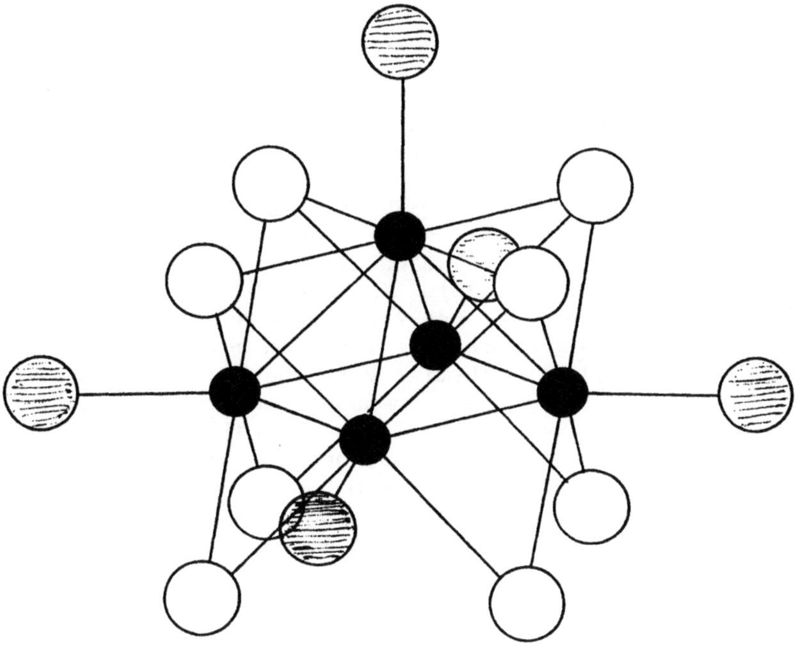

Figure 5. Structure of the $Mo_5Cl_{13}^{2-}$ cluster anion.

HEXANUCLEAR CLUSTERS

Recent developments have served to make this one of the most interesting areas of metal cluster chemistry. Until about 1970 the study of $M_6X_{12}^{n+}$(M=Nb,Ta,W; X=F,Cl,Br,I) and $M_6X_8^{n+}$(M=Nb,Mo, W; X=Cl,Br,I) clusters dominated published work. These studies established that bonding within the octahedral clusters of either type was clearly delocalized and that bonding MCE counts of 14 to 16 were favored for $M_6X_{12}^{n+}$, and 24 MCE's were dominant for $M_6X_8^{n+}$ clusters. Rough MO treatments were in substantial agreement with these facts in that 8 bonding M-M MO's were predicted for $M_6X_{12}^{n+}$, and 12 bonding MO's for $M_6X_8^{n+}$ clusters (26). In O_h symmetry the array of MO's involved in the M-M interactions were derived as:

$$M_6X_{12} \; : \; (a_{1g}+t_{1u}+t_{2g}+a_{2u})^b \; (t_{2g}+e_g+t_{2u}+t_{1u}+e_u+t_{1u})^*$$

$$M_6X_8^{n+} : \; (a_{1g}+t_{1u}+t_{1g}+t_{2u}+e_g)^b \; (e_g+t_{2u}+a_{2g}+t_{2g}+t_{1u})^*$$

For the Nb and Ta clusters, $M_6X_{12}^{n+}$, the magnetic properties (43) provide support for a_{2u} as the HOMO for n=2+ or 3+, with 16 and 15 MCE, respectively. The weakly bonding character of this orbital, combined with an absence of any structural reorganization of the cluster upon change in MCE, account for the reversible redox processes and stable derivatives observed among the three oxidation states. Representative examples of structurally well-characterized compounds are listed in Table IV. It can be seen that the change in M-M bond distance with MCE count clearly supports removal of bonding electrons in the series $M_6X_{12}^{2+} \rightarrow M_6X_{12}^{3+} \rightarrow M_6X_{12}^{4+}$. The derivatives of $M_6X_{12}^{n+}$(M=Nb or Ta) are uniformly intensely colored with dipole-allowed electronic transitions occurring over the region 350 to 1000 nm (44).

Renewed interest in M_6X_{12} cluster chemistry has been kindled by the discovery of the lower scandium and zirconium halides Sc_7Cl_{12}, Zr_6I_{12} and Zr_6Cl_{15}(10). The occurrence of these new compounds evoke interest because they possess only 9, 12 and 9 MCE's respectively, and hence have very low net M-M bond orders. Extension of metal cluster chemistry to the Group III and IV transition elements, where such compounds were previously unknown, considerably broadens the field and raises the prospect that even elements of the lanthanide and actinide series will be included. Because of their lower M-M bond orders and highly reducing character such compounds probably will be unable to survive in solution, but as stable crystalline phases they may exhibit interesting or useful solid state properties.

Similarly, renewed interest has been aroused in the M_6X_8 cluster compounds by the discovery and proliferation of the "Chevrel phases" (51). Until 1970 nearly all examples of the M_6X_8 clusters were among the halides. The only exceptions to the $M_6X_8^{4+}$ clusters of Mo(II) and W(II) having 24 MCE count were with the mixed-valence cluster $(Nb_6I_8)I_3$ (52,53) and the compounds containing the $W_6Br_8^{6+}$ unit, viz. $(W_6Br_8)Br_6$ (54), $(W_6Br_8)Br_4(Br_4)_{2/2}$ (54,55) and $(W_6Br_8)Br_2(Br_4)_{4/2}$ (54). Then in 1971 the first results of an extensive investigation of binary and ternary molybdenum chalcogenide phases were published by the group at Rennes, France (56). Because many of the subsequent compounds discovered in these investigations proved to be superconductors, some with extraordinarily high critical field values (57), both chemists and physicists have been attracted to the area, and remarkable growth has occurred. Only a few representative examples of these compounds which have been structurally well-characterized are included in Table V.

Among the M_6X_8 clusters there is an excellent correlation between the M-M bond order (and bond distance) calculated from the MCE count and the observed M-M bond distances. This shows that the nonmetal atoms in the M_6X_8 clusters play a less important role then in the M_6X_{12} clusters in determining the strength of the M-M bonds. The geometric arrangement of the M_6X_8 cluster is such that the M atoms can approach to within M-M single bond distance without displacement from the center of the surrounding square of X atoms, even for large X. Also, in the absence of M-X π-bonding, the orbitals used for M-X bonds and M-M bonds are essentially completely separable; the $M-X^i$ and $M-X^a$ bonds[2] may be constructed from dsp^3 hybrids and the M-M bonds then are left with nearly

Table IV. Hexanuclear Cluster Compounds of M_6X_{12} Type

Compound	Ref.	MCE[a]	Av.d(M-M),Å	n[b]	D(n),Å[c]
$(Nb_6Cl_{12})Cl_2$	45	16	2.915	0.667	2.79
$(Nb_6F_{12})F_3$	46	15	2.80	0.625	2.81
$(Me_4N)_3[(Nb_6Cl_{12})Cl_6]$	47	15	2.967(5)	0.625	2.81
$(Me_4N)_2[(Nb_6Cl_{12})Cl_6]$	1	14	3.018(2)	0.583	2.83
$(Ta_6I_{12})I_2$	48	16	2.90	0.667	2.79
$(Ta_6Cl_{12})Cl_3$	49	15	2.925	0.625	2.81
$H_2[(Ta_6Cl_{12})Cl_6]\cdot 6H_2O$	50	14	2.962(2)	0.583	2.83
Zr_6I_{12}	10	12	3.200(2)	0.500	3.09
$(Zr_6Cl_{12})Cl_3$	10	9	3.207(4)	0.375	3.16
$Sc[Sc_6Cl_{12}]$	10	9	3.219(2)	0.375	3.13

[a] Total number of metal-based cluster electrons

[b] Formal bond order, n=MCE/24

[c] M-M bond length calculated from D(n)=D(1)-0.6 log n as given by Pauling, with D(1)=2.684(Nb), 2.686(Ta), 2.908(Zr) and 2.878Å(Sc).

Table V. Some Mixed-Valence Clusters of the Type M_6X_8

Compound	Ref.	MCE[a]	Av.d(M-M),Å	n[b]	D(n),Å[c]
$(Nb_6I_8)I_3$	53	19	2.85	0.792	2.75
$Cs(Nb_6I_8)I_3$	59	20	2.826	0.833	2.73
$(W_6Br_8{}^{6+})Br_4(Br_4)_{2/2}$	55	22	2.64(2)	0.917	2.63
Mo_6S_8	60	20	2.780	0.833	2.64
Mo_6Se_8	2	20	2.760	0.833	2.64
$PbMo_6S_8$	61	22	2.702	0.917	2.62
$LaMo_6S_8$	51	23	2.687	0.958	2.60
$Cu_{3,66}Mo_6S_8$	62	(23.66)	2.670	0.986	2.60

[a] Metal-based cluster electron count

[b] Formal bond order, n=MCE/24

[c] M-M bond length calculated from D(n)=D(1)-0.6 log n as given by Pauling.

pure d-character. Thus the X-atoms may influence the M-M bonding mainly through the effective charge imparted to the metal atoms and its consequence on the effective radius of the d-orbitals. Through the latter effect, the shortest M-M distance for a given MCE count should be prompted when X is most electronegative, although the comparison of Mo_6S_8 and Mo_6Se_8 appears to contradict this.

The conductivity of the Chevrel phases is thought to arise from a broadening of these cluster MO's derived from the metal d-orbitals into narrow conduction bands caused by the approach of metal atoms in neighboring clusters (58). This in turn is caused by the low X/M ratio and the interlinking of clusters via bridging of a nonmetal atom within one cluster to the exo position of a metal atom in a neighboring cluster, as represented in the formula $Mo_6S_8{=}(Mo_6S_2{}^{i}S_{6/2}{}^{i\text{-}o})S_{6/2}{}^{i\text{-}o2}$. Among these compounds the Mo-Mo intercluster distance varies from 3.084A for Mo_6S_8 to 3.271A for $PbMo_6S_8$ (51). The superconducting properties depend in a critical way upon both the MCE count and the intercluster Mo-Mo distance. Optimally the MCE count should be such that the Fermi level falls in a region of the d-bond with a high density of states, and the intercluster Mo-Mo distance should be sufficiently large that the d-bonds remain narrow. For example, although Mo_6S_8 has a favorable MCE count the intercluster distance is too short and the solid is not a superconductor. When the d-bond becomes filled, as would be the case for phases where the MCE count is 24, the compounds are semiconductors. These relations are supported by the recent theoretical work published by Anderson, Klose and Nohl (58).

COMPOUNDS WITH CONDENSED CLUSTER SYSTEMS

Only recently has it become apparent that a vast new area of metal cluster chemistry exists, viz. the extension from discrete cluster units into systems of condensed clusters where individual cluster units are welded together by sharing of metal atoms on corners, edges or faces of the cluster polyhedra. Although the structures of many compounds with such condensed cluster systems were known earlier, their structures were not viewed from this perspective until recently. An interesting paper on metal-rich compounds by Simon (63) in 1976 articulated this perspective very effectively. Since then important new compounds have been discovered which tend to confirm that metal clusters can be condensed, perhaps one step at a time, to generate families of compounds related by the number of condensed units. Concrete examples with a limited number of condensed units now are known among both organometallic and nonorganometallic clusters, e.g. $[Pt_3(CO)_3]_n{}^{2-}$ (n=2 to 5) (64) and Mo_9Se_{11} (as contained in the new high field superconducting compound $In_{\sim3}Mo_6Se_8 \cdot Mo_9Se_{11}$) (65).

The Mo_9Se_{11} cluster unit in the latter compound may be considered as derived from the fusion of two Mo_6Se_8 units by removal of three facial Mo atoms and the four attached Se atoms of one unit and then inserting the second unit, after removal of one capping Se atom, such that the two units share a common face. Conceptually this can be formulated by the equation $2Mo_6Se_8 \longrightarrow Mo_9Se_{11}+Mo_3Se_5$. Next step in the condensation by building on the Mo_9Se_{11} unit would be $Mo_9Se_{11}+Mo_6Se_8 \longrightarrow Mo_{12}Se_{14}+Mo_3Se_5$. From this we can see that a chain can be built from stacked Mo_3Se_3 units with the end members capped by an extra Se atom. The general formula would be $M_{3n}X_{3n+2}$. Amazingly enough compounds in this series are now known with n=2,3,4 and ∞, in all cases ternary

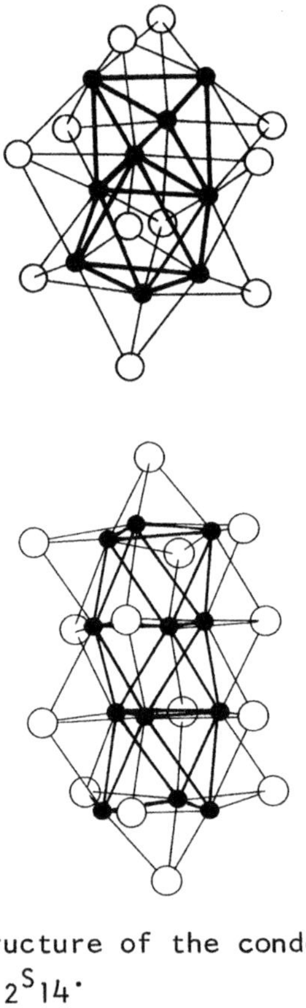

Figure 6. Structure of the condensed cluster units Mo_9S_{11} and $Mo_{12}S_{14}$.

molybdenum chalcogenides. Figure 6 depicts the Mo_9S_{11} and $Mo_{12}S_{14}$ cluster units found in the compounds $K_2Mo_6S_8 \cdot Mo_9S_{11}$ and $K_4Mo_6S_8 \cdot M_{12}S_{14}$, respectively (66). In the very recent communication of these results the new compounds, $M_2Mo_6X_6$(M=Rb,Cs with X=S; M=Na,K,Tl with X=Se), representing the infinite chain structure, also were reported (66).

A second mode of condensation of octahedral clusters involves the fusion of adjacent units along common edges to form either linear chains or zig-zag chains. In the linear mode, which is the only one to be discussed here, adjacent octahedra share opposite (parallel) edges. Condensation of two M_6X_8 or M_6X_{12} units in this way would produce $M_{10}X_{14}$ or $M_{10}X_{18}$, respectively. Condensation of _n_ such units would produce $M_{4n+2}X_{6n+2}$ or $M_{4n+2}X_{6n+6}$. At the present time discrete cluster units of this type are unknown for n>1. Many examples do exist however for the infinite chains. Among the compounds having chains derived from M_6X_8 units are $Gd_2Cl_3(=Gd_2Gd_{4/2}Cl_4Cl_{4/2})$ (6) and Y_2Cl_3 (67). Two examples with

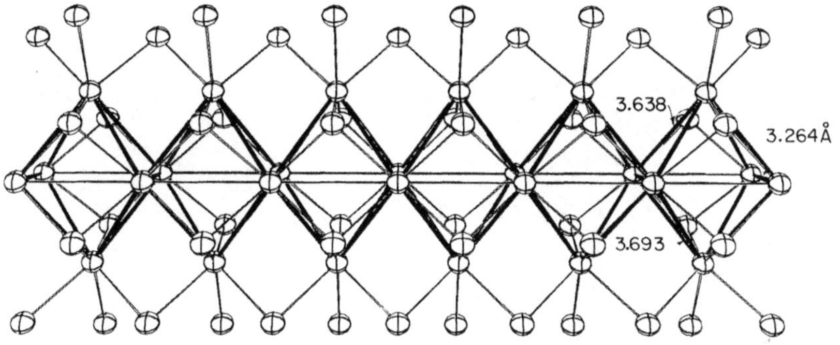

Figure 7. Structure of the infinite chains in Y_2Cl_3.

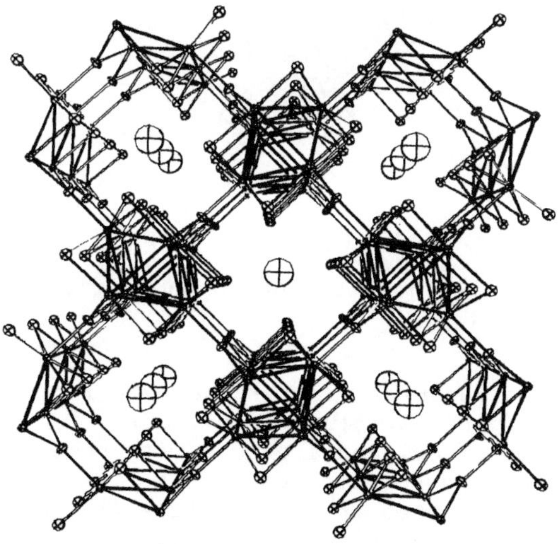

Figure 8. A view of the infinite anion chains and interlinking of the chains to form channels for the Na^+ ions in $NaMo_4O_6$.

chains constructed from condensed M_6X_{12} units are $Sc_5Cl_8 = ScCl_2^+(Sc_2Sc_{4/2}Cl_2Cl_{8/2}^-)$ (9) and $NaMo_4O_6 = Na^+(Mo_2Mo_{4/2}O_2O_{8/2}^-)$ (68).

Figures 7 and 8 depict these chains for Y_2Cl_3 and $NaMo_4O_6$, respectively. The interesting compound $Sc_7Cl_{10} = ScCl_2^+(Sc_6Cl_8^-)$ (69) can be viewed as a further condensation of two chains of the type in Gd_2Cl_3 such that "double chains" are formed by sharing of common edges along the parallel chains as shown in Fig. 9.

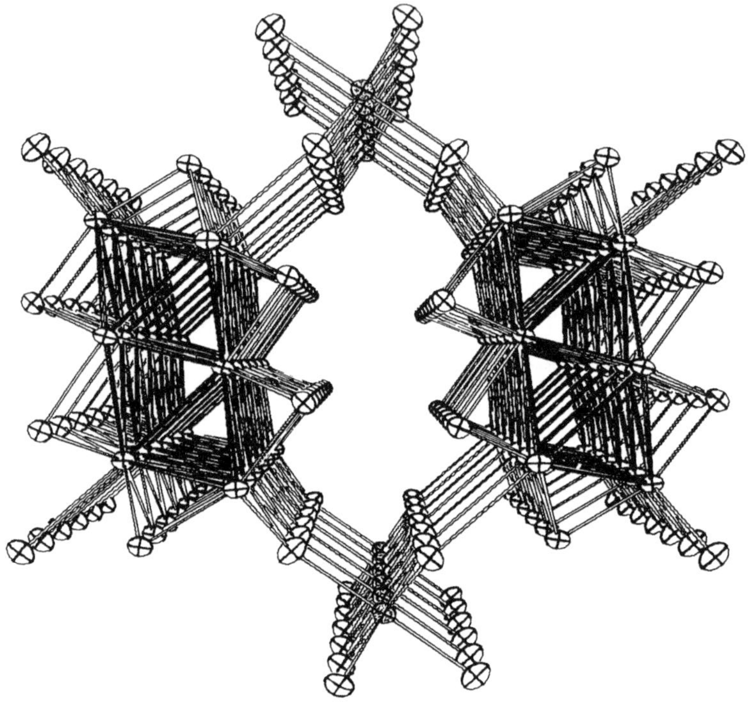

$(ScCl_2^+)(Sc_6Cl_8^-)$

Figure 9. A view down the anion and cation chains in $ScCl_2^+(Sc_6Cl_8^-)$.

Linear chain structures also are derived from the sharing of atoms at opposite apices of cluster polyhedra. The sharing of opposite apex atoms of octahedra occurs with some frequency in metal-rich chalcide and pnictide compounds. The simplest of these structures is that observed for Ti_5Te_4 (8) which is the prototype of many compounds having the composition M_5X_4, with M a metal from Groups IV, V or VI and X chalcide or pnictide. This structure is best illustrated by viewing the contents of the tetragonal unit cell along the c-axis, as shown in Fig. 10. Because all atoms occupy positions with z-coordinates of either 0 or

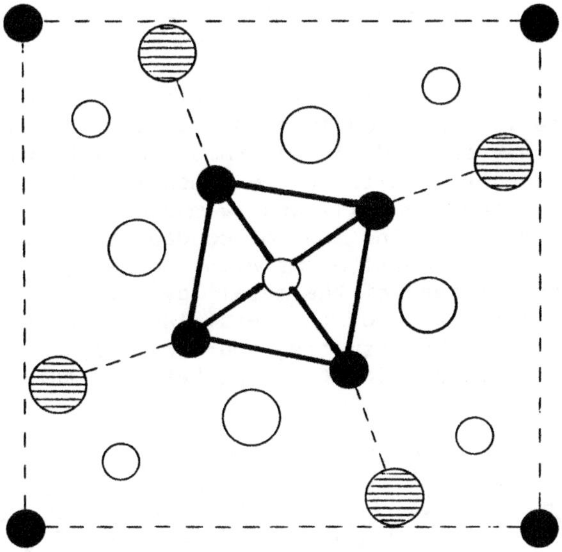

Figure 10. A projection down the c-axis of the Ti_5Te_4 structure. Small circles Ti, large circles Te; filled circles at Z=0, open circles at Z=½.

½ the structure is easily visualized. One readily distinguishes octahedral units of metal atoms centered at the corners and body center, each surrounded by an array of nonmetal atoms arranged exactly like that in the M_6X_8 cluster units. We note also that the apical metal atoms shared between adjacent octahedral units in the chain become enclosed in the center of eight metal atoms disposed at the corners of a cube, as in the bcc metals. Finally we note that the individual chains are knit together by bridging of each nonmetal atom in one cluster chain to the exo position of a metal atom in an adjacent chain. There are vacant octahedral coordination sites at the face centers of the unit cell which may be occupied by metal ions in some cases. The compound V_6As_4 has exactly such structure (70) and is better formulated $V(V_5As_4)$.

Factors important to the stability of these chain structures are poorly understood. Whereas it can be stated with certainty that metal-metal bonding with delocalized bond formation is essential to these structures it is not known if gaps dictating specific electron/metal ratios exist in the electronic bands. In most known examples however the Pauling bond order per metal-based electron, averaged over all metal-metal bonds of the repeat unit, fall in the range 0.8-1.0 (71) indicating that all such electrons are indeed utilized for bonding purposes. One can also see that in these chain structures the spacing of metal atoms and nonmetal atoms along the chain must be the same. This is a severe requirement since optimal M-M bonding in the chain direction will dictate stability only with nonmetal atoms within a narrow range of vanderWaals radii and with given valence.

There are many other conceptual variants of the modes of cluster condensation which could be discussed. In the space available here only one additional example will be given. Condensation of octahedral units in two dimensions by edge sharing is illustrated in the Group III and IV monohalides, prototype ZrCl (72) or ZrBr (73). The mode of condensation results in two close-packed layers of metal atoms with the halogen atoms, one over each triangular face of the metal layers, forming commensurate layers on either side of the metal layers, as shown in Fig. 11. The stacking of the resulting four-layer slabs then accounts for the slight difference in structure between ZrCl and ZrBr.

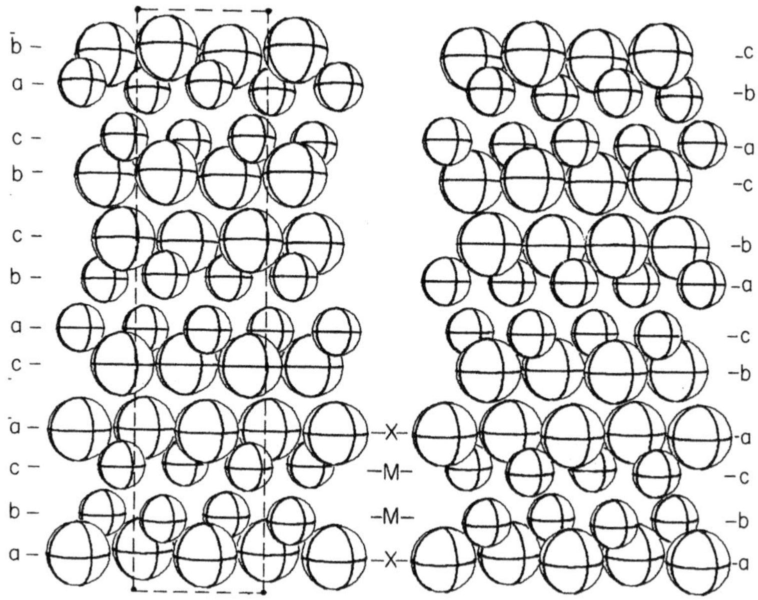

Figure 11. The close-packed layer structures (X-Zr-Zr-X) of ZrBr on the left, and ZrCl on the right. Note the different stacking sequence of the layers in the two cases.

Further condensation as the M/X ratio increases generates the so-called metal-rich compounds with often complex structures. These are worthy of separate treatment as given in the review by Franzen (74).

NOTES

1. Formal bond order for metal-metal bonds in cluster species is here defined as one-half the net number of metal-based, bonding electrons divided by the total number of metal-metal bonds within the cluster. Usually the number of metal-metal bonds in a cluster corresponds to the number of edges in the metal atom polyhedron.

2. In this notation i indicates a ligand atom which occupies an inner bridging position of the cluster, and 0 indicates a ligand in an outer or exo position; see Schafer, H., and von Schrering, H. G.: 1964, Angew, Chem. 76, pp. 833-849.

REFERENCES

(1) Koknat, F. W. and McCarley, R. E.: 1972, Inorg. Chem. 11, pp. 812-816.

(2) Bars, O., Guillevic, and Grandjean, D.: 1973, J. Solid State Chem. 6, pp. 335-339.

(3) Templeton, J. L., Jacobson, R. A., and McCarley, R. E.: 1977, Inorg. Chem. 16, pp. 3320-3328.

(4) Broll, A., Simon, A., vonSchnering, H. G., and Schafer, H.: 1969, Z. Anorg. Allg. Chem. 367, pp. 1-18

(5) Stensvad, S., Helland, B. J., Babich, M. W., Jacobson, R A., and McCarley, R. E.: 1978, J. Am. Chem. Soc. 100, pp. 6257-6258.

(6) Lokken, D. A., and Corbett, J. D.: 1973, Inorg. Chem. 12, pp. 556-559.

(7) Jodden, K., vonSchnering, H.G., and Schafer, H.: 1975, Angew. Chem. 87, pp. 594-595.

(8) Gronvold, F., Kjekshus, A., and Raaum, F : 1961, Acta Cryst. 14, pp. 930-934.

(9) Poeppelmeier, K. R., Corbett, J. D.: 1978, J Am. Chem. Soc. 100, pp. 5039-5044.

(10) Corbett, J. D., Daake, R L., Poeppelmeier, K. R , and Guthrie, D. H.: 1978, J. Am. Chem. Soc. 100, pp. 652-654.

(11) Converse, J. G., Hamilton, J. B. and McCarley, R. E.: 1970, Inorg. Chem. 9, pp. 1366-1372.

(12) Cotton, F. A. and Lippard, S. J.: 1965, Inorg. Chem. 4, pp. 59-65.

(13) Schafer, H., Laumanns, R., Krebs, B., and Henkel, G.: 1979, Angew, Chem. 91, pp. 343-344.

(14) Bennett, K. G., Caulton, K. G. and Cotton, F. A.: 1969, Inorg. Chem. 8, pp. 1-6.

(15) Cotton, F. A., Eiss, R. and Foxman, B. R.: 1969, Inorg. Chem. 8, pp. 950-957.

(16) Abraham, F , Trehoux, J., and Thomas, D.: 1979, J. Solid State Chem. 29, 73-79.

(17) Sleight, T. P., Hare, C. R., and Sleight, A. W.: 1968, Mater. Res. Bull. 3, pp. 437-444.

(18) Cotton, F. A., Frenz, B. A., Pedersen, E. and Webb, T. R.: 1975, Inorg. Chem. 14, pp. 391-398.

(19) Shive, L. W. and Cotton, F. A.: 1975, Inorg. Chem. 14, pp. 2032-2035.

(20) Bennett, M. J., Cotton, F. A. and Walton, R. A.: 1968, Proc. Roy. Soc. (London) A303, pp. 175-

(21) Cotton, F. A., Shive, L. W. and Stults, B. R.: 1976, Inorg. Chem. 15, pp. 2239-2243.

(22) Katovic, V. Templeton, J. L., Hoxmeier, R. J. and McCarley, R. E.: 1975, J. Am. Chem. Soc. 97, pp. 5300-5302.

(23) Bertrand, J. A., Cotton, F. A., and Dollase, W. A.: 1963, Inorg. Chem. 2, pp. 1166-1171.

(24) Ansell, G. B. and Katz, L.: 1966, Acta Crystallogr. 21, pp 482-485; McCarroll, W. H., Katz, L., and Ward, R.: 1957, J. Am. Chem. Soc. 79, pp. 5410-5414.

(25) Bino, A., Ardon, M., Maor, I., Kaftory, M. and Dori, Z.: 1976, J. Am. Chem. Soc. 98, pp. 7093-7095.

(26) Cotton, F. A. and Haas, T. E.: 1964, Inorg. Chem. 3, pp. 10-17.

(27) King, R. B.: 1972, Progr. Inorg. Chem. 15, pp. 314-321.

(28) Pauling, L.: The Nature of the Chemical Bond, 3rd ed., 1960, Cornell University Press, Ithaca, N.Y., pp. 400-404.

(29) Cotton, F. A.: 1964, Inorg. Chem. 3, pp. 1217-1220.

(30) Schafer, H., vonSchnering, H. G.: 1964, Angew, Chem. 76, pp. 833-849.

(31) Bino, A., Cotton, F. A., and Dori, Z.: 1979, Inorg. Chim. Acta 33, pp. 133-134.

(32) Bino, A., Cotton, F. A., and Dori, Z.: J. Am. Chem. Soc. 100, pp. 5252-5253.

(33) Simon, A., and vonSchnering, H. G.: 1966, J. Less-Common Metals 11, pp. 31.

(34) Mattes, R., and Mennemann, K.: 1977, Z. Anorg. Allg. Chem. 437, pp. 175-182.

(35) Katovic, V., Templeton, J. L., and McCarley, R. E.: 1976, J. Am. Chem. Soc. 98, pp. 5705-5706.

(36) Berg, J. M., Hodgson, K. O., and Holm, R. H.: 1979, J. Am. Chem. Soc. 101, pp. 4586-4593, and references therein.

(37) Vahrenkamp, H.: 1977, Structure and Bonding 32, pp. 1-56.

(38) Perrin, C., Chevrel, R., and Sergent, M.: 1975, C. R. Acad. Sc. Paris 281C, pp. 23-25.

(39) Vandenberg, J. M. and Brasen, D.: 1975, J. Solid State Chem. 14, pp. 203-208; Perrin, C , Chevrel, R., Sergent, M.: 1975, C. R. Acad. Sc. Paris 280C, pp. 949-951.

(40) Gall, R. S., Chu, C. T.-W., and Dahl, L. F.: 1974, J. Am. Chem. Soc. 96, pp. 4019-4023; Trinh-Toan, Teo, B. K., Ferguson, J. A., Meyer, T. J., and Dahl, L. F.: 1977, ibid. 99, pp. 408-416.

(41) Averill, B. A., Herskovitz, T., Holm, R. H., and Ibers, J. A.: 1973, J. Am. Chem. Soc. 95, pp. 3523-3534.

(42) Schafer, H., vonSchnering, H. G., Tillack, J., Kuhnen, F., Wohrle, H., and Baumann, H.: 1967, Z. Anorg. Allg. Chem. 353, pp. 281-310.

(43) Converse, J. G. and McCarley, R. E.: 1970, Inorg. Chem. 9, pp. 1361-1366.

(44) Fleming, P. B. and McCarley, R. E.: 1970, Inorg. Chem. 9, pp. 1347-1354, and references therein.

(45) Simon, A., vonSchnering, H. G., Wohrle, H., and Schafer, H.: 1965, Z. Anorg. Allg. Chem. 339, pp. 155-170.

(46) Schafer, H., vonSchnering, H. G., Niehues, K.-J., Nieder-Vahrenholz, H. G.: 1965, J. Less Common Metals 9, pp. 95-104.

(47) Koknat, F. W. and McCarley, R. E.: 1974, Inorg. Chem. 13, pp. 295-300.

(48) Bauer, D., vonSchnering, H.G., and Schafer, H.: 1965, J. Less Common Metals 8, pp. 388-401.

(49) Bauer, D., and vonSchnering, H.G.: 1968, Z. Anorg. Allg. Chem. 361, pp. 259-276.

(50) Thaxton, C. B., and Jacobson, R. A.: 1971, Inorg. Chem. 10, pp. 1460-1463.

(51) An excellent review of the Chevrel phases is given by Yvon, K.: 1978, Current Topics in Materials Science 3, pp. 55-129.

(52) Simon, A., vonSchnering, H. G., and Schafer, H.: 1967, Z. Anorg. Allg. Chem. 355, pp. 295-310; Kust, M. A., Corbett, J. D., and Friedman, R. M.: 1968, Inorg. Chem. 7, pp. 2081-2086.

(53) Bateman, L. R., Blount, J. F. and Dahl, L. F.: 1966, J. Am. Chem. Soc. 88, pp. 1082-1084.

(54) Schafer, H , and Siepmann, R.: 1968, Z. Anorg. Allg. Chem. 357, pp. 273-288.

(55) Siepmann, R., and vonSchnering, H. G.: 1968, Z. Anorg. Allg. Chem. 357, pp. 289-298.

(56) Chevrel, R., Sergent, M., and Prigent, J.: 1971, J. Solid State Chem. 3, pp. 515-519.

(57) Odermott, R., Fischer, O. Jones, H., and Bongi, G.: 1974, J. Phys. C7, pp. L13-15; Fischer, O.: 1976, Europhysics News 7, pp. 1-4.

(58) Andersen, O. K., Klose, W. and Nohl, H.: 1978, Phys. Rev. B17, pp. 1209-1237.

(59) Imoto, H. and Corbett, J. D., to be published.

(60) Chevrel, R., Sergent, M., and Prigent, J.: 1974, Mat. Res. Bull. 9, pp. 1487-1498.

(61) Marezio, M., Dernier, P. D., Remeika, J. P., Corenzwit, E., and Matthias, B. T.: 1973, Mat. Res. Bull. 8, pp. 657-668; Guillevic, P. J., Lestrat, H., and Grandjean, D.: 1976, Acta Cryst. B32, pp. 1342-1345.

(62) Yvon, K., Paoli, A., Flugiger, R., and Chevrel, R.: 1977, Acta Cryst, B33, pp. 3066-3072.

(63) Simon, A.: 1976, Chem. Unserer Zeit 10, pp. 1-9.

(64) Chini, P., Longoni, G., and Albano, V. G.: 1976, Adv. Organometal. Chem. 14, pp. 285-344.

(65) Seeber, B., Decroux, M., Fischer, O., Chevrel, R., Sergent, M., and Gruttner, A.: 1979, Solid State Commun. 29, pp. 419-423.

(66) Potel, M., Chevrel, R., Sergent, M., Decroux, M., and Fischer, O.: 1979, C. R. Acad. Sc. Paris 288, pp. C429-432.

(67) Mattausch, H. J., Hendricks, J. B., Eger, R., Corbett, J. D., and Simon, A.: to be published.

(68) Torardi, C. and McCarley, R. E.: 1979, J. Am. Chem. Soc. 101, pp. 3963-3964.

(69) Poeppelmeier, K. R., and Corbett, J. D.: 1977, Inorg. Chem. 16, pp. 1107-1111.

(70) Berger, R.: 1977, Acta Chem. Scand. A31, pp. 287-291.

(71) Corbett, J D.: to be published.

(72) Adolphson, D. G., and Corbett, J. D.: 1976, Inorg. Chem. 15, pp. 1820-1823.

(73) Daake, R. L. and Corbett, J. D.: 1977, Inorg. Chem. 16, pp. 2029-2033.

(74) Franzen, H. F.: 1978, Prog. Solid State Chem. 12, pp. 1-39.

HETEROPOLY BLUES

Michael T. Pope

Department of Chemistry, Georgetown University
Washington, D. C. 20057 USA

Polyoxotungstate, -molybdate and -vanadate anions in which the metal atoms occupy sites of approximately C_{4v} symmetry with one terminal oxygen each, are reducible in reversible one- or two-electron steps to mixed-valence polyanions (heteropoly blues). Variable temperature electron spin resonance (ESR) spectroscopy of the reduced anions reveals that the added electrons are trapped at W(V), Mo(V) centers at temperatures between 50 and 120K depending upon the structure of the anion. Other spectroscopic studies of heteropoly blues are reviewed. Certain polyoxotungstates may be reduced beyond the 'blue' stage to give isostructural complexes with firmly trapped W(IV) + W(VI) valences. The latter complexes contain metal-metal bonds and may be reduced even further to give fully delocalized species with an average metal oxidation state of +10/3.

INTRODUCTION

Heteropoly blues are a class of mixed valence complexes which are obtained by reduction of isopoly- and heteropoly vanadates, -molybdates and -tungstates (1). Although some of these polyoxoanions give colloidal or insoluble mixed valence oxides, so-called "molybdenum-" and "tungsten blues", we restrict the term heteropoly blue to well-defined crystalloidal molecular or ionic species. Even with this restriction the field of heteropoly blues encompasses several hundred complexes, for a single polyanion may give rise to a dozen or more mixed valence species. Most, but by no means all, of the reduced anions are intensely blue, hence the generic name.

D. B. Brown (ed.), Mixed-Valence Compounds, 365-386.

Berzelius is generally credited with the first description and analysis of a heteropoly compound, ammonium 12-molybdophosphate, in 1826, although it was not until Marignac's discovery of the tungstosilicates in 1862 that such complexes were recognized as a distinct class. During the subsequent 50-60 years there was considerable activity in the field, and many of the currently known types of polyanions were first synthesized and analyzed during this period (2). Reliable characterization of such complex species was not of course possible until more recently, but later work using modern physical methods has in most cases confirmed the stoichiometries reported and attests to the skills of the earlier chemists. The terms isopoly- and heteropoly acids came to denote anions with one or more types of positive-valent atoms respectively, thus $12WO_3 \cdot 9H_2O$ (modern formula $H_6[H_2W_{12}O_{40}] \cdot 5H_2O$) an isopoly acid; $P_2O_5 \cdot 18MoO_3 \cdot 36H_2O$ ($H_6P_2Mo_{18}O_{62} \cdot 33H_2O$) a heteropoly acid. The reduction of such species to heteropoly blues was observed at an early date and this behavior was soon adopted for analytical purposes (3), although the composition and structure of these substances was then unknown. The first isolation and analysis of crystalline heteropoly blues was made in 1920 (4), fourteen years before the first X-ray structure determination of a heteropoly complex by Keggin (5).

STOICHIOMETRY

Those heteropoly blues that have been the most intensely studied are derived from heteropoly anions with the so-called α-Keggin (5,6) structure, $XM_{12}O_{40}^{n-}$ ($M=Mo^{VI}$, W^{VI}), see Figure 1.

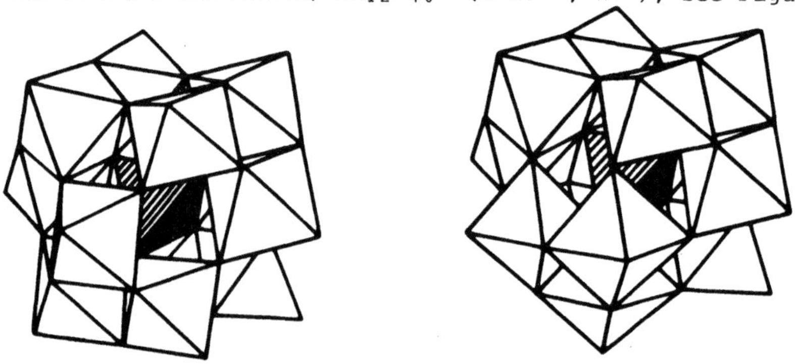

α **β**

Figure 1. Structures of α- and β-$XM_{12}O_{46}^{n-}$ heteropoly anions as arrangements of MO_6 octahedra. Note rotation of lower M_3O_{13} unit by 60° in β-isomer.

Most analytical applications of heteropoly blues are based on the formation and subsequent reduction of molybdates with this structure. The structure is seen to consist of four M_3O_{13} groups of three edge-shared MO_6 octahedra, each group linked to the next M_3O_{13} by corner-sharing and the whole arrangement enclosing a corner-sharing central XO_4 tetrahedron. As in all polyanion structures the M-atoms are strongly displaced from the centers of their MO_6 octahedra towards the terminal unshared oxygen atoms. The α-Keggin structure has overall T_d symmetry. A β-isomer of C_{3v} symmetry in which one M_3O_{13} group has been rotated by 60° (Figure 1) is also known (7). Although thermodynamically less stable than the α-isomer for the oxidized anions, the β-structure appears to be stabilized upon reduction (8). Rates of isomerization are generally very low for the tungstates and can be retarded for the molybdates by working in nonaqueous or mixed solvents.

The redox chemistry of these anions has been elucidated by numerous electrochemical studies (9). The behavior of α- and β-$SiMo_{12}O_{40}{}^{4-}$ (this, and other Keggin formulae, will be abbreviated, e.g. $SiMo_{12}$) are used to illustrate the general pattern for Keggin structure anions. Polarograms of both isomers in acidic solutions show three reversible two-electron waves, the half-wave potentials of which vary with pH in such a way as to correspond to the addition of a proton for each electron. Thus the reductions may be written (10)

$$\alpha\text{-}SiMo_{12}{}^{4-} \longrightarrow \alpha\text{-}H_2SiMo_{12}{}^{4-} \longrightarrow \alpha\text{-}H_4SiMo_{12}{}^{4-} \longrightarrow \alpha\text{-}H_6SiMo_{12}{}^{4-}$$

$$\text{or} \quad O\alpha \longrightarrow II\alpha H_2 \longrightarrow IV\alpha H_4 \longrightarrow VI\alpha H_6$$

where Roman numerals denote the number of electrons added. A parallel series of reductions at potentials that are slightly more positive, may be written for the β isomer.

The reduced heteropoly acids H_6SiMo_{12} (II) and H_8SiMo_{12} (IV) can be isolated by controlled potential electrolysis. Since the reduced anions are more stable against alkaline decomposition than are the oxidized anions it is possible to map the redox behavior of the $SiMo_{12}$ structure between pH 0 and 14 from polarograms of II or IV, see Figure 2. As the pH increases the two-electron waves successively split into two pH-independent one-electron waves, the first above pH 2.4, the second above pH 9.5, and the third in 0.1 M LiOH.

Other Keggin anions, molybdates and tungstates, exhibit analogous polarographic behavior corresponding to the addition, in reversible one- or two-electron steps (depending upon pH), of four or six electrons. Differences between individual anions arise because (a) tungstates have more negative reduction potentials than molybdates, (b) tungstates are less strongly protonated

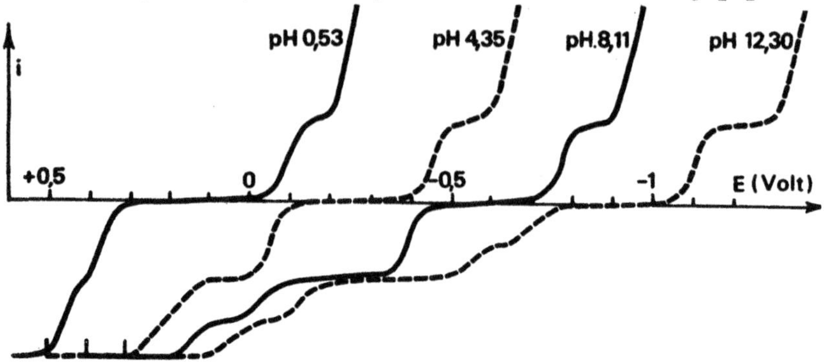

Figure 2. Polarograms of the four-electron blue of β-$SiMo_{12}$ at different pH. (Ann.Chim., 1969, 4, p. 368).

than analogous molybdates and splitting of the two-electron waves occurs at lower pHs, e.g. SiW_{12} undergoes 1,1 and 2 electron reductions at pH 0 (9), and (c) there is an electrostatic effect of the anion charge upon the reduction potentials, see Figure 3.

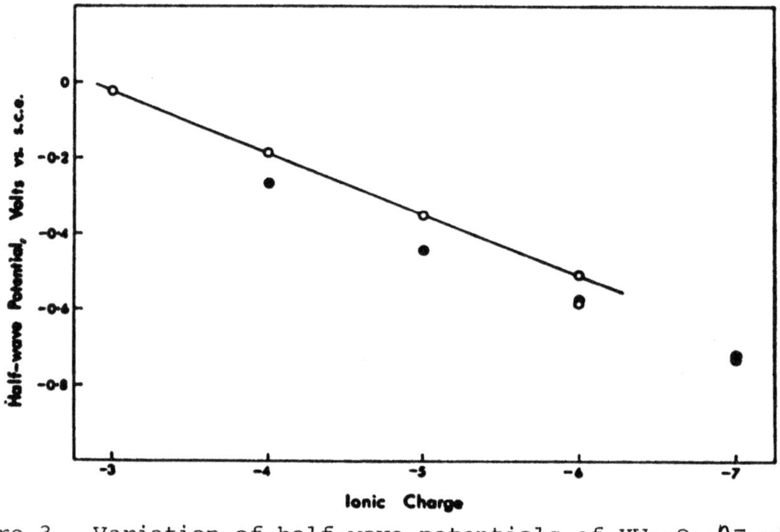

Figure 3. Variation of half-wave potentials of $XW_{12}O_{40}^{n-}$ with ionic charge. X=P(V), Si(IV), Fe(III), Co(II) and H_2: O and ●, first and second one-electron reductions respectively. (Inorg. Chem., 1966, 7, p. 1253).

At the six-electron level there appears to be a discontinuity in the reduction process which is especially marked for the tungstates. Subsequent reductions, of up to 26 more electrons in one case (11), occur by polarographically irreversible multielectron steps. The resulting species, generally brown or red, are still quantitatively oxidizable to the original Keggin structures. The nature of these highly reduced species will be discussed in Section (e) below.

STRUCTURAL CONSIDERATIONS

Each MO_6 octahedron in the Keggin structure has one terminal (unshared) oxygen atom. The metal atom therefore occupies a site of approximately C_{4v} symmetry as in mononuclear oxocomplexes such as $WOCl_5^-$, $VO(H_2O)_5^{++}$. In some other polyanion structures the MO_6 octahedra have two (cis) terminal oxygens as in $MoO_2F_4^{2-}$. The fact that cis dioxo complexes are known only for d^0 metal ions, whereas monoxo (C_{4v}) complexes may contain d^0, d^1, or d^2 central atoms led Pope to propose that only polyanions with one terminal oxygen per metal (Type I structures) would undergo reversible one-electron reductions to form heteropoly blues (12). The implication of this proposal is that the electron is added to a metal orbital (d_{xy} or b_2 in C_{4v} symmetry) which is formally nonbonding or weakly π-antibonding, see Figure 4. No such low-lying

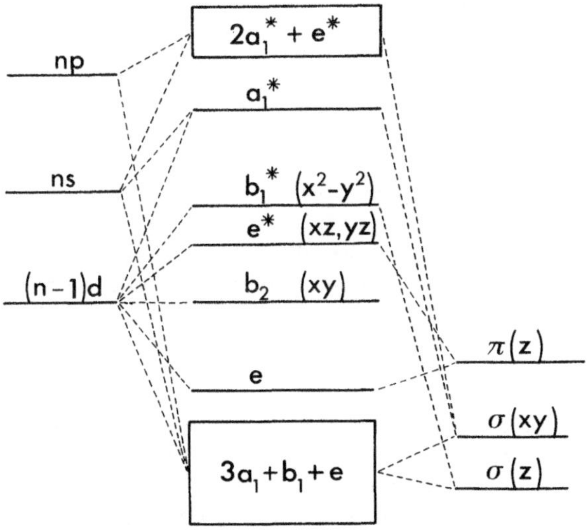

Figure 4. Simplified Molecular Orbital Scheme, neglecting equatorial π-bonding, for MOL_5 complexes of C_{4v} symmetry. For d^1 metals the unpaired electron occupies the b_2 orbital.

orbital is available in the cis dioxo case, and no examples are known of polyanions based on cis dioxo octahedra (Type II structures) which are reducible to isostructural heteropoly blues.[1]

Other polyanions which, according to the above criterion, should be reducible to heteropoly blues are illustrated in Figure 5(a)-(f). Each of these structures will be briefly considered here. (a) The α- and β-isomers of $X_2M_{18}O_{62}^{6-}$ (X=P, As: M=Mo,W) have long been known to give rise to heteropoly blues with 2,4 and 6 electrons (4,13). Evidence for the existence of the odd reduction states is available for the tungstates but curiously not for the molybdates. An explanation for this has been offered (14). (b) Recent work has shown that $Mo_6O_{19}^{2-}$ and $W_6O_{19}^{2-}$ undergo two reversible one-electron reductions in non-aqueous solvents (15,16). A number of mixed (V-W and Nb-W) polyanions of this structure are reducible also (17). (c) Decatungstate, $W_{10}O_{32}^{4-}$, alias polytungstate-Y, undergoes two one-electron reductions to blue species (18). (d) There is as yet no convincing evidence for heteropoly blues based on the decavanadate structure $V_{10}O_{28}^{6-}$. Mixed valence species written as decavanadates (19) now appear to involve more complex structures (20). Direct reduction of $V_{10}O_{28}^{6-}$ to an isostructural 'blue' if it is possible, will probably have to be carried out in a non-aqueous solvent to restrict hydrolysis. (e) The anion $MnV_{13}O_{38}^{7-}$ has been chemically reduced (21) to isomorphous $MnV^{IV}V_{12}O_{38}^{8-}$. (f) No reduction of $XW_{10}O_{36}^{8-}$ heteropoly anions has yet been reported (22). A possible explanation for this may lie in the fact that eight of the ten tungsten atoms have effectively cis dioxo environments[2] as a result of low bond orders for the X-O bonds (23).

ELECTRONIC STRUCTURES OF HETEROPOLY BLUES

The bulk of the physical evidence to be reviewed below supports a model for heteropoly blues in which the first added electrons are weakly trapped on individual metal atoms. To the extent that each metal atom in a reducible polyanion occupies a site of C_{4v} symmetry, the extra electron occupies a b_2 ("d_{xy}") orbital as in $MoOCl_5^{2-}$. Electron delocalization in heteropoly blues is then viewed as a combination of two effects, a thermally-activated hopping process from one metal atom to the next, and a ground-state delocalization gsd (24), presumably involving π-bonding through bridging oxygens from the reduced metal atom to its neighbors. The degree of gsd is relatively small, but its existence is necessary to account for the non-zero intensities of intervalence charge transfer (IVCT) optical absorption bands. The energies of IVCT bands (E_{op}) are related approximately to an activation energy for the corresponding thermal electron hop (E_{th}) by $E_{op} \approx 4E_{th}$ (25). Other factors being equal, E_{th} will be diminished as gsd becomes more significant, see Figure 6.

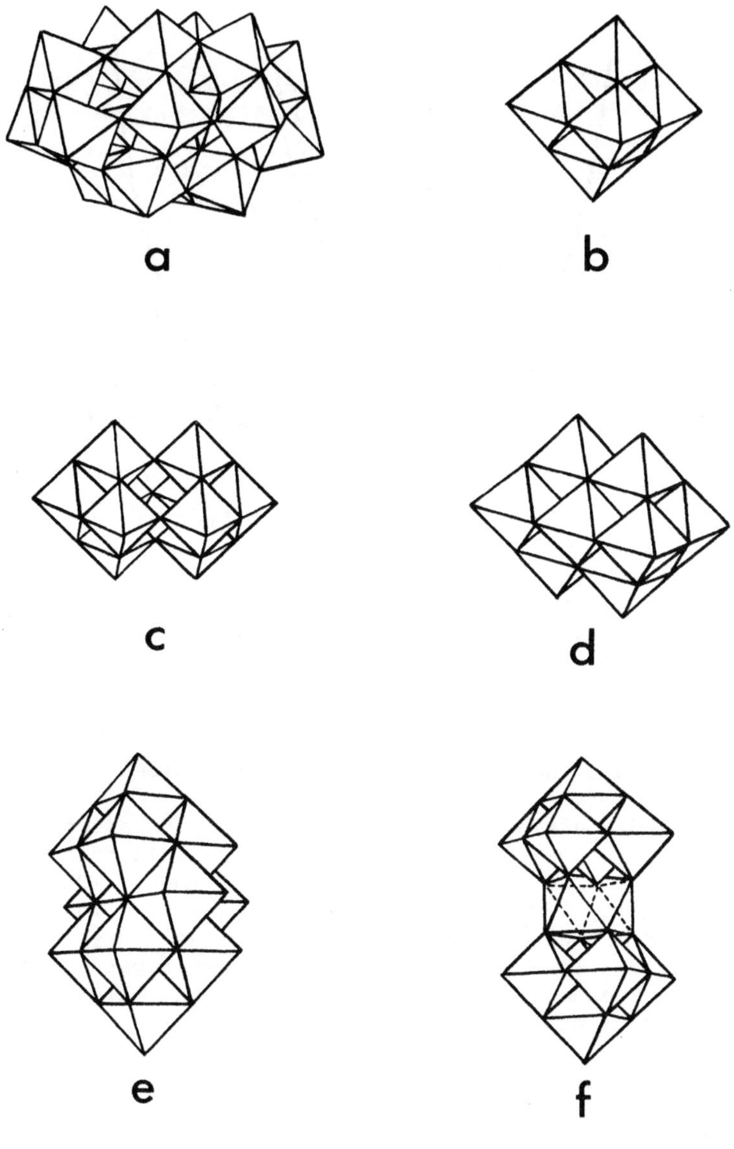

Figure 5. Structures of some Type I polyanions. (a) α-$X_2M_{18}O_{62}^{6-}$ (the β-isomer has one edge-shared M_3O_{13} group rotated by 60°); (b) $M_6O_{19}^{2-}$; (c) $W_{10}O_{32}^{4-}$; (d) $V_{10}O_{28}^{6-}$; (e) $MnV_{13}O_{38}^{7-}$; (f) $XW_{10}O_{36}^{8-}$.

The more highly reduced polyanions with six or more added electrons appear to have strongly localized valences, see section (e).

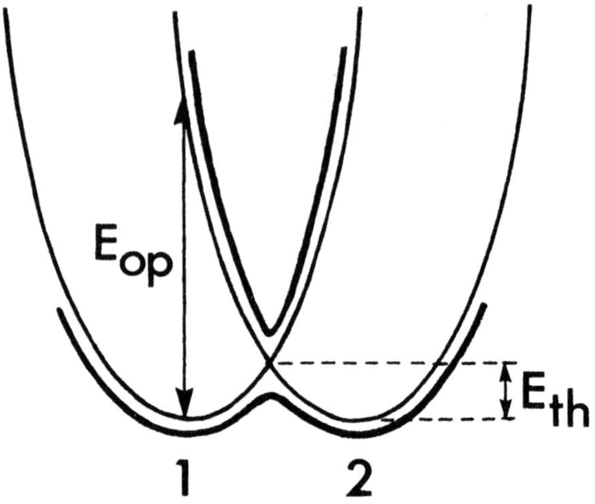

Figure 6. Schematic representation of potential energy surfaces of atoms 1 and 2 which can undergo valence interchange. Optical and thermal charge transfer barriers shown are for the limit of no resonance interaction (ground state delocalization) between 1 and 2. Bold lines show exaggerated effects of a small degree of such delocalization. (See Mayoh, B. and Day, P.: 1972, J.Amer.Chem.Soc., 94, p. 2885).

(a) Electron Spin Resonance of One-Electron Blues

Before discussing the results of ESR spectroscopy of heteropoly blues it seems advisable to review the main features of the ESR spectra of axial oxocomplexes of C_{4v} symmetry MOL_5^{n-}. A schematic energy level diagram has been presented in Figure 4. Taking the M-O bond as the z-axis, the d-levels are ordered as shown, with d_{xy} lowest. The latter can participate in π-bonding with the four equatorial ligands. Following Ballhausen and Gray (26) the appropriate LCAO-MO orbitals are

$$|B_1\rangle = \beta_1|d_{x^2-y^2}\rangle - \beta_1'|\phi_{b_1}\rangle$$

$$|\beta_1\rangle = \beta_1|d_{x^2-y^2}\rangle - \beta_1'|\phi_{b_1}\rangle$$

$$|E\rangle = \varepsilon|d_{xz}, d_{yz}\rangle - \varepsilon'|\phi_e\rangle$$

where the ligand orbitals ϕ are group orbitals of the necessary symmetry. For d^1 metals the unpaired electron occupies the b_2 orbital and the ESR parameters have been calculated to be (27)

$$g_{\parallel} = 2.0023 - [2\frac{(2\lambda_M\beta_2\beta_1 - \lambda_L\beta_2'\beta_1')}{\Delta E(B_2 \to E_1)}$$

$$\times (2\beta_2\beta_1 - 2\beta_1\beta_2' S_{b_2} - 2\beta_2\beta_1' S_{b_1} - \beta_1'\beta_2') \quad (1)$$

$$g_{\perp} = 2.0023 - [\frac{2\lambda_M\beta_2\varepsilon}{\Delta E(B_2 \to E)} (\beta_2\varepsilon - \beta_2\varepsilon' S_e - \varepsilon\beta_2' S_{b_2})] \quad (2)$$

$$-\frac{A_{\parallel}}{P} = \kappa\beta_2^2 + \frac{4}{7}\beta_2^2 + 2.0023 - g_{\parallel} + \frac{3}{7}(2.0023 - g_{\perp})$$

$$+ [\frac{6}{7}\frac{\lambda_M\beta_2\varepsilon}{\Delta E(B_2 \to E)} (\beta_2\varepsilon' S_e + \varepsilon\beta_2' S_{b_2})]$$

$$+ [2\frac{(2\lambda_M\beta_2\beta_1 - \lambda_L\beta_2'\beta_1')}{\Delta E(B_2 \to B_1)} (2\beta_2\beta_1' S_{b_1} + 2\beta_1\beta_2' S_{b_2} + \beta_1'\beta_2')] \quad (3)$$

$$-\frac{A_{\perp}}{P} = \kappa\beta_2^2 - \frac{2}{7}\beta_2^2 + \frac{11}{14}(2.0023 - g_{\perp})$$

$$+ \frac{11}{7}[\frac{\lambda_M\beta_2\varepsilon}{\Delta E(B_2 \to E)}](\beta_2\varepsilon' S_e + \varepsilon\beta_2' S_{b_2}) \quad (4)$$

where λ_M, λ_L are spin-orbit coupling parameters for metal and ligand atoms respectively, S_{b_2}, S_{b_1} and S_e are overlap integrals, $P = 2.0023\beta_e g_n \beta_n \langle r^{-3}\rangle$, and κ^2 is the isotropic Fermi contact term. As Che etal (15) have pointed out, a greater range of values is observed for $g_{\parallel}$ than for $g_{\perp}$ so that while $g_{\perp} > g_{\parallel}$ for most complexes of Mo(V) with oxygen ligands, $g_{\parallel} > g_{\perp}$ for species $MoOBr_5^{2-}$ with larger λ_L values. The possibility of inversion of the relative magnitudes of $g_{\parallel}$ and $g_{\perp}$ implies that situations may arise in which $g_{\parallel} \approx g_{\perp}$ even for complexes of axial symmetry. If overlap and λ_L terms are omitted, equations (1)-(4) may be reduced and combined to give

$$g_{\parallel} = 2.0023 - \frac{8\beta_2^2\beta_1^2\lambda}{\Delta E(B_2 \to B_1)} \quad (5)$$

$$g_{\perp} = 2.0023 - \frac{2\beta_2^2\varepsilon^2\lambda}{\Delta E(B_2 \to E)} \quad (6)$$

$$\beta_2^2 = \frac{7}{6}\left(\frac{A_\perp - A_\parallel}{P}\right) - \frac{7}{6}(2.0023 - g_\parallel) + \frac{5}{12}(2.0023 - g_\perp) \quad (7)$$

where (7) yields an approximate value of the MO-coefficient for the equatorial π-bonds. Application of equations (5) and (6) requires that the ligand field transitions $B_2 \rightarrow B_1$ and $B_2 \rightarrow E$ have have been observed and correctly assigned. This is not always possible, especially for heteropoly blues, the spectra of which (see below) contain intense IVCT bands.

The simplest and most symmetrical heteropoly blue studied to date is the brown (!) polyanion $Mo_6O_{19}^{3-}$ (Fig. 5b) prepared by controlled potential electrolysis of $(Bu_4N)_2Mo_6O_{19}$ in dimethylformamide (15). The room temperature ESR spectrum consists of a single line (width 62 G) with no detectable hyperfine structure from the magnetic isotopes of molybdenum $^{95,97}Mo$, I = 5/2, 25.3%. At 77K however the spectrum is fully resolved into axial components, with $g_\perp = 1.930$; $g_\parallel = 1.919$; $A_\perp = 34.5G$; $A_\parallel = 80.5G$. Since no exchange broadening was observed, it was concluded that the hopping frequency for Mo^V-O-$Mo^{VI} \rightarrow Mo^{VI}$-O-$Mo^V$ was much lower than ~5×10^7 Hz at 77K. Absence of hyperfine structure at 300K suggests that electron-hopping is rapid at the higher temperature. These data may be compared with those for the Keggin anion α-PMo^VW_{11} (24) in which the electron must be trapped on the molybdenum atom. Although the site symmetry of Mo in the Keggin structure is strictly only C_s, an axial ESR spectrum was observed, with parameters similar to those of $Mo_6O_{19}^{3-}$: $g_\perp = 1.939$; $g_\parallel = 1.913$; $A_\perp = 37.9G$; $A_\parallel = 81.8G$. However, as expected, the hyperfine structure was still observable in the isotropic solution spectrum at room temperature.

The anions $Mo_6O_{19}^{3-}$ and $PMoW_{11}$ therefore illustrate two different degrees of valence trapping. At room temperature, hopping electron delocalization is rapid and complete for $Mo_6O_{19}^{3-}$ and is presumably non-existent for $PMoW_{11}$. A 'covalent' gsd (see above) which places some of the unpaired electron density on the neighboring Mo or W atoms, is determined by the relative energies, symmetries, and overlap of the orbitals involved. In polyanions, where metal-metal separations are 3.3-3.7Å it appears that gsd occurs via a π-interaction of the d_{xy} orbital and bridging oxygen atoms. The magnitude of this interaction is reflected in the value of the MO-coefficient β_2 which can be estimated from ESR data by means of equation (7). Values of β_2 obtained in this way[3] are: $Mo_6O_{19}^{3-}$, 0.92; $PMoW_{11}O_{40}^{4-}$, 0.87. The difference between these two values may indicate that gsd is slightly greater in $PMoW_{11}$ than in $Mo_6O_{19}^{3-}$. Inspection of the two structures (Figs. 1, 5b) shows that on the grounds of symmetry alone, there are better possibilities for d_{xy}-pπ-d_{xy} overlap between the *corner-shared* MO_6 octahedra in the Keggin structure than between the edge-shared octahedra in either structure.

The ESR data for $PM\overset{V}{o}Mo_{11}O_{40}^{4-}$ demonstrate that the unpaired electron is less firmly trapped than in $Mo_6O_{19}^{3-}$. Polycrystalline spectra for dilute crystals of $(Bu_4N)_4PMo^VMo_{11}O_{40}$ in $(Bu_4N)_4SiW_{12}O_{40}$ between 6 and 300K revealed no hyperfine structure above ca 50K (28). At lower temperatures, apparently isotropic spectra with six hyperfine lines (a~61G) were resolved. The appearance of an isotropic spectrum was difficult to understand since although it is possible for $g_{\parallel} \approx g_{\perp}$ (see above) it would be highly unusual for $A_{\parallel} = A_{\perp}$. The problem has recently been clarified by a remeasurement of the spectrum in frozen dimethylformamide solution which provided better resolution than before (29). A very slight rhombic distortion of the central line is observed. Virtual superposition of some of the parallel and perpendicular hyperfine lines, with $A_{\parallel} \approx 2A_{\perp}$ accounted for the earlier spectrum. Since the rhombic component of the spectrum is very small[4] and is not even detectable in the hyperfine structure, the following axial parameters can be determined, $g_{\perp} = 1.9465$; $g_{\parallel} = 1.9458$; $A_{\perp} = 27.2G$; $A_{\parallel} = 60.7G$. From these, using equation (7), $\beta_2 = 0.79$, a value which implies significantly more gsd than in $Mo_6O_{19}^{3-}$. The ESR spectrum of isotopically-enriched $PMo^VMo_{11}O_{40}^{4-}$ (98% ^{95}Mo, I=5/2) showed essentially the same hyperfine spectrum at low temperatures but with broader lines which, it is believed, indicate unresolved superhyperfine interactions with the neighboring ^{95}Mo nuclei (28).

Since each MoO_6 octahedron in PMo_{12} has two edge-shared neighbors (as in $Mo_6O_{19}^{2-}$) and two corner-shared neighbors, the difference both in gsd and the temperature at which hyperfine structure first becomes visible for these two polyanions suggests that there is both more facile electron hopping and greater gsd between the corner-shared octahedra than between the edge-shared octahedra of the Keggin structure. This conclusion would be consistent with the considerations of orbital overlap mentioned above, and would indicate that electron-hopping in a heteropoly blue with the Keggin structure is described by two different activation energies, one for corner-shared and one for edge-shared octahedra.

In an attempt to verify and/or refine this model, current measurements are being made on mixed-metal Keggin anions such as PMo_2W_{10} and PMo_3W_9. Although normal syntheses of such anions produce mixtures of stereoisomers (30) detectable by multiple ^{31}P NMR resonances (31), it has been possible to isolate single isomers of PMo_2W_{10} ($\delta(^{31}P)=13.9$) and PMo_3W_9 ($\delta=13.1$) in which the MoO_6 octahedra are corner-shared (32). At 77K the electron in $PMo^VMo^{VI}W_{10}$ is trapped according to ESR hyperfine structure, but in $PMo^VMo_2^{VI}W_9$ at the same temperature no hyperfine structure is observed, cf PMo^VMo_{11}.

A number of polytungstate one-electron blues have been studied by X-band ESR spectroscopy between 10 and 300K (33). In every case, the spectrum is gradually resolved into three lines as the temperature is lowered, indicating a rhombic environment for the unpaired electron. The spectral variation with temperature could be satisfactorily simulated with linewidth as the single variable parameter. Values of g are given in Table 1.

Table 1. ESR Parameters for Reduced Heteropoly Tungstates with α-Keggin[a] and α-Dawson[b] Structures

Anion	<g>	g_x	g_y	g_z
$PW_{12}O_{40}^{4-}$	1.827	c	c	c
$SiW_{12}O_{40}^{5-}$	1.821	1.85	1.822	1.79
$GeW_{12}O_{40}^{5-}$	1.814	1.848	1.816	1.779
$BW_{12}O_{40}^{6-}$	1.817	1.854	1.823	1.773
$H_2W_{12}O_{40}^{7-}$ d	1.805	1.852	1.805	1.759
$P_2W_{18}O_{62}^{7-}$	1.858	1.906	1.854	1.814
$As_2W_{18}O_{62}^{7-}$	1.854	1.905	1.849	1.810

a See Figure 1
b See Figure 5a
c Spectrum appears slightly rhombic at 14K but reliable g-values are not resolvable, see Figure 7.
d This anion contains two protons in the central tetrahedron.

These results are entirely consistent with those observed for PMo_{12} and imply a trapping of the electron on a single tungsten atom at low temperatures. The large spread of g-values for each anion is a consequence of the large spin-orbit coupling parameter (ca 2700 cm^{-1}) which also contributes to the linewidths (20-30G per component at 10K). No hyperfine structure from ^{183}W, I=1/2, 14.4%, could be detected, but this is not unexpected in view of the relatively broad spectra and the abundance of the isotope. The variation in resolution of the rhombic components, see Fig. 7 for example, may indicate differences in the firmness of electron trapping as a result of structural distortions (33).

Unlike the Keggin (1:12) structure the α-Dawson (2:18) structure has two types of metal sites: six of type A and twelve of type B (Fig. 5a). The type A octahedra have a similar environment to the octahedra in the Keggin structure, and the very different g-values for the reduced Dawson anions in Table 1 suggest

Figure 7. X-band ESR spectra of aqueous-glycerol glasses of one-electron blues of PW_{12} and H_2W_{12}. (Inorg.Chem., 1976, 15, p. 2551).

that the unpaired electron is trapped on a type B metal atom. Such a conslusion is reinforced by recent studies of α-$P_2W_{17}VO_{62}^{7-}$ and α-$P_2W_{17}MoO_{62}^{6-}$ both of which exist as two isomers with V and Mo in A and B sites (34). The reduction potentials ($V^V \rightarrow V^{IV}$; $Mo^{VI} \rightarrow Mo^V$) of the B-substituted isomers are more positive (by 0.2V for the Mo pair, 0.1V for the V pair) than those for the A-substituted isomers. In the polytungstate blues, trapping of the electron, as revealed by the resolution of a rhombic ESR spectrum, occurs at higher temperatures for the 2:18 polyanions (ca 100K) than for the Keggin anions (<77K).

(b) Two-Electron Blues

ESR spectra of frozen solutions (77K) of α-$XMo^V_2Mo_{10}$ (X=P, As, Si, Ge) and β-$SiMo^V_2Mo_{10}$ have been reported (35-38). An isotropic line, g=1.945-1.952, was found for the α-isomers, and an axial spectrum (no g-values reported) for β-$SiMo_{12}$. From the latter result it was concluded that the electrons were trapped on the MoO_6 octahedra adjacent to the rotated Mo_3O_{13} group, see Fig. 1. A similar conclusion had been drawn from considerations of the electronic spectra of α- and β-isomers (39). That ESR spectra are observed at all for the two-electron blues implies incomplete antiferromagnetic coupling of the two electrons, i.e., a triplet state. From the relative intensities of the signals of the one- and two-electron blues Rabette etal (35) concluded that the electrons were ~60% paired in α-$H_6SiMo^V_2Mo_{10}$ at 77K. Kazanskii etal (38) report that the spectrum of $H_5PMo^V_2Mo_{10}$ is, by a factor of 10^6, less intense than that of the corresponding silicate. These results might indicate contamination with one-electron blues rather than a paramagnetic two-electron blue. Observation of unbroadened NMR spectra (38) for the two-electron blues of $SiMo_{12}$ (^{17}O) and PMo_{12} (^{17}O, ^{31}P) suggests that these species are diamagnetic. In the tungstate series, the central protons in H_2W_{12} which are detected by ^1H NMR at 6.2 ppm are also seen in the two-electron blue (5.5 ppm) but not in the paramagnetic one-electron blue (40).

Kazanskii etal (36-38,41) have reported ^{17}O and ^{31}P NMR spectra and X-ray photoelectron spectra of two-electron blues of H_5PMo_{12} and H_6SiMo_{12}. According to ^{17}O-NMR the two electrons are evenly distributed over the twelve Mo atoms at 60°C, a result which places a lower limit of ca 10^3 sec^{-1} for the rate of electron hopping. The photoelectron spectra, with a characteristic time-scale of ca 10^{18} sec^{-1} could be deconvoluted to reveal Mo^{6+} and Mo^{5+} centers in the anticipated ratio of 10:2. The ^{17}O and ^{31}P spectra (38) of reduced PMo_{12} showed changes consistent with an $\alpha \rightarrow \beta$ isomerization first reported by Fruchart and Souchay (42).

(c) Electronic Spectra

Spectra of oxidized polyanions have oxygen-to-metal charge transfer bands in the near ultraviolet which diminish in intensity as the anion is reduced (40). The energies of these bands have been correlated with the reduction potentials for a series of polyanions of the same charge (43), a result which implies that the electrons added upon reduction of the polyanions occupy the acceptor levels of the charge-transfer transitions.

The absorption spectra of the reduced anions generally show several broad bands throughout the visible and near infrared and are often profoundly influenced by the degree of protonation of the anions (see Fig. 8). Attempts have been made to interpret such spectra in terms of ligand field transitions of Mo(V), W(V) or V(IV) centers, and intervalence charge transfer (IVCT). The lowest energy band in all heteropoly blues occurs at 8-10kK and is assigned as an IVCT transition since (i) it disappears upon removal of mixed valence character, e.g. upon reduction of $SiMo^VMo^{VI}W_{10}$ to $SiMo^V_2W_{10}$ and (ii) in mixed metal polyanions such as $V^{IV}W_5O_{19}{}^{4-}$ the band appears at higher energy and its position can be related to optical electronegativity differences between 'donor' (reduced) and 'acceptor' atoms (39,44). The intensity of

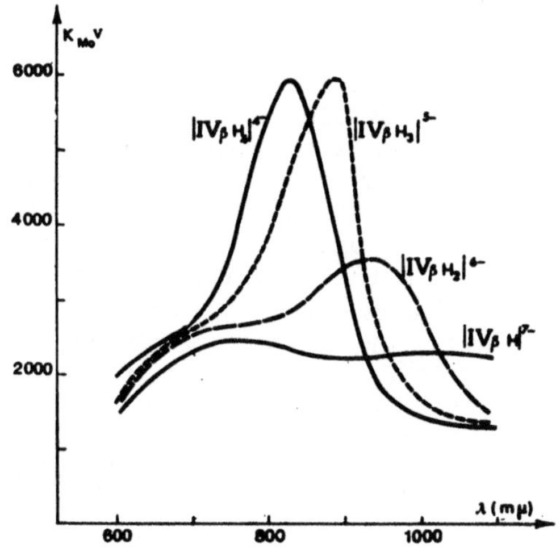

Figure 8. Absorption spectra of four-electron blue of β-$SiMo_{12}$ with different degrees of protonation. (Ann.Chim., 1969, 4, p. 367).

the 8-kK band can vary significantly, e.g. ε_{max} = 280 $M^{-1}cm^{-1}$ for $W^VW_5O_{19}{}^{3-}$, 1600 $M^{-1}cm^{-1}$ for $PW^VW_{11}O_{40}{}^{4-}$, and this can be attributed to differences in delocalization of electron density from the reduced atom to the neighboring acceptors as described in section (a). Characteristic spectra of unprotonated Keggin anions are shown in Fig. 9. In most cases three bands are

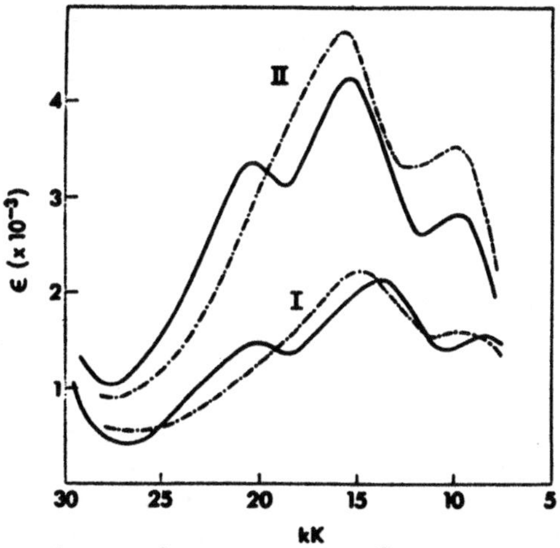

Figure 9. Absorption spectra of unprotonated one- and two-electron blues of PW_{12} (solid lines) and CoW_{12} (broken lines). (Inorg.Chem., 1970, 9, p. 664).

observed at ca 8-, 14- and 20kK (18kK for molybdates). Upon reduction to the two electron blue the spectra double in intensity and the 8- and 14kK bands undergo a blue shift of some 1-2kK. Based on the spectra of several simple and substituted Keggin anions, Fruchart etal (39) have made the following provisional assignments. The band at 8kK: IVCT between edge-shared MO_6 octahedra ('intra' group transition), and the band at 14kK: IVCT between corner-shared octahedra ('extra' group transition - note that this band is not present in the spectrum of reduced $W_6O_{19}^{2-}$ which has no corner-shared octahedra). The band at 20(18)kK, the position of which is unaffected by further reduction, was assigned as a d-d transition ($^2B_2 \rightarrow {}^2B_1$ in C_{4v}) of W(V) or Mo(V). The anomalously high intensity of this band was considered to be caused by intensity-stealing from the nearby IVCT band and by the reduction of symmetry of the M(V) center from C_{4v} to C_s. Although these assignments, and the corresponding heteronuclear IVCT (e.g. $Mo^V \rightarrow W^{VI}$) transitions in substituted polyanions are internally self consistent, it has recently been shown that assumptions made regarding the structures of certain substituted polyanions were incorrect. For example, the complex β-$SiMo_3W_9$, originally believed to be that isomer with the three MoO_6 octahedra in an edge-shared group, is now known to have the MoO_6 octahedra linked by corners[5]. Since the spectrum of β-$SiMo^VMo_2W_9$ has the 8kK band, it is clear that such a band cannot solely be identified by an 'intra'-group IVCT. It is possible that both 'intra' and 'extra' transitions in Keggin anions occur in the 8-10kK region. The 'intra' transition is expected to be less allowed than the 'extra' transition on account of the smaller gsd between edge-shared octahedra, and the 'intra' IVCT band could be hidden by a more intense 'extra' band. However this modification does not take into account the intense band at 14kK and it is evident that more information is required before a satisfactory understanding of heteropoly blue spectra is reached.

(d) Magnetic Properties

Very few magnetic measurements of heteropoly blues have been reported. Magnetic moments of ca 1.4 BM at 37° were found for one-electron blues of PW_{12}, SiW_{12} and H_2W_{12}; the corresponding two electron blues were reported to be diamagnetic (40). A value of 0.6 BM has been reported for the two-electron blue of $SiMo_{12}$ at 'room temperature' (36). Variable-temperature magnetic measurements of the latter complex would be desirable in view of its supposed antiferromagnetic nature. A recent study of $SiV_3^{IV}W_9$ in which the VO_6 octahedra share corners shows antiferromagnetic coupling with $J \sim 50\ cm^{-1}$ (45).

The question arises as to why the polarograms of reducible polyanions, especially molybdates, show an enhanced stability for heteropoly blues with even numbers of electrons. Magnetic exchange energies are too small to account for this behavior and it seems that deformation energies associated with valence-trapping and subsequent protonation must be considered (39).

(e) Highly-Reduced Heteropoly Anions

Reduction of Keggin anions beyond the six-electron stage results in a modification of their polarograms. This behavior has been observed for SiW_{12}, BW_{12}, H_2W_{12} and $SiMo_{12}$. The case of H_2W_{12} has been studied in some detail by Launay (46) who concludes that the initially formed six-electron blue undergoes valence-isomerization.

$$H_2W_6^{V}W_6^{VI} \text{ (blue)} \longrightarrow H_2W_3^{IV}W_9^{VI} \text{ (brown)}$$

It is proposed that the brown complex contains the three W(IV) atoms in an edge-shared W_3O_{13} group of the Keggin structure, the three terminal oxygens of this group being converted to OH groups and the tungstens participating in metal-metal bonds. The brown six-electron product may be further reduced to twelve-, eighteen- and twenty-four-electron species, each group of the Keggin structure being in turn reduced to $W_3^{IV}O_{13}$. Oxidation of the amphoteric 24-electron product restores the original polyanion without decomposition. Evidence for W(IV) and W(VI) in the reduced species has been provided by X-ray photoelectron spectra, see Fig. 10 for example (47). The 24-electron product,

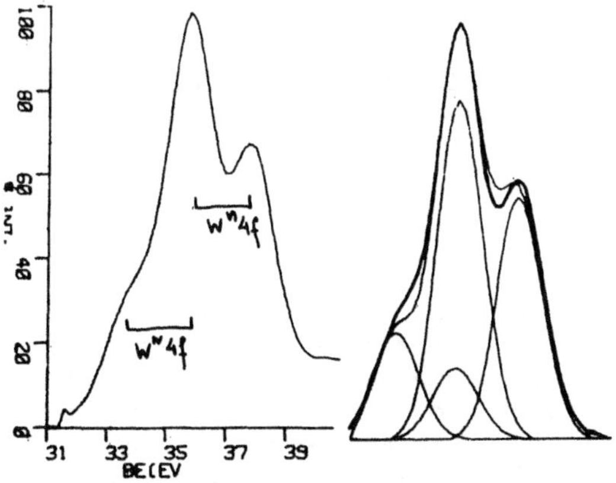

Figure 10. X-ray photoelectron spectrum of $Rb_4H_8[H_2W_3^{IV}W_9^{VI}O_{40}]$ in W(4f) region. Left: experimental spectrum. Right: deconvoluted spectrum. (Chem.Phys.Lett., 1977, 51, p. 244).

isolated from 1 M HCl as a cation, undergoes a slow transformation to an isostructural species with one proton (instead of two) in the central tetrahedral cavity of the Keggin structure. The new complex may be reoxidized to give $HW_{12}O_{40}{}^{7-}$ which is slowly converted to $H_2W_{12}O_{40}{}^{6-}$; both species have been characterized by

^1H NMR (48). The deprotonation of the 24-electron product is presumed to be caused by an electrostatic influence of the reduced tungsten atoms which have moved closer together to allow metal-metal bonding in the W_3O_{13} groups. Cotton has described an MO scheme for an M_3L_{13} metal-metal bonded cluster (49). Three bonding-, one non-bonding-, and five antibonding orbitals were obtained from the t_{2g} orbitals of the metals. In $W_3^{IV}O_{13}$ groups the bonding levels are filled, and Launay has shown that the 24-electron product of H_2W_{12} can accept eight more electrons which presumably occupy the non-bonding levels of the four W_3O_{13} groups.

The other Keggin anions show analogous behavior but cannot be reduced to the same extent as $H_2W_{12}O_{40}{}^{6-}$. This is attributed to increased electrostatic repulsions between the central atoms B(III), Si(IV), and the W_3^{IV} groups.

CONCLUSIONS

This article has attempted to summarize the main features of the chemistry of heteropoly blues as they are known at present. In terms of the Robin-Day classification (50) of mixed valence substances, heteropoly blues can exmplify all types: Class I, strong valence trapping as in $H_2W_3^{IV}W_9^{VI}O_{40}H_3{}^{9-}$; Class II, weak valence trapping as in $PMo^VMo_{11}O_{40}{}^{4-}$; Class III, complete valence mixing in the 32-electron reduction product of $H_2W_{12}O_{40}{}^{6-}$. Clearly there are many gaps in our understanding of these complexes, but they offer an exceptional opportunity to study the ramifications of electron delocalization, electron mobility and magnetic exchange processes in metal-oxide environments.

ACKNOWLEDGEMENTS

The hospitality of Professor Y. Jeannin during a recent sabbatical leave is gratefully acknowledged, as are valuable discussions with Drs. J.P. Launay, R. Massart and their colleagues at the University of Paris VI. The author's current research on heteropoly blues is supported by a grant from the National Science Foundation. Figures 2 and 8 are reprinted with permission from Masson et Cie, Paris; Figures 3, 7 and 9 with permission from Inorganic Chemistry, Copyright by the American Chemical Society; Figure 10 with permission from the North-Holland Publishing Company, Amsterdam.

NOTES

1. Protonation of one of the terminal oxygens of a cis dioxo group would in effect yield a reducible moiety with a single remaining M-O double bond. However such terminal oxygens are not known to be susceptible to protonation under normal conditions. Photolysis of molybdate solutions containing isopropylammonium cations has yielded crystals of a mixed valence octamolybdate in which it may be postulated that hydrogen atoms from the cation had been transferred to cis $Mo^{VI}O_2$ groups yielding $Mo^{V}O(OH)$ centers. (Isobe, M., Marumo, F., Yamase, T., and Ikawa, T.: 1978, Acta Cryst. B34, p 2728; Yamase, T., 1978, J.Chem.Soc., Dalton Trans., p. 283).

2. Oxygen-17 NMR chemical shifts for $XW_{10}O_{36}{}^{8-}$ indicate bond orders appropriate for cis WO_2 groups. (Kazanskii, L.P., Fedotov, M.A., and Spitsyn, V.I.: 1977, Dokl.Akad.Nauk SSSR, 233, p 152; 234, p. 1376).

3. The calculations were made using $P = -55 \times 10^{-4} cm^{-1}$, the value appropriate for Mo^{3+}. (See for example, McGarvey, B.R.: 1966, Transit.Metal Chem., 3, p. 89).

4. From 35-GHz ESR spectra of a partially-reduced single crystal of $Ca_3(PMo_{12}O_{40})_2 \cdot 58H_2O$ at 4.2K, $g_x = 1.956$, $g_y = 1.948$, $g_z = 1.939$. (Zaitov, M.M.: 1975, Zh.Strukt. Khim., 16, p 366: Russ.J.Inorg.Chem., 20, p. 1726).

5. The complexes are prepared directly from β-$SiW_9O_{34}{}^{9-}$. The crystal structure of the sodium salt of this anion (Y. Jeannin, private communication) reveals a β-Keggin unit from which the three corner-shared WO_6 octahedra opposite to the rotated W_3O_{13} group have been removed.

REFERENCES

1. Weakley, T.J.R.: 1974, Struct.Bonding(Berlin), 18, pp. 131-176; Evans, H.T., Jr.: 1971, Perspec.Struct.Chem., 4, pp. 1-69; Kepert, D.L.: 1972, *The Early Transition Metals*, Academic Press, New York, N.Y., pp. 46-62, 288-304.

2. Rosenheim, A.: 1921, *Abeggs Handbuch der Anorganische Chemie*, Hirzel, Leipzig, 4, I(2), pp. 977-1064.

3. Osmond, F.: 1887, Bull.Soc.Chim., 47, p. 745.

4. Wu, H.: 1920, J.Biol.Chem., 43, p. 189.

5. Keggin, J.F.: 1934, Proc.Roy.Soc., A144, pp. 75-100.

6. Recent determinations include Brown, G.M., Noe-Spirlet, M. R., Busing, W.R., and Levy, H.A.: 1977, Acta Cryst., 33A, pp. 1038-46; Strandberg, R.: 1977, Acta Cryst., 33B, pp. 3090-6; D'Amour, H., and Allmann, R.: 1976, Z.Kristallogr., 143, pp. 1-13.

7. Matsumoto, K.Y., Kobayaski, A., and Sasaki, Y.; 1975, Bull.Chem.Soc.Japan, 48, pp. 3146-51.

8. See for example, Souchay, P. and Contant, R.: 1967, C.R. Acad.Sci., Ser.C, 265, p. 723.

9. Pope, M.T. and Varga, G.M., Jr.; 1966, Inorg.Chem., 5, p. 1249; Massart, R. and Hervé, G., 1968, Rev.Chim.Minérale, 5, p. 501.

10. Massart, R.: 1968, Ann.Chim., 3, p. 507; 1969, Ann.Chim., 4, pp. 285-96, 365-70, 441-52.

11. Launay, J.P.: 1976, J.Inorg.Nucl.Chem., 38, p. 807.

12. Pope, M.T.: 1972, Inorg.Chem., 11, p. 1973.

13. Contant, R. and Fruchart, J.M.: 1974, Rev.Chim.Minérale, 11, p. 123; Papaconstantinou, E. and Pope, M.T.: 1967, Inorg.Chem., 6, p. 1152.

14. Garvey, J.F. and Pope, M.T.: 1978, Inorg.Chem., 17, p. 1115.

15. Che, M., Fournier, M., and Launay, J.P.: 1979, J.Chem.Phys., in press.

16. Boyer, M. and LeMeur, B.: 1975, C.R.Acad.Sci., Ser.C, 281, p. 59.

17. Flynn, C.M., Jr. and Pope, M.T.: 1973, Inorg.Chem., 12, p. 1626; Dabbabi, M., Boyer, M., Launay, J.P., and Jeannin, Y.: 1977, J.Electroanal.Chem., 76, p. 153.

18. Boyer, M.: 1971, J.Electroanal.Chem., 31, p. 441; Termes, S.C. and Pope, M.T.: 1978, Inorg.Chem., 17, p. 500.

19. Ostrowetsky, S.: 1964, Bull.Soc.Chim.France, pp. 1012, 1018.

20. For example $HV_{19}O_{49}^{8-}$ (Johnson, G.K.: 1977, Ph.D. Thesis, University Missouri - Columbia).

21. Flynn, C.M., Jr. and Pope, M.T.: 1970, J.Amer.Chem.Soc., 92, p. 85.

22. Termes, S.C. and Pope, M.T.: 1978, Transition Met.Chem., 3, p. 103.

23. Iball, J., Low, J.N., and Weakley, T.J.R.: 1974, J.Chem. Soc., Dalton Trans., p. 2021.

24. Altenau, J.J., Pope, M.T., Prados, R.A., and So, H.: 1975, Inorg.Chem., 14, p. 417.

25. Hush, N.S.: 1967, Prog.Inorg.Chem., 8, p. 391.

26. Ballhausen, C.J. and Gray, H.B.: 1962, Inorg.Chem., 1, p. 111.

27. DeArmond, K., Garrett, B.B., and Gutowsky, H.S.: 1965, J.Chem.Phys., 42, p. 1019.

28. Prados, R.A., Meiklejohn, P.T., and Pope, M.T.: 1974, J.Amer.Chem.Soc., 96, p. 1261.

29. Launay, J.P., Sanchez, C., Pope, M.T., Livage, J., and Fournier, M.: Manuscript in preparation.

30. Pope, M.T. and Scully, T.F.: 1975, Inorg.Chem., 14, p. 953.

31. Pope, M.T., O'Donnell, S.E., and Prados, R.A.: 1976, Adv. Chem.Ser., 150, p. 85.

32. Mossoba, M.M.: Unpublished results.

33. Prados, R.A. and Pope, M.T.: 1976, Inorg.Chem., 15, p. 2547.

34. Contant, R. and Ciabrini, J.P.: 1977, J.Chem.Res.(S), p. 222; Rocchiccioli-Deltcheff, C. and Thouvenot, R.: 1979, Spectrosc.Lett., 12, p. 127; Acerete, R., Harmalker, S.P., Hammer, C.F., Pope, M.T., and Baker, L.C.W.: 1979, Chem. Commun., In press.

35. Rabette, P., Ropars, C., and Grivet, J.P.: 1967, C.R.Acad. Sci., Ser.C, p. 153.

36. Dorokhova, E.N. and Kazanskii, L.P.: 1976, Dokl.Akad.Nauk SSSR, 229, p. 622.

37. Kazanskii, L.P.: 1978, Izvest.Akad.Nauk SSSR, Ser.Khim., p. 274.

38. Kazanskii, L.P., Fedotov, M.A., Potapova, I.V., and Spitsyn, V.I.: 1979, Dokl.Akad.Nauk SSSR, 244, p. 372.

39. Fruchart, J.M., Hervé, G., Launay, J.P., and Massart, R.: 1976, J.Inorg.Nucl.Chem., 38, p. 1627.

40. Varga, G.M., Jr., Papaconstantinou, E., and Pope, M.T.: 1970, Inorg.Chem., 9, p. 662.

41. Spitsyn, V.I., Potapova, I.V., and Kazanskii, L.P.: 1978, Dokl.Akad.Nauk SSSR, 243, p. 426.

42. Fruchart, J.M. and Souchay, P.: 1968, C.R.Acad.Sci., Ser.C, 266, p. 1571.

43. Kazanskii, L.P.: 1975, Izvest.Akad.Nauk SSSR, Ser.Khim., p. 499.

44. So, H. and Pope, M.T.: 1972, Inorg.Chem., 11, p. 1441.

45. O'Connor, C.J. and Mossoba, M.M.: Unpublished results.

46. Launay, J.P.: 1976, J.Inorg.Nucl.Chem., 38, p. 807.

47. Kazanskii, L.P. and Launay, J.P.: 1977, Chem.Phys.Lett., 51, p. 242.

48. Launay, J.P., Boyer, M., and Chauveau, F.: 1976, J.Inorg. Nucl.Chem., 38, p. 243.

49. Cotton, F.A.: 1964, Inorg.Chem., 3, p. 1217.

50. Robin, M.B. and Day, P.: 1967, Adv.Inorg.Chem.Radiochem., 10, 247-422.

LINEAR CHAIN MIXED-VALENCE SYSTEMS WITH DIRECT METAL-METAL INTERACTIONS

H. J. Keller

Anorganisch-Chemisches Institut der Universität Heidelberg, D-6900 Heidelberg 1,/GFR
Im Neuenheimer Feld 270

1. INTRODUCTION

1.1 Scope of the paper

This article describes the preparation and structural chemistry of a very restricted class of solid transition metal compounds. Only mixed valence solids containing infinite linear arrays of nonbridged transition metal ions will be discussed. The paper is limited to these materials, because high and anisotropic conductivities in linear molecular solids can only be expected from directly interacting systems (1d metals).

The enormous variation in chemical composition and physical properties of mixed valence compounds have been reviewed and classified earlier (1-3). These complete summaries of the "mixed valence" (m.v.) literature up to 1966 are mainly concerned with ligand bridged systems in solutions or threedimensionally linked solids. Some of the more recent reviews on linear chain metal compounds (4-14) include, in part, a discussion of directly interacting chain systems - with emphasis on their unusual physical properties. Two of the latter reviews (13 + 14) deal explicitly with directly interacting metal chain systems. The main concern in these papers, however, are not mixed valence, but integral oxidation state compounds. The aim of this article is not to review all these published data but to emphasize that a new class of compounds with very special properties has emerged.

D. B. Brown (ed.), Mixed-Valence Compounds, 387-411.

1.2 Formal oxidation numbers

A few words about the concept of formal oxidation numbers: Its use is a prerequisite in mixed valence chemistry. Before assigning different oxidation states to certain metal ions in a compound, one has to "count" electrons and ascribe formal oxidation numbers to each of these ions. The rules are simple to handle (15). Whether the central metal is actually in an oxidation state as calculated by the formal oxidation number rules is a totally different question, which can only be answered after extensive physical investigations.

In directly interacting systems which are discussed here the interactions between the metal ions are usually so strong that the materials can be assigned to class III B of the Robin-Day scheme (1) even if the metal sites are not exactly identical. As pointed out by Allen and Hush (2) in the strong interaction limit a band structure description of the solid is more appropriate. Assigning certain integral oxidation numbers to the different metal ions is not very meaningful then. In these cases the average number per metal atom is important, because it tells something about the band filling. Here the concept of integral or nonintegral oxidation states (better integral or nonintegral formal oxidation numbers), IOS and NIOS (14) might be helpful, though interband overlap can lead to partially filled bands and high electrical conductivity in IOS compounds, (semimetal), especially under pressure.

2. COMPOUNDS, STRUCTURES AND PREPARATION

Up to now only the chemical elements of the cobalt, nickel and copper triad (group VIII and Ib of the Periodic Table) have been found in directly interacting linear chain mixed valence systems.

VIII			Ib
^{26}Fe	^{27}Co	^{28}Ni	^{29}Cu
^{44}Ru	^{45}Rh	^{46}Pd	^{47}Ag
^{76}Os	^{77}Ir	^{78}Pt	^{79}Au

Since stacks of directly interacting complex molecules can only be formed by two-coordinate or planar three- and four-coordinate species, those ions which favourably occur in these geometries are principally the best candidates for linear structures. But only a few of the known compounds contain certain ions at different or nonintegral oxidation levels.

2.1 The cobalt group

The two heavier elements of the cobalt group form directly interacting linear chain systems preferrably with "strong" ligands like isonitriles and carbon monoxide. Some of these materials are potential mixed valence systems, but only a few of them have been proven to be partially oxidized Class III B (or NIOS) materials.

2.1.1 Cobalt compounds

To the authors best knowledge directly interacting, linear chain mixed valence cobalt complexes have not been characterized so far. There is one report of a partially oxidized cobalt complex with the ligand phthalocyanine (16). The detailed analysis of the spectroscopic data on the nickel derivative showed that these complexes are mainly interacting through their ligands (see section 2.2.1.2). Though there are only limited physical data on the cobalt compound, it can be assumed that this system is interacting through its ligands also.

There is one other general problem in assigning direct interactions in linear chain metal complexes, which is especially evident in planar complexes crystallizing in columnar stacks. As shown in figure 1, a

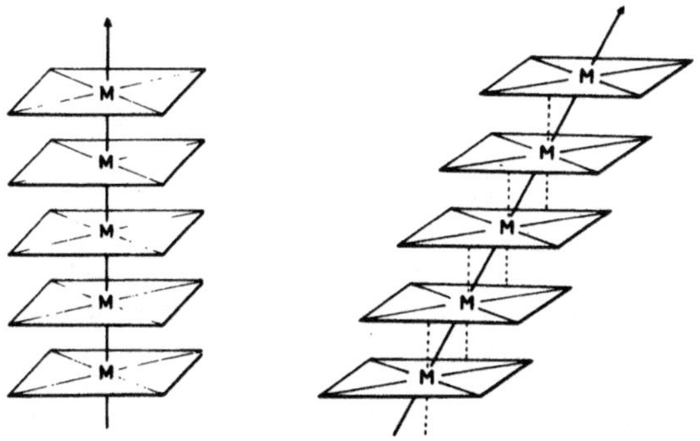

Figure 1: (a) Directly interacting "M-M" stack.
(b) Bridged "M-L-M" stack.
Slight slipping of molecules converts (a) into (b).

change from a direct to a bridged arrangement can be brought about just by slipping the molecules by a small amount or in alternative description, by allowing angles between the molecular plane and the stacking direction which differ from 90°. The definition "direct interaction" clearly excludes structures like that observed for bis(diaminoglyoximato)cobalt(II)diaminoglyoxim (17), which contains stacks of planar complex molecules but no direct interactions between the metal centers. As in different other planar cobalt(II) complexes this structure clearly can be regarded as a slipped arrangement of planar molecules giving rise to metal-ligand interactions between neighboring molecules only. This kind of arrangement has previously been described as a "M-L-M" stack (18). Only "M-M" arrangements are discussed here.

2.1.2 Linear chain rhodium solids

Two types of linear complexes are mentioned here, although a mixed valence state has not been proven for these materials so far.

(i) Tetrakis(isonitrile)rhodium(I) salts.
(ii) Dicarbonylchelatorhodium(I) compounds.

2.1.2.1 Tetrakis(isonitrile)rhodium(I) Salts

The surprising colors of these materials in solid state and solution aroused the interest of chemists immediately following their preparation (19). Nevertheless an unambiguous answer to the question of whether directly interacting linear chain NIOS solids exist in this class of compounds and a description of their properties is still lacking. There is one report on the "partially oxidized" tetrakis(vinylisonitrile)rhodium- and tetrakis(ethylisonitrile)rhodium-chlorides, tetrafluoroborates and perchlorates with a few structural data (20). The partial oxidation was assumed from the results of the elemental analysis. But since many of these compounds crystallize with solvent molecules, the analytical data alone are not conclusive in my opinion.

More recent investigations on these compounds (21-23) show that polynuclear species exist in solutions and in the solid state but the full structure of an "infinite" isonitrile material is still missing. Crystallization of solutions which contain oxidized species leads to a trinuclear mixed valence compound

(22 b) which can be regarded as an intermediate in the Aufbau process to higher aggregates (figure 2) but no infinite chain lattice has been characterized so far.

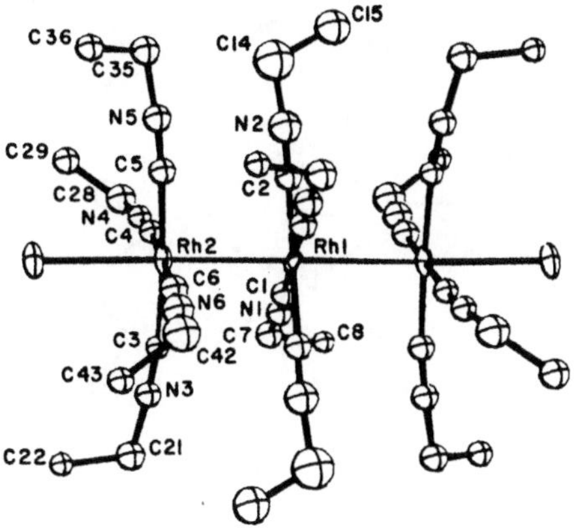

Figure 2: Trinuclear mixed valence tetrakis(benzyliso-nitrile)diiodorhodium(I,III)trication (only the first carbon atom of each phenyl group has been shown). An intermediate in the chain Aufbau processes? [with permission from ref. (22b)].

2.1.2.2 Dicarbonylchelatorhodium(I) complexes

Numerous solids of general stoichiometry cis-$Rh(CO)_2L_2$ or cis-$Rh(CO)_2ClL_1$ (with L_2 = bidentate mono-anion, e.g. acetylacetonate, L_1 = monodentate neutral donor, e.g. aniline or toluidine) with very anisotropic physical properties have been prepared (24,25). Some of them could be identified structurally (26-28) and different linear chain structures with short direct metal-metal contacts were found. A typical example being the acetylacetonatodicarbonylrhodium(I) (figure 3). Though quite short metal-metal bonds occur, there is no clear evidence whether at least a few of these directly interacting materials can be regarded as partially oxidized. Since there are many planar cationic species in rhodium(I) chemistry, since rhodium(II) (d^7) can form quite stable compounds in appropriate (sterically hindered) environment and since finally rhodium(I) can easily be oxidized by different chemicals, or undergo oxidative addition reactions the synthesis of infinite directly interacting materials should be no principal problem.

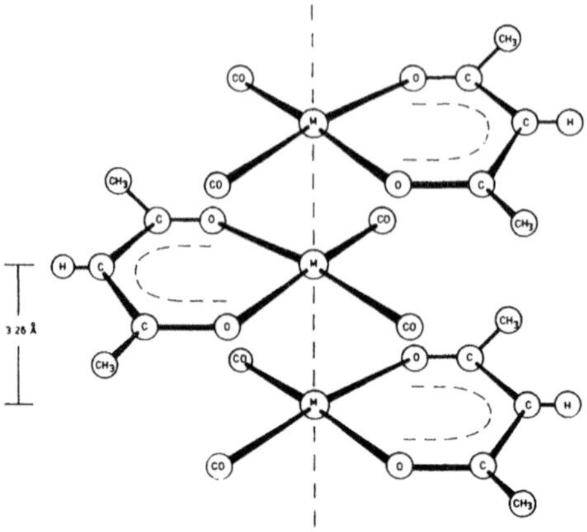

Figure 3: Schematic structure of dicarbonylacetylacetonatorhodium(I).

2.1.3 Iridium compounds

There is an extensive "mixed valence" chemistry of linear chain iridium compounds. One of these complexes ($Ir(CO)_3Cl$) has also been discussed as a mixed valence solid and an IOS material. The dicarbonylchelatoiridium(I) derivatives of the above mentioned rhodium complexes (2.1.2.2) are known but with a few exceptions they have not been characterized structurally.

2.1.3.1 Dicarbonyldihaloiridates(I,III)

These complexes were first obtained in 1958 by Malatesta (29), who described the remarkable color of the solutions (varying with concentration) and of the solid potassium salts of the chloro and bromo derivative. The nonintegral oxidation number was ascertained later (30). A wide variety of compounds with different cations (even TTF) and varying degree of partial oxidation has been obtained later (9,11,33-35) and more convenient routes for the preparation of these mixed valence materials have been found (31,32). Considerable effort has been made to characterize the solids (33-35). Due to the severe difficulties during the crystallization of these compounds (34), a full single crystal X-ray structure is not available, but there are some Guinier and single crystal data which clearly suggest that the solids contain linear stacks of planar $[Ir(CO)_2X_2]^-$ moieties with short metal-metal contacts (34) (figure 4). So this seems to be a very promising class of one-dimensional metals, if the crystallization problems can be overcome.

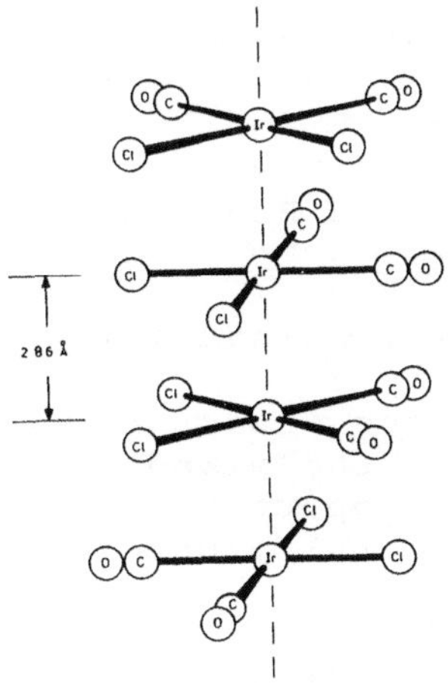

Figure 4: Schematic drawing of $(Ir(CO)_2X_2)^-$ stacks in the "Malatesta" compounds.

2.1.3.2 Tricarbonylchloroiridium(I)

This well-known compound (36,37), which serves as a starting material for most low-valent metalorganic iridium species (38), has later been claimed as a mixed valence (Class III b) material by different groups (39,40). Though some doubts about the non-stoichiometry of this compound arose (41,42) the structural proof of the stoichiometry $Ir(CO)_3Cl$ was not obtained before 1977 (43). The structure is very similar to the schematic drawing of the $[Ir(CO)_2X_2]^-$ chains. Staggered planar $Ir(CO)_3Cl$ units are stacked on top of each other in the crystals leading to a metal-metal distance of 2.844(1) Å only. The unusual physical properties of this IOS compound could be explained by strong metal-metal interactions, leading to overlap of fully occupied and empty bands (semimetal) (43). This compound shows, that partial oxidation is not a prerequisite for high conductivity in the strongly and directly interacting transition metal compounds.

2.2 The Nickel group

The distinction between the directly interacting M-M-forms and the indirectly interacting M-L-M forms is often subtle in the many planar "IOS" complexes of the ions Ni(II), Pd(II), Pt(II) (18). All the mixed valence materials in this group, however, contain the directly interacting M-M modifications. All three elements have an extensive mixed valence chemistry of directly interacting linear chain compounds.

2.2.1 Nickel compounds

2.2.1.1 Partially oxidized Bis(α,ß-dionedioximato)-nickel complexes

2.2.1.1.1 Bis(diphenylglyoximato)nickel(II,IV)

As in all other cases discussed in this review, the first preparation of the mixed valence species occured long before the importance and impact of these materials as an experimental probe for physical models were recognized.

Bis(diphenylglyoximato)nickel iodide can be chosen as a typical example for the "normal" history of conducting one-dimensional mixed valence solids. In 1950 Edelman reported (44) the isolation of the compounds Ni $C_{28}H_{22}N_4O_4Br$ and Ni $C_{28}H_{22}N_4O_4I$. Both were obtained by oxidation of bis(diphenylglyoximato)nickel(II) ($Ni(dpg)_2$) with the appropriate molecular halogens in organic solvents. The latter compound was described as glistening bronze crystals. Because of its simple stoichiometry (one additional halogen atom per complex unit after oxidation) the compound was declared to contain nickel with a formal oxidation number of three. Since this is very unusual for nickel complexes, which occur mainly in the +2 state, the compound was reinvestigated by Simek in 1962. He reported (45) the magnetic data of $Ni(dpg)_2Br$ and $Ni(dpg)_2I$ and concluded that the solid actually contains nickel(II). In 1967 Foust and Soderberg (46) recognized the stacked structure of the complex molecules and proposed that the material should be regarded as "inclusion-like molecular complexes stabilized by charge-transfer interaction" (46). Some years later the occurence of linear triiode ions in the channels of the lattice was suggested (47). It was proposed that these linear triiodide ions form chains parallel to the metal stacks very similar to the chains in the well-known blue starch-polyiodide compounds (48). The existence of triiodide ions instead of single charged iodide ions would lead to a formal oxidation number of 2.33 for the central nickel ions. In 1975 the full three-dimensional X-ray structure of the compound was solved (49) (figure 5). The iodine

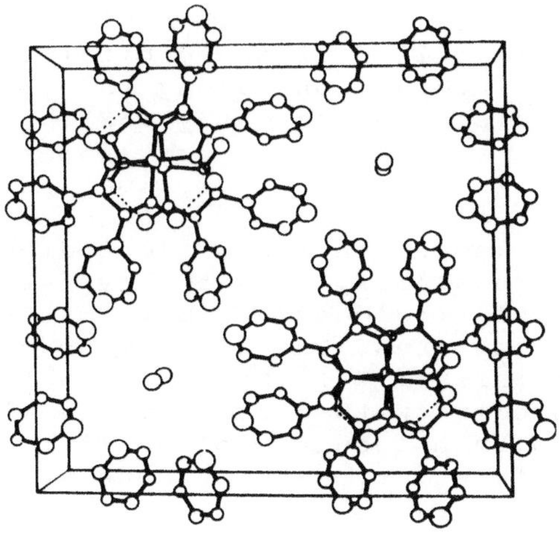

Figure 5: Structure of the partially oxidized Ni(dpg)$_2$I (with permission from reference (49)).

positions in the channels were disordered. On the basis of Resonance Raman and ^{129}I-Mößbauer data it was concluded that the channels contain triiodide ions leading again to the "typical" formal oxidation number of 2.33 for the nickel. Since the compound can loose considerable amounts of halogen without changing its structure, the exact chemical composition of the solid under different conditions was investigated (50). A very detailed, recent structural and physical (optical, RR, electrical conductivity ^{129}I-Mößbauer) study of this compound (51) finally solved most of the remaining problems: The solid was recognized as semi-conducting, linear chain mixed valence crystals containing I_5^- ions. This gives a formal oxidation number of 2.20 for the nickel ion. Therefore, only a comparatively small change in formal oxidation number occurs upon oxidation of the nickel(II) parent compound with molecular iodine. Though the overall physical properties of Ni(dpg)$_2$ and Ni(dpg)$_2$I are very different indeed (51) the central metal ion "feels" nothing about the changes in its environment: Linewidth and binding energies in the X-ray photoelectron spectra (XPS) of Ni(dpg) and Ni(dpg)I do not differ significantly from each other (51). Whether this result suggests that the charge is removed mainly from ligand orbitals (as in the phthalocyaninatonickel complexes, see 2.2.1.2) is still an open question. At the present time it is not known whether the charge transport in these semiconducting systems occurs through the metal-metal contacts or via π-orbital overlap between the partially filled ligand orbitals. The problem of ligand-ligand interactions along the chains of these solids has been discussed recently (52).

2.2.1.1.2 Bis(1.2-benzoquinonedioximato)nickel(II,IV)

Of the many known stacked bis(α,ß-dionedioximato)-nickel(II) compounds (53), only a very few can be oxidized with molecular halogens to m.v. materials without destroying the complexes. One of them is bis(1.2-benzoquinonedioximato)nickel(II) ($Ni(bqd)_2$) which gives golden lustreous needles with stoichiometry $Ni(bqd)_2 \cdot 0.5$ I upon oxidation with molecular iodine (54). The lattice contains two different types of channels. (figure 6).

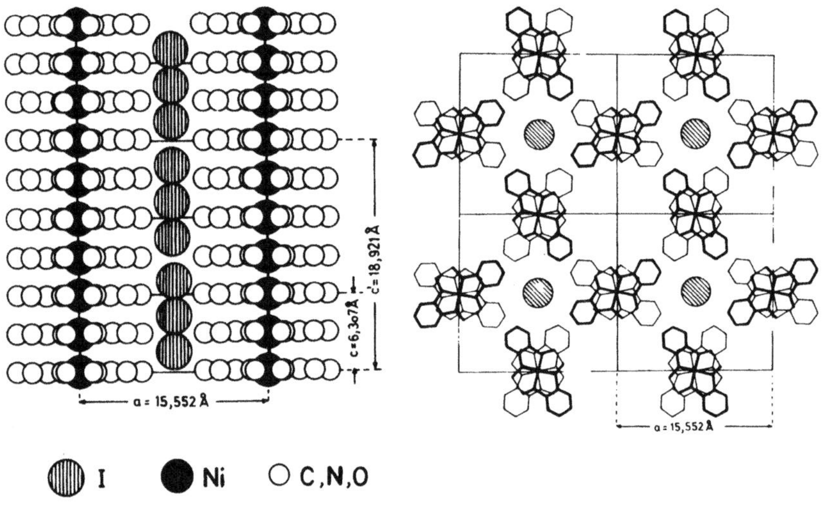

Figure 6: Structure of $Ni(bqd)_2I_{0.5}$, (a) projection onto the ac-plane, (b) projection onto the ab-plane (with permission from ref. (54b)).

The smaller channels are filled with linear triiodide ions while the larger ones were proposed to be empty. The occurence of triiodide ions was confirmed by detailed X-ray analysis of the diffuse layer lines (54) and by ^{129}I-Mößbauer and Resonance Raman data (55). The overall stoichiometry of $Ni(bqd)_2 \cdot 0.5$ I gives a formal oxidation number of 2.16 for the nickel ions. Additional solvent molecules were found in the iodine-free channels of the structure recently (56) and the solid obtained from toluene was reformulated as $Ni(bqd)_2 \cdot 0.52\ I \cdot 0.32\ C_7H_8$.

Another partially oxidized material (stoichiometry $Ni(bqd)_2 \cdot 0.018\ I$) was published in the same paper (56) and a formal oxidation number of 2.018 for the nickel was proposed. Whether this compound actually can be regarded as mixed valence material or could be understood as a pure and so far unknown $Ni(bqd)_2$ phase, which is made accessible by crystallization in the presence of iodine, is not fully understood at this time.

2.2.1.2 Phthalocyaninatonickel-iodide

Phthalocyaninatonickeliodide, (NiPcI), is a representative of a large class of "molecular metals" which can be obtained by oxidation of the MPc complexes (with the central metals M = Fe, Co, Ni, Cu, Zn, Pt and of the "free" ligand phthalocyanine) with molecular iodine (16,57). The structure of the nickel compound closely resembles those discussed under headings 2.2.1.1: Parallel to the stacks of the π-interacting complex molecules linear arrays of triiodide ions are found (RR and ^{129}I-Mößbauer data). The stoichiometry can be formulated as $NiPc(I_3)_{0.33}$, therefore.

While the question of conduction mechanism in $Ni(Bqd)_2\ 0.5\ I \cdot xS$ is still open (56), e.s.r. data indicate that the electron in the MPc materials is removed from a ligand π-orbital (58). This result suggests that the conduction is achieved through partially filled bands made up from ligand-ligand-π-interactions and <u>not</u> through the metal spine. The calculation of formal oxidation numbers (of 2.33) for the central metal ions is especially questionable in these materials, therefore.

Single crystals of Ni Pc I show <u>metallic</u> d.c.-conductivity (16,58) down to 90° K with a comparatively long mean free path, of 1.0 - 2.3 intermolecular spacings at room temperature. These increase to a value of 4 - 9 at the maximum conductivity around 90° K. A sharp metal-insulator transition takes place at 90° K but the degree of partial oxidation remains unaltered.

2.2.1.3 Porphyrinatonickel compounds

Several "mixed valence" porphyrinato complexes of nickel have been studied recently (57,59). Tetrabenzoporphyrinatonickeliodide (NiTBPI) which is indeed very similar to NiPcI in molecular and crystal structure again crystallizes as $NiTBP(I_3)_{0.33}$ and the crystal structure is almost indistinguishable from $NiPc(I_3)_{0.33}$ (metal complex stacks parallel to I_3^- chains) and has a similar conductivity ($\sigma_{\parallel}$ = 550 $\Omega^{-1}cm^{-1}$ for NiPcI and $\sigma_{\parallel}$ = 350 $\Omega^{-1}cm^{-1}$ for TBPNiI at room temperature). Alkylsubstitution at the ligand increases the Ni-Ni spacing considerably and as expected from simple steric considerations, the conductivity of the mixed valence systems $Ni(OMTBP)(I_3)_{0.33}$ ($OMTBPH_2$ = 1,4,5,8,9,12,13,16-octamethyltetrabenzporphyrin), ($\sigma_{\parallel}$ = 10 $\Omega^{-1}cm^{-1}$ at room temperature) is quite lower (59). The transition from a metallic conductor to a semiconductor is quite smooth in the latter compound. The differences in physical properties brought about by slight changes in the molecular design of the stacked species suggests that a wide range of desired physical properties can be synthesized at will (57).

2.2.2 Palladium complexes

2.2.2.1 Bis(α-ß-dionedioximato)palladium(II,IV)iodides

The polyiodides of the nickel species described in 2.2.1.1.1 and 2.2.1.1.2 can be obtained with the central metal palladium as well. Structures and physical properties of the partially oxidized compounds are very similar to the pertinent nickel derivatives.

2.2.2.1.1 Bis(diphenylglyoximato)palladium iodide

The compound was reported first by Foust and Soderberg (46) who determined the unit cell data from single crystal precession photographs and the chain structure was proposed. The material was investigated in detail later (47,50,51) and the close resemblace between the nickel and palladium analogue in structure and physical properties was recognized. The compound forms a semiconducting, quasi one-dimensional solid (51).

2.2.2.1.2 Bis(1.2-benzoquinonedioximato)palladium iodide

$Pd(bqd)_2 \cdot 0.5$ I was reported in 1974 and the close resemblance to the nickel derivative was mentioned several times after the first preparation of the material (54). The full structure of the compound was not determined until very recently (56). The solid contains additional solvent molecules in the larger channels and again triiodides in the smaller ones. Crystallization from 1.2-dichlorobenzene yields a compound with stoichiometry $Pd(bqd)_2 \cdot I_{0.5} \cdot 0.52\ C_6H_4Cl_2$ (56).

2.2.2.1.3 Bis(glyoximato)palladium-iodide

While the nickel derivative of the unsubstituted glyoximato ligand could not be stabilized in a mixed valence solid, $Pd(Gl)_2I$ was obtained in 1975 (60). The structure consists of cation stacks leaving only one type of channel (figure 7) filled with triiodide ions.

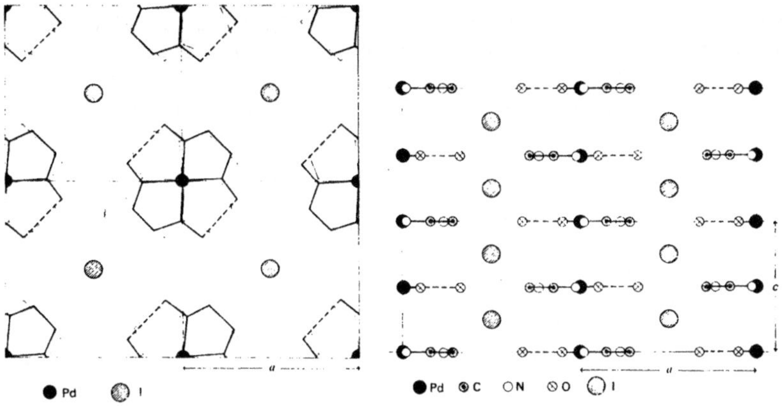

Figure 7: Structure of $Pd(Gl)_2I$, (a) projection onto the ab-plane, (b) projection onto the ac-plane (with permission from ref. (61)).

The latter information, though not confirmed by RR or ^{129}I-Mößbauer data, is contained in the intensities and distances of the diffuse layer lines in the X-ray photographs (61). The overall stoichiometry of $Pd(Gl)_2I$ again leads to a formal oxidation number of 2.33 for the palladium ions.

Comparing the oxidation numbers in all the known linear polyiodides of transition metal complexes, it appears that two factors control the amount of oxidation:

(i) Steric considerations (type of channels, ligands etc.).

(ii) Redox potential of the complex molecule.

It, therefore, can be expected, that a wide variety of new m.v. nickel and palladium derivatives can be obtained by following the above mentioned routes, mainly by using macrocyclic ligands with four nitrogen donors.

2.2.3 Platinum compounds

2.2.3.1 Tetracyanoplatinates(II,IV)

The evaluation of the structures of different, directly interacting, mixed valence platinum complexes was the starting point for the physicist's interest in one-dimensional metals. After the initial interest generated by Krogmann (4), hundreds of papers and many reviews have appeared (5-12) dealing especially with "KCP": Dipotassium-tetracyanoplatinate(II,IV)bromid-trihydrate. Many other mixed valence tetracyanoplatinates(II,IV) have been found and characterized during the last few years. A survey of the more recently synthesized materials and their structures is found in (62) and (63).

The preparative chemistry of these compounds is quite old, the first period ranging between 1846 - 1912 (64-67). During this time the compound "KCP" was mentioned as a double salt with composition 5 $K_2Pt(CN)_4 \cdot K_2Pt(CN)_4Br_2$. A. Werner (68) pointed out, that platinum in the two different oxidation states +2 and +4 is responsible for the unusual properties of these materials. The field was revived end of the sixties with detailed single crystal X-ray investigations (69).

Most of these complexes contain varying amounts of water which is very often statistically spread out over several lattice positions. Some of these sites are additionally occupied by counterions in a statistical fashion. This makes the X-ray analysis very difficult and gives additional problems in the interpretation of the physical data. A typical example is again KCP, which was firstly claimed to crystallize in the space group P4/mmm (69) later shown to be correctly assigned to space group P4 mm (70). These problems make the anhydrous compounds very preferrable because especially simple structures occur in these cases (63) (figure 8).

Without any doubt the tetracyanoplatinates(II,IV) are the best investigated and most important class of m.v. or partially oxidized (NIOS) linear chain solids. They were the first real one-dimensional metals available and many of the relevant physical theories (Peierls, Kohn, Fröhlich e.g.) have been tested in experiments on these compounds.

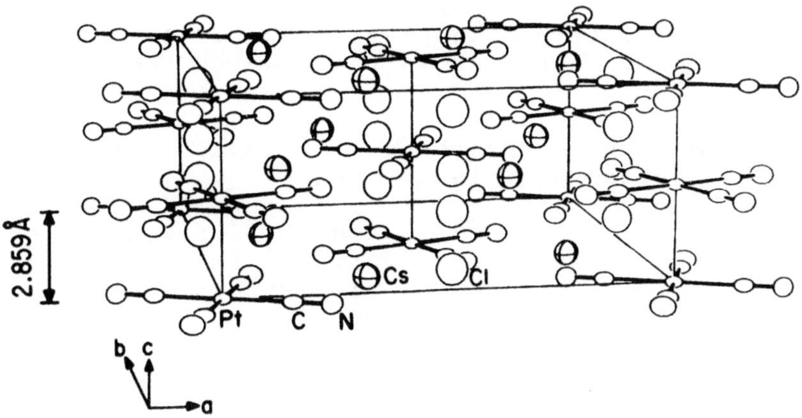

Figure 8: A perspective view of the unit cell of anhydrous $Cs_2(Pt(CN)_4)Cl_{0.3}$ (with permission of ref. (72)).

The physical properties (5-12,71) as well as their chemical preparation and structural peculiarities (4, 62,63,72) have been discussed in detail during the past few years. Even though these are the physically most interesting compounds, the discussion in this paper can be kept short in view of the extensive recent literature.

2.2.3.2 Bis(oxalato)platinates(II,IV)

Compared to the above (2.2.3.1) mentioned compounds the bis(oxalato)platinates(II,IV) are much less crystallizable and much less investigated, therefore. There are only a few detailed results for these materials. Again, after initial interest at the end of the last century (73,74), Werner recognized the mixed valence character of the solids (68), and again the starting point for all the physical investigations were papers written by Krogmann (75,76). Up to know the full X-ray structures of three different bis(oxalato)platinates(II,IV) have been published: $Mg_{0.82}Pt(C_2O_4)_2\cdot 5.3H_2O$ (76), $K_{1.6}Pt(C_2O_4)_2\cdot 1.2H_2O$ (77, 78) and $Co_{0.83}Pt(C_2O_4)_2\cdot 6H_2O$ (79). There are common features in this class of compounds:

(i) Though the interchain Pt-Pt distances are quite short (in the range of 2.85 Å), the conductivity of the compounds is low compared with KCP or other tetracyanoplatinates (II,IV). They do not behave metallic in the studied temperatur range (79b).

(ii) All the compounds so far investigated are of the "cation deficient" type.

(iii) The anions are stacked in a staggered fashion and the torsion angle varies between 45° and 60°, which seems to maximize the intermolecular ligand HOMO-LUMO overlap (80,81).

The magnesium and the cobalt analogue are isomorphous compounds (figure 9) and one of them contains a paramagnetic cation (79). A systematic comparison of physical and chemical properties of these very closely related materials seems to be possible now and it will help in the thorough understanding of the physical behaviour of this special compounds.

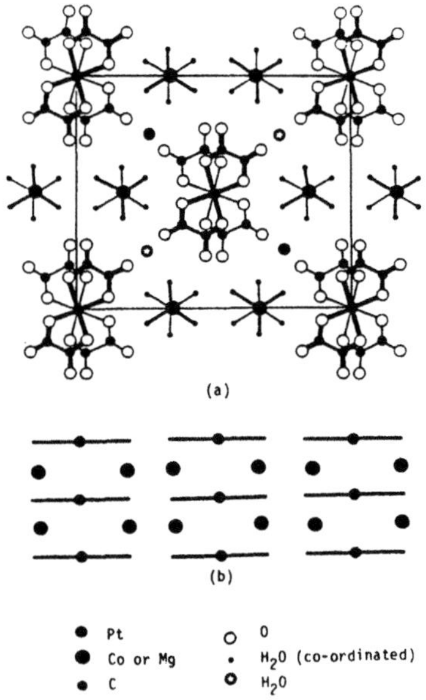

Figure 9: Crystal structure of $M_{0.8}(Pt(C_2O_4)_2)\cdot 6H_2O$ (M = Co, Mg),
(a) projection parallel to, and
(b) projection perpendicular to, Pt chain (the oxalate groups are represented by thick lines (b); with permission from ref.(79b)).

2.2.3.3 Partially oxidized Magnus Green Salt

One other ill-defined mixed valence linear chain platinum compound with possibly direct metal-metal interactions should be mentioned because it shows the way to new materials: the partially oxidized Magnus Green Salt (MGS) (82). MGS is a mixed stack platinum(II) solid, which contains tetraammineplatinum(II) cations and tetrachloroplatinate(II) anions in a arrangement as shown in figure 10 (83). Stacked planar platinum(II)

Figure 10: Schematic drawing of the MGS structure (Ⓝ represents ammonia).

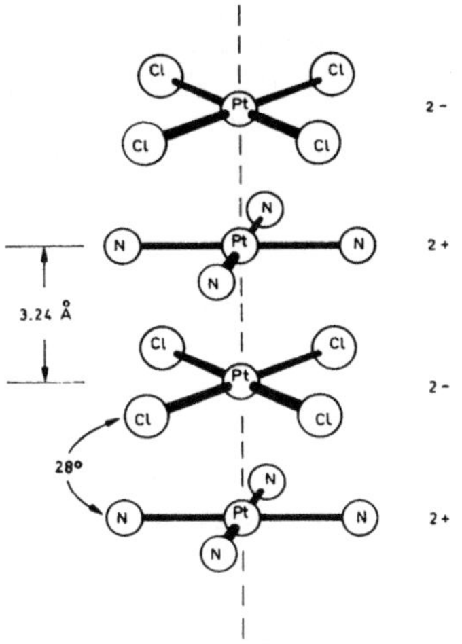

complexes have in some cases been converted to mixed valence solids by oxidation (see e.g. 2.2.3.1 and 2.2.3.2). It is obvious, therefore, that this procedure can be used for synthesizing other m.v. solids. In our opinion this is mainly a problem of solvent and there should be a wide variety of new compounds accessible by working in appropriate solvents like SO_2 (84).

MGS can be oxidized in sulfuric acid to a dark purple compound with a remarkable metallic lustre (82). It was suggested that the compound is a linear chain mixed valence solid of stoichiometry $Pt_6(NH_3)_{10}Cl_{10}(SO_4)_4$. Some years later it was found that the polycrystalline sample has more or less the same structure as MGS and behaves metallic (85). This would be the first example of a mixed stack one-dimensional metal.

As pointed out earlier, a mixed stack material with metal ions in different crystal fields could become metallic if the bandwidth of the partially filled conduction band would be larger than the differences in site energies along the stacks (86). In the organic molecular metals, the bandwidth is only of the order of 0.1 eV and slight distortions in the regular arrangement of the molecules along the stack lead to a breakdown of the conduction mechanism. In the m.v. inorganics, (especially KCP) the bandwidth was estimated to be of the order of 2.0 eV (71). This means that even crystallographically remarkable differences in site energy could be overcome as long as the overlap along the chains is strong enough. One could interpret the reported metallic conductivity of the partially oxidized MGS as the first example in which the bandwidth overcomes the site energy differences.

But the mentioned structural assignment (85) is at variance with recent Resonance Raman data (87). These hint to halide bridged linear chains in partially oxidized MGS. A M-X-M structure on the other hand would clearly rule out the observed high conductivity of the material. More investigations seem to be necessary to solve these problems since the so far reported physical results are in clear contradiction.

The search for new one-dimensional inorganic materials should nevertheless include partially oxidized mixed stack materials because strong metal-metal interactions could overcome even remarkable differences in crystal field energies.

2.2.3.4 Bis(diphenylglyoximato)platinum(III)-perchlorate

This compound stands as a typical example for the problems encountered in defining "mixed valence" compounds in directly interacting IOS systems. It can be obtained by oxidation of bis(diphenylglyoximato)platinum(II) with iodine and after exchange of the halide by perchlorate ions (88). A projection of the structure down the stacking axis is shown in figure 11. An exactly equidistant arrangement of the platinum ions along the chains gives a formal platinum(III) state, but slight distortions of the platinum chain would lead to a mixed valence platinum(II,IV) description of the solid. The missing EPR-signal is consistant with the second model. In any case it is an IOS material.

$Pt(dpg)_2ClO_4$ additionally is an example for the wide range of formal oxidation numbers which can be obtained by selecting appropriate metal ions, ligands, counter ions and preparative conditions.

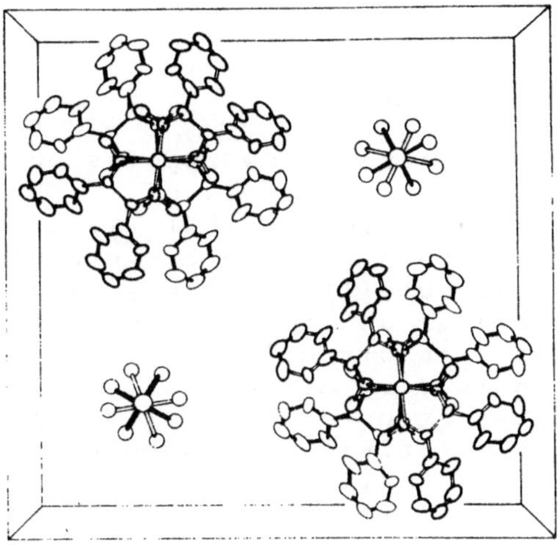

Figure 11: Projection of the structure of Pt(dpg)ClO_4 parallel to the stacking axis (c-axis), (with permission from ref. (88)).

2.3 The Copper group

Copper, silver and gold occur in linear two-coordinate environment in their monovalent state. Copper(II) and gold(III) compounds are often found four-coordinated and planar. Both geometries should allow stacking with direct metal-metal interactions principally, but compared with the d^8 metal ions (nickel(II), palladium(II), platinum(II), rhodium(I) and iridium(I)) only a very few directly interacting linear chain structures are known.

2.3.1 Copper complexes

A few planar copper(II) compounds form linear arrays with direct metal-metal bonds (89) but none of them could be classified as a m.v. material and all of them have quite weak intermolecular interactions. Mixing solutions of positively charged planar copper(II) complexes with negatively charged linear $(CuX_2)^-$ species might lead to the desired compounds.

2.3.2 Silver compounds

The only silver compound which could be mentioned in this connection is the bronze-colored Ag_2F. It has a high conductivity and strong <u>two</u>-dimensional metal-

metal interactions. Double-layers of silver atoms alternate with layers of the fluoride counter ions (90).

2.3.3 Gold complexes

There are quite a few reports on chain compounds of gold with different types of direct interactions along the metal chains (91,92). In order to synthesize m.v. gold materials in a more systematic way one has to remind that gold(I) is mainly linear, gold(III) in most cases planar and four coordinated. All mixed valence solids with directly interacting ions have to contain gold in very different crystal fields, therefore. One well-known structurally characterized compound of this type (93) - bis(2.3-butanedionedioximato)gold(III)-dichloroaurate(I) (figure 12) - shows only minor intermolecular interactions (yellow crystals).

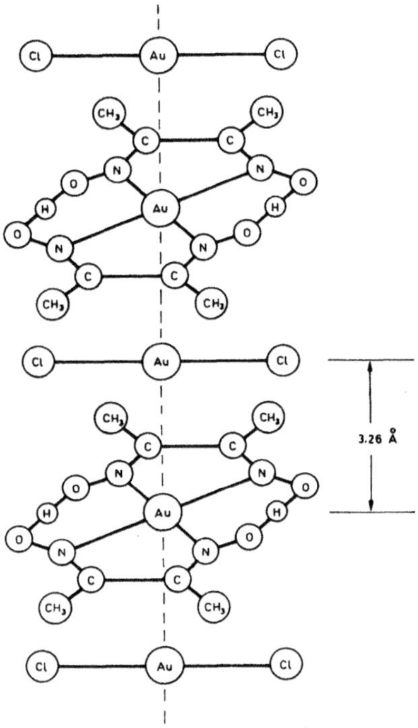

Figure 12:

Schematic drawing of the structure of bis(2.3-butanedionedioximato)gold(III)-dichloroaurate(I).

Stronger interactions could possibly achieved, if the differences in site energies between Au(I) and Au(III) could be reduced or if the interionic interactions could be strengthened (by applying high pressure e.g.). The preparative recipe again is to mix linear $(\overset{I}{Au}X_2)^-$ species with cationic and planar $(\overset{III}{Au}L_4)^+$ complexes.

3. CONCLUSION

A new class of materials with strong direct metal-metal interactions along linear chains has emerged during the last few years. Most of the highly conducting ones are partially oxidized NIOS (Underhill) or Class III b (Day-Robin) but some of the interesting ones are clearly IOS solids. Most of the directly interacting compounds with metal-metal distances below 2.9 Å and with small ligands behave like one-dimensional metals. If larger ligands are used the conductivity drops (decreasing intermolecular M-M overlap). This drop can also be interpreted as a "dilution" effect (94) if M-M distances remain short (e.g. bis(oxalato)platinates (II,IV)). Ligands with extended π-systems give larger ($>$ 3.2 Å) M-M distances but high conductivities if intermolecular interactions through the partially oxidized ligand π-system are possible. (Macrocyclic compounds).

4. LITERATURE

(1) Robin, M.B., and Day, P.: 1967, Adv. Inorg. Chem. Radiochem. 10, pp. 247-463.

(2) Allen, G.C., and Hush, N.S.: 1967, Progr.Inorg. Chem. 8, pp. 357-389.

(3) Hush, N.S.: 1967, Progr.Inorg.Chem. 8, pp. 391-444

(4) Krogmann, K.: 1969, Angew.Chem. 81, pp. 10 ff., 1969, Int.Ed.Engl. 8, pp.35 ff.

(5) "Extended Interactions between Metal Ions in Transition Metal Complexes", ACS Symposium Series, Vol. 5, 1974, ed. L.V.Interrante, Washington, D.C.

(6) "Low-dimensional Cooperative Phenomena", NATO-ASI Series B, Vol. 7, 1975, ed. H.J.Keller, Plenum Press, N.Y.

(7) "One-dimensional Conductors", 1975, Lecture Notes in Physics, Vol. 34, Springer Verlag, ed. H.G. Schuster.

(8) "Chemistry and Physics of One-dimensional Metals", 1977, NATO-ASI Series B, Vol. 25, Plenum Press, N.Y., ed. H.J.Keller

(9) Miller, J.S., and Epstein, A.J.: 1976, Prog.Inorg. Chem. 20, pp. 1 ff.

(10) "Inorganic Compounds with Unusual Properties", 1976, Adv. in Chem. Series 150, ed. R.B.King

(11) "Synthesis and Properties of Low-dimensional Materials", 1978, Ann.N.Y.Acad.Sci. 313, ed.J.S.Miller and A.J.Epstein.

(12) "Molecular Metals", NATO-Conference Series VI, Vol. 1, Plenum Press, N.Y., 1979, ed.W.A.Hatfield
(13) Thomas, T.W., and Underhill, A.E.: 1972, Chem. Soc.Rev. 1, pp. 99 ff.
(14) Underhill, A.E. in ref. (6), pp. 287-297
(15) See for example: Campbell, J.A.: 1970, "Chemical Systems, Energetics, Dynamics, Structure",
(16) a) Petersen, J.L., Schramm,C.S.,Stojakovic,D.R., Hoffman,B.M., and Marks,T.J.: 1977, J.Amer. Chem.Soc., 99, pp. 286
b) Marks,T.J., in ref. (11), pp. 609-614.
(17) Bekaroglu, Ö, Sarisaban, S. Koray, A.R., and Ziegler, M.L.: 1977, Z.Naturforsch. 32b, pp. 387-392
(18) Endres, H., Keller, H.J., Lehmann, R., Poveda, A., Rupp, H.H., van de Sand, H.: 1977, Z.Naturforsch. 32b, pp. 516 ff.
(19) Dart, J.W., Lloyd, M.K., Mason, R., and McCleverty, J.A.: 1973, J.Chem.Soc.Dalton, pp. 2046 ff.
(20) Gordon II, J.G., Williams, R., Hsu, C.H., Cuellar, E., and Somoano, R.: 1978, ref. (11), pp. 580
(21) Miskowski, V.M., Nobinger, G.L., Kliger, D.S., Hammond, G.S., Lewis, N.S., Mann, K.R., and Gray, H.B.: 1978, J.Amer.Chem.Soc. 100, pp. 485 ff.
(22) a) Balch, A.L., and Olmstead, M.M.: 1976, J.Amer. Chem.Soc. 98, pp. 2354
b) Balch, A.L., and Olmstead, M.M.: 1979, J.Amer. Chem.Soc. 101, pp. 3128
(23) Endres, H., Gottstein, N., Keller, H.J., Martin, R., Rodemer, W., and Steiger, W.: 1979, Z.Naturforsch. 34b, pp. 827-833
(24) Bonati,F., and Ugo, R.: 1967, J.Organometal. Chem. 7, pp. 167-180
(25) Ugo, R., La Monica, G., Cenini, S., and Bonati, F.: 1968, J.Organometal.Chem. 11, pp. 159
(26) Cozens, R.J., Murray, K.S., and West, B.O.: 1971, J.Organometal.Chem. 27, pp. 399-407
(27) Bailey, N.A., Coates, E., Robertson, G.B., Bonati, I., and Ugo, R.: 1967, J.Chem.Soc.Chem.Commun. pp. 1041
b) Hug, F., and Skapski, A.C.: 1974, J.Cryst. Molec.Struct. 4, pp. 411
(28) De Haven, P.W., and Goedken, V.L.: 1979, Inorg. Chem. 18, pp.827
(29) Malatesta, L., and Angoletta, M.: 1958, J.Inorg. Nucl.Chem. 8, pp. 273
(30) Malatesta, L., and Canziani, F.,: 1961, J.Inorg. Nucl.Chem. 19, pp. 81-93

(31) a) Chernyaev, I.I., and Novozhenyuk, Z.M.: 1966, Russ.J.Inorg.Chem. 11, pp. 1398.
b) Forster, D.: 1969, Inorg.Nucl.Chem.Letters, 3, pp. 433-436.
(32) Cleare, M.J., and Griffith, W.B.: 1969, J.Chem. Soc. A1969, pp. 372-380 and 1970, J.Chem.Soc. A 1970, pp. 2788-2794.
(33) Ginsberg, A.P., Koepke, J.W., Cohen, R.L., and West, K.W.: 1976, Chem.Phys.Letters, 38, pp. 310
(34) Krogmann, K.: 1977, in ref. (8), pp. 229.
(35) Rosencwaig, A., Ginsberg, A.P., and Koepke, J.W.: 1976, Inorg.Chem. 15, pp. 2540 and references cited therein.
(36) Hieber, W., and Lagally, H.: 1940, Z.Anorg.Allg. Chem. 245, pp. 321.
(37) Hieber,, W., Lagally, H., and Mayr, H.: 1941, Z.Anorg.Allg.Chem. 246, pp. 138
(38) Fischer, E.O., and Brenner, K.S.: 1962, Z.Naturforsch. 17b, pp. 774.
(39) Krogmann, K., Binder, W., and Hausen, H.D.: 1968, Angew. Chem. 80, p. 844 and 1968, Angew.Chem. Int.Ed. 7, p. 812.
(40) Ginsberg, A.P., Cohen, R.L., DiSalvo, F.J., and West, K.W.: 1974, J.Chem.Phys. 60, pp. 2657.
(41) Gitzel, W., Keller, H.J., Lorentz, R., and Rupp, H.H.: 1973, Z.Naturforsch. 28b, pp. 161-163.
(42) a) Keller, H.J.: 1975, in ref. (6), pp. 315
b) Endres, H., Keller, H.J., Lehmann, R., van de Sand, H., Vu, D., and Poveda, A., in ref. (11), pp. 627-632
(43) a) Reis, A.H., jr., Hagley, V.S., Peterson, S.W.: 1977, J.Amer.Chem.Soc. 99, pp. 4184.
b) Reis, A.H., jr., and Peterson, S.W., in ref. (11), pp. 571-578.
(44) Edelman, L.E.: 1950, J.Amer.Chem.Soc. 72, pp. 5765.
(45) Simek, M.: 1962, Collection Czech.Chem.Commun. 27, pp. 337.
(46) Foust, A.S., and Soderberg, R.H.: 1967, J.Amer. Chem.Soc., 89, pp. 5507.
(47) Keller, H.J., and Seibold, K.: 1971, J.Amer.Chem. Soc., 93, pp. 1309.
(48) Teitelbaum, R.C., Ruby, S.L., and Marks, T.J.: 1978, J.Amer.Chem.Soc. 100, pp. 3215 and ref. cited therein.
(49) Gleizes, A., Marks, T.J., and Ibers, J.A.: 1975, J.Amer.Chem.Soc. 97, pp. 3545.
(50) Miller, J.S., and Griffiths, C.H.: 1977, J.Amer. Chem.Soc. 99, pp. 749.

(51) Cowie, M., Gleizes, A., Grynkewich, G.W., Webster Kalina, D., Mc.Clure, M.S., Scaringe, R.P., Teitelbaum, R.C., Ruby, S.L., Ibers, J.A., Kannewurf, C.R., and Marks, T.J.: 1979, J.Amer.Chem. Soc. 101, pp. 2921.

(52) Miller, J.S.: 1977, Inorg.Chem. 16, pp. 957.

(53) Keller, H.J., and Lehmann, R.: 1979, Z.Naturforsch. 34b, pp. 245 and references cited therein.

(54) Moroni, W., and Nöthe, D.: 1974, Inorg.Nucl.Chem. Letters, 10, pp. 467.

(54b) Endres, H., Keller, H.J., Moroni, W., and Weiß, J.: 1975, Acta Cryst., B31, pp. 2357.

(54c) Endres, H., Keller, H.J., Mégnamisi-Bélombé, M., Moroni, W., Pritzkow, H., Weiß, J., and Comès, R.: 1976, Acta Cryst., A 32, pp. 954.

(55) Marks, T.J., Webster, D.F., Ruby, S.L., and Schultz, S.: 1976, J.Chem.Soc.Chem.Commun., pp.444.

(56) Brown, L.D., Webster Kalina, D., McClure, M.S., Schultz, S., Ruby, S.L., Ibers, J.A., Kannewurf, C.R., and Marks, T.J.: 1979, J.Amer.Chem.Soc., 101, pp. 2937.

(57) Hoffman, B.M., Phillips, T.E., Schramm, C.J., and Wright, S.K., in ref. (12), pp. 393 and references cited therein.

(58) Schramm, C.J., Stojakovic, D.R., Hoffman, B.M., and Marks, T.J.: 1978, Science, 200, pp. 47.

(59) Phillips, T.E., and Hoffman, B.M.: 1977, J.Amer. Chem.Soc., 99, pp. 7734.

(60) Endres, H., Keller, H.J., and Lehmann, R.: 1975, Inorg.Nucl.Chem.Letters, 11, pp. 769.

(61) Endres, H., Keller, H.J., Lehmann, R., and Weiß, J.: 1976, Acta Cryst., B32, pp. 627.

(62) Stucky, G.D., Schultz, A.J., and Williams, J.M.: 1977, Ann.Rev.Mater.Sci., 7, pp. 301.

(63) Williams, J.M., and Schultz, A.J., in ref. (12), pp. 337-368.

(64) Knop, W., and Schnedermann, G.: 1846, J.prakt. Chem., 37, pp. 462.

(65) Hadow, E.A.: 1861, Quart.J.Chem.Soc., 14, pp.104.

(66) Wilm, Th.: 1888, Ber.dtsch.chem.Ges., 21, pp.1436.

(67) Levy, L.A.: 1912, J.Chem.Soc.(London), pp. 1081.

(68) Werner, A: 1896, Z. Anorg. Chem., 12, pp. 46

(69) Krogmann, K.: 1968, Z.Anorg.Allg.Chem., 358, pp. 67.

(70) Williams, J.M., Petersen, J.L., Gerdes, H.M., and Peterson, S.W.: 1974, Phys.Rev.Letters, 33, pp. 1079.

(71) Zeller, H.R.: 1973, Adv.Solid State Phys., 13, pp. 31.

(72) Brown, R.K., and Williams, J.M.: 1979, Inorg. Chem., 18, pp. 1922 and references cited therein.
(73) Söderbaum, H.G.: 1886, Bull.Soc.Chim.France, 45, pp. 188.
(74) Werner, A.: 1899, Z.Anorg.Allg.Chem., 21, pp.377.
(75) Krogmann, K., and Dodel, P.: 1966, Chem. Ber., 99, pp. 3402, pp. 3408.
(76) Krogmann, K.: 1968, Z.Anorg.Allg.Chem., 358, pp. 97.
(77) Reis, A.H., jr., Peterson, S.W., and Lin, S.C.: 1976, J.Amer.Chem.Soc., 98, pp. 7839 and in ref. (11), pp. 566.
(78) Kobayashi, H., Shirotani, I., Kobayashi, A., and Sasaki, Y.: 1977, Solid State Commun., 23, pp.409.
(79) a) Schultz, A.J., Underhill, A.E., and Williams, J.M.: 1978, Inorg.Chem., 17, pp. 1313.
b) Underhill, A.E., and Wood, D.J., in ref. (12), pp. 377.
(80) Miller, J.S.: 1976, Inorg.Chem., 15, pp. 2357.
(81) a) Reis, A.H., jr.: 1977, Inorg.Nucl.Chem.Letters, 13, pp. 231.
b) Reis, A.H., and Peterson, S.W., in ref. (11), pp. 560.
(82) Gitzel, W., Keller, H.J., Rupp, H.H., and Seibold, K.: 1972, Z.Naturforsch., 27b, pp. 365.
(83) Atoji, M., Richardson, J.W., and Rundle, R.E.: 1957, J.Amer.Chem.Soc., 79, pp. 3017.
(84) Buse, K.D., Keller, H.J., and Nöthe, D.: 1976, Z.Naturforsch., 31b, pp. 194.
(85) Tsujikawa, I., Kubota, R., Enoki, T., Miyajima, S., and Kobayashi, H.: 1977, J.Phys.Soc.Japan, 43, pp. 1459.
(86) Soos, Z.G., and Keller, H.J., in ref. (8), pp. 391-412.
(87) Papavassiliou, G.C., and Theophanides, T.: 1979, Z.Naturforsch., 34b, pp. 986.
(88) Endres, H., Keller, H.J., van de Sand, H., and Vu Dong: 1978, Z.Naturforsch., 33b, pp. 843.
(89) Bartkowski, R.R., and Morosin, B.: 1972, Phys.Rev., B 6, pp. 4209.
(90) Terauchi, H., Sakamoto, N., and Shirotani, I.: 1975, J.Phys.Soc.Japan, 38, pp. 595.
(91) Lawton, S.L., Rohrbaugh, W.J., and Kokotailo, G.T.: 1972, Inorg.Chem., 11, pp. 2227.
(92) Hesse, R., and Jennische, P.: 1972, Acta Chem. Scand., 26, pp. 3855.
(93) Rundle, R.E.: 1954, J.Amer.Chem.Soc., 76, pp.3101.
(94) Miller, J.S.: 1974, J.Amer.Chem.Soc., 96, pp.7131.

THE VERWEY TRANSITION REVISITED

John B. Goodenough

Inorganic Chemistry Laboratory, South Parks Road, Oxford OX1 3QR

Abstract: Magnetite, the ferrospinel $Fe^{3+}[Fe^{2+}Fe^{3+}]O_4$, is ferrimagnetic with atomic moments characteristic of localized, high-spin configurations with orbital-momentum quenched. An unactivated electrical conductivity of low electron mobility u leaves open the question whether the mobile d electrons on the octahedral sites are itinerant ($u = e\tau/m^*$) or polarons ($u = eD_o/kT$) having no activation energy for a jump between molecular orbitals of Fe-Fe pairs. Below a sharp Verwey transition at about 120K, the mobile charges are ordered; but the structural evidence is still inadequate to clarify whether this is a complex ordering of small polarons or the stabilization of a charge-density wave.

INTRODUCTION

The word "magnet" is derived from the ancient lodestone of Magnesia; particles of lodestone were known to attract or repel each other according to their relative orientations. As early as 121 B.C. the Chinese had discovered that lodestone could attract iron, and the first compasses used powered lodestone on the needle. Lodestone is now known as magnetite, Fe_3O_4. Magnetite has also found a place in the biological world. Certain bacteria manufacture small particles - just large enough to sustain a spontaneous magnetization - that are oriented perpendicular to their body axis to allow orientation with respect to the Earth's magnetic field [1].

Magnetite has not only a sizeable spontaneous magnetization M_s at room temperature, but also a high electrical conductivity σ. In 1931, Okamura [2] observed a reversible (no thermal hysteresis)

D.B. Brown (ed.), Mixed-Valence Compounds, 413-425.

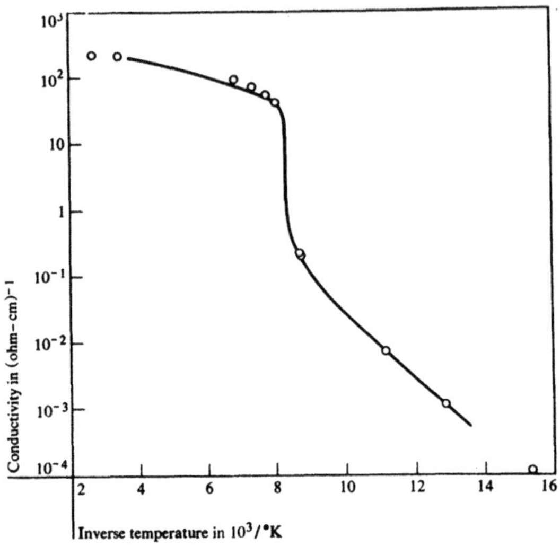

Figure 1. Conductivity vs temperature for a single crystal of Fe_3O_4

sharp discontinuity in the conductivity at about 120K, see Fig.1. The specific-heat anomaly associated with the transition has the Λ shape typical of an order-disorder transition [3], and Verwey [4] proposed that the transiton reflected an ordering of Fe^{2+} and Fe^{3+} ions over sites that are crystallographically equivalent at higher temperatures, inequivalent at lower temperatures. Magnetic data, see below, seemed to confirm this prediction, and the transition has become known as the Verwey transition occurring at the Verwey temperature $T_v \approx 120K$. In fact, T_v varies somewhat from sample to sample: $119 \lesssim T_v \lesssim 123K$.

Magnetite belongs to the $M^{2+}Fe_2{}^{3+}O_4$ family of oxides that crystallize with the cubic structure of spinel, $MgA\ell_2O_4$: space group $O_h{}^7$-Fd3m [5]. The cubic unit cell, illustrated in Fig.2(a), contains eight formula units: 8 $A[B_2]O_4$, where the A sites are tetrahedral and the B sites are octahedral. The oxygen atoms form a close-packed cubic array, the A sites consist of 1/8 of the tetrahedra ordered to form a diamond-lattice subarray, and the B sites are ½ the octahedra ordered so as to give intersecting strings of edge-shared octahedra along the <110> directions. The A-O bond lengths are adjustable through an oxygen u-parameter; $u = 3/8$ for perfect close packing of the oxygens, but generally a $u>3/8$ is found. Variation of u permits modulation of the trigonal component of the crystal field found at the B sites. The primitive unit cell, Fig.2(b), contains two molecular units, and each of the four B sites has its trigonal axis along a different <111> direction.

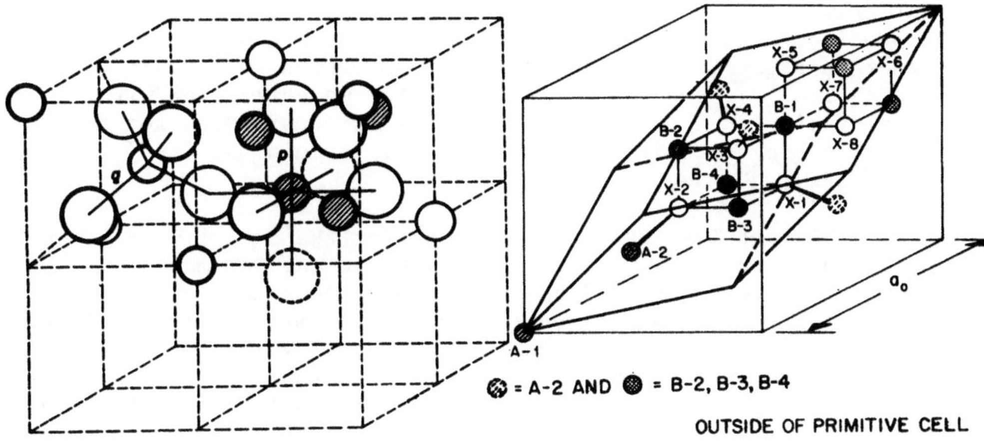

Figure 2. Unit cell of spinel: (a) cubic (b) primitive

A normal 2-3 spinel has the A sites occupied by the M^{2+} ions, as in $Zn[Fe_2]O_4$; an inverse 2-3 spinel has the A sites occupied by trivalent ions, as in $Fe[NiFe]O_4$; and a mixed spinel has both valence states on the A sites, as in $Mg_xFe_{1-x}[Mg_{1-x}Fe_{1-x}]O_4$. Since it is difficult to distinguish by x-rays ions of similar weight, determination of ion distributions is made indirectly by magnetization measurements - with the aid of the Néel hypothesis - and directly by neutron diffraction and by Mössbauer or NMR measurements.

MAGNETIZATION AND TRANSPORT STUDIES

The Verwey model of the transition in magnetite requires that at $T>T_V$ the cubic spinel is inverse: $Fe^{3+}[Fe^{2+}Fe^{3+}]O_4$. Indirect evidence for this, aside from theoretical prediction based on crystal-field stabilization energies, first came from magnetization measurements together with the 1948 Néel [6] hypothesis that in the inverse ferrospinels the A-site atomic moments are aligned antiparallel to the B-site atomic moments: $M_s = |M_B - M_A|$. Néel had observed that at high temperatures the ferrospinels and certain antiferromagnetic oxides exhibited a temperature-dependent magnetic susceptibility described by a Curie-Weiss law $\chi = C/(T-\theta)$, but with a Weiss constant $\theta<0$. The negative θ could be straightforwardly interpreted as signalling antiferromagnetic - as against ferromagnetic - coupling. Measurements of the Curie constant C were consistent with high-spin atomic moments, so the small M_s measured implied an antiferromagnetic coupling between two crystallographically distinguishable subarrays having unequal sublattice magnetizations M_A and M_B. In the case of magnetite, Néel ferrimagnetic ordering would yield a spin-only magnetization $M_s = 6\mu_B$ for the normal configuration $Fe^{2+}[Fe_2^{3+}]O_4$ and an $M_s = 4\mu_B$ for an inverse configuration $Fe^{3+}[Fe^{2+}Fe^{3+}]O_4$. The measured values are $4.0\mu_B < M_s < 4.1\mu_B$ [7], and the variation of the magnetization with composition in the systems $Zn_xFe_{3-x}O_4$ is consistent, for smaller x, with Néel ordering and a structural formula $Zn_x^{2+}Fe_{1-x}^{3+}[Fe_{1-x}^{2+}Fe_{1+x}^{3+}]O_4$ [8,9].

The fact that the atomic moment for the Fe^{2+} ion is the spin-only value, $\mu_{Fe^{2+}} \simeq 4.0\mu_B$, implies that the angular-momentum contribution has been quenched by the trigonal component of the crystalline field and/or by delocalization of the minority-spin electron outside the closed half-shell. (The majority-spin d electrons will be referred to as α-spin electrons, the minority-spin electron the β-spin electron.) The trigonal field would change the ground state from $^5T_2$ to $^5A_1$. Compounds containing localized, high-spin $3d^6$ configurations at octahedral-site Fe^{2+} ions exhibit a substantial orbital contribution to the atomic moment if only cubic crystalline fields are present above the magnetic-ordering temperature; below the magnetic-ordering temperature a Jahn-Teller distortion that preserves the orbital angular momentum introduces a giant magnetostriction and magnetic anisotropy [10]. These latter features are noticeably absent from magnetite.

Direct confirmation of Néel ferrimagnetic order in magnetite came with the advent of neutron diffraction [11], and a careful measurement of the magnetization change on passing through T_V gave a $\Delta M_s/M_s \approx 0.1$ % [12], showing that the atomic moments and the magnetic order are not significantly changed by the electron ordering below T_V.

In 1928, Heisenberg [13] pointed out the quantum-mechanical origin of magnetic spin-spin interactions and formulated the direct-exchange Hamiltonian between spins $\mathbf{S}_i$ and $\mathbf{S}_j$ located at positions $\mathbf{R}_i$ and $\mathbf{R}_j$ as

$$\mathcal{H}_{ij} = -J_{ij}(\mathbf{S}_i \cdot \mathbf{S}_j) \quad (1)$$

$$J_{ij} = J_{ij}^D = (J_e - \alpha^2 J_c)/(1-\alpha^4) \quad (2)$$

where J_e and J_c are the well known exchange and coulomb integrals and α is the overlap integral for interacting electrons in orbitals ψ_i and ψ_j. For orthogonal orbitals, α = 0 and $J_{ij}^D > 0$ makes the spin-spin coupling ferromagnetic.

In 1934, Kramers [14] pointed out that electron-transfer excited states may couple nonorthogonal orbitals. For localized-electron configurations, this interaction, known as superexchange, gives an expression similar to Equation (1), but with

$$J_{ij} = J^{S}_{ij} = -(2b^2_{ij}/4S^2U)\underset{\sim}{S}_{ij}\cdot\underset{\sim}{S}_j \tag{3}$$

for the interaction between half-filled orbitals and

$$J_{ij} = J^{S}_{ij} = +(2b^2_{ij}J^{intra}/4S^2U^2)\underset{\sim}{S}_i\cdot\underset{\sim}{S}_j \tag{4}$$

for the interaction between a half-filled and an empty orbital, or a filled and a half-filled orbital. These expressions follow directly from the conservation of spin during charge transfer, the definition of the electrostatic correlation splitting U shown pictorially in Fig.3(a), the definition of the spin-independent electron-transfer matrix element (or resonance integral)

$$b_{ij} = (\psi_i, \mathcal{H}'\psi_i) \approx \varepsilon_{ij}(\psi_i, \psi_j) \tag{5}$$

where $\mathcal{H}'$ is the perturbation of the electron potential energy at $\underset{\sim}{R}_j$ as a result of the presence of the atom at $\underset{\sim}{R}_i$, and the character of the maxtrix describing rotation of an electron spin through an angle θ

$$\alpha = \alpha \cos(\theta/2) + \beta \sin(\theta/2) \tag{6}$$

$$\beta = -\alpha \sin(\theta/2) + \beta \cos(\theta/2)$$

From second-order perturbation theory,

$$\mathcal{H}_{ij} = -t_{ij}{}^2/U = -b_{ij}{}^2\sin^2(\theta/2)/U \tag{7}$$

if the interacting orbitals are half-filled; the Pauli exclusion

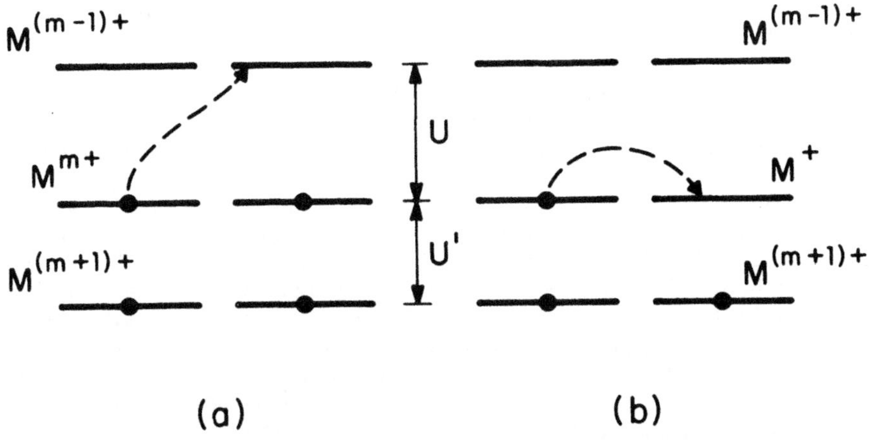

Figure 3. Electron-transfer excited states for (a) super-exchange and (b) double exchange

principle only permits creation of spin-paired polar states, so the interaction is antiferromagnetic and $t_{ij} = b_{ij}\sin(\theta/2)$. Equation (4) follows from third-order perturbation theory; in this case electrons of either spin may be transferred, but the transfer that maximizes the atomic spin on the acceptor cation is stabilized relative to its alternative by the factor J^{intra}/U, where J^{intra} is the intraatomic-exchange (direct-exchange) stabilization on the acceptor ion. Therefore this interaction is ferromagnetic.

In 1951, Zener [15] pointed out that mixed valency on equivalent lattice sites can give rise to a ferromagnetic coupling known as double exchange. As indicated in Fig.3(b), electron transfer takes place by direct tunneling, so first-order perturbation theory gives

$$\mathcal{H}^{D}_{ij} = -Nc(1-c)zt_{ij} = -Nc(1-c)zb_{ij}\exp(-\varepsilon_p/2kT)\cos(\theta_{ij}/2) \quad (8)$$

where Nc is the density of mobile electrons (low-valence cations) and (1-c)z is the number of like nearest neighbors to which it can jump (high-valence cations). If $b_{ij}<h\omega_R$, where ω_R is the Restrahl frequency, the lattice has time to relax about a mobile electron between jumps, thereby trapping it in a local deformation. An electron "dressed" in its local deformation is called a small polaron with a trapping energy ε_p. A thermal energy is required to overcome this polaron energy before tunneling to the neighboring site can occur, which introduces the exponential term in Eq.(8). If $\varepsilon_p\neq 0$, the time between electron jumps may be longer than the relaxation time for a spin; in this case, the mobile electron (polaron) "forgets" the spin orientation of the neighbor from which it last jumped, and the double-exchange vanishes. This is why the mixed-valance fluorides K_xFeF_3, for example, exhibit no double-exchange contribution [16].

In magnetite, three interatomic-exchange interactions contribute: a 135° A-O-B superexchange interaction dominated by half-filled σ-bonding orbitals, a ca. 90° B-O-B interaction dominated by a direct-exchange interaction between nearly orthogonal, half-filled σ-bonding orbitals (overlap over the O^{2-} ions is associated with covalently mixed anion-p_σ orbitals), and a double-exchange B-B interaction between B-site Fe^{2+} and Fe^{3+} ions. The A-O-B superexchange interaction is antiferromagnetic whereas the B-O-B and B-B interactions are ferromagnetic. Therefore all three interactions cooperate to produce the Néel ferrimagnetic ordering: ferromagnetic A and B subarrays coupled antiferromagnetically. Measurement of the spin-wave dispersion [17] has confirmed ferromagnetic coupling among the B-site iron atoms of Fe_3O_4.

The Curie temperature T_c is proportional to the toal exchange-energy parameter J, which is dominated by the A-O-B superexchange contribution J^S. The transfer energy b_{ij} increases with the overlap integral $\alpha\equiv(\psi_i,\psi_j)$ whereas the electrostatic energy U decreases; therefore J^S increases unambiguously with increasing overlap so long as the perturbation expansion leading to Eqs. (7) and (3) is convergent. In this localized-electron limit, we must anticipate a $dT_c/dP>0$, where P represents hydrostatic pressure. Where the spontaneous magnetization is due to itinerent-electron magnetism, we obtain a $dT_c/dP<0$ [18]. Samara and Giardini [19] have reported a $dT_c/dP = 2.05 \pm 0.1$ K/kbar for magnetite, which confirms the localized character of the σ-bonding electrons, which belong to the majority-spin (α-spin) half-shell. However, it does not follow that the β-spin electrons are also localized. At a high-spin d^6 cation, the maximum distinction exists between the atomic potentials experienced by α-spin and β-spin electrons; the former are stabilized by strong intraatomic exchange whereas the latter is not [20]. Therefore, the magnetization data are not able to distinguish whether the β-spin-electron mobilities should be described by a diffusion model or by an itinerant-electron model.

In the diffusion model, the mobility $u = eD/kT$ is proportional to the diffusion coefficient D. Above room temperature, the conductivity is essentially temperature-independent, indicating that any $D = D_o$ cannot have an activation enthalphy for an electron jump. In the itinerant-electron model, the mobility $u = e\tau/m^*$ is proportional to the ratio e/m^* of electron charge to mass as well as the mean free time τ between scattering collisions. In the former model, distinguishable electrons tunnel from site to site (or between Fe-Fe bands). In the latter the electrons are indistinguishable; they have a momentum $p = \hbar \tilde{k}$, and Fermi-Dirac statistics defines a Fermi surface.

Ambiguity about the character of the β-spin electrons does not prevent construction of the schematic energy diagram of Fig.4. From ESR measurements of transition-metal impurities dilutely doped in TiO_2, it is known that a $U\approx3$eV separates octahedral $Fe^{2+}:d^6$ and $Fe^{3+}:d^5$ energies [21]. Moreover, the inverse ion distribution of Fe_3O_4 requires that the A-site $Fe^{2+}:d^6$ level lie above the B-site $Fe^{2+}:3d^6$ level; this requirement is consistent with the shorter A-O bond lengths, greater covalent mixing destabilizing more the antibonding d orbitals [22]. At temperature $T>T_v$, the B-site $Fe^{2+}:d^6$ band is half-filled; this band must be split in two by a crystallographic distortion at the semiconducting temperatures $T<T_v$.

Mössbauer measurements at $T>T_v$ could readily distinguish the A-site Fe^{3+} signal from the B-site signal by a 1:2 difference in intensity, but they were unable to distinguish different B-site iron atoms [23, 24]. An internal field $H_{int}\approx500$kOe and an isomer shift relative to α-Fe I.S.$\approx$0.5mm/s is characteristic of a tetrahedral-site Fe^{3+} ion; the remaining $H_{int}\approx470$kOe and

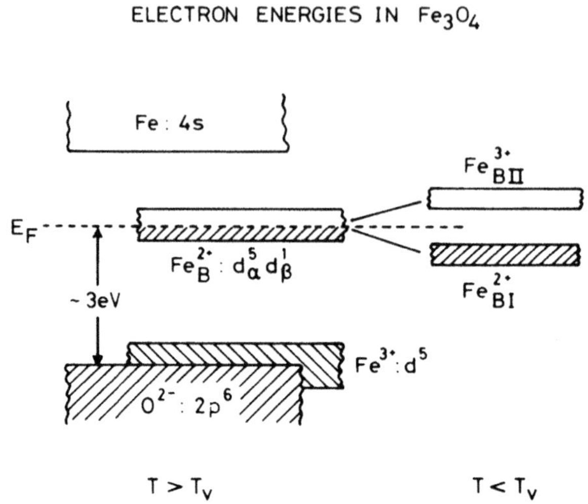

Figure 4. Schematic electron-energy diagram for Fe_3O_4

I.S.≈0.9mm/s indicates that the β-spin electrons are shared equally by the B-site ions over the time 10^{-8}s required to probe an iron nucleus. Therefore any β-spin-electron jump frequency is $\tau_p^{-1}>10^8$/s. However, small-polaron jumps have a fequency requirement $\tau_p^{-1}<10^{12}$/s, and small-polaron jump frequencies in the range $10^8<\tau_p^{-1}<10^{12}$ would be consistent with a jump enthalpy $2\varepsilon_p = 0$.

Transport data supply evidence that any small polaron at $T>T_V$ cannot be a distinguishable Fe^{2+} ion. Nicolau et al [25] have measured the Seebeck coefficient θ and electrical conductivity σ as a function of composition for the system $Fe^{3+}[Ni^{2+}_{1-x}Fe^{2+}_xFe^{3+}]O_4$. For small polarons jumping between N sites,

$$\theta = \frac{k}{e}\{\ell n\left(\frac{1-c}{c}\right)+ a\} \tag{9}$$

where c is the concentration of carriers of charge e and a is a transport term that can generally be neglected in these oxides. If the electrons jump between B sites, Eq. (9) becomes

$$\theta \approx (k/e)\ell n([Fe_B^{3+}]/[Fe_B^{2+}]) \tag{10}$$

The electrical conductivity is

$$\sigma = [Fe^{2+}]eu = ([Fe^{2+}]e^2D_o/kT)\exp(-\varepsilon_p/2kT) \tag{11}$$

The data on polycrystalline samples having x<0.4 gives a temperature-independent θ for all samples, indicating a mobile $[Fe^{2+}]$ independent of temperature. Moreover, θ is described quantitatively by Eq. (10) for $x<0.2$, but begins to deviate from Eq. (10) for $x>0.2$. In addition, the activation energy $2\varepsilon_p$ obtained from linear $\ell n(\sigma T)$ vs T^{-1} plots decreases monotonically from 0.5eV for smallest x to about 0.05eV for $x = 0.38$. The sharp decrease in polaron-stabilization energy ε_p with increasing polaron concentration appears to reflect the influence of polaron-polaron interactions, and the deviations from Eq. (10) for $x>0.2$ signals a delocalization inconsistent with distinguishable Fe^{2+} ions in high-temperature Fe_3O_4. Polarons, if they exist, must jump between molecular-cluster orbitals, the statistics from Eq. (9) having Fe-Fe pairs [26].

Siemons [27] has measured the Hall mobility u_H in single-crystal Fe_3O_4 below 300K. He obtained linear plots vs 1/T for $T<T_v$ and $T>T_v$, a discontinuity and a minimum occurring at T_v. The magnitudes were in the range $0.05<u_H<0.5$ cm^2/Vsec, about what would be predicted for $u = eD_o/kT$. However, the slope of u_H vs 1/T is negative in the range $T_v<T<300K$; but this could reflect short-range ordering above T_v. An itinerant-electron model would have a $u = e\tau/m^*$ with $\tau>\hbar/w_b$ and a bandwidth $w_b \approx \hbar(\pi/a_o)^2/2m^*$, which corresponds to a $u>0.4$cm^2/Vsec for a free-electron effective mass. With a heavier effective mass, the observed magnitudes of u_H could just be at the lower range of possibility for itinerant electrons.

In conclusion, the magnetic and transport data leave ambiguous the question whether the β-spin electrons are itinerant. They establish the localized character of the α-spin electrons, the lack of any small-polaron stabilization energy ε_p for the β-spin electrons, and a delocalization of the β-spin electrons at least to the extent of jumping between Fe-Fe pair orbitals rather than ionic Fe^{2+} orbitals. We turn now to evidence from the low-temperature structural data.

LOW TEMPERATURE STRUCTURE

The crystal structure below a cooperative, diffusionless transition generally offers a sure guide to the character of the electronic ordering that drives the transition. For example, the lifting of an accidental atomic degeneracy by a cooperative intraatomic Jahn-Teller electron ordering leaves each cation in the center of its interstice; the splitting of a narrow itinerant-electron band by the formation of a charge-density wave changes the periodicity of the structure so as to introduce a smaller Brillouin zone that just contains the occupied electron states. Nearly stoichiometric FeS, for example, exhibits a semiconductor-metal transition at a temperature T_α. For $T>T_\alpha$, FeS has the hexagonal $B8_1$ structure of NiAs; for $T<T_\alpha$ the close-packed basal planes of edge-shared octahedral sites exhibit the triangular

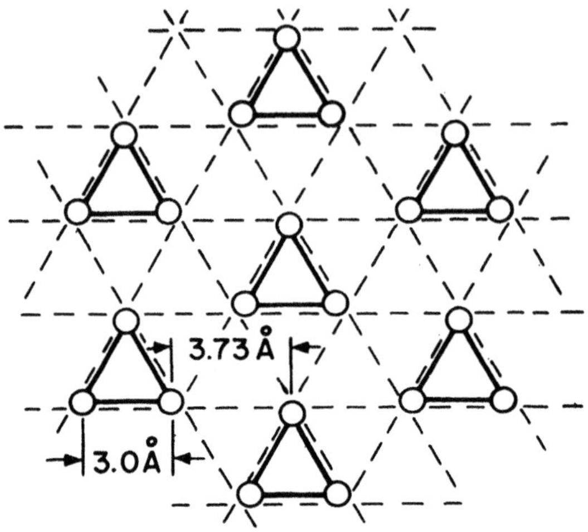

Figure 5. Triangular cation clustering in basal planes of low-temperature FeS

clustering of Fe^{2+} ions shown in Fig.5 [28]. As in Fe_3O_4, the α-spin electrons of the high-spin Fe^{2+} ions appear to be localized; but the low-temperature distribution clearly indicates that the β-spin electrons are itinerant [29]. This example not only illustrates the significance of low-temperature structural information; it also demonstrates that localized and itinerant d electrons may coexist on the same iron ions if the localized electrons belong to a high-spin half-shell and the itinerant electrons have different spin. This condition applies, of course, to the situation in Fe_3O_4.

Verwey [30], assuming that the low-temperature structure reflected a simple ordering of Fe^{2+} and Fe^{3+} ions, postulated an ordering into alternate (001) planes, thus defining a c-axis. Such an ordering would place Fe^{2+} ions in $[1\bar{1}0]$ rows and Fe^{3+} ions in [110] rows of edge-shared octahedra, thereby creating unequal a and b axes within a basal plane. Therefore, Verwey ordering would produce an orthorhombic structure.

Although early x-ray powder measurements suggested rhombohedral symmetry [31], magnetic-anisotropy measurements on single crystals — first grown by Smiltens [32] — clearly indicate that low-temperature Fe_3O_4 is magnetically orthorhombic. Bickford [33] first demonstrated this with ferrimagnetic resonance and strain-gauge measurements; Calhoun [34] confirmed it with torque-magnetometer measurements. Moreover, Abrahams and Calhoun [35] reported an orthorhombic x-ray structure at 78K with a_o = 5.912, b_o = 5.945,

c_o = 8.388 Å at 78K, and Hamilton [36] claimed to confirm the structure with neutron diffraction. At the close of the 1950's, the Verwey model appeared to be established.

The initial difficulty with the model was manifest in attempts to interpret low-temperature Mössbauer data [24], but the first serious questioning of the model came only in 1968 [37] when electron diffraction of synthetic single crystals at 77K revealed extra spots that indicated a doubling of the unit cell proposed by Verwey. Shirane et al [38] subsequently remeasured single-crystal Fe_3O_4 with neutron diffraction; they confirmed the need for a larger unit cell, but were unable to obtain a satisfactory structural refinement. This result was followed by an x-ray diffraction study of single-crystal Fe_3O_4 that also confirmed the extra reflections and showed that the symmetry should be reduced at least to monoclinic with lattice parameters: a_m = 11.888, b_m = 11.847, c_m = 16.773 Å and $\beta = 89.76_2{}^o$ [39]. However, Rado and Ferrari [40] found a nonlinear mangetoelectric effect in low-temperature Fe_3O_4; this effect requires loss of a center of symmetry and may require a triclinic space group.

It is now clear that the original ordering proposed by Verwey is incorrect even though the crystal is magnetically orthorhombic. This observation has introduced differing theoretical views: one considers the complexity of the low-temperature structure a manifestation of the itinerant character of the β-spin electrons, but without indicating how the new periodicity best stabilizes occupied states at the expense of empty states [41]; another attempts to interpret the data in terms of a complex ordering of distinguishable Fe^{2+} and Fe^{3+} ions [38, 42]. A third alternative, not seriously explored, is that the complexity of the low-temperature structure manifests retention of some β-spin electrons in Fe-Fe pairs and some in discrete Fe^{2+}, Fe^{3+} couples. Such an order would be characteristic of strong polaron-polaron interactions and some delocalization, but without a transition to the itinerant-electron state.

Whatever the structural model finally chosen, it must take account of the low-temperature Mössbauer and NMR data. These two types of data are complementary: the NMR signals give greater accuracy of the values of the hyperfine fields at ^{57}Fe, but the Fe^{2+}-ion signals are weak relative to the Fe^{3+}-ion signals. Mössbauer data, on the other hand, allow quantitative comparisons of the signals from Fe^{2+} and Fe^{3+} ions. In addition, a larger isomer shift of the Fe^{2+} relative to the Fe^{3+} lines permits better resolution of the spectra. Both NMR and Mössbauer data exhibit no splitting of the tetrahedral Fe^{3+} spectrum below T_v, but a splitting of the octahedral spectrum into at least four [43-46], but perhaps considerably more, distinguishable iron atoms [47, 48]. The spectra are sensitive to the direction of the applied field and exhibit considerable complexity. Apparently the orbital momentum remains quenched on all ions, but a site identified as Fe^{2+}-II has a large quadrupole splitting with an electric-field gradient having its unique axis parallel to the c-axis. For this ion the β-spin charge density appears to occupy a d_{xy} orbital. Another site, identified as Fe^{2+}-I, exhibits a smaller quadrupole field oriented at right angles to the c-axis. As many as five different Fe^{3+} ions are said to be resolved [47, 48]. This degree of complexity would seem to point to stabilization of a charge-density wave at low temperatures, characteristic of itinerant β-spin electrons.

REFERENCES

1. R. Blakemore, Science 190, 377 (1975)
2. T. Okamura, Sci. Rep. Tahoku Univ. 21, 231 (1931)
3. R.W. Millar, JACS 51, 215 (1929); M. Matzui, S. Tado, and S. Chikazumi, J. Phys. Soc. Japan 43, 47 (1977)
4. E.J.W. Verwey, Nature 144, 327 (1939); E.J.W. Verwey and P.W. Haaymann, Physica 8, 979 (1941)
5. W.H. Bragg, Phil. Mag. 30, 305 (1915); S. Nishikawa, Proc. Tokyo Math. Phys. Soc. 8, 199 (1915)
6. L. Néel, Annals de Phys. 3, 138 (1948)
7. J.B. Goodenough, Magnetism and the Chemical Bond (Interscience and John Wiley, New York, 1963) pp.122
8. C. Guillaud, J. Physique Rad. 12, 239 (1951)
9. E.W. Gorter, Philips Res. Repts. 9 295, 321, 403 (1954)
10. J.B. Goodenough, Phys. Rev. 171, 466 (1968)
11. C.G. Shull, E.O. Wollan, and W.C. Koehler, Phys. Rev. 84, 912 (1951)
12. S. Umemura and S. Iida, J. Phys. Soc. Japan 40, 679 (1976)
13. W. Heisenberg, Z. Physik 49, 619 (1928)
14. H.A. Kramers, Physica 1, 182 (1934)
15. C. Zener, Phys. Rev. 81, 440 (1951)
16. R. De Pape, Compt. Rendus Acad. Sci. Paris 260, 4527 (1965)
17. H. Watanabe and B.M. Brockhouse, Phys. Letters 1, 189 (1962)
18. J.B. Goodenough, J.M. Longo, and J.A. Kafalas, Mater. Res. Bull. 3, 471 (1968)
19. G.A. Samaro and A.A. Giardini, Phys. Rev. 186, 577 (1969)
20. J.B. Goodenough, Mater. Res. Bull. 6, 967 (1976)
21. K. Mizushima, M. Tanaka, A. Asai, S. Iida, and J.B. Goodenough, J. Phys. Chem. Solids (in press)
22. J.B. Goodenough, J. Solid State Chem. 1, 349 (1970)
23. R. Banminger, S.G. Cohen, A. Marinov, S. Ofer, and E. Segal, Phys. Rev. 122, 1447 (1961)
24. K. Ono, Y. Ishikawa, A. Ito, and E. Hirahara, J. Phys. Soc. Japan 17, Sup. B-1, 125 (1962)
25. P. Nicolau, I. Bunget, M. Rosenberg, and I. Belciu, IBM J. Res. Develop. 1970, 248 (May)
26. J.B. Goodenough, Mater. Res. Bull. 5, 621 (1970)

27. W.J. Siemons, IBM J. Res. Develop. 1970, 245 (May)
28. E.F. Bertaut, Bull. soc. franc. minéral et crist. 79, 276 (1956)
29. J.B. Goodenough, J. Appl. Phys. Suppl. 33, 1197 (1962)
30. E.J.W. Verway, P.W. Haaymann, and F.C. Romeijn, J. Chem. Phys. 15, 181 (1947)
31. N.C. Tombs and H.P. Rooksby, Acta Cryst. 4, 474 (1951); H.P. Rooksby and B.T.M. Willis, Acta Cryst. 6, 565 (1953)
32. J. Smiltens, J. Chem. Phys. 20, 990 (1952)
33. L.R. Bickford, Jr, Phys. Rev. 78, 449 (1950); Revs. Mod. Phys. 25, 75 (1953)
34. B.A. Calhoun, Phys. Rev. 94, 1577 (1954)
35. S.C. Abrahams and B.A. Calhoun, Acta Cryst. 6, 105 (1953); 8, 257 (1955)
36. W.C. Hamilton, Phys. Rev. 110, 1050 (1958)
37. T. Yamada, K. Suzuki, and S. Chikazumi, Appl. Phys. Lett. 13, 172 (1968)
38. G. Shirane, S. Chikazumi, J. Akimitsu, K. Chiba, M. Matsui, and Y. Fujii, J. Phys. Soc. Japan 39, 949 (1975)
39. J. Yoshida and S. Iida, J. Phys. Soc. Japan 42, 230 (1977)
40. G.T. Rado and J.M. Ferrari, Phys. Rev. B12, 5166 (1975)
41. J.R. Cullen and F. Callen, J. Phys. (Paris) C1-1971 2, 1110 (1971)
42. S. Iida, K. Mizushima, M. Mizoguchi, S. Umemura, and Y. Yoshida, J. Appl. Phys. 49, 1455 (1978)
43. M. Rubinstein and D.W.Forester, Solid State Commun. 9, 1675 (1971); M. Rubinstein, G.H. Stauss, and F.J. Brumi, A.I.P. Conf. Proc. 10, 1384 (1972)
44. N.M. Kovtun and A.A. Shamyakov, Solid State Commun. 13, 1345 (1973)
45. R.S. Hargrove and W. Kündig, Solid State Commun. 8, 303 (1970)
46. J. Mada and S. Iida, J. Phys. Soc. Japan 39, 1627 (1975); 42, 1184 (1977)
47. M. Mizaguchi, J. Phys. Soc. Japan 44, 1501, 1512 (1978)
48. S. Umemura and S. Iida, J. Phys. Soc. Japan 47, 458 (1979)

POLYMETALLIC CENTERS IN BIOLOGY

Stephen J. Lippard

Department of Chemistry, Columbia University,
New York, New York 10027

ABSTRACT

Three classes of polymetallic centers in biology are systematically reviewed. Tight clusters having metal-metal bonding interactions occur in iron-sulfur proteins and in the enzyme nitrogenase. The geometric and electronic structures of these clusters are consistent with their assignment as mixed valence compounds. Most prevalent are the $Fe_4S_4^*$ clusters having distorted cubic geometries. They exhibit reversible redox reactions that are essential to their biological function. Ligand bridged systems, including the oxygen transport proteins hemerythrin and hemocyanin, superoxide dismutase from erythrocytes, and possibly cytochrome oxidase, constitute a second broad class of biological polymetallic centers. The value of having two or more metal atoms working in concert to achieve an enzymatic function is discussed. Finally, brief mention is made of a third class, comprised of loose assemblies such as the multifunction redox enzymes, allosteric proteins, and membrane bound systems, that are more complex and pose interesting problems for future study.

INTRODUCTION

The importance of metal ions in biology is now widely recognized. The purpose of this article is to outline the various classes of polymetallic centers that occur and to discuss how several metal ions working together can perform a biological function better than a single metal center.

D. B. Brown (ed.), Mixed-Valence Compounds, 427-450.

If one defines mixed valence compounds as those containing the same element in more than one oxidation state, then the best studied examples thus far in biology are the iron-sulfur proteins (1). These proteins are the principal members of the first class of polymetallic centers to be discussed, the tight clusters. A number of proteins are known to contain two metal ions bridged by one or more common ligands. While not mixed valence in the sense of having mixed oxidation states, the metal ions in these proteins are known or thought to undergo redox reactions that alter their oxidation states. Ligand-bridged systems constitute the second broad class of polymetallic centers in biology to be surveyed. The third class encompasses loose assemblies such as multi-subunit proteins or membrane bound systems where the metal centers are physically remote from one another but manage cooperatively to bring about a biological transformation.

TIGHT CLUSTERS

Iron-Sulfur Proteins

The chemical characteristics of simple iron-sulfur proteins are summarized in Table I (1,2). They are all low molecular weight, readily crystallizable proteins. Rubredoxin contains iron coordinated to four sulfur atoms contributed by four cysteine amino acids. The resulting distorted tetrahedral structure is not a tight cluster and will not be considered further. The remaining proteins listed in Table I all have the nFe-nS* stoichiometry, n = 2,4,8, where S* represents sulfide ion that is readily released as H_2S upon acidification. X-ray crystallographic studies have revealed the presence of one or two $Fe_4S_4^*$ units in proteins where n = 4 or 8, respectively (3). The structure of the $Fe_4S_4^*$ unit is that of a distorted cube, the geometry of which is shown in Figure 1. Four cysteine sulfur atoms bind the cube to the protein. The structure of the iron-sulfur core in a 2Fe-2S* protein has not yet been determined by x-ray crystallography although, as will be mentioned later, there is strong circumstantial evidence to support the existence of the $Fe_2S_2^*$ unit with a specific geometry.

Although the biological functions of the iron-sulfur proteins listed in Table I have not been fully elucidated, they are all believed to participate in electron transfer reactions. The $Fe_nS_n^*$ units are capable of transferring one electron reversibly to various substrates. Proteins containing two $Fe_4S_4^*$ units function as two-electron carriers. The manner by which two such units communicate with one another at a center-to-center distance of 12 Å in the 8Fe-8S* proteins is not known. The mechanism of communication may share some of the features to be outlined later in our discussion of ligand bridged systems and loose assemblies.

TABLE I. Chemical Characteristics of Simple Iron-Sulfur Proteins

PROTEIN CLASS	SOURCE	M.W.	# Fe	# Cys	# S^*	$E^{0'}$	# elns.
($1Fe-0S^*$)							
Rubredoxin	Clostridium pasteurianum	6,000	1	4	0	-0.057 V	1
($2Fe-2S^*$)							
Plant Ferredoxin	Spinach; algae	11,500	2	5	2	-0.4 V	1
Adrenodoxin	Mammals	13,000	2	5	2	-0.34 V	1
Putidaredoxin	Pseudomonas putida	12,000	2	6	2	-0.23 V	1
($4Fe-4S^*$)							
HiPIP	Chromatium	10,000	4	4	4	+0.35 V	1
Bacterial Ferredoxin	Bacillus polymyxa	9,000	4	4	4	-0.37 V	1
	Desulfovibrio gigas	6,000	4	6	4	-0.32 V	1
($8Fe-8S^*$)							
Bacterial Ferredoxin	Azotobacter vinelandii	14,200	8	8	8	-0.42 V	2
	Chromatium	9,600	8	8	8		2
	Clostridium; Micrococcus	6,000	8	8	8	-0.4 V	2

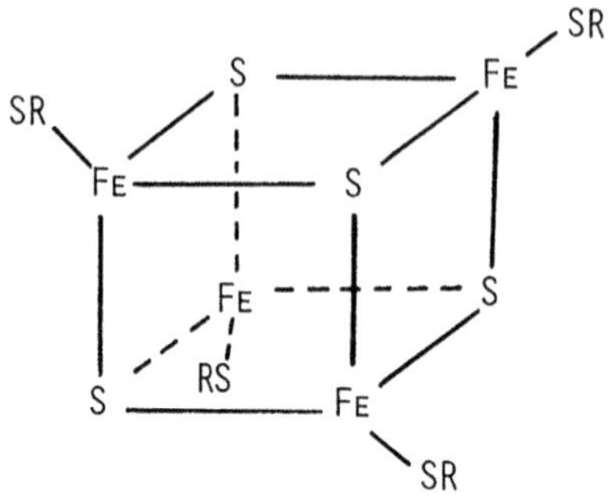

PROTEIN	DISTANCES (Å)		
	Fe - Fe	Fe - S*	Fe - S(cys)
Reduced HiPIP	2.74	2.32	2.22
Oxidized HiPIP	2.81	2.26	2.20
Ferredoxin	2.85	2.30	2.19

FIGURE 1. Geometry of the $Fe_4S_4^*$ cubes in three iron-sulfur proteins.

An intriguing feature of the data presented in Table I is the difference between the redox potential of the high potential iron protein (HiPIP) from chromatium, $E^{0'} = +0.35$ V, and the potentials of other proteins known to contain the $Fe_4S_4^*$ cluster, $-0.32 \text{ V} \leq E^{0'} \leq -0.42$ V. This difference is adequately explained by the "three state hypothesis" (4) in which the oxidation level in reduced HiPIP is postulated to be the same as in the oxidized bacterial ferredoxins. Before outlining this hypothesis, however, we must first discuss the synthesis and properties of model compounds for the iron-sulfur protein cores.

In extensive studies, Holm and his co-workers have prepared and characterized $Fe_2S_2^*$ and $Fe_4S_4^*$ clusters that exhibit many of the physical properties of the iron-sulfur proteins (3,5). Comparison of the structure of one of these complexes, $[Fe_4S_4(SCH_2Ph)_4]^{2-}$ (Fig. 2), determined by Ibers, with that of the cube in ferredoxin (Fig. 1) reveals many similarities. From the knowledge of the charge on the model compounds (not available from studies of the proteins) and through comparisons of the

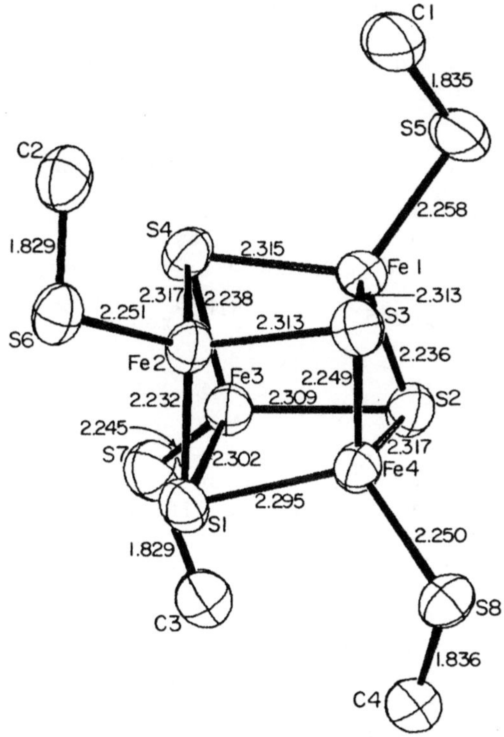

FIGURE 2. Structure of the $[Fe_4S_4(SCH_2Ph)_4]^{2-}$ cluster. Phenyl groups are omitted. The average Fe•••Fe distance is 2.747 Å. Reproduced with permission from J. Am. Chem. Soc., 95, 3523 (1973).

electron spin resonance, Mössbauer, optical, and NMR spectra it was possible to assign the oxidation levels of the clusters in the proteins. Thus the $[Fe_4S_4(SR)_4]^{2-}$ model compounds, oxidized ferredoxin (Fd_{ox}), and reduced high potential iron protein $(HiPIP)_{red}$ were all shown to have the same iron oxidation level (3,5). Formally this level corresponds to two Fe(II) and two Fe(III) centers, but the electrons must be delocalized over the entire cluster since no physical technique was found that could distinguish two different, trapped valence species. Voltammetric studies showed that electrons could be reversibly added to the cubes. The species $[Fe_4S_4(SR)_4]^{n-}$ with n = 1,2, or 3 have either been isolated and structurally characterized or detected electrochemically.

The model compounds provide a valuable key to understanding the variation in redox potentials of the $Fe_4S_4^*$ clusters in the iron-sulfur proteins. As outlined on the top of the following page, the proteins were postulated (4) to exist in three states (a-c). In ferredoxin, the low potential redox reaction occurs between states (a) and (b). In HiPIP, the high potential (b) $\rightleftarrows$ (c) redox couple is observed. The different polarity and

$$[Fe_4S_4(SR)_4]^{4-} \rightleftarrows [Fe_4S_4(SR)_4]^{3-} \rightleftarrows [Fe_4S_4(SR)_4]^{2-} \rightleftarrows [Fe_4S_4(SR)_4]^{-}$$

$$Fd_{red} \rightleftarrows Fd_{ox} \rightleftarrows Fd_{superox}$$

$$HiPIP_{supered} \rightleftarrows HiPIP_{red} \rightleftarrows HiPIP_{ox}$$

(a) (b) (c)

hydrophobicity of the surrounding amino acid milieu is presumably responsible for the stabilization of the different redox couples in the two protein systems. In support of this idea, Cammack (6) was able to "super-reduce" HiPIP with dithionite in 80% DMSO/H_2O solutions and observe ESR and optical spectra that were very similar to those of Fe_{red} and the $[Fe_4S_4(SR)_4]^{3-}$ model compounds. There is also some evidence for the existence of a "super-oxidized" ferredoxin. It is therefore clear from the combined model and protein studies that states (a-c) exist in the 4Fe-4S* and 8Fe-8S* proteins and are responsible for their redox activities and potentials.

Although x-ray structural results are not yet available for the 2Fe-2S* proteins, physical studies (ESR, Mössbauer, etc.) show their iron-sulfur cores to be well approximated by $[Fe_2S_2(SR)_4]^{2-}$ model compounds (5). The crystal structure of one such compound is shown in Figure 3. Reversible electron transfer reactions to form the 3- and 4- anions were established electrochemically and it seems quite likely that the 3- $\rightleftarrows$ 2- equilibrium in the model compounds corresponds to the $Fd_{red} \rightleftarrows Fd_{ox}$ redox process in the plant ferredoxins, adrenodoxins, and putidaredoxin (Table I). The fine tuning of the potentials among these $Fe_2S_2^*$ proteins is quite likely the result of their differing amino acid contents and sequences (see ref. 6 for further discussion). The potentials of both the $Fe_nS_n(SR)_n$ model compounds and the proteins are known to depend upon solvent and the nature of the R group.

Nitrogenase

Nitrogenase is the enzyme system by which bacteria on the nodules of green plants convert dinitrogen (N_2) from the air into

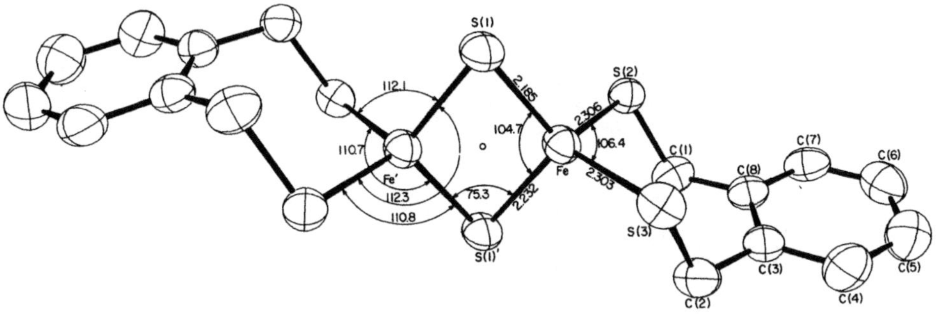

FIGURE 3. Structure of bis[o-xylyl-α,α'-dithiolato-μ_2-sulfido-diferrate(III)]. Reproduced with permission from Proc. Natl. Acad. Sci. U.S.A., 70, 2429 (1973).

ammonia. The composition of a typical nitrogenase is displayed in Table II (7). As can be seen from the table there are two proteins, an iron protein and an iron-molybdenum protein, both of which have subunit polypeptide chains. The iron protein was shown by core extrusion studies to have one $Fe_4S_4^*$ cluster, similar to those found in the simpler iron-sulfur proteins. Core extrusion involves chemical unfolding of the protein, extraction of the $Fe_4S_4^*$ cluster with a thiol such as benzenethiol, and subsequent identification of the cluster as a 2-Fe or 4-Fe unit by spectroscopic methods (5). Core extrusion studies also show that four $Fe_4S_4^*$ clusters exist in the iron-molybdenum protein. The remaining two molybdenum and 14-16 iron atoms comprise a component of the iron-molybdenum protein that can be extracted chemically (8). This extract is called the iron-molybdenum cofactor, or FeMo-co. At present its structure is unknown, although several groups are trying to crystallize it for x-ray diffraction studies.

Using x-ray absorption spectroscopy (see ref. 9 for a review), Hodgson, Mortenson, and their co-workers were able to eliminate the possible existence of Mo=O units in the iron-molybdenum protein and the FeMo-co and proposed the two structures shown below as possible models for the metallic core (10). The

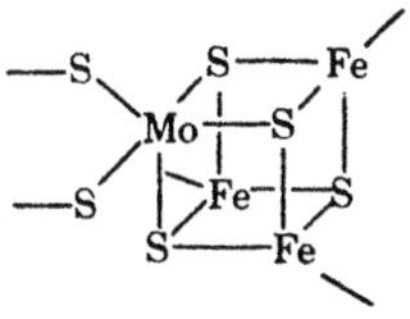

I

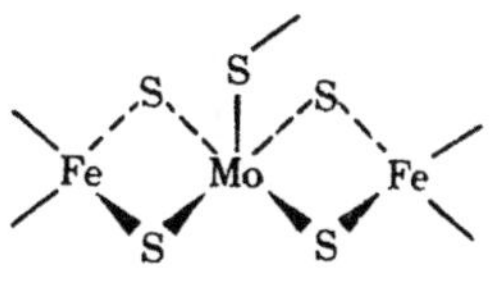

II

TABLE II. Composition of a Typical Nitrogenase System.

Azotobacter vinelandii : two proteins : a MoFe protein and an Fe protein

	MOLECULAR WEIGHT		METAL AND SULFIDE COMPOSITION (g-atom/mol)		
	Holoprotein	Subunits ($\alpha_2\beta_2$)	Mo	Fe	S*
MoFe Protein	220,000 - 240,000	56,000(2) 60,000(2)	2	30 - 32	>24
	Core extrusion shows 4 Fe_4S_4 clusters.				
Fe Protein	64,000	33,000	-	4	4
	Core extrusion shows one Fe_4S_4 cluster.				

structure on the left bears an obvious resemblance to the $Fe_4S_4^*$ cluster in the iron-sulfur proteins, having one of the iron atoms in the cube replaced by molybdenum. Subsequent work by Holm (11), Garner (12), and their collaborators has led to several compounds containing the $Fe_3MoS_4^*$ core, one synthetic route to which is shown in Figure 4. The structure (11) of one such complex, the $[Mo_2Fe_6S_8(SEt)_9]^{3-}$ ion, is shown in Figure 5. In this anion there are two $Fe_3MoS_4^*$ cubes bridged by ethane thiolate ligands. As Figure 4 reveals, this $Fe_3MoS_4^*$ unit occurs in all the compounds thus far structurally characterized. Formally these cubes have two Fe(III), one Fe(IV), and a Mo(III) or Mo(IV) centers; that is, they are mixed valent clusters. While it has thus far not been possible to demonstrate any chemical reactivity of the Mo center in these clusters toward N_2 or other substrates (they do undergo reversible electron transfer reactions, however), they are beautiful new compounds that may ultimately prove of value in understanding the FeMo-co in nitrogenase.

At present our understanding of nitrogenase is far from complete. It is possible that nature accomplishes the six electron reduction of N_2 to NH_3 by pumping in electrons through the $Fe_4S_4^*$ unit of the iron protein, storing them in the $Fe_4S_4^*$ clusters of the iron-molybdenum protein, then transferring them to dinitrogen which is activated by binding at the molybdenum atom of an $Fe_3MoS_4^*$ or similarly related cluster. Clearly mixed valence compounds play an important role in this biologically intriguing system.

Platinum Blue -- An Antitumor Drug

Although our discussion focuses on polymetallic centers that occur in nature, we digress for a moment to mention a class of mixed valent clusters that are potentially useful as drugs for the treatment of cancer. The compounds are the platinum blues (13), which, as shown in Figure 6, may be obtained in the reaction of the widely used antitumor drug (14) cis-$[(NH_3)_2PtCl_2]$, with pyrimidine bases. The structure of a crystalline platinum blue, synthesized from cis-$[(NH_3)_2PtCl_2]$ and α-pyridone (15), is shown in Figure 7. This mixed valence cluster has four platinum atoms with an average formal oxidation state of 2.25 arranged in a linear array. It is stabilized by metal-metal bonding, the bridging α-pyridonate ligands, and hydrogen bonding interactions. Platinum blues are reported to have antitumor activity with lower kidney toxicity than the parent drug. Their mechanism of action is unknown, but the pyrimidine base has been shown not to accompany the platinum atom when platinum uracil blue binds to DNA (16). Further discussion of platinum blues will be given in a subsequent lecture by the author at this workshop and the

SYNTHESIS OF MIXED VALENT Fe-Mo CLUSTERS

$(R'_4N)_2(MS_4) + FeCl_3 + 10\ RSH + NaOMe \xrightarrow[25^0]{MeOH\ (EtOH)}$ SEVERAL PRODUCTS AS TETRAALKYLAMMONIUM SALTS

M = Mo, W R = Et, Ph, Bz

$M_2Fe_6S_9(SR)_8^{3-}$ X-RAY; FORMALLY A $Mo^{III}Fe_3S_4^0$ AND A $Mo^{IV}Fe_3S_4^+$ CORE

$M_2Fe_6S_8(SR)_9^{3-}$ X-RAY; FORMALLY TWO $Mo^{III}Fe_3S_4^0$ CORES

$M_2Fe_7S_8(SR)_{12}^{3-}$ X-RAY*; FORMALLY TWO $Mo^{III}Fe_3S_4^0$ CORES; CENTRAL Fe^{III}

$M_2Fe_7S_8(SR)_{12}^{4-}$ NO X-RAY YET; PRESUMABLY CENTRAL Fe^{II}

CUBES HAVE FORMALLY 2 Fe(III) AND 1 Fe(II) CENTERS.

CLUSTERS EXHIBIT REVERSIBLE ELECTRON TRANSFERS.

PRODUCTS WERE SEPARATED BY FRACTIONAL CRYSTALLIZATION.

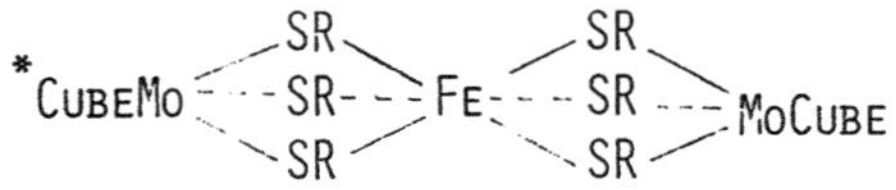

FIGURE 4. Synthesis of mixed valent iron-molybdenum clusters (ref. 11 and R.H. Holm, private communication).

TABLE III. Comparison of Some Properties of the Oxygen Binding Pigments (reproduced with permission from ref. (17)).

	HEMOGLOBIN	HEMERYTHRIN	HEMOCYANIN
Metal	Fe	Fe	Cu
Metal : O_2	Fe : O_2	2Fe : O_2	2Cu : O_2
Oxidation state of metal in the deoxy protein	Fe^{II}	Fe^{II}	Cu^{I}
Coordination of metal	**Porphyrin**	**Protein side chains**	Protein side chains
No. subunits	4	8	Variable
Mol. wt.	65,000	108,000	400,000 - 9,000,000
Color			
Oxygenated	Red	Violet-pink	Blue
Deoxygenated	Red-blue	Colorless	Colorless

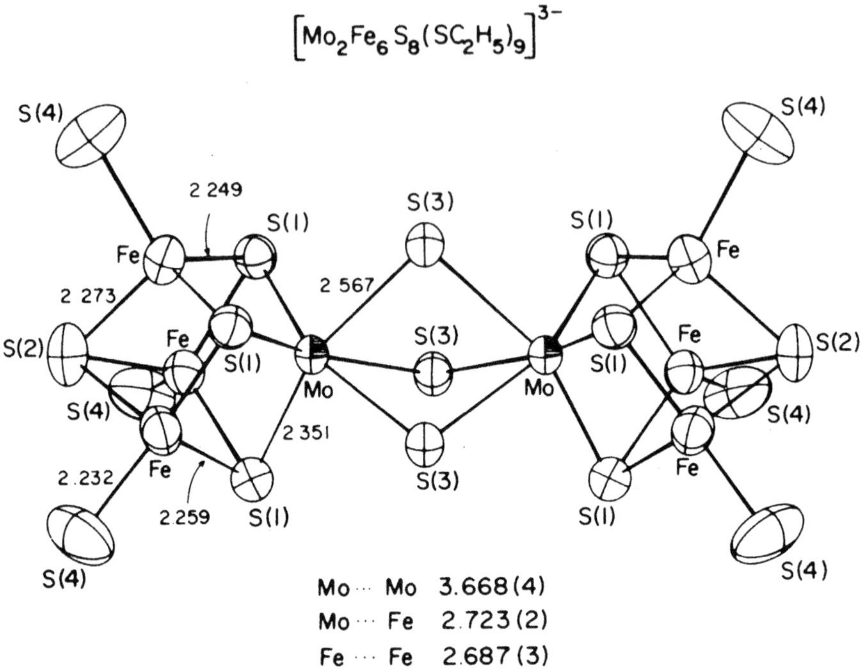

FIGURE 5. Structure of $[Mo_2Fe_6S_8(SEt_9)]^{3-}$ omitting the nine ethyl groups of the bridging and terminal thiolate ligands. Reproduced with permission from ref. 11.

interested reader is referred to the literature for details (13-16).

LIGAND BRIDGED SYSTEMS

Oxygen Transport Proteins

We are all familiar with hemoglobin, the protein containing iron porphyrin that transports oxygen in the blood of most major phyla in the animal kingdom. Perhaps less familiar are two proteins, hemerythrin (17) and hemocyanin (18), that transport oxygen in some invertebrates, specifically in worms and in arthropods and molluscs, respectively. Table III reveals that binuclear iron and copper centers occur in these proteins. These centers are believed to react with oxygen in a reversible oxidative addition reaction as shown in Figure 8.

X-ray diffraction results are available for two hemerythrin (17) proteins isolated from different species. Both proteins studied were in the met-form which means that the iron atoms were

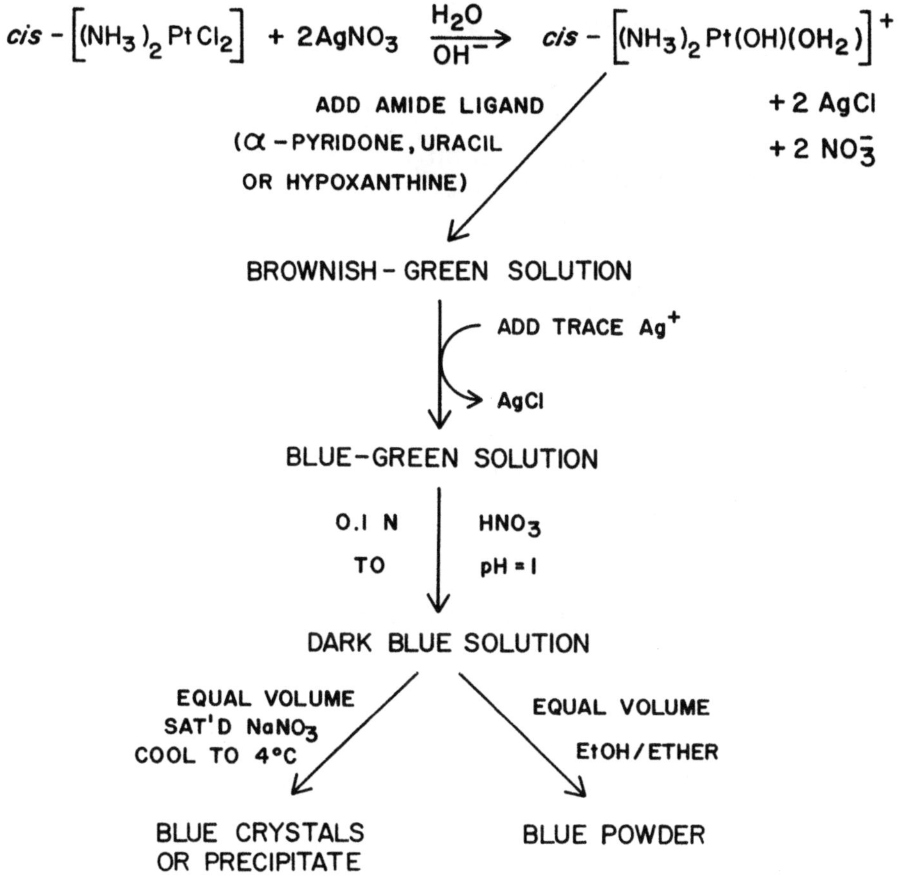

FIGURE 6. Synthetic route to the cis-diammineplatinum blues.

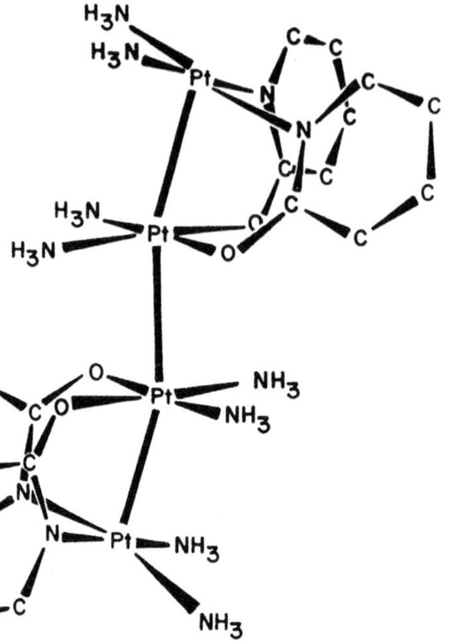

FIGURE 7. Structure of the $[Pt_4(NH_3)_8(C_5H_4NO)_4]^{5-}$ anion. Double bonds in the α-pyridone rings are not illustrated. The inner and outer Pt-Pt distances are 2.88 and 2.77 Å, respectively.

oxidized to the ferric state without binding oxygen. The structures are shown in Figure 9. Although the two crystallographic models appear to be different, one structure was carried out on azidomet- and the other on aquomethemerythrin, which may account for the discrepancy. Both active sites contain bridging ligands and it is likely that a bridged, bimetallic center exists in the reduced form of the protein. It is believed that oxygen is

HEMERYTHRIN: $Fe^{II} \cdots Fe^{II} + O_2 \rightleftharpoons Fe^{III} \cdots O_2^{2-} \cdots Fe^{III}$

HEMOCYANIN: $Cu^{I} \cdots\cdot Cu^{I} + O_2 \rightleftharpoons Cu^{II} \cdots O_2^{2-} \cdots Cu^{II}$

FIGURE 8. Bimetallic centers for oxygen transport in biology.

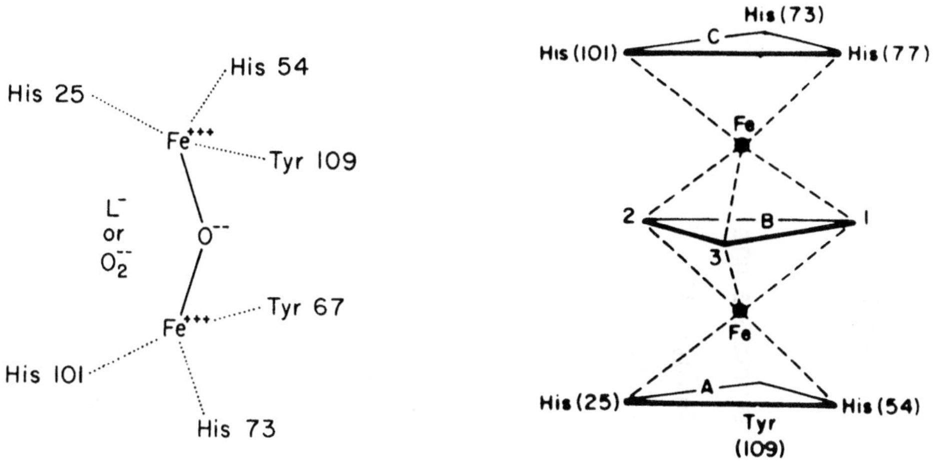

FIGURE 9. Active site structures of azidometmyohemerythrin (left) and aquomethemerythrin (right). Reproduced with permission from reference 17.

prepared for transport in these proteins by a two-electron reduction to the peroxide state (Figure 8). The ligand bridged binuclear diiron(II) center nicely provides the required reducing power and also serves to coordinate and transport the reduced ligand. Two models proposed for the geometry of the active center in oxyhemerythrin are shown below.

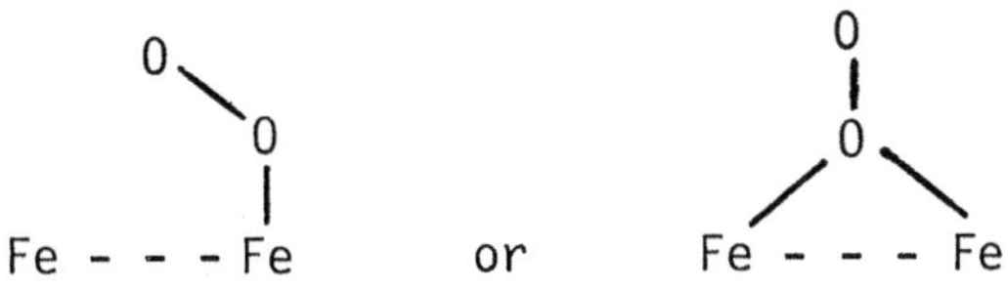

The structure of the bimetallic center in hemocyanin (18) has not been determined by x-ray crystallography and, owing to the high molecular weight of this protein (Table III), may not be available for some time.

Superoxide Dismutase

Another protein in which a ligand bridged bimetallic center occurs at the active site is the superoxide dismutase present in the red blood cells of mammals (19). Properties of the bovine

BOVINE ERYTHROCYTE SUPEROXIDE DISMUTASE (BESOD)

SOURCE: BOVINE ERYTHROCYTES

SUBUNITS: TWO IDENTICAL SUBUNITS

MW: 31,400

COLOR: BLUE-GREEN; λ_{MAX} 680nm

METALS: $2\,Cu^{2+} + 2\,Zn^{2+}$

ESR: $g_{\parallel} = 2.265$; $A_{\parallel} = 142$ G

REDOX POTENTIAL; $E^{\circ\prime} = 0.42$V (pH dependent)

REACTION CATALYZED: $2O_2^- + 2H^+ \longrightarrow O_2 + H_2O_2$

RATE CONSTANT: $k = 2.37 \times 10^9$ $M^{-1}sec^{-1}$ at 25°, $4.8 < pH < 9.5$

PROPOSED REACTION MECHANISM:

$$ECu(II) + O_2^- \rightleftharpoons ECu(I) + O_2$$

$$ECu(I) + O_2^- + 2H^+ \rightleftharpoons ECu(II) + H_2O_2$$

FIGURE 10. Properties of bovine erythrocyte superoxide dismutase.

erythrocyte enzyme, BESOD, are shown in Figure 10. BESOD contains two identical subunits, each of which has a copper and a zinc atom bridged by the imidazolate ring of a histidine residue (Figure 11) (20). The enzyme functions to scavenge the

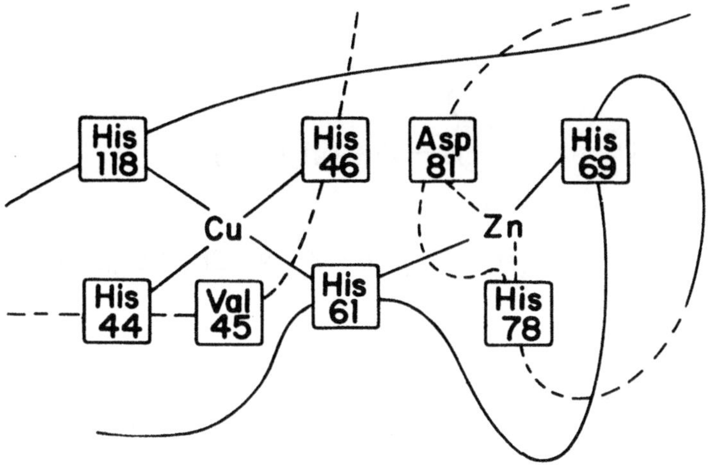

FIGURE 11. Schematic diagram of the active site structure of BESOD showing the coordinated amino acids and partial tracing of the peptide chain. Reproduced with permission from reference 21.

superoxide ion, O_2^-, that is presumed to be a toxic by-product of oxygen metabolism. It does so through a reversible redox shuttle involving the copper center (Fig. 10). Since copper is the active catalyst in this protein, the role of the zinc has been a subject of considerable study. One possibility, discussed in detail elsewhere (21), is that coordination of zinc to the histidine ligand attenuates the pK_a of the other imidazolate ring nitrogen atom such that its binding preference is Cu(II) > H^+ > Cu(I). This preference could be manifest in a bridge splitting mechanism for BESOD activity, as outlined in Figure 12.

Details of the enzyme mechanism, particularly with regard to the role of the bridging imidazolate ligand and the zinc ion, are being sought through model chemistry. Imidazolate bridged dicopper(II) complexes have been synthesized by the author and his co-workers (22) and by others (23) that show many of the features of Cu_4BESOD, in which copper replaces zinc in

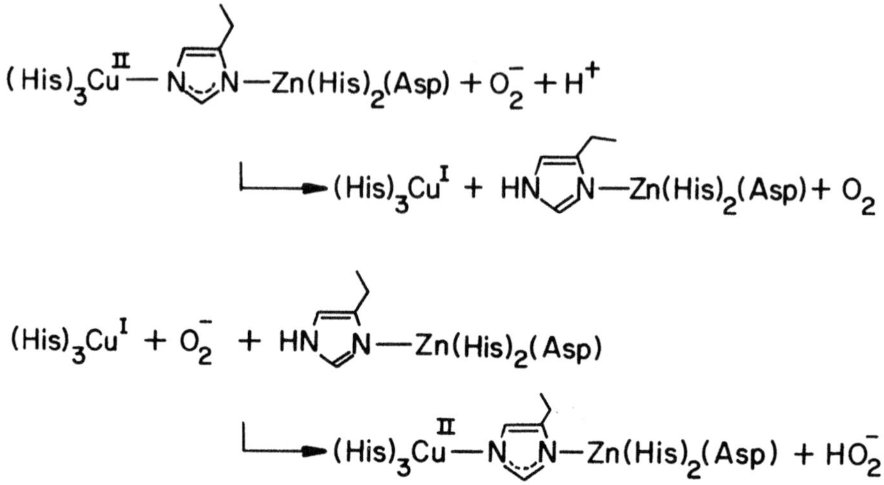

FIGURE 12. Possible bridge-splitting mechanism for bovine superoxide dismutase action. Reproduced with permission from reference 21.

Cu_2Zn_2BESOD. The structure of one such model complex (24) is presented in Figure 13.

A histidine bridge between heme iron and copper has also been proposed (25) for cytochrome c oxidase, an enzyme that will be discussed later. Although BESOD is the only protein where a histidine bridge has been rigorously established, enough is known about the properties of imidazolate bridged bimetallic complexes to predict that further examples will be found in biology. If so, there will be additional opportunity to learn how nature takes advantage of two metals in electronic and structural communication with one another through the intervention of a bridging ligand.

Metal Storage Proteins - Ferritin

The storage and transport of metal ions in biology is a fascinating story that is only beginning to unfold. Polymetallic centers are involved in at least one system and probably others will be identified. Ferritin (26) is an iron storage protein in mammals that contains 24 subunits. The approximate composition of its iron core is $FeO(PO_3H_2)\cdot 8Fe(OH)_3$. The iron is readily

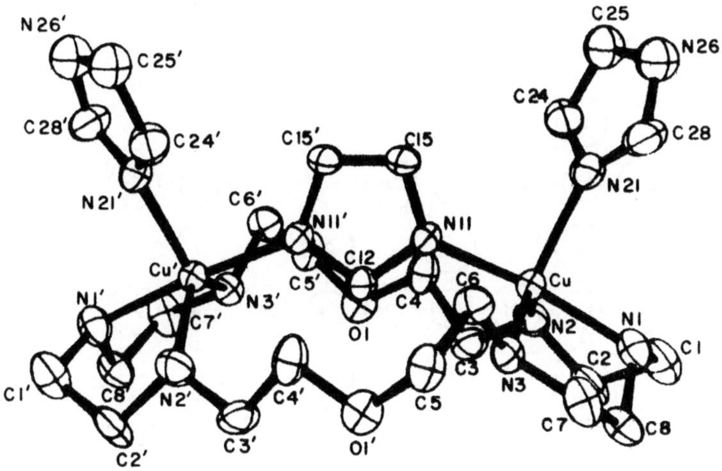

FIGURE 13. Structure of the imidazolate bridged dicopper(II) ion incorporated into a circular cryptate macrocycle. Reproduced with permission from reference 24.

removed from the protein upon reduction to iron(II), consistent with its function as a storage protein. An interesting model for the iron core in ferritin, proposed by Spiro and Saltman (27), is shown in Figure 14. As may be seen, a polymetallic center involving bridging oxide and hydroxide ligands is postulated.

LOOSE ASSEMBLIES

The discussion thus far has dealt with polymetallic centers in direct contact either through metal-metal bonding or a bridging ligand. There are numerous examples in biology of more complex systems in which more remotely organized metal atoms cooperate to catalyze a specific reaction or perform a needed function. While these loose assemblies are not as well characterized they are worthy of brief discussion because they pose some interesting problems for future investigations.

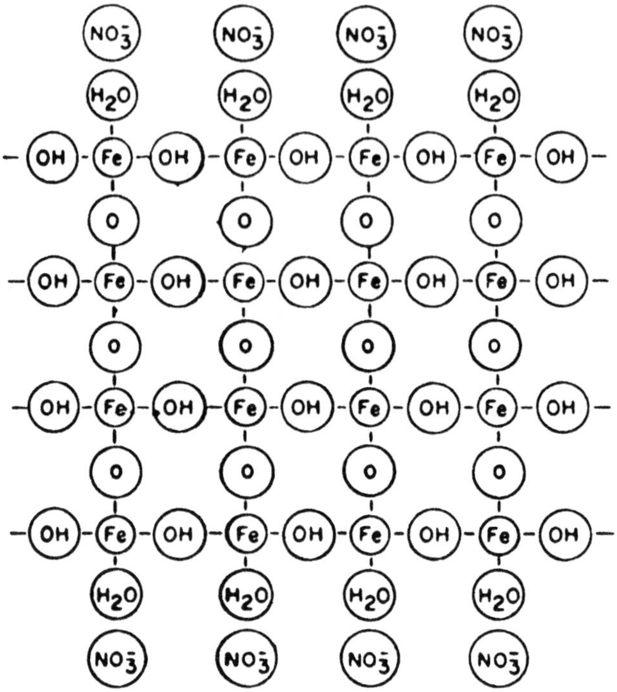

FIGURE 14. Structural model proposed for the iron core of ferritin. Reproduced with permission from Adv. Chem. Ser., 100, 365 (1971).

Multifunction Redox Enzymes

Table IV lists the properties of several copper containing proteins that catalyze the reduction of oxygen to water (28). This four electron redox reaction is achieved in biology by enzymes that have four one electron redox centers, either Cu(I)/Cu(II) as in the case of laccase, ascorbate oxidase, and ceruloplasmin, or two such copper centers and two heme iron centers, as in cytochrome c oxidase. The latter is coupled to the cellular production of ATP in a process known as oxidative phosphorylation. It is estimated that cytochrome c oxidase is responsible for more than 90% of the oxygen consumption in the biosphere.

The multifunction redox copper enzymes have been postulated to reduce O_2 to H_2O in two 2-electron steps (28). It can be seen from the oxygen potentials (Table IV) that O_2^- is thermodynamically uphill from O_2, suggesting that any mechanism for the catalytic reduction of oxygen would proceed more rapidly by

TABLE IV. Multifunction Redox Enzymes - Summary of Properties

PROTEIN	SOURCE	MW	METAL CONTENT	FUNCTION
Laccase	lacquer tree fungi	110,000	4 Cu Type 1 Cu Type 2 Cu 2 Type 3 Cu	$O_2 \longrightarrow H_2O$ catalysis "Blue oxidase"
Ascorbate oxidase	cucumber, squash	130,000	8 Cu	Blue oxidase
Ceruloplasmin	animal serum	160,000	8 Cu	Blue oxidase
Cytochrome c oxidase	mitochondria	lipoprotein	2 Cu 2 heme Fe	oxidative phosphorylation

OXYGEN POTENTIALS

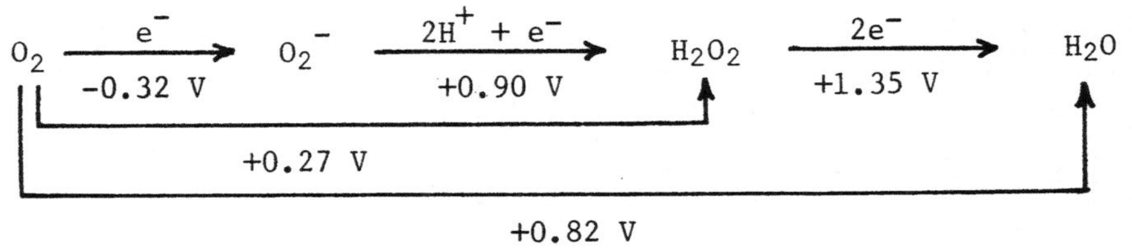

multielectron transfer steps. Nature has evolved proteins to perform this task and it remains for us to determine how they work.

Allosteric Proteins

Allosterism is a term used to explain the cooperative interactions found to occur among the components of a complex system (29). Hemoglobin is one such system, as is hemocyanin. In these proteins, binding of oxygen to a heme iron or binuclear dicopper center in one subunit of the protein facilitates binding of oxygen to the metals in adjacent subunits. This cooperativity, which has been extensively discussed (30), arises through changes induced in the coordination environment and surrounding polypeptide chain of the metal centers.

Membrane Bound Systems

In the mitochondria of cells there is a mechanism of electron transport that involves a number of metalloproteins in constant communication. These are the cytochromes that, together with cytochrome c oxidase, catalyze the electron transfer from organic substrates to oxygen and produce ATP through oxidative phosphorylation. How do these proteins communicate with one another? What are the mechanisms of electron transfer? Why has a polymetallic system of this degree of complexity evolved to carry out the desired biological task? Attempts to answer these and related questions are at the forefront of modern bioinorganic chemisty.

REFERENCES

Note: These references are meant to give the student a lead into the relevant literature and not to provide a comprehensive bibliography. Review articles are cited wherever possible.

1. Lovenberg, W., ed., "Iron-Sulfur Proteins," Academic Press, New York, Vols. I (1973), II (1973), and III (1977).

2. Orme-Johnson, W.H.: 1973, Ann. Rev. Biochem. 42, p. 159.

3. Holm, R.H.: 1975, Endeavor 34, p. 38.

4. Carter, C.W., Jr., Kraut, J., Freer, S.T., Alden, R.A., Sieker, L.C., Adman, E., and Jensen, L.H.: 1972, Proc. Natl. Acad. Sci. 69, p. 3526.

5. Holm, R.H.: 1977, Acc. Chem. Res. 10, p. 427.

6. Lippard, S.J.: 1973, Acc. Chem. Res. 6, p. 282.

7. A review by W.H. Orme-Johnson and L.C. Davis of nitrogenase systems may be found in Volume III of reference 1. Other data mentioned here come from recent work by E.I. Stiefel, R.H. Holm, and their co-workers (private communications).

8. Shah, V.K. and Brill, W.J.: 1977, Proc. Natl. Acad. Sci. USA 74, p. 3249.

9. Cramer, S.P. and Hodgson, K.O.: 1979, Progr. Inorg. Chem. 25, p. 1.

10. Cramer, S.P., Hodgson, K.O., Gillum, W.O., and Mortenson, L.E.: 1978, J. Am. Chem. Soc. 100, p. 3398.

11. Wolff, T.E., Berg, J.M., Hodgson, K.O., Frankel, R.B., and Holm, R.H.: 1979, J. Am. Chem. Soc. 101, p. 4140.

12. Christou, G., Garner, C.D., Mabbs, F.E., and Drew, M.G.B.: 1979, J.C.S. Chem. Comm. 91.

13. Barton, J.K. and Lippard, S.J.: 1978, Ann. N.Y. Acad. Sci. 313, p. 686.

14. For numerous articles and reviews, see J. Clin. Hematol. Oncol. (Wadley Medical Bulletin), Vol. 7 (1977).

15. Barton, J.K., Szalda, D.J., Rabinowitz, H.N., Waszczak, J.V., and Lippard, S.J.: 1979, J. Am. Chem. Soc. 101, p. 1434.

16. Bauer, W., Gonias, S.J., Kam, S.K., Wu, K.C., and Lippard S.J.: 1978, Biochemistry 17, p. 1060.

17. Kurtz, D.M., Jr., Shriver, D.F., and Klotz, I.M.: 1977, Coord. Chem. Revs. 24, p. 145.

18. Lontie, R. and Witters, R., in Eichhorn, G.I., ed., "Inorganic Biochemistry," Ch. 12 (1973).

19. Fridovich, I.: 1975, Ann. Rev. Biochem. 44, p. 147.

20. Richardson, J.S., Thomas, K.A., Rubin, B.H., and Richardson, D.C.: 1975, Proc. Natl. Acad. Sci. USA 72, p. 1349.

21. Lippard, S.J., Burger, A.R., Ugurbil, K., Valentine, J.S., and Pantoliano, M.W.: 1977, Adv. Chem. Ser. 162, p. 251.

22. O'Young, C-L., Dewan, J.C., Lilienthal, H.R., and Lippard, S.J.: 1979, J. Am. Chem. Soc. 100, p. 7291.

23. Haddad, M.S. and Hendrickson, D.N.: 1978, Inorg. Chem. 17, p. 2622.

24. Coughlin, P.K., Dewan, J.C., Lippard, S.J., Watanabe, E-i., and Lehn, J-M.: 1979, J. Am. Chem. Soc. 101, p. 265.

25. Palmer, G., Babcock, G.T., and Vickery, L.E.: 1976, Biochemistry 7, p. 2206.

26. Crichton, R.R.: 1973, Angew. Chem. Internat. Ed. 12, p. 578.

27. Brady, G.W., Kurkjian, C.R., Lyden, E.F.X., Robin, M.B., Saltman, P., Spiro, T., and Terzis, A.: 1968, Biochemistry 7, p. 2185.

28. Malmström, B.G.: 1977, Adv. Chem. Ser. 162, p. 173.

29. Monod, J., Changeux, J.P., and Jacob, E.: 1963, J. Mol. Biol. 6, p. 306.

30. Stanley, H.E., Bansil, R., and Herzfield, J.: 1978, Metal Ions Biol. Syst. 7, p. 311.

MIXED-VALENCE IN THE ORGANIC SOLID STATE

James J. Mayerle

IBM Research Laboratory
San Jose, California 95193

ABSTRACT: A review of mixed valence in the organic solid state is given, with the intention of demonstrating the interplay of the various energies that stabilize the phenomenon. Since high electrical conductivity is one of the most spectacular results of mixed valence, those factors that lead to it will be discussed. Particular attention is paid to three classes of materials: the TTF halides, TTF-TCNQ and derivatives, and the newly discovered, highly conducting TTF-tetrahalo-*p*-benzoquinones.

INTRODUCTION

Mixed valence is usually thought of by inorganic chemists in terms of isolated, non-interacting molecules containing two or more metal atoms in different oxidation states, some examples being the biferrocene derivatives described by Cowan and co-workers (1) and the iron-sulfur clusters found in electron-transfer proteins and their synthetic analogs (2,3). In addition to these molecular species there occur in the solid state, and only in the solid state, materials in which the metal atoms in different oxidation states are located in different molecules or are part of large networks of atoms in which there are no discrete molecules. An example of the former is $K_2Pt(CN)_4{\cdot}Br_{0.3}{\cdot}3H_2O$ (4), in which the average oxidation state of the Pt atom is +2.3, while examples of the latter are the tungsten bronzes (5). In these extended systems, the unique properties brought about by the mixed valence are conferred not on individual molecules but on the solid as a whole. These

D. B. Brown (ed.), Mixed-Valence Compounds, 451-473.

macroscopic properties, such as electrical conductivity, are among the most spectacular effects of non-integral valence.

The materials cited above are examples of atomic mixed valence, i.e., the non-integral oxidation states can, to a large extent, be identified with a single element (usually a metal atom) within the molecule or solid. In contrast, the molecules to be discussed in this chapter do not contain metal atoms. Furthermore, the non-integral valence (or electrical charge) is associated not with a single atom but with the entire molecule, i.e., these are examples of compounds exhibiting molecular mixed valence. If the crystalline structure of the material permits, the solid state properties (and this type of mixed valence also occurs only in the solid state) resulting from the mixed valence can be quite unusual for organic materials.

Since the discovery of the unusual electrical properties of TTF-TCNQ (6), there has been a tremendous quantity of work aimed at gaining an understanding of this unique material. From this has emerged a recognition (7,8) of the importance of molecular mixed valence in the organic solid state. It remains, however, a phenomenon that is of interest mainly because of the properties that result from it.

The main thrust of this review deals not with the high conductivity of certain classes of organic materials, but with mixed valence in organic solid state chemistry. It is a fact, however, that many and certainly the most interesting of the materials to be discussed are highly conducting. As a sub-theme, then, we will also review those conditions necessary for high conductivity in organic solids, which includes mixed valence (8), and at the end will present some of our more recent work in this direction.

Although TTF-TCNQ is the most familiar of the organic mixed-valence materials, we will not begin with a discussion of it. Rather, the crystal chemistry of the TTF halides will be thoroughly reviewed since, as shown by Scott and co-workers (9) and by Torrance and Silverman (10), many of the fundamental features of organic mixed valence are so simply and beautifully illustrated in these materials. We will then proceed to the highly conducting 1:1 salts of TCNQ and describe how their mixed valence has been explained (10) in terms similar to those discussed for the TTF halides. In the process, we will outline the conditions requisite for high conductivity in this type of material, one of them being mixed valence, and illustrate guidelines (8,11) for the synthesis of new classes of such materials. Finally, a new class of highly conducting organic materials, the TTF-tetrahalo-*p*-benzoquinones, synthesized on the basis of those guidelines, will be described (11).

We hope to demonstrate in this review a common thread that unifies the field of organic mixed valence. Although the physics of the organic solid state is both complex and controversial, we hope to show that by keeping in mind a few considerations one can understand many of the general features of the area and, in the process, gain an appreciation for the beauty of the field.

THE TTF HALIDES

The TTF halides, the most fully characterized of which are the bromides, represent an extremely interesting and important class of organic mixed valence materials, both from a chemical and crystallographic (12,13) point of view. They are especially important to us in that they demonstrate all the features associated with mixed valence in the organic solid state.

TTF reacts with halogens (9) and pseudohalogens (14) to yield salts having anion:TTF ratios of 2:1 and 1:1 as well as various phases in which the anion:TTF ratio is less than unity. The phases formed by the halide salts are summarized (9) in Fig. 1. Throughout this paper, we shall refer to the quantity ρ, the degree of charge-transfer from a donor to an acceptor or, equivalently, the fractional charge on an organic moiety. Assuming that any non-oligomerized halogen atoms in these

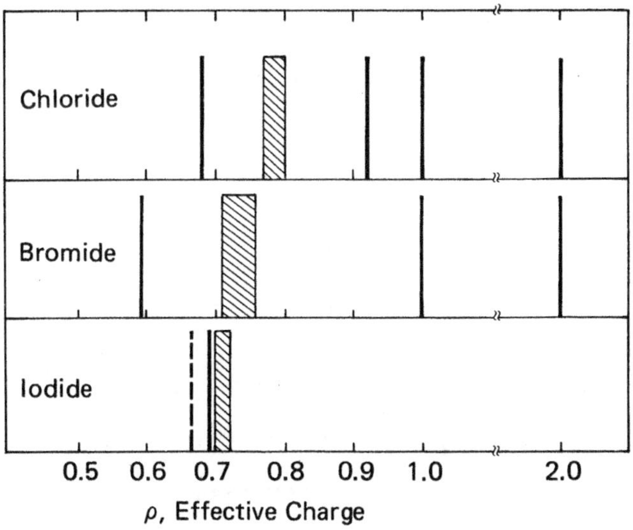

Figure 1. Summary of the phases observed for the TTF halides, plotted as a function of ρ (from Ref. 9).

materials exist as fully charged ions, one can see from Fig. 1 that there are basically two types of TTF-halide salts, those in which the charge on the TTF molecular ion is integral and those in which the charge, ρ, is less than one (and is equal to the halogen content), i.e., the materials of composition $TTF{\cdot}X_\rho$ ($\rho<1$) contain TTF molecules having a non-integral charge associated with them. They are examples of organic molecular mixed valence materials.

The structure (13) of a typical member of this family of compounds is illustrated in Fig. 2. In this particular example, $TTF{\cdot}Br_{0.79}$, the TTF moieties stack on top of each other in an eclipsed fashion, forming infinite columns of uniformly spaced, partially oxidized molecules parallel to the c axis of the monoclinic unit cell. The bromide ions are also arrayed in columns parallel to the c axis. However, as one can see in Fig. 2(b), the repeat distance along the stacking axis is different for the cations and anions. That is, the structure

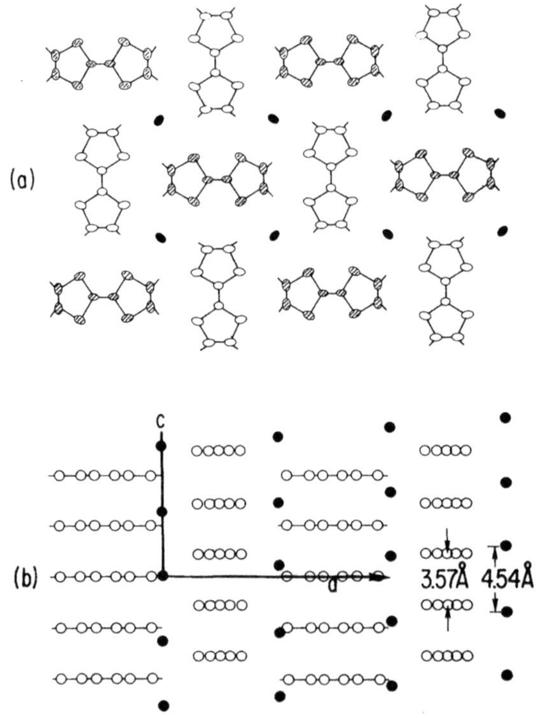

Figure 2. Crystal structure of $TTF\text{-}Br_{0.79}$ viewed (a) down the c axis and (b) down the b axis (from Refs. 10 and 13).

consists of two semi-independent monoclinic lattices, one for the TTF^+ ions and one for the Br^- ions. The two lattices have identical dimensions in the plane (a*b*) perpendicular to the stacking ($\underset{\sim}{c}$) direction. In the stacking direction, however, the lattice constants are not only unequal, they are incommensurate, i.e., they are not simple multiples (or fractions of one another). Since the a*b* nets of the two lattices are identical, one can obtain the chemical composition and therefore the degree of charge transfer, ρ, directly from the crystallographic data according to the equation $\rho = 3.57/\underset{\sim}{c}_{Br}$, where 3.57 corresponds to the $\underset{\sim}{c}$ axis length of the TTF subcell and c_{Br} is the same for the bromide sublattice. The point to be emphasized here is that the Br spacing is dependent on the stoichiometry while the TTF spacing is essentially independent of it.

In attempting to understand the TTF halides, what needs to be explained are the stabilization of the mixed valence state and why the stoichiometries, at least those of the ordered phases, are limited to $0.7 < \rho < 0.8$. A partial explanation for the stoichiometry range of the TTF halides is that the absolute value of the electrostatic binding energy of the crystal is at a maximum in this region (9,10). In the formation of the $TTF{\cdot}X_\rho$ lattice, the TTF molecule becomes partially oxidized. More precisely, $\rho \times 100\%$ of the TTF molecules become oxidized to the TTF radical cation. To achieve this oxidation requires an energy proportional to (I-A), where I is the ionization potential of the TTF molecule and A is the electron affinity of the halogen atom. This energy consumption must be compensated or the TTF halides would not exist. The most likely source of energy compensation is the electrostatic (Madelung) energy gained in forming the ionic lattice.

Torrance and Silverman (10) have calculated the net electrostatic binding energy, $E_B(\rho)$, defined in Eq. (1),

$$E_B(\rho) = -\rho[|E_M(\rho)| - (I-A)] \qquad (1)$$

as a function of ρ. E_M is the attractive Madelung energy per ionized TTF^+-Br^- pair. Their results are shown in Fig. 3. As can be clearly seen, there is an energy minimum in the region of stoichiometries actually observed. This can be explained as follows. Starting with a lattice of neutral TTF molecules, spaced as they are in the TTF halides, one can begin to add Br atoms, creating TTF^+ molecular ions. As this occurs, the crystalline binding energy begins to increase, the result of the creation of sheets of positive and negative charge in planes perpendicular to the stacking axis. At some value of ρ the absolute value of the binding energy reaches a maximum. Beyond this point repulsion between like charges along the stacking

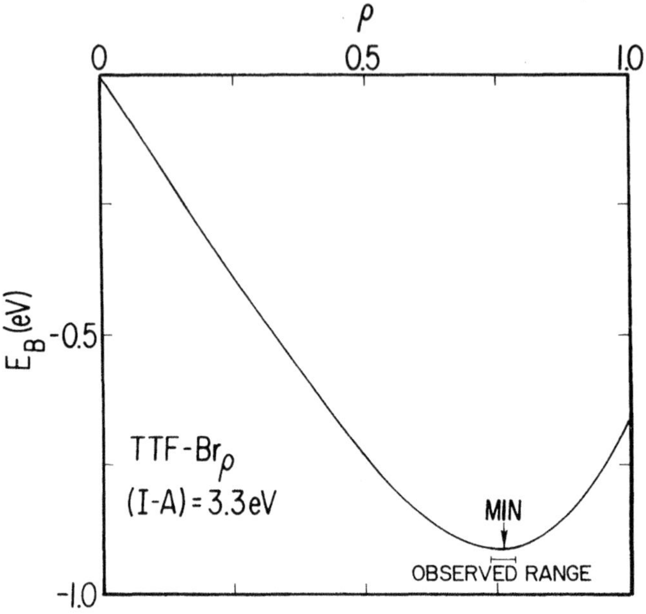

Figure 3. Results of Madelung calculations for TTF-Br$_\rho$ showing minimum in electrostatic binding energy near observed composition range (from Ref. 10).

axis begins to dominate and the net binding energy decreases with added bromine. There are sources of stabilization in addition to the electrostatic binding energy, most notably that gained by delocalizing the positive charges over the TTF stacks, the band energy. Although this may be the dominant attractive force along the stack, its magnitude is small in salts of this type, on the order of 0.15 eV. This is significantly smaller than the Madelung energy, the magnitude of which is on the order of several volts. However, the addition of this quantity to the binding energy does shift the calculated minimum to slightly lower values of ρ. Experimental data for the Cl^- and I^- salts are consistent with this explanation, the degree of TTF oxidation shifting as the electron affinity of the halogen atom changes. This has the effect of changing the (I-A) term in Eq. (1), reducing it for Cl^- and increasing it for I^-. The significance of these calculations (10) is that they show that classical electrostatic forces are primarily responsible for mixed valence in the TTF halides, in that Coulomb repulsion along the stack is reduced by forming salts in which the Br:TTF ratio (ρ) is less than 1:1. We point out that, in some ways, these materials are similar to the family of mixed valence inorganic salts represented by $K_2Pt(CN)_4 \cdot Br_{0.3} \cdot 3H_2O$ (4). In these materials the average oxidation state of the Pt atom is always near +2.3. Messmer and Salahub (15) suggest that Madelung arguments similar to the above may partially account for the stoichiometry of these salts, although agreement is not general on this point.

An instructive comparison can be made between the TTF halides, in which a mixed valence state is energetically favored, and the iodide salt of tetramethyl-p-phenylenediamine, in which the 1:1 mono-valence salt is formed. There seem to be two reasons why the 1:1 stoichiometry is preferred. The first is that TMPD is a better donor than TTF, the result of which is that the (I-A) term in Eq. (1) is smaller and it requires less energy to ionize the lattice. Another difference between the $TTF \cdot X_\rho$ ($\rho<1$) salts and TMPD•I is the manner in which the cations stack in the crystal. In the TTF salts, the donor molecules are nearly eclipsed (13) whereas in TMPD•I (16) the cations are slipped relative to each other. That is, the molecular plane of TMPD makes an angle of ~34.8° with the stacking axis. The net effect of this slippage is that, although the interplanar spacing between molecules is the same, the molecular centers (or areas of like charge) are further away from each other and Coulomb repulsion of like charges along the stacking direction is reduced, allowing for an increased value of ρ.

Examination of solid state optical properties is a useful method for qualitatively demonstrating the presence of mixed valence. Since we know from their stoichiometry that the TTF halides are of mixed valence, examination of their optical spectra should be quite instructive and should be helpful in interpreting the spectra of charge-transfer complexes which are thought to have non-integral valence but which is not evident from their stoichiometry. An additional feature of the mixed valence TTF halides that makes them particularly useful as models is their crystalline structure. Since the molecular planes are perpendicular to the stacking axis, the charge-transfer excitations along the stack do not mix with the intramolecular excitations, which are polarized in the molecular plane. This makes interpretation of the spectra more straight-forward than is usually the case for charge-transfer salts. More often, the normals to the molecular planes make an angle of 20-30° with the stacking axis, thereby mixing the charge-transfer excitation with the intramolecular transitions and complicating the interpretation of the spectra.

Figure 4 shows the solid state optical spectra (17) of both a 1:1 and a mixed-valence TTF halide as well as the solution spectra of the TTF radical cation and its dimer. Aside from a slight energy shift, the solution spectra of TTF^+ and $(TTF^+)_2$

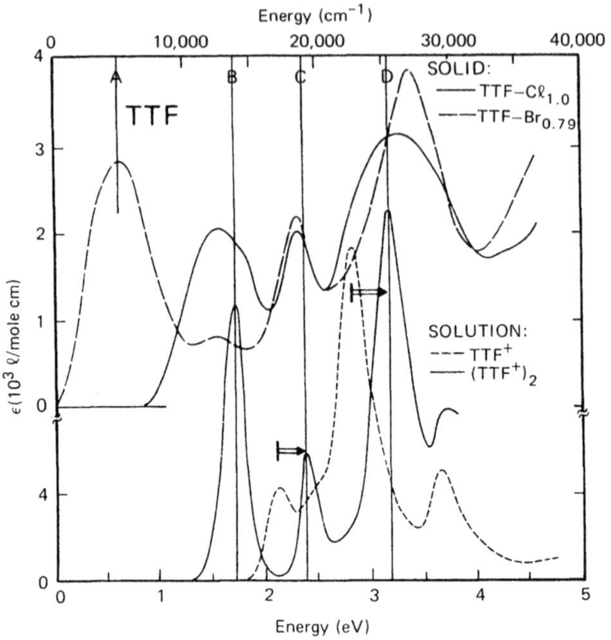

Figure 4. Summary of optical data showing the evolution of the TTF spectrum from monomer and dimer in solution to monovalent and mixed valent salt in the solid state (from Ref. 17).

are quite similar except that an additional band, B, occurs in the dimer spectrum at ~1.7 eV. Bands C and D are intramolecular transitions of TTF^+ whereas B is assigned (17) to an intermolecular charge-transfer transition between the TTF^+ moieties of the dimer. As for the solid state spectra, one of the most remarkable features is the similarity between the spectrum of the 1:1 salt and that of the solution dimer. The structure (9) of the 1:1 salt consists essentially of TTF cations dimerized in an eclipsed configuration and relatively isolated from other dimers, at least in the direction perpendicular to the molecular planes. As can be clearly seen, its spectrum consists of the two TTF^+ intramolecular transitions and the charge-transfer intermolecular transition between the constituents of the dimer. The second feature of the solid state spectra to note is that the spectrum of the mixed-valence salt exhibits all the features of the solution dimer, namely two intramolecular transitions and the charge-transfer transition between adjacent cations. In addition, it exhibits a low energy band at ~0.7 eV. As with transition B, this peak (A) is polarized along the stacking axis of the complex and is assigned (17) to an intrastack mixed-valence charge-transfer transition between neutral TTF and an adjacent cation (TTF°, TTF^{+}→TTF^{+}, TTF°). The presence of this band, characteristic of materials such as TTF-TCNQ, for which other methods have independently shown the presence of a mixed-valence state, is a very useful diagnostic for the detection of non-integral valence in the organic solid state. A localized picture of bands A and B is shown at the top of Fig. 7.

In summarizing the TTF halides, two points should be made. The first is that the occurrence and gross features of mixed valence in these materials can be understood using very simple, but elegant, arguments (9,10). The second is that these same considerations will provide insight (10) into the occurrence of non-integral valence in the more complicated, stoichiometric charge-transfer salts such as TTF-TCNQ.

TCNQ SALTS

As with most of the materials to be discussed in this brief review, interest in the TTF-TCNQ family of compounds was generated not because of their mixed valence but because of the unique electrical properties that non-integral valence, combined with their crystalline structures, confer upon them. In fact, only after these materials had been studied for several years was the importance of mixed-valence to the electrical properties recognized (7).

There are two features of this class of charge-transfer complex that lead to their mixed valence, the redox properties of the constituent molecules and the manner in which they arrange themselves in the solid state. The structure (18) of a representative member of this class, HMTTF-TCNQ, is shown in projection in Fig. 5. The structure consists of stacks of HMTTF and stacks of TCNQ arrayed along the crystallographic b axis. This self-stacking (hereafter segregated stacking) of donors and acceptors, as opposed to the more common mixed stacking of donor and acceptor molecules in the same stack, is one of the common features of highly conducting organic charge-transfer salts.

A characteristic structural feature of these materials is a short crystallographic axis (~4Å) parallel to the stacks, indicating that the molecules are uniformly spaced in this direction. In the case of HMTTF-TCNQ, the perpendicular distance between donor planes is 3.57Å, whereas the interplanar separation of the TCNQ planes is 3.23Å. To maintain a commensurate lattice of anions and cations in the face of this apparent discrepancy,

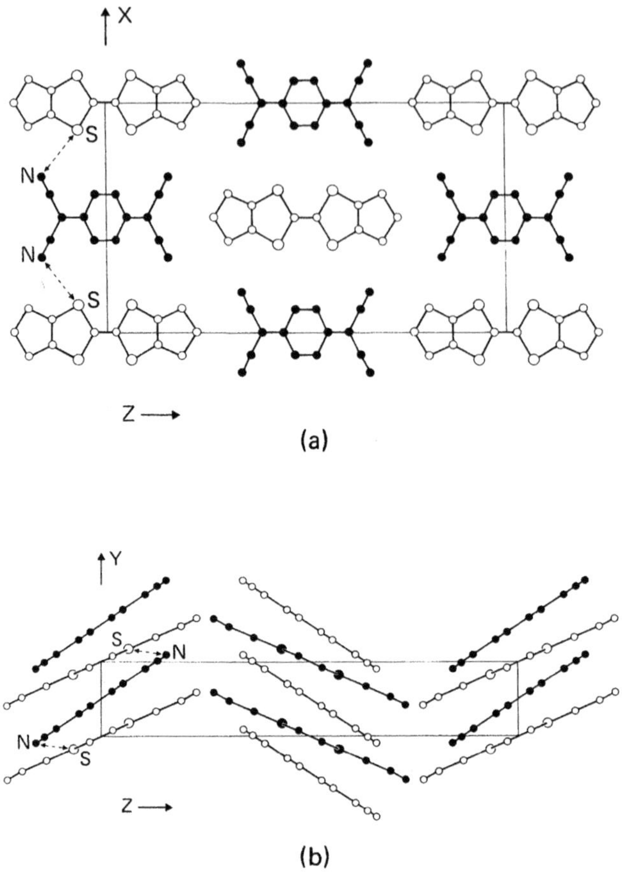

Figure 5. Projection of the structure of HMTTF-TCNQ onto (a) the plane (010) perpendicular to the stacking axis and (b) a plane (100) containing the stacking axis (from Ref. 18).

the molecules stack not in an eclipsed configuration but in a staggered one, Fig. 6. That is, the molecules pack such that the normals to their molecular planes are not parallel to the stacking axis but make an angle with it, the angles being different for donor and acceptor. For the structure shown in Fig. 5, this angle is 23.8° for the donor molecules and 34.2° for TCNQ. The TCNQ-TCNQ overlap is such that the exocyclic double bond of one lies over the quinone ring of the molecule adjacent to it along the stack. This has been shown to be an energy minimum for TCNQ and can be explained (19) in terms of the nodal

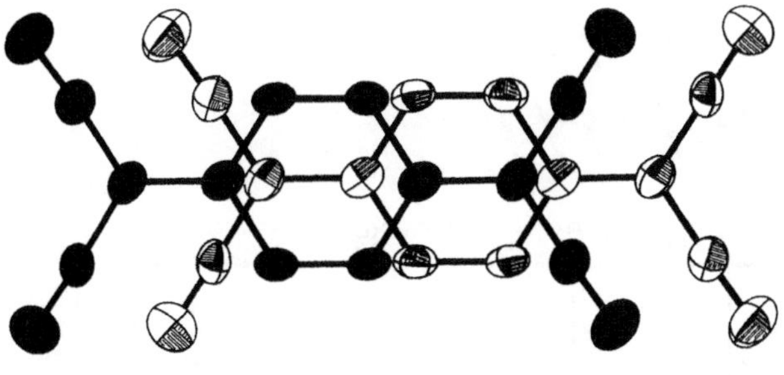

Figure 6. The slipped overlap of the TCNQ molecule typically found in its highly conducting salts.

pattern of the lowest unoccupied molecular orbital of the molecule.

Several groups of workers have performed electrostatic binding energy calculations on this class of materials (10,20). A common result of this work is that the calculated electrostatic binding energy defined in Eq. (1) does not, by itself, account for the stability of these materials. That is, the Madelung energy gained in forming the ionic lattice is not great enough to offset the energy necessary to completely ionize the constituent molecules. It has been suggested (10) that one way of increasing the electrostatic binding energy would be to increase the effective distance between like (repulsive) charges along the stacking direction. As discussed earlier, this is accomplished in the TTF halides by the formation of mixed-valence salts of the type $TTF{\cdot}X_{\rho}$, where $\rho<1$, i.e., by making the average charge on each TTF moiety less than unity. For charge-transfer salts in which both donor and acceptor are large organic molecules, the formation of such non-stoichiometric materials crystallizing with incommensurate lattices would likely be energetically unfavorable. However, the increased effective distance between like charges along the stack can be brought about if the transfer of charge from donor to acceptor is incomplete. That is, in contrast to the TTF halides, where the charge of each halide is unity and partial charge-transfer is realized by the formation of non-stoichiometric materials, in the TTF-TCNQ family of compounds the donor and acceptor molecules are both partially charged (mixed-valence) and the materials are stoichiometric.

Qualitative evidence for incomplete charge-transfer in TTF-TCNQ can be seen in the solid state optical spectrum of the material shown, along with that of a few other TCNQ salts, in Fig. 7 (7). The structures of all the materials represented in

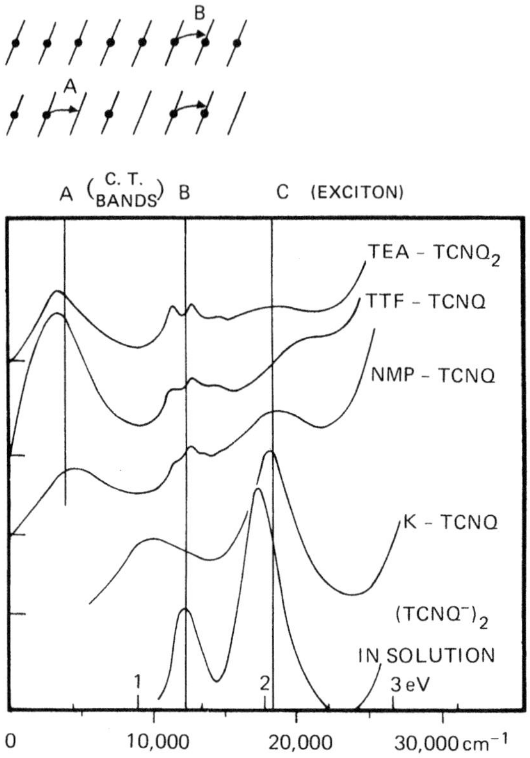

Figure 7. Powder absorption spectra of various TCNQ salts (from Ref. 7).

Fig. 7 are composed of segregated stacks of TCNQ. The simplest spectrum is that of the potassium salt. Peaks C and D have been assigned (7) to intramolecular transitions while peak B, polarized along the stacking axis, has been interpreted as being an intermolecular charge-transfer transition between adjacent, fully ionic TCNQ moieties. The most prominent feature of the spectra of the remaining materials is the presence of a band at ~0.3 eV, also polarized parallel to the stacking axis, that is absent in the spectrum of K•TCNQ. This is the mixed-valence intermolecular transition, seen also in the spectra of the mixed-valence TTF halides (17). Clearly the TCNQ moiety in the non-stoichiometric (but commensurate) acridinium, tetraethylammonium, and cesium salts is of non-integral valence. The presence of the band at ~0.3 eV implies the same is true of the 1:1 TCNQ salts of TTF, TSeF, and NMP.

In addition to the qualitative evidence for mixed-valence found in the optical spectrum, quantitative evidence can be obtained from methods that directly determine ρ, the degree of charge transfer. TTF-TCNQ undergoes a transition from a conducting to an insulating state at 59°K. This transition is characterized not by a gross structural change, but by a subtle periodic modulation of the lattice due to a coupling of the conduction electrons with the lattice. This shows up in the form of satellite x-ray reflections (or diffuse sheets) surrounding the Bragg peaks. Because the structure is sinusoidally modified, the Bragg peaks (caused by the average structure) remain essentially unchanged and new, weak peaks caused by the slight modulation grow in near them. Since these satellite peaks are the result of an interaction between the conduction electrons and the lattice, they are located at positions that depend on the filling of the conduction band, i.e., on ρ. Comes and Kagoshima (21) have determined that for TTF-TCNQ, ρ=0.59. By another method, the numerical integration of x-ray diffraction amplitudes, Coppens (22) arrived at an estimate for ρ of 0.48-0.60 for TTF-TCNQ. Another method of indicating incomplete charge transfer in TCNQ salts, by studying the field dependence of the proton NMR second moment, was used

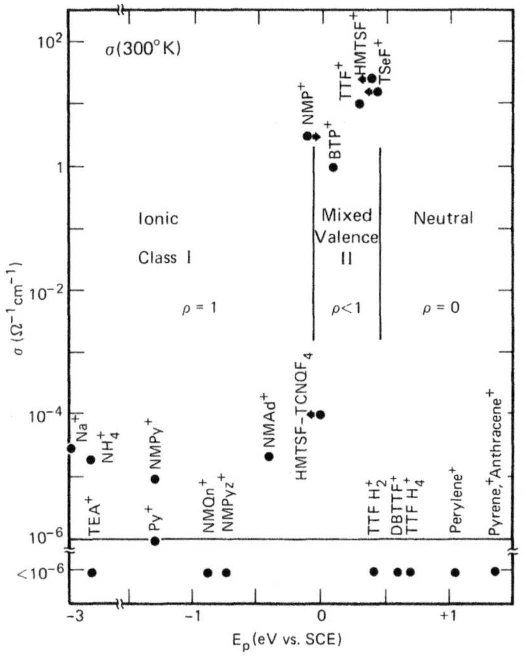

Figure 8. The powder conductivity at 300°K of various TCNQ salts plotted as a function of the reduction potential of the cation (from Ref. 8).

by Butler, Wudl and Soos (23) to show that there are unpaired electrons on ~6% of the NMP sites in NMP-TCNQ, implying that $\rho=0.94$ in this highly conducting 1:1 salt. X-ray diffuse scattering places the value of ρ for NMP-TCNQ at 0.67 (24).

The effect of ρ on the electrical properties of organic charge transfer salts is illustrated in Fig. 8, where the conductivity of a number of 1:1 TCNQ salts is plotted as a function of the reduction potential of the cation (8). The easily reduced cations on the far right side of the plot represent relatively poor donors. Their complexes with TCNQ are neutral ($\rho\simeq 0$). Because of this, the complexes would be expected to have mixed, rather than segregated, stacks and, consequently, low conductivity. On the left side of the plot are cations that are difficult to reduce. Their complexes with TCNQ are fully ionic ($\rho=1$) and they, too, exhibit low conductivity. Only over a narrow range of cation reduction potentials are highly-conducting complexes obtained. On the basis of the optical and x-ray measurements noted above, the charge-transfer in these materials has been shown to be non-integral, i.e., they are all mixed-valence materials.

The effect of ρ on the conductivity of the TCNQ salts can be visualized in the following manner. For electrical conduction to occur, electrons must move along the stack from one TCNQ to another. In the case of the fully ionic salts, this process can be described by the disproportionation $TCNQ^- + TCNQ^- \rightarrow TCNQ^\circ + TCNQ^=$, which involves the creation of a TCNQ dianion. The Coulomb repulsion between the two electrons on the dianion make the energy involved prohibitive, so the ionic salts are insulators. In the mixed-valence salts, the electrons can move along the stack by the process $TCNQ^\circ + TCNQ^- \rightarrow TCNQ^- + TCNQ^\circ$. The energy of this process is small and the mixed-valence materials are highly conducting. This localized picture is not entirely realistic. A more rigorous explanation is given by the Hubbard band model. In the case of the fully ionic salts, the TCNQ conduction band is filled and the resulting materials are Mott insulators. In the mixed-valence salts, the conduction band is only partially filled, the electrons can move freely, and the conductivity is quite high.

The effect of mixed-valence is dramatically illustrated by the difference in the electrical properties of HMTTF-TCNQ and HMTTF-$TCNQF_4$ (25). Although the two salts are isostructural (25), their electrical behavior is very different. HMTTF-TCNQ has a room temperature conductivity along the stacking axis of $\sim 400\Omega^{-1}\cdot cm^{-1}$ with metallic temperature dependence (18). In contrast, HMTTF-$TCNQF_4$ is a semiconductor with room temperature conductivity seven orders of magnitude less than that of the TCNQ salt. This difference in conductivity can be explained by the difference in the electrochemical properties of the acceptors. The reduction potential of $TCNQF_4$ is ~0.5 volts more negative than that of TCNQ. It is a much stronger acceptor. Thus, the charge-transfer in its salt with HMTTF is complete and the material is an insulator. The weaker acceptor TCNQ forms a salt with HMTTF in which $\rho=0.72$, i.e., it forms a mixed-valence salt that, consequently, exhibits metallic conductivity. This same divergence of behavior has been previously seen in the isostructural salts of HMTSeF with TCNQ and $TCNQF_4$ (26). This suggests that by careful consideration of the electrochemical properties of the donor and acceptor, one can choose pairs of molecules such that incomplete charge-transfer should result (8).

THE TTF-TETRAHALO-*p*-BENZOQUINONES

Incomplete charge-transfer is only one of the requirements for high conductivity. Another is that the molecules must crystallize in separate (segregated) stacks of donors and acceptors. Choosing pairs of molecules that satisfy this condition is more difficult than selecting pairs that have the proper match of redox potentials. In our attempts to synthesize new classes of highly conducting organic materials, however, we have developed guidelines (11,27), to be discussed now, that indicate whether segregated stacking might occur.

It is usually stated that the vast majority of organic charge-transfer compounds crystallize in stacks of alternating donor and acceptor molecules, and that materials such as TTF-TCNQ which crystallize in segregated stacks are the rare exception. It is indeed true that most organic charge-transfer complexes crystallize with mixed stacks, but it is also true that in the vast majority of such complexes the degree of charge-transfer is very small. Therein lies the problem. One of the stabilizing features of these materials, in solution as dimers or in the solid state, is the Mulliken charge-transfer interaction between donor and acceptor. This type of interaction is weak between two neutral donors or two neutral acceptors. In one case, the charge-transfer interaction would involve placing electron density in the lowest unoccupied molecular orbital (LUMO) of the donor while in the other it would involve taking electron density from the highest occupied molecular orbital (HOMO) of the acceptor. No substantial stabilization occurs in either case. Thus, those materials with neutral ground states will always crystallize with alternate stacks.

If a significant fraction of the donors and acceptors are ionized, then examples of both mixed and segregated stacking can be found. For the mixed stacked materials, the charge-transfer stabilization still exists, being a "back charge-transfer" from the negatively charged acceptor to the now positively charge donor. However, a mode of charge-transfer now exists that can stabilize segregated stacked materials. This involves charge-transfer from a negatively charged acceptor to another neutral or negatively charged acceptor and likewise for the donor molecules. So, if both modes of stacking are possible, what is it that leads to the formation of one phase over another? In choosing donors and acceptors that we though would lead to segregated stacked materials, we have formulated still-evolving guidelines (11,27) based partially upon the symmetry of the frontier orbitals that would be involved in the stabilizing Mulliken charge-transfer interaction of the alternate stacking phase, i.e., the HOMO of the donor and the LUMO of the acceptor. The strategy is to inhibit this interaction with the hope that segregated stacking will then result. Before proceeding further, I wish to state that until further proof is forthcoming, this is a guideline only, however appealing.

The guideline, stated quite simply, is the following: if the symmetries of the frontier orbitals are equivalent, the resulting complex will have mixed stacks. However, if the symmetries of the frontier orbitals are different, orbital overlap between donor and acceptor may be inhibited and segregated stacks may result. This is illustrated in Fig. 9 and explained more fully in the text.

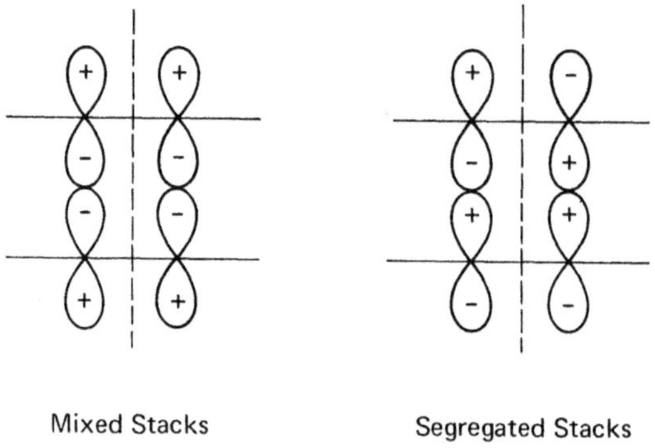

Figure 9. Schematic view of donor-acceptor orbital overlap showin; the situation leading to (a) alternate stacking and (b) segregated stacking.

Figure 9 shows schematically a donor-acceptor overlap for two situations. In 9(a), the symmetry of the donor's HOMO matches that of the acceptor's LUMO in that both possess a mirror plane perpendicular to the molecular axis. Net attractive HOMO/LUMO overlap occurs essentially in the entire region between the two molecular planes, maximizing the charge-transfer stabilization, and leading to mixed stacking. In 9(b), on the other hand, the symmetries (with respect to an imaginary mirror plane perpendicular to the molecular planes) are different. One has mirror symmetry while one does not. Such being the case, the wave functions are in phase in only half the region between the molecular pairs and are out of phase in the other half. Stated differently, any net stabilizing overlap is compensated by a destabilizing overlap on the other side of the imaginary mirror plane. The net orbital overlap is zero for an eclipsed molecular overlap. In such a case, two events may occur. By sliding and rotating relative to each other, the donor/acceptor pair can possibly obtain a non-eclipsed configuration in which there is significant net positive overlap and, because of this, form mixed stacks. Or, if this is not possible, segregated stacks may result.

Concrete examples can be obtained by examining two ionic salts of TCNQ, that with the donor TTF and that with the donor N,N,N',N'- tetramethyl-*p*-phenylenediamine, TMPD. The wave functions of the frontier orbitals of these molecules, along with that of the tetrahalo-*p*-benzoquinones, are shown schematically in Fig. 10, where black and white circles denote opposite phases of the wave function. As can be seen, the HOMO of TTF possesses a mirror plane (with regard to the phase of the wave function) perpendicular to the molecular plane whereas

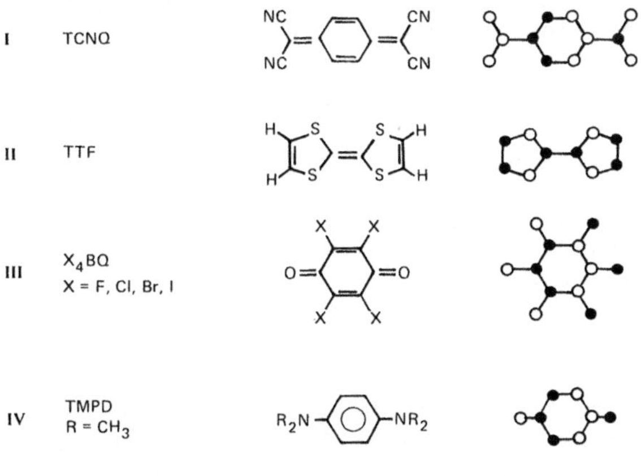

Figure 10. Schematic illustrations of the wave functions of TTF, TMPD, TCNQ and the tetrahalo-*p*-benzoquinones. Black and white circles denote opposite phases of the wave function.

the HOMO of TMPD and the LUMO of TCNQ do not possess this symmetry element. Because the symmetries of the frontier orbitals of TMPD and TCNQ are similar, that molecular pair should have significant net positive overlap and should form ionic mixed stacks. This is what is found (28). Because of the symmetry mismatch of their frontier orbitals, orbital overlap between TTF and TCNQ should be inhibited and, according to the guideline, one expects the possibility of segregated stacks. This, too, is what is found. Symmetry mismatch does not, however, guarantee the formation of segregated stacks. Thus, TMTSF, which forms a 1:1 complex with TCNQ that has segregated stacks, also forms a 1:1 mixed stacked complex with TCNQ in which the overlap between donor and acceptor is essentially eclipsed (29). As will be discussed below, TTF and the tetrahalobenzoquinones also yield 1:1 phases of both mixed and segregated stacks.

In describing this guideline, we have not stated whether the effect is a thermodynamic one, related to the overall energy differences between mixed and segregated phases, or a kinetic one, in which the growth of one phase is preferred over the growth of another. In fact, because of the small contribution of the Mulliken charge-transfer energy to the total energy of an ionic crystal, it seems unreasonable to expect it to be the determining factor insofar as which phase is preferred. This leads one to suspect that, if orbital symmetry does play a role in phase selection, it does so by favoring the growth of one phase over another. The effect is a kinetic one. Briefly, the argument would be the following. When a donor-acceptor pair form in solution, the energy that binds them together is the Mulliken charge-transfer interaction as well as the electrostatic energy of the possibly ionized pair. If the orbital symmetry permits significant overlap, the pair stays together and becomes a nucleating site for a mixed stack crystalline phase. If, however, the orbital symmetry permits little overlap, the donor-acceptor pair flies apart. Hence, the formation of acceptor-acceptor and donor-donor pairs becomes competitive with donor-acceptor pairs, creating the possibility of both segregated and alternate stacking phases, the formation of one versus the other perhaps now depending on the growth conditions.

As we have seen, two conditions must be fulfilled to achieve high conductivity in organic charge-transfer salts. First, the constituent molecules must crystallize in segregated stacks of donors and acceptors and, secondly, the degree of charge-transfer must be less than unity. As discussed above, comparison of the redox potentials of the donor and acceptors (8) as well as consideration of the symmetry properties of their frontier orbitals (11,27,30) are useful in selecting pairs of molecules that will yield highly conducting complexes. In our search for new classes of highly conducting materials, we have employed these considerations and have selected (11) the TTF family of donors and the tetrahalo-*p*-benzoquinone family of acceptors as being two types of materials whose electrochemical and orbital symmetry properties are such that the likelihood of their charge-transfer complexes fulfilling the two above-mentioned criteria are significant.

The LUMO of the tetrahalo-*p*-benzoquinones (X_4BQ) is shown schematically in Fig. 10. The point to note is that the LUMO of X_4BQ, like that of TCNQ, is not symmetric with respect to a mirror plane perpendicular to the long molecular axis. One would expect, therefore, a significant probability that the charge-transfer complexes of TTF and X_4BQ would crystallize in segregated stacks, assuming that the complexes were ionic. The expectation of an ionic ground state is quite reasonable, since the tetrahalo-*p*-benzoquinones are relatively strong electron acceptors, although not quite as strong as TCNQ.

In practice, TTF and its tetramethyl derivative (TMTTF) combine with the tetrahalo-*p*-benzoquinones to give rise to a complex systems of materials (11,27,30). Nearly every combination of donor and acceptor yields at least two distinct phases of 1:1 stoichiometry. Although we have not fully characterized each of these compounds, it has become apparent that this is an extremely interesting class of materials, among its members being the first examples (11) of highly conducting organic charge-transfer salts that do not contain TCNQ or one of its derivatives as the acceptor. The values of the pellet conductivites of the most highly conducting phases of several of these salts are listed in Table I. Of the TTF salts, only that of fluoranil exhibits high conductivity. All of the TMTTF-tetrahalo-*p*-benzoquinones, however, have pellet conductivities only slightly less than that of the TCNQ salt.

Table I. Values of the pressed powder conductivities at 300°K for the TTF and Tetramethyl-TTF salts of the Tetrahalo-*p*-benzoquinones. For comparison, those of TCNQ are included.

σ(300°K)	X=F	X=Cl	X=Br	TCNQ
TTF	10	8×10^{-4}	6×10^{-4}	70
TMTTF	7	20	1	35

The member of this family of complexes that we have been able to best characterize is TMTTF-bromanil, a structural projection of which is shown in Fig. 11. The complex crystallizes in the triclinic space group $P\bar{1}$ with unit cell parameters a=15.754, b=17.520, c=3.939, α=93.64, β=82.84, γ=70.54 and Z=2. As we had anticipated, the donors and acceptors crystallize in separate (segregated) stacks.

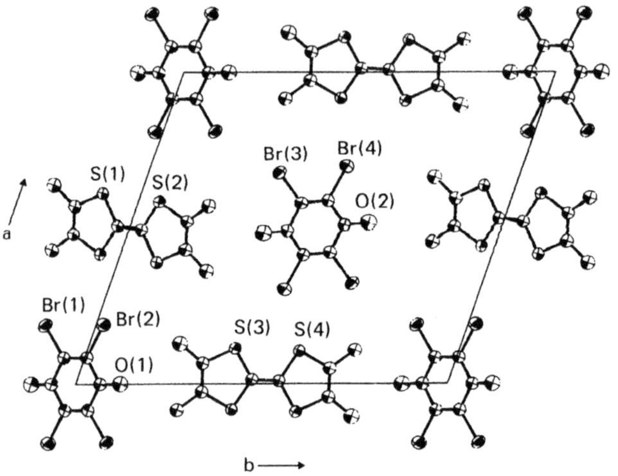

Figure 11. The (001) projection of TMTTF-bromanil, showing the relationships between the segregated stacks of TMTTF and bromanil (11).

Within these stacks, the molecules are uniformly spaced, as shown by the short c crystallographic axis. The structure bears a striking resemblance to that of TMTTF-TCNQ (31), considering that the acceptors are different molecules. The overlap of the TMTTF molecules is virtually identical in the two structures. So, too, is the interplanar spacing of ~3.54Å between the donor molecules. The overlap between the bromanils is analogous to that between the TCNQ molecules in TMTTF-TCNQ in that the exocyclic C=O bond of one bromanil lies over the quinoid ring of the bromanil molecule adjacent to it along the stack. This ring-double bond overlap, shown in Fig. 6 for TCNQ, is common to all of the highly conducting TCNQ salts although, in those materials, the exocyclic double bond is a C=C rather than a C=O bond. A significant difference between TMTTF-TCNQ and TMTTF-bromanil, however, is that in the former the perpendicular distance between the molecular planes of the acceptors is 3.29Å whereas in the latter the corresponding distance is 3.39Å.

Although the structures of TMTTF-bromanil and TMTTF-TCNQ are quite similar, their electrical properties differ significantly. The single crystal conductivity of the bromanil salt at 300°K is $\sim 5\Omega^{-1}\cdot cm^{-1}$, with an activation energy of 0.11 eV (11), whereas the room temperature conductivity of TMTTF-TCNQ is $\sim 350\Omega^{-1}\cdot cm^{-1}$ and shows metallic behavior. The relatively large interplanar acceptor-acceptor spacing in TMTTF-bromanil, presumably caused by the bulk of the bromine atoms, may be partially responsible for the lower conductivity. Orbital overlap, hence band formation, must be diminished relative to the more closely spaced TCNQ molecules in TMTTF-TCNQ. For this reason, the attainment of single crystals of the fluoranil salts seems a noteworthy goal since the lesser bulk of the fluorine atoms should allow closer contact of the acceptor molecules. This problem of poor overlap is not limited to the organic solid state. The relatively poor conductivity and semiconducting behavior of those partially oxidized platinum chain complexes containing bulky ligands such as dioximes is probably due to the poor overlap of the Pt d_z^2 orbitals, the result of the large intermolecular spacing. In contrast, the Pt-Pt spacing in Krogmann's salt, which contains the relatively small cyanide ligand, is much less (2.85Å) and the conductivity at room temperature is $750\text{-}1000\Omega^{-1}\cdot cm^{-1}$ (4).

Qualitative evidence for mixed-valence in TMTTF-bromanil can be obtained from its optical spectrum. In addition to the molecular transitions of the constituent molecules, there is a broad, intense infrared band centered near 4000 cm^{-1} that we (11) ascribe to a mixed-valence charge-transfer transition characteristic of salts such as TTF-TCNQ, which are known to be of non-integral valence. In addition, Pecile and co-workers have recently determined the degree of charge transfer, ρ, in TMTTF-bromanil to be ~0.60 (32). Thus these materials, which represent a new class of organic materials that exhibit mixed valence in the solid state, demonstrate that TCNQ salts are not unique in this respect.

SUMMARY

In this paper, we have briefly reviewed mixed valence in the organic solid state, limiting the discussion to three prototypical classes of such materials in an attempt to show that simple considerations allow one to understand on a crude level a great deal about the phenomenon. These considerations, exemplified in the TTF halides and the 1:1 TCNQ salts, allow one to make reasonable guesses as to what other materials might demonstrate mixed valence. Indeed, these considerations led us to choose TTF and the tetrahalo-*p*-benzoquinones as classes of donors and acceptors whose electrochemical and orbital symmetry properties were such that their charge-transfer complexes would be likely to exhibit mixed valence. Such has been the case. The TTF-tetrahalo-*p*-benzoquinones have been found (11) to yield materials that are mixed valence and are the first examples of highly conducting 1:1 charge-transfer salts that do not contain TCNQ or one of its derivatives as the acceptor molecule.

ACKNOWLEDGMENT

I thank J. B. Torrance for a critical reading of the manuscript as well as for illuminating discussions of the physics of the organic solid state.

REFERENCES

1. D. O. Cowan, C. LeVanda, J. Park and F. Kaufman, Accts. Chem. Res. 6, 1 (1973).
2. B. A. Averill, T. Herskovitz, R. H. Holm and J. A. Ibers, J. Amer. Chem. Soc. 95 3523 (1973).
3. J. J. Mayerle, S. E. Denmark, B. V. DePamphilis, J. A. Ibers and R. H. Holm, J. Amer. Chem. Soc. 97, 1032 (1975).
4. J. M. Williams and A. J. Schultz, Molecular Metals, Nato Conf. Ser. VI, Vol. I, W. E. Hatfield, ed., Plenum, New York, 1978, p. 337 and references therein.
5. M. B. Robin and P. Day, Advan. Inorg. Chem. Radiochem. 10, 247 (1967).
6. J. Ferraris, D. O. Cowan, V. V. Walatka, Jr. and J. H. Perlstein, J. Amer. Chem. Soc. 95, 948 (1973); L. B. Coleman, M. J. Cohen, D. J. Sandman, F. G. Yamagishi, A. F. Garito and A. J. Heeger, Solid State Commun. 12, 1125 (1973).
7. J. B. Torrance, B. A. Scott and F. B. Kaufman, Solid State Commun. 17, 1369 (1975).
8. J. B. Torrance, Accts. Chem. Res. 12, 79 (1979).
9. B. A. Scott, S. J. LaPlaca, J. B. Torrance, B. D. Silverman and B. Welber, J. Amer. Chem. Soc. 99, 6631 (1977).
10. J. B. Torrance and B. D. Silverman, Phys. Rev. B15, 788 (1977).
11. J. B. Torrance, J. J. Mayerle, V. Y. Lee and K. Bechgaard, J. Amer. Chem. Soc. 101, 4747 (1979).
12. C. K. Johnson and C. R. Watson, J. Chem. Phys. 64, 2271 (1976).
13. S. J. LaPlaca, P. W. R. Corfield, R. Thomas and B. A. Scott, Solid State Commun. 17, 635 (1975).
14. F. Wudl, J. Amer. Chem. Soc. 97, 1962 (1975).
15. R. P. Messmer and D. R. Salahub, Phys. Rev. Letts. 35, 533 (1975).

16. J. L. de Boer, A. Vos and K. Huml, Acta Crystallogr. B24, 542 (1968).
17. J. B. Torrance, B. A. Scott, B. Welber, F. B. Kaufman and P. E. Seiden, Phys. Rev. B19, 730 (1979).
18. R. L. Greene, J. J. Mayerle, R. Schumaker, G. Castro, P. M. Chaikin, S. Etemad and S. J. LaPlaca, Solid State Commun. 20, 943 (1976).
19. B. D. Silverman, J. Chem. Phys. 70, 1614 (1979).
20. R. M. Metzger and A. N. Bloch, J. Chem. Phys. 63, 5098 (1975); A. J. Epstein, N. O. Lipari, D. J. Sandman and P. Nielson, Phys. Rev. B13, 1569 (1976); V. E. Klymenko, V. Ya Krivnov, A. A. Duchinnikov, I. I. Ukrainsky and A. F. Shvets, Sov. Phys. - JETP, 42, 123 (1976).
21. R. Comes, Chemistry and Physics of One-Dimensional Metals, NATO Adv. Study Inst. Series: Series B, Physics, Vol. 25, H. J. Keller, ed., Plenum, New York, 1977, pp. 315-339 and references therein.
22. P. Coppens, Phys. Rev. Letts. 35, 98 (1975).
23. M. A. Butler, F. Wudl and Z. G. Soos, Phys. Rev. B12, 4708 (1975).
24. J. P. Pouget, S. Megtert and R. Comes, Lecture Notes in Physics, Quasi One-Dimensional Conductors I, S. Barisic, A. Bjelis, J. R. Cooper and B. Leontic, eds., Springer-Verlag, Berlin (1979), pp. 14-27.
25. J. B. Torrance, J. J. Mayerle, K. Bechgaard, B. D. Silverman and Y. Tomkiewicz, to be submitted for publication.
26. M. E. Hawley, T. O. Poehler, T. F. Carruthers, A. N. Bloch, D. O. Cowan and T. J. Kistenmacher, Bull. Am. Phys. Soc. 23, 424 (1978).
27. J. J. Mayerle, J. B. Torrance and J. I. Crowley, Acta. Crystallogr. in press.
28. A. W. Hanson, Acta. Crystallogr. 19, 610 (1965).
29. T. J. Kistenmacher, private communication.
30. J. B. Torrance and J. J. Mayerle, unpublished results.
31. T. E. Phillips, T. J. Kistenmacher, A. N. Bloch, J. P. Ferraris and D. O. Cowan, Acta. Crystallogr. B33, 422 (1977).
32. R. Bozio, C. Pecile, J. B. Torrance and J. J. Mayerle, work in progress.

PHOTOELECTRON SPECTROSCOPY OF MIXED-VALENCE COMPOUNDS

P.A. Cox, R.G. Egdell and A.F. Orchard

Inorganic Chemistry Laboratory,
South Parks Road, Oxford (U.K.)

The applications of X-ray photoelectron spectroscopy in the investigation of the electronic structures of mixed-valence compounds are briefly surveyed. The experimental and theoretical problems that arise are discussed.

INTRODUCTION

Photoelectron spectroscopy (PES) is yet another technique that has been used to study mixed-valence materials and, like any other spectroscopic method, it has its inherent advantages and disadvantages when applied to a particular problem. The technique is by now familiar to most physical scientists, so a brief description of the technique should suffice as an introduction (1).

PES involves subjecting the sample to (essentially) monochromatic radiation of sufficient photon energy $h\nu$ to cause ionization, and measuring the kinetic energy distribution of the emitted photoelectrons. In general, the photon energy $h\nu$ will be large enough to ionize more than one species of electron, each separate ionization event being subject to the Einstein relation,

$$E_k = h\nu - I_k \qquad (1)$$

where E_k and I_k are the kinetic energy (KE) and ionization energy (IE), respectively, of electron species $\underline{k}$. Loosely bound electrons are ejected with a high KE, and more tightly bound electrons with a lower KE. If there are, say, three distinct species of

D.B. Brown (ed.), Mixed-Valence Compounds, 475-502.

electron (that is, three occupied energy levels or bands well separated from each other) accessible at the incident photon energy selected, then the photoelectron (PE) spectrum will consist of three separate bands (fig.1). The structure of PE spectra is of course direct evidence for the discrete nature of electronic energy levels, and at the simplest level of approximation the ionization energies I_K are to be identified with minus the SCF orbital energies ε_k. This is known as Koopman's rule (2): it is based on the naive assumption that there is no relaxation of orbital wave functions in the ionised state (the 'frozen orbital approximation'), together with the neglect of electron correlation effects. However, PE spectra should strictly be interpreted in terms of transitions to states of the <u>ionised</u> system, rather than in terms of the orbital electronic structure of the neutral system.[3]

Fig. 2 shows the essential components of a PE spectrometer. Photoionization takes place in a central UHV chamber, the photoelectrons emitted by the sample being analysed in an adjacent chamber containing an electron dispersion system, commonly an electrostatic deflection device. The PE spectrum is normally recorded in 'differential' form - that is as a plot of $^{di}/_{dE}$ vs. electron kinetic energy E, where <u>i</u> is the photocurrent through the analyser. This yields directly the energy distribution n(E) (fig.1), modified only by commonly negligible analyser transmission effects. It is of course essential to contrive a high vacuum pathway for the photoelectrons because they readily suffer inelastic scattering by gas molecules.

A variety of photon sources is available. Most commonly employed are line sources - for example, hard u.v. lines such as He I ($h\nu$ = 21.2 eV) from gas discharges or soft X-rays such as MgKα (1253.6 eV) and Al Kα (1486.6 eV) emitted by electron-bombarded metal targets. Alternatively, where the facility presents itself, good use can be made of synchroton radiation which, with sufficiently flexible monochromators, provides a continuously tunable flux of ionising photons ($h\nu$ = 10 - 250 eV, say). Now most applications of PES to mixed valence compounds have concerned <u>core</u>, rather than valence electrons, which necessitates the use of X-rays ($h\nu \gtrsim$ 100 eV, say). Accordingly we restrict our considerations to this aspect of PES.

The core electron ionization energies of a given atom are approximately constant, within narrow limits, and characteristic of that atom - indeed, sufficiently characteristic that there is usually no problem in identifying the various signals in a complex X-PE spectrum. Core level PES (X-PES) therefore furnishes us with an analytical method; and hence the familiar acronym ESCA (Electron Spectroscopy for Chemical Analysis) (4,5). At the same time, however, the core electron binding energies of an atom

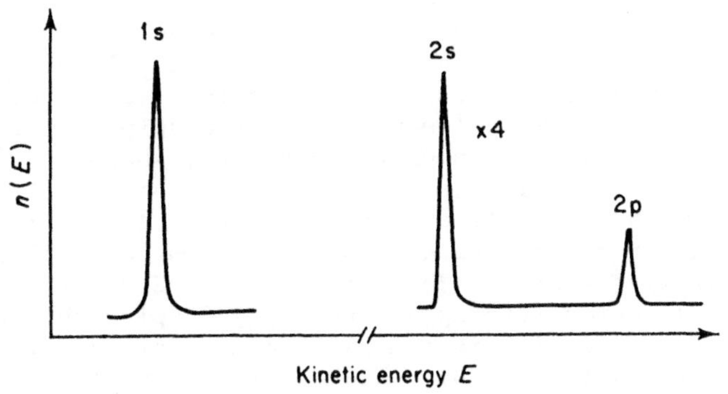

Fig. 1 PE spectrum of neon ($1s^2 2s^2 2p^6$) excited by X-rays

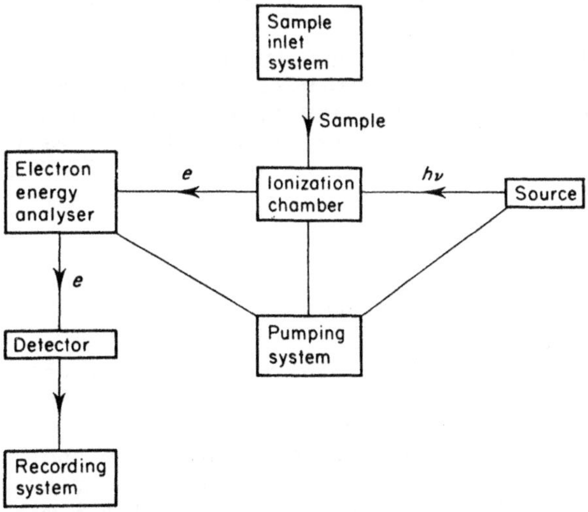

Fig. 2 Block diagram of a photoelectron spectrometer

usually vary somewhat (and measurably) with change of chemical environment - that is, with change in the electrostatic potential experienced by the core electrons. This chemical shift effect was first observed in the thiosulphate anion $S_2O_3^{2-}$, the X-PE spectrum of which shows a 'doubling' of the sulphur core signals (4). Thus, the sulphur 2p region of the spectrum, shown in fig. 3, exhibits two well-resolved signals with a separation of 5.9 eV, as compared with peak half-widths of 2.1 eV. In contrast, the X-PE spectrum of the sulphate anion has a single sulphur 2p signal of similar width. (In both cases the S 2p signals show incipient resolution into $2p_{1/2}$ and $2p_{3/2}$ spin-orbit components; such multiplet splitting is a normal feature of PE signals from core subshells having $\ell > 0$). Elementary arguments lead us to assign the higher binding energy peak of the thiosulphate anion to the central sulphur atom, and the other to the terminal sulphur. The equal intensity of the two signals reflects in a simple way the stoichiometry of the anion. Fig. 4 provides a similar example, the nitrogen 1s signals of the azide ion N_3^-. Observe that the relative intensities of the two component signals (2:1) again faithfully reflect the composition of the anion. This direct correspondence between signal intensity and the stoichiometric proportion of the atom in question is usually maintained in such cases.

The species considered above are of course technically mixed-valency compounds (but of non-metallic elements) and, historically, the X-PES results on such systems held out great promise for the application of the technique in the field of mixed-valence chemistry of the metallic elements. This expectation has been partly borne out by events but, for reasons that should become clear below, only in part. It is as well, before we launch into case studies, to recognise some general limitations of PES and certain special difficulties that may be encountered in the study of mixed-valence compounds.

LIMITATIONS OF THE TECHNIQUE

1. Surface sensitivity. The mixed-valence compounds we are concerned with are solids, and PES is a highly surface-sensitive technique. Whereas the exciting photons penetrate thousands of Ångstroms into the bulk of the sample, the short attenuation length of an electron means that photoelectrons can in the main escape from only a thin surface layer of the material, perhaps 20 Å thick or even less. It is essential therefore that a 'clean' sample surface is presented to the incident photons. Unfortunately however, this is often difficult to ensure (for a variety of possible reasons). A common problem is that the sample has an 'impurity' surface layer of different composition from the bulk; and that this cannot be removed without the risk of chemical damage to the underlying material.

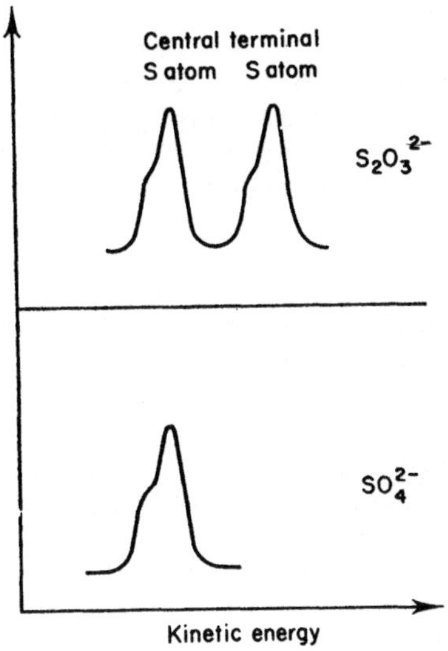

Fig. 3 Sulphur 2p signals in the X-PE spectra of $Na_2S_2O_3$ and Na_2SO_4.

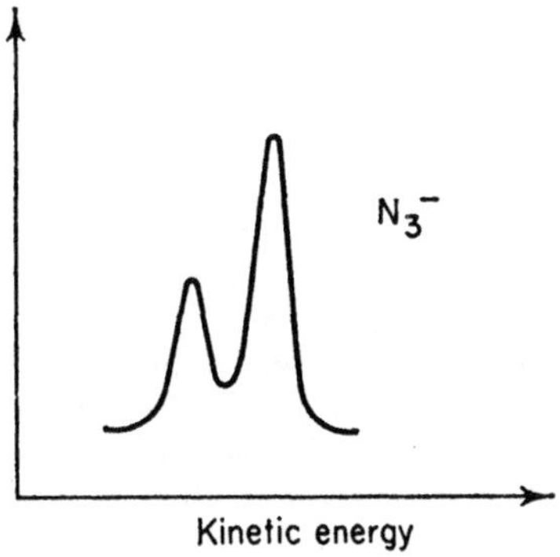

Fig. 4 Nitrogen 1s signals in the X-PE spectrum of NaN_3.

2. In situ sample decomposition. Some materials cannot withstand UHV conditions, because of the out-gassing of volatile components. A simple case in point is provided by the Tl(I)/Tl(III) chlorides, $Tl(TlCl_4)$ and $Tl_3(TlCl_6)$, which lose chlorine even in moderate vacuum. Another is 'KCP', $K_2Pt(CN)_4Br_{0.30}.3H_2O$, which out-gasses H_2O, with consequent collapse of the crystal lattice (6).

A further problem is that many compounds are susceptible to 'radiation damage' on exposure to the exciting X-rays (7). (Actually the destructive agency is almost certainly the secondary electron flux present). We illustrate this point later.

3. Limitations on resolution in X-PES. The resolution achievable in X-PES is severely limited by two factors: (a) X-ray lines have quite large inherent half widths, typically about 1 eV; (b) X-ray induced photoelectrons have high kinetic energies and, the analyser having a fixed resolving power $R = E/\Delta E$ there is a substantial instrumental contribution to the line width of the signal. Various effects peculiar to the solid state may make an additional contribution to PE line widths. The upshot is that core PE signals are quite broad, with half widths of 1.5 eV or greater. The situation can be improved by monochromatisation of the X-rays (e.g. to a band-pass of 0.3 eV), but even with this expedient the sharpest X-PE line thus far observed from a solid is of 0.4 eV half-width.

All this matters because the chemical shift effect is usually less pronounced for metallic elements than for non-metals. Thus, while the sulphur 2p ionization energy can vary by as much as 10 eV with chemical environment (the range for SH_2 - SF_6), the variation encountered for metallic elements is often as little as 1 eV, and rarely more than 3 eV.

4. Theoretical ambiguities. The application of X-PES to mixed-valence compounds may also be confounded by certain theoretical problems which can occasion ambiguity in the interpretation of data. Among these is the not always predictable effect of site potential on an observed core electron binding energy. At the simplest level of approximation the magnitude of the chemical shift in a mixed-valence compound can be expressed as (5)

$$\Delta I = I_A - I_B = (\varepsilon_B - \varepsilon_A) + e(V_B - V_A) \qquad (2)$$

where the subscripts A and B distinguish the non-equivalent atoms. The quantities V_A and V_B are the electrostatic potentials experienced by the atoms at their respective sites, while ε_A and ε_B are the core electron orbital energies in the atomic valence state. It is important to appreciate that the 'non-local' term, (V_B-V_A), is an important ingredient of the chemical shift

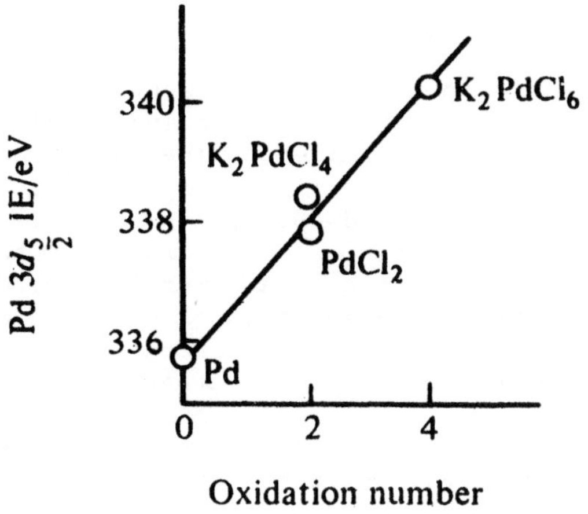

Fig. 5 The variation of the Pd $3d_{5/2}$ binding energy with oxidation state in some palladium compounds.

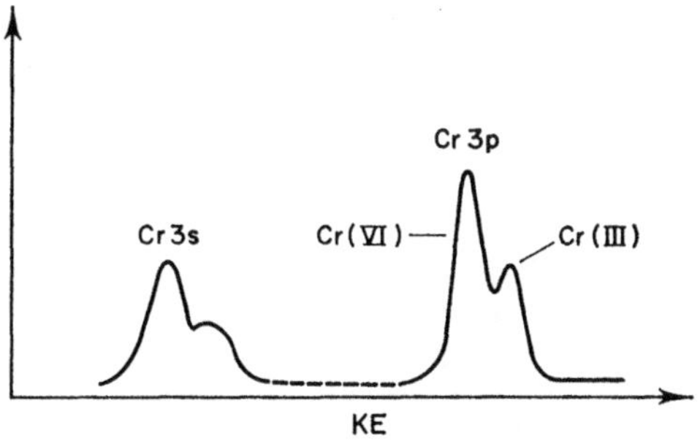

Fig. 6 Chromium 3p and 3s structure in the X-PE spectrum of KCr_3O_8(low resolution).

effect; and, unfortunately, it is very difficult to calculate reliably.

When different compounds of an element are compared, there may be additionally a differential charging problem to contend with: an insulating solid sample acquires a surface charge during the PES measurements, and this might vary significantly from one compound to another. This and the site potential effect may combine to yield unexpected results: for example, PbO and PbO_2 are reported to have essentially the same metal 4f binding energies (8). (Or was the PbO_2 sample largely decomposed to PbO at the surface ?) Certainly, the once compelling correlations of the kind shown in Figure 5 proved to be the exception rather than the rule.

A further potential source of confusion is the fact that some elements (in a single oxidation state) show core PE signals with multiplet structure beyond that explicable on the basis of spin-orbit coupling. We will return to this problem later.

But let us now turn to consider some specific examples where PES has yielded chemically interesting results. It is convenient to deal first with classes I and II of the Robin-Day scheme (9). (To remind the reader, these are mixed-valence compounds in which the two oxidation states are represented by crystallographically distinct metal atoms. Class II compounds are distinguished from those of class I by showing, spectroscopically or otherwise, evidence of a 'significant' interaction between the two sites).

COMPOUNDS OF CLASSES I & II

We begin with the simple example of KCr_3O_8, a Cr(III,VI) compound containing Cr^{3+} and CrO_4^{2-} ions (class I). Even at quite low resolution, the X-PE spectrum shows two sets of chromium core signals (Fig. 6), these having an intensity ratio (2:1) consistent with the known stoichiometry.[10] This is a particularly favourable case, however, since both oxidation states are quite stable and the difference core PE signals well apart. The limited resolution of X-PES frequently precludes the observation of well separated PE signals from the different oxidation states. Thus the U(IV) and U(VI) core PE signals are incompletely resolved in the spectrum of U_3O_8 (another class I compound) excited by achromatic X-rays (Fig. 7);(11) and the mixed-valence character is not at all evident unless great care is taken in sample preparation (12). Similar remarks can be made concerning the class II compounds Cs_2SbCl_6 (Fig.10) below), containing Sb(III) and Sb(V) in equal proportion (13).

The metal 4f spectra of two familiar class II materials are reproduced in Fig. 8. In both cases the complex form of the 4f signals can be understood in terms of the superposition of two

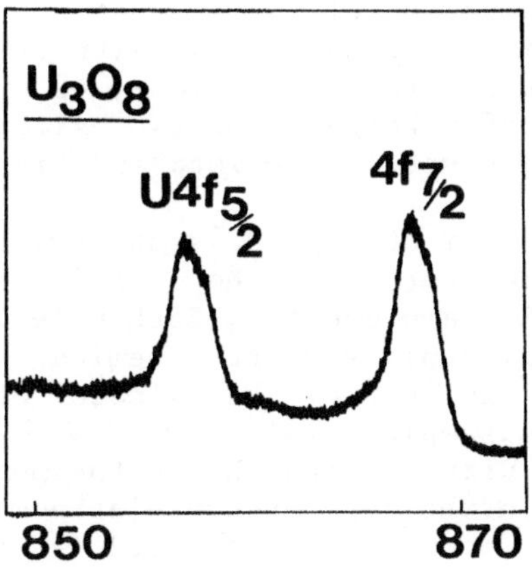

Fig. 7 Uranium 4f structure in the X-PE spectrum of U_3O_8.

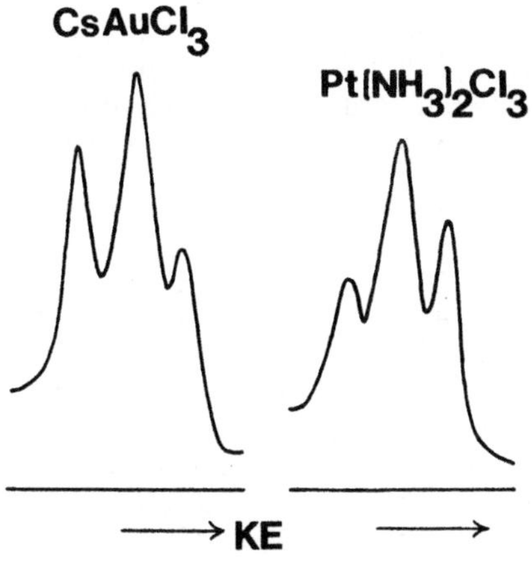

Fig. 8 The 4f signals of gold and platinum in the X-PE spectra of $CsAuCl_3$ and $Pt(NH_3)_2Cl_3$.

$4f_{5/2,7/2}$ doublets, their separation being comparable to the spin-orbit splitting. The potential interest of PES measurements on such compounds lies in the possibility, with the necessary supporting electronic structure calculations, of quantifying the degree of charge-transfer between the two oxidation states. However, there has been as yet little work in this direction.

The Pt(II,IV) system of Fig. 8 is otherwise of interest as furnishing a good illustration of the occasionally destructive consequences of X-PES measurements. Both Pt(en)Cl_3 and the Au(I, III) compound $CsAuCl_3$, rapidly suffer chemical reduction by the secondary electrons issuing from the 'window' of the X-ray source or created within the sample itself. (7). (Indeed both spectra in Fig. 8 were recorded with indecent haste the moment the X-rays had been switched on). In the case of the platinum system, a mixed-valence compound can apparently be created (at the surface) starting with a sample of the Pt(IV) compound, Pt(en)Cl_4 (Fig. 9). The reduced surface phase obtained after some 18 hours exposure appears to be a Pt(II) compound. Similar observations have been made on the Sb(III,V) compound Cs_2SbCl_6 (Fig. 10), and its Sb(V) analogue $CsSbCl_6$.[13a] There one sees also the chemical damage, again reduction, that can be caused by ion bombardment, once regarded as a viable means of 'cleaning' sample surfaces.

It is worth noting in passing that X-PES is singularly well adapted to the investigation of mixed-valence <u>surface</u> phases and to the demonstration of mixed-valence character in thin films. The technique has been applied to, for example, u.v.-coloured (blue) WO_3 films, with the conclusion that this reduced state is probably a disordered hydrogen tungsten bronze, H_xWO_3 (14). (cf. work on the sodium bronzes, Na_xWO_3, referred to below).

As foreshadowed earlier, mixed-valence materials involving a comparatively unstable higher oxidation state may be seriously depleted of that oxidation state at the surface. Thus Sb(III,V) bromides such as $(NH_4)_2SbBr_6$ appear in X-PES measurements to consist only of Sb(III) (6). A particularly enigmatic example is the Pb(II,IV) oxide, Pb_3O_4 (class I). The Pb(II) and Pb(IV) sites are utterly different crystallographically, yet the compound shows just one set of Pb 4f signals, these having a comparable linewidth to that observed in PbO and PbO_2 (8). These observations remain a mystery at the present time, though there have been tentative explanations (8). Perhaps both Pb_3O_4 and PbO_2 have surface phases of essentially the composition PbO. That is conjecture, which could be evaluated by SIMS (Secondary Ion Mass Spectroscopy) or LEED studies on single crystal samples. A similar problem arises with Sb_2O_4 and $BaBiO_3$, both class I (III,V) systems. But in these latter cases there is tangible evidence that the surface phases predominantly involve the III oxidation state (15).

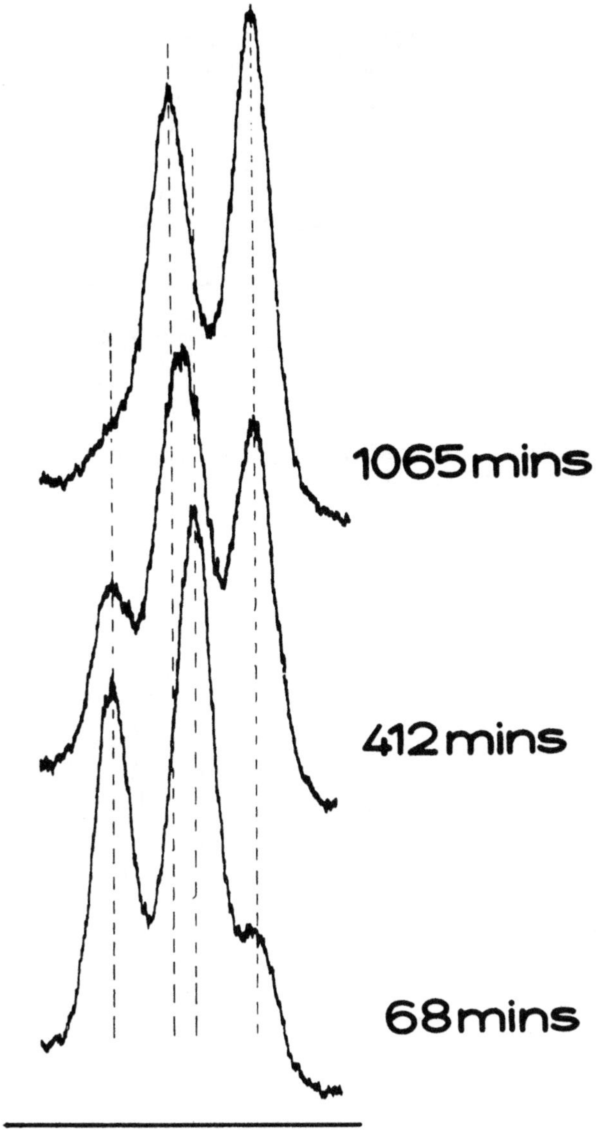

Fig. 9 The platinum 4f region of the X-PE spectrum of $Pt(en)Cl_4$ as a function of exposure to X-rays.

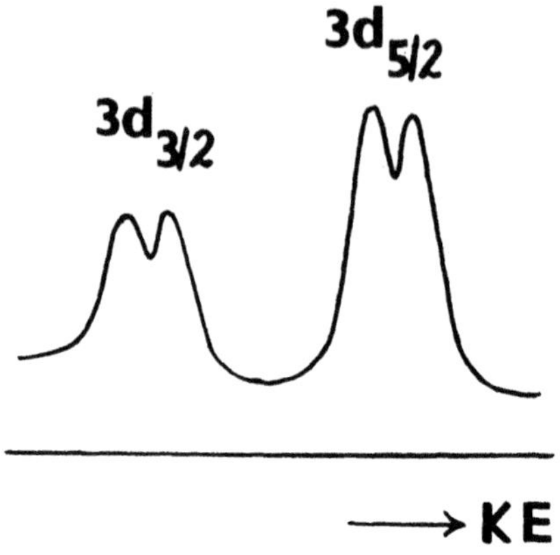

Fig. 10 The antimony 3d structure in the X-PE spectrum of Cs_2SbCl_6.

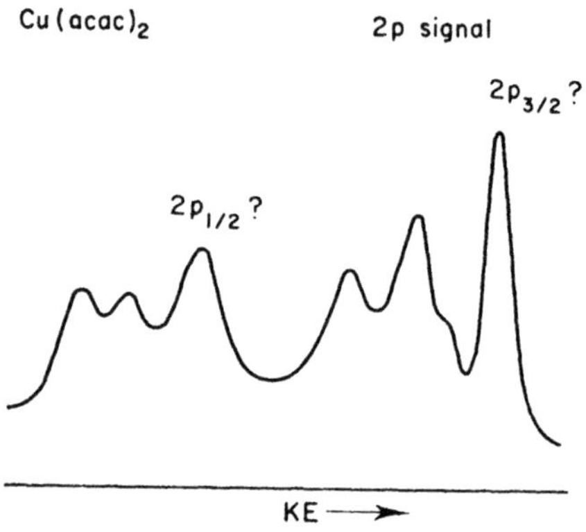

Fig. 11 The copper 2p region of the X-PE spectrum of copper bis-acetylacetonate.

The Pb_3O_4 case history is a disturbing one, and it is not unique. Consider, for example, the lithium/titanium spinel phase $Li_{1+x}Ti_{2-x}O_4$ $(0 < x < 1/3)$, a Ti(III,IV) system with a III/V ratio of $(1-3x)/(1+2x)$. When $x = 0.15$, the spinel is an insulator with Ti(III)/Ti(IV) $\approx 1/3$. After prolonged heat treatment in vacuo, He I spectra showing a weak Ti 3d band can be obtained, yet the Ti 2p signal in the X-PE spectrum is a simple doublet closely similar to that encountered in TiO_2 (16). The reason for this is unclear as yet.

SATELLITE STRUCTURE : A DIGRESSION

Thus we sometimes fail to see two PE signals when we had expected to; but what is perhaps more disconcerting, as regards the application of X-PES to mixed-valence compounds, is the not infrequent observation of multi-peak structure where we can be quite certain that only one oxidation state is present. Take Fig. 11, for example, which is quite typical of copper(II) compounds (17). Instead of a simple (and well-resolved) $2p_{3/2,1/2}$ spin-orbit doublet, the copper 2p spectrum shows a complex group of at least seven PE signals; in addition to the well-separated $2p_{3/2,1/2}$ doublet there are apparently several additional or 'satellite' lines. Fig. 12 gives two more illustrations, drawn from Lanthanoid chemistry (18). A superficial consideration of such spectra clearly might lead one to conclude that the materials contained more than one metal oxidation state. However, satellite lines in core PE spectra are not normally as intense in relation to the 'parent' line as is found in the above examples.

To understand the origins of these satellite lines, consider the PE spectrum of helium in Fig. 13 (19). In addition to the dominant, primary signal representing formation of the He^+ ion in its ground state (1s), one observes also weak signals corresponding clearly to processes such as

$$He(1s^2) + h\nu \longrightarrow \left\{ \begin{array}{l} He^+ (2s) \\ He^+ (2p) \end{array} \right. + e$$

- that is, to production of He^+ in its excited states. Such transitions, forbidden in the simple, one-electron picture of photoionization, may formally be described in terms of concerted events such as

$$He(1s^2) + h\nu \longrightarrow He^+ (1s) + e(E_1) \qquad (1)$$

$$He^+(1s) + e(E_1) \longrightarrow \left\{ \begin{array}{l} He^+ (2s) \\ He^+ (2p) \end{array} \right. + e(E_2) \qquad (2)$$

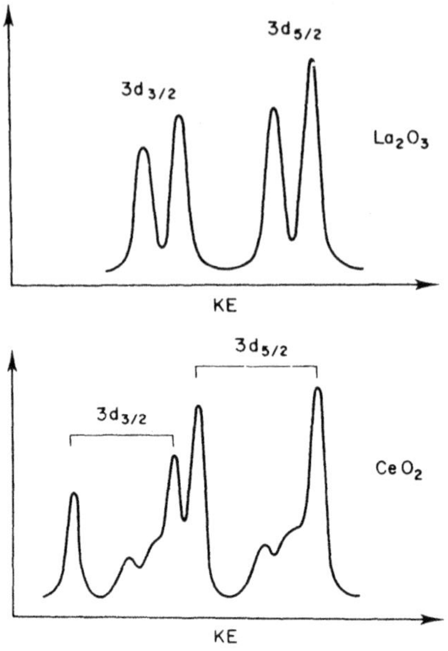

Fig. 12 The 3d signals in the X-PE spectra of La_2O_3 and CeO_2.

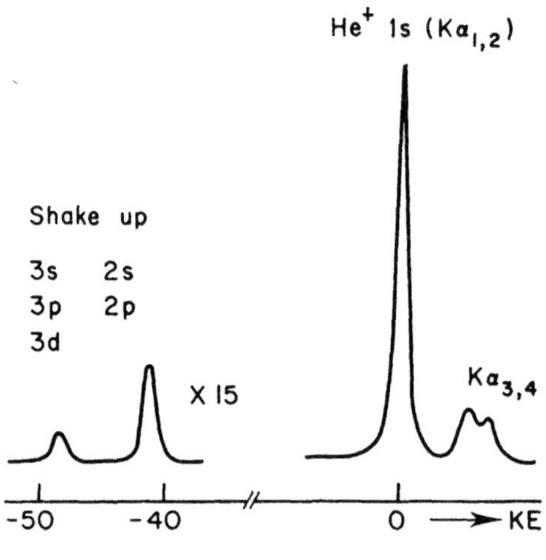

Fig. 13 The X-PE spectrum of helium gas showing shake up satellites.

where the second step represents the excitation 1s ⟶ 2s, 2p through an energy loss $E_1 - E_2$ on the part of the outgoing photoelectron. A coupled excitation of this type is frequently referred to as a shake-up process, (5) and the additional PE signal (occurring at a lower KE E_2) termed a shake-up satellite of the main core PE signal at KE = E_1. The satellite structure in Figs. 11 and 12 can be interpreted on a similar basis. In each case the satellites are thought to be due to shake-up processes involving O 2p ⟶ metal charge-transfer. One piece of evidence supporting this assignment is the fact that the 2p spectra of Cu(I) compounds, where the metal *d* subshell is full, are free of satellite lines (20).

One simple model for the production of satellite lines makes use of the so-called sudden approximation (5). Ejection of a core electron is considered to take place instantly, the other electrons being 'frozen' in their ground state distributions. Formally, the result of this sudden process can be described as $\hat{a}_i.\Psi_o(N)$, where $\hat{a}_i$ is the operator annihilating an electron in orbital *i* and $\Psi_o(N)$ is the N-electron, ground state wave function. This state - we might call it the Koopmans' state - is not a stationary state of the ionised, (N-1)-electron system. Rather it is a mixture of the real states of the ion, $\Psi_j(N-1)$, in which the orbitals have relaxed:

$$\hat{a}_i\Psi_o(N) = \sum_j c_{ij}\Psi_j(N-1)$$

The expansion coefficients c_{ij} are simply the overlap integrals

$$c_{ij} = \langle \Psi_j(N-1) | \hat{a}_i\Psi_o(N) \rangle$$

The probability that the ion is produced in state *j*, when the hole is in orbital *j* rather than *i*, is then given by

$$P_{ij} = |c_{ij}|^2$$

Clearly, in this model, the shake-up processes i ⟶ j are subject to a monopole selection rule: that is, the states *j* must be of the same symmetry as the Koopmans' state. However, other, non-monopole transitions become possible in more elaborate models which incorporate many-electron (correlation) effects.

CLASS III COMPOUNDS

It remains for us to discuss the PES of mixed-valence compounds belonging to class III of the Robin-Day scheme. In compounds of this category the metal atoms occupy exactly equivalent sites, each apparently representing a hybrid, non-integral oxidation state. Class III mixed-valence compounds are further distinguished according to whether they involve discrete polynuclear units (class III A) or have extended, 'infinite' crystal lattices (class III B); thus the compounds $[Nb_6Cl_{12}]Cl_n$, where n = 2,3 or 4, are examples of class III A, while the system $La_{1-x}Sr_xVO_3$ is representative of class III B. It is of course this aspect of mixed-valence chemistry, which spawns a great variety of materials having interesting electronic properties, in which there has been, and continues to be, the most intense activity.

Let us take class III A materials first.

CLASS III A SYSTEMS

Most applications of PES in this context have been directed towards answering the question: Are the electron densities at the metal sites in fact all equal, as their geometrical equivalence would imply? On the face of it, PES is singularly well adapted to dealing with this problem. If there is indeed charge equalization, then there should be just a single set of the characteristic metal core signals in the X-PE spectrum; if, on the other hand, there is an unequal distribution of electron density among the ostensibly equivalent metal atoms, then (resolution permitting) more than one set of metal core PE signals should be observed. However, it is not as simple as that.

We limit our discussion to a single case history, that of the Creutz-Taube complexes (21)

$$[(H_3N)_5Ru\text{-}pyr\text{-}Ru(NH_3)_5]^{n+} \qquad \underline{A}$$

where pyr = pyrazine and n = 4, 5 or 6, and analogues such as

$$[Cl(dpy)_2Ru\text{-}pyr\text{-}Ru(dpy)_2Cl]^{(n-2)+} \qquad \underline{B}$$

This will enable us to make most of the relevant points. The cases n = 4 and 6 are of course simply Ru(II) and Ru(III), respectively, while n = 5 represents a mixed-valence situation.

The X-PE spectra of these binuclear systems have been measured by Citrin (22); the structure observed in the ruthenium 3d region is reproduced in Fig. 14. Unfortunately, the Ru 3d signals (the most intense of the metal signals) are strongly overlapped by the intense, composite carbon 1s signal of the bipyridyl ligand (system B), the bridging pyrazine molecule and the p-toluenesulphonate counter anion. It was necessary therefore to 'deconvolute' the spectra (as indicated in the figure), frequently an unreliable procedure. As subjects for X-PES, the

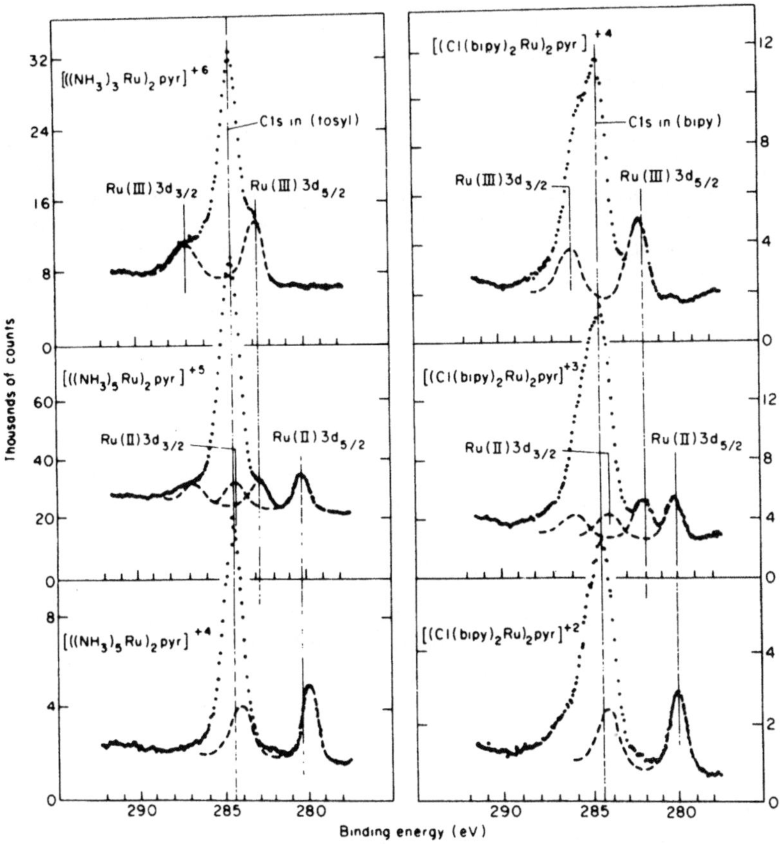

Fig. 14 The ruthenium 3d signals in the X-PE spectra of the Creutz-Taube complexes (A) and a related system (B) - see text.

materials are also less than ideal because of the risk of surface contamination, radiation damage and (in the case of system A) outgassing of ammonia in the high vacuum (5×10^{-9} torr in these experiments). Nevertheless, Citrin's results are apparently reproducible and meaningful, there being evident two sets of Ru 3d signals (each a spin-orbit doublet) in the spectra of the mixed-valence compounds (n = 5). It was concluded that these complexes are Ru(II,III) species with 'trapped valencies'; that is, they were assigned to class II, rather than class III A, of the Robin-Day scheme.

However, it has been pointed out by Hush (23) that this interpretation may not be correct: it is conceivable that two sets of Ru 3d signals can be observed even if there is perfect charge delocalization in the binuclear system. Let us recast Hush's arguments in terms of the sudden approximation (*vide supra*). We consider two, equivalent sites, labelled *a* and *b*, and denote the lower oxidation state by X, and the higher by Y ($= X^+$). Assuming complete charge equalization, the ground state of the system may be described by

$$|0\rangle = \frac{1}{\sqrt{2}}(|X_a, Y_b\rangle + |Y_a, X_b\rangle)$$

Now the ionization of a core electron - of atom *a*, say - may be imagined to produce suddenly the (non-stationary) state

$$|K_a = \frac{1}{\sqrt{2}}(|X_a^*, Y_b\rangle + |Y_a^*, X_b\rangle)$$

where * indicates the presence of a core hole. But, because of relaxation in response to the electrostatic field of the core hole at site *a*, the ionised system cannot sustain a symmetrical charge distribution. There will be, first, a state

$$|f_1\, a\rangle = N(|\bar{X}_a^*, Y_b\rangle + \lambda|\bar{Y}_a^*, X_b\rangle)$$

where $N = (1 + \lambda^2)^{-1/2}$. The notation $\bar{X}$, $\bar{Y}$ signifies rescaled atomic orbitals. Secondly, orbital relaxation combined with charge-transfer will yield a final state

$$|f_2\, a\rangle = N(\lambda|\bar{X}_a^*, Y_b\rangle - |\bar{Y}_a^*, X_b\rangle)$$

This latter state may be viewed as the result of charge-transfer, *b* → *a*, stimulated by the creation of a core hole in atom *a* (a shake-up process). The relative probabilities of reaching these two states of the ionised system are approximately

$$P_2/P_1 = (1 - \lambda)^2/(1 + \lambda)^2 \qquad (3)$$

If $\lambda = 1$, corresponding to negligible orbital relaxation, the probability P_2 of producing the charge transfer state is zero: but as λ decreases towards zero the transition probabilities P_2 and P_1 tend to become equal. Of course, in the limit $\lambda = 0$, the final states $|f_1\rangle$ and $|f_2\rangle$ are completely localised.

Thus, a genuine class III A mixed valence system may well have an X-PE spectrum with multi-peak metal core signals. Singlet core PE signals will be observed only when orbital relaxation in the ionised system is slight. It is unfortunate but, in general, PES cannot be straightforwardly applied to demonstrate the prescence of trapped oxidation states. If core PES is to contribute in this context, we need a thorough-going theory of photoionization in polynuclear systems, one incorporating relaxation effects at least, with which to analyse the spectroscopic data. It will almost certainly also be necessary to 'ring the chemical changes', exploring the effects of systematic chemical substitutions.

The model we have used above is perhaps not well suited to quantitative applications, but let us nonetheless explore its implications a little further, and estimate the energy separation of the two PE signals that could arise on core ionization in a class III A binuclear system. Neglecting orbital rescaling, the energies of the final states relative to the neutral system are

$$E_1 = N^2 \{-\varepsilon_c + \lambda^2(-\varepsilon_c + J) - \gamma(1-\lambda)^2 - 2\lambda\rho\}$$

and

$$E_2 = N^2 \{\lambda^2(-\varepsilon_c) + (-\varepsilon_c + j) - \gamma(1+\lambda)^2 + 2\lambda\rho\}$$

where:

ε_c is the orbital energy of the relevant core subshell in X (including extra-atomic potentials);

$J = \langle \phi_a c_a || \phi_a c_a \rangle - \langle \phi_b c_a || \phi_b c_a \rangle$, the difference between one centre and two centre coulomb integrals for the core and valence orbitals involved;

$$\gamma = \langle X_a Y_b | \hat{H} | Y_a X_b \rangle = \langle \phi_a | \hat{h}_{eff} | \phi_b \rangle + 2\langle \phi_a c_a || \phi_b c_a \rangle ;$$

and $\rho = \langle \phi_a c_a || \phi_b c_a \rangle$.

($\underline{c}$ denotes the core orbital, and ϕ the occupied valence orbital in X_2; $||$ indicates a matrix element of the electron repulsion $\sum e^2/4\pi\varepsilon_o r_{ij}$). The difference in energy between the two final states, $|f_1 a\rangle$ and $|f_2 a\rangle$, is therefore

$$\Delta E = E_2 - E_1 = (1 + \lambda^2)^{-1}\{ (1 - \lambda^2) J - 4 (\gamma - \rho) \} \quad (4)$$

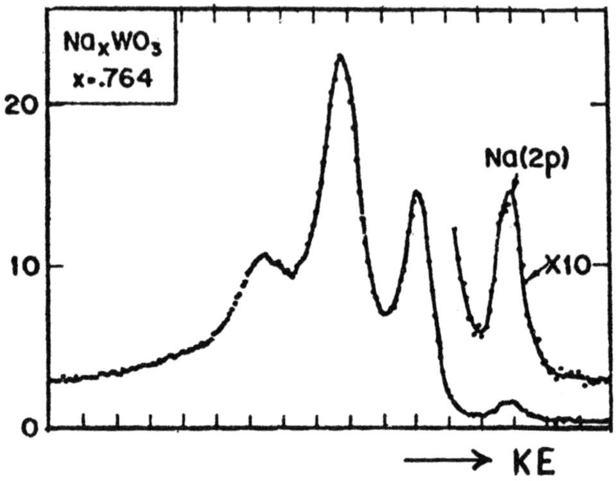

Fig. 15 The tungsten 4f region of the X-PE spectrum of the bronze $Na_{0.764}WO_3$.

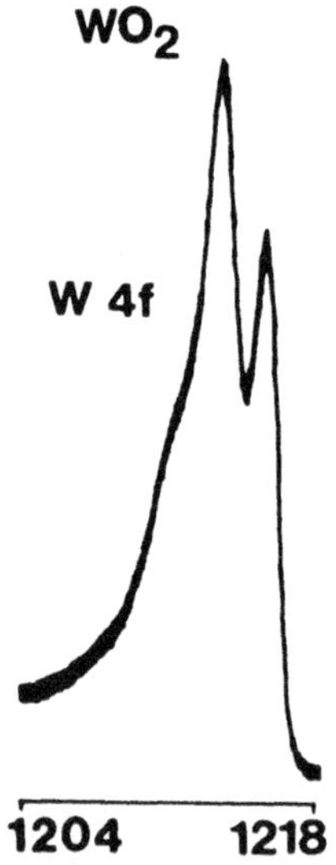

Fig. 17 Tungsten 4f structure in the X-PE spectrum of WO_2.

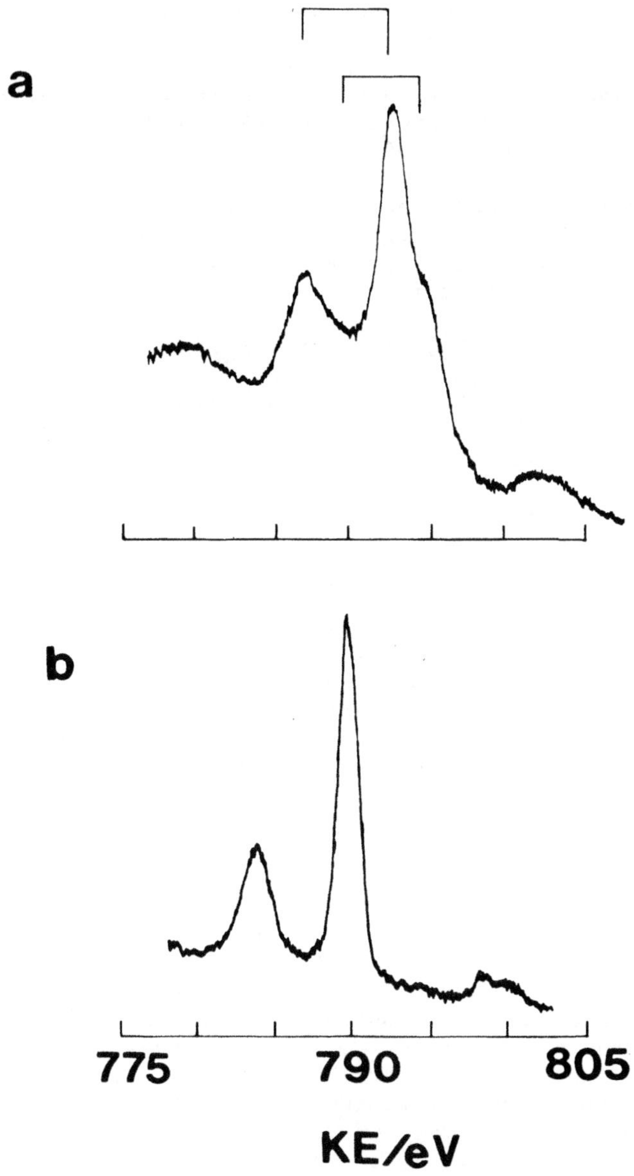

Fig. 16 The titanium 2p structure in the X-PE spectra of (a) LiT_2O_4 , and (b) $Li_{4/3}Ti_{5/3}O_4$.

In the limit of strong relaxation, $\lambda \rightarrow 0$ (and P_2/P_1), the two core PE signals should have an energy separation $\Delta E \approx J$; in the other limit, corresponding to negligible relaxation, $\lambda \rightarrow 1$ and $\Delta E \rightarrow 2\lambda(\gamma-\rho) \approx 2\lambda\gamma$, but of course the shake-up probability P_2 must then tend to zero. The expressions (4) and (3) are formally equivalent to the results of Hush (23).

It should be added that, in a class II binuclear system, a total of four PE signals might be observed for a given core ionization process - two from each of the trapped oxidation states. Suppose the ground state is

$$|0\rangle = N_o(|X_a, Y_b'\rangle + \lambda_o|Y_a, X_b\rangle)$$

(where $N_o = (1 + \lambda_o)^{-1/2}$) and that a core electron is ejected at site a; then, as before, two PE signals can in principle arise, but in this case with relative intensities

$$P_2/P_1 = (\lambda-\lambda_o)^2/(1+\lambda\lambda_o)^2$$

Ionization of a core electron at site b instead can produce two further signals which in general will occur at different KE from the first two. But to the best of the authors' knowledge, no class II mixed-valence system has yet shown more than the doubling of characteristic core PE signals. This is somewhat surprising if only because it is not at all uncommon, as we pointed out earlier, to observe a multi-peak core signal in the X-PE spectrum of a single-valence compound.

COMPOUNDS OF CLASS III B

Core level PE spectra of class III B compounds are in general complex. For example, Fig. 15 shows the tungsten 4f signals in the X-PE spectrum of a sample of vacuum-cleaved $Na_{0.764}WO_3$, a tungsten bronze (24). In contrast to the simple $^2F_{7/2,5/2}$ doublet found in the case of WO_3, one observes in the W(VI,V) compound a complicated spectral profile. A similar situation pertains in the lithium/titanium spinels, $Li_{1+x}Ti_{2-x}O_4$. The end member with $x = \frac{1}{3}$, a compound containing only Ti(IV), yields a simple $^2P_{3/2,1/2}$ doublet in its titanium 2p spectrum, whereas the metallic end member, $LiTi_2O_4$ ($x = 0$), a Ti(III,IV) mixed-valence material, displays complex structure only superficially interpretable in terms of two overlapping spin-orbit doublets (Fig. 16) (16).

Interpretation of data of this kind is still controversial. However, enlarging upon the ideas of Wertheim, the present authors consider that core PE signals from class III B mixed-valence compounds should, for the most part, be interpreted along the following lines (25). Class III B materials are commonly metallic, but the conduction band is frequently of quite narrow width. In these circumstances, ionization of a core subshell can often produce a local electrostatic perturbation strong enough to disengage one or more valence orbitals of the ionized atom from the conduction band. The resulting, localised atomic state will lie below the Fermi surface in energy and the site in question will behave as an electron 'trap', akin to an impurity atom doped into the lattice. Because of inter-electronic repulsion the maximum permitted occupancy of the trapped state will not necessarily be the same as the degeneracy of the valence subshell involved. In the case of the tungsten bronzes, for example, it is likely that the localised t_{2g}(5d) level can accommodate only one electron. Two different final states are then accessible upon photoionization, corresponding to the alternative situations in which the trap (1) becomes occupied, and (2) remains empty. The former possibility resembles a Re(VI) ion doped into the lattice and the latter an impurity Re(VII) ion. In materials having a greater density of conduction electrons, screening of the electron-electron repulsion at the localised site may permit trap occupancies greater than one. According to a model advanced by Kotani et al. (26) occupied trap states should give rise to asymmetric peaks situated on the low binding energy side of the lifetime-broadened PE signals associated with the empty localised state.

Many difficulties attend the quantitative application of these ideas but they nonetheless provide a persuasive picture of the physical processes involved in core ionization in narrow-band metals. In particular, the model is in accord with the itinerant nature of the valence electrons in most class III B compounds and with the charge-equivalence of atoms in the initial state of the system. Electron localization and the emergence of distinct valence states occur only in the final state of the photoionization process.

It should be apparent from the above discussion that the complex, multi-peak core PE structure expected of narrow-band metallic materials may well be observed in systems having a single, well-defined oxidation state. The probabilities of reaching the different final states are difficult to guage quantitatively, but distinct states differing by one electron in the occupancy of a trap site should be accessible on core ionization in a variety of materials. In accord with this expectation, we have shown recently that compounds such as the rutile-series dioxides (MoO_2, WO_2, etc.) and related perovskite systems ($SrMoO_3$ etc.) do in fact display complex core PE structure consistent with the ideas outlined above (25,27). Consider, for example, the molybdenum 3d signals in the X-PE spectrum of WO_2, reproduced in Fig. 17. Many workers have, wrongly we believe, taken data of this sort as an indication of the presence of mixed-valence, impurity phases at the sample surfaces.

We should not fail to point out, however, that some of these authors have questioned the interpretation that we (following Wertheim) favour for the core PE spectra of narrow-band metals. For example, de Angelis and Schiavello (28) argue that the 4f structure in the X-PE spectra of the tungsten bronzes is best understood in terms of overlapping peaks associated with different metal oxidation states, localised in the neutral system. They like many other workers, have carried out curve-fitting analyses in which the experimental PE spectra were synthesised from component signals attributed to a variety of extant oxidation states - W(VI), W(V) and W(IV), in the case of the tungsten bronzes. We regard such procedures as highly questionable, for a variety of reasons. Most of all, we would dispute such an interpretation because it is based on the assumption that the time-scale of PES is short enough to permit the detection of atoms in different instantaneous oxidation states arising from local fluctuations in the itinerant electron density. This is encouraged by the widespread, but erroneous view that the effective time-scale of PES is determined by the escape time for photoelectrons - about 10^{-18} sec. In fact, as in other spectroscopic techniques, the relevant timescale, when seeking to distinguish states whose energies differ by ΔE, is $\hbar/\Delta E$ or, typically, around 3×10^{-16} sec. for states differing in energy by 2 eV. We explore this point further in the Appendix.

To summarise: as with compounds belonging to class IIIA of the Robin Day scheme, the crystallographic equivalence of metal atoms in the initial state of the system does not preclude the observation of multi-peak structure in the core PE spectra of narrow-band metallic compounds.

CONCLUDING REMARKS

That core PES has something to offer mixed-valence chemistry is unquestionable; but, for a variety of reasons, both experimental and theoretical, the interpretation of core PE spectra from mixed-valence compounds is rarely straightforward. The scope of the technique can be extended by improvements in sample-handling and preparation procedures, and by wider exploitation of monochromatised X-rays. Advances in theory, especially as regards final state effects in the spectra of class III compounds, will also hopefully remove any of the existing ambiguities in interpretation. But it is likely in the main, that the usefulness of PES in mixed-valence chemistry will continue to depend on its application in conjunction with other physical techniques.

APPENDIX: The time-scale in X-PES

The idea of a time-scale associated with a spectroscopic measurement is one which is most frequently encountered with magnetic resonance techniques such as NMR (29). In this case, the time-scale is given by $\hbar/\Delta E$, where ΔE is the energy difference observed for a nucleus in two positions - i.e. the chemical shift. Then if the nucleus exchanges between the two sites, we have two limiting cases: in fast exchange, where the residence time at either site is much less than $\hbar/\Delta E$, a single line is observed in the spectrum; in slow exchange, on the other hand, the residence time is long compared with $\hbar/\Delta E$, and two distinct lines are observed.

If this notion can be extended to high energy PES measurements, it would seem that the important quantity determining the time-scale would not be, as has been claimed, the absolute core binding energy, or the kinetic energy of the ejected electron. What is important is the difference in core binding energies - for example, between two oxidation states of an element. Since this difference is typically of the order of 2 eV, this would lead to a time-scale of around 3×10^{-16} s, which is rather different from the 10^{-18} s often quoted (28,30).

It is interesting to examine these ideas a little more fully using the model originally due to Hush (23), which is discussed in the text. The difference between the core IEs in the localised limit is J: this is the difference in the coulomb interaction between the core hole and the valence electron on the two different atoms. Thus the time-scale, according to our present definition, is $\hbar/J$. On the other hand, the rate of exchange of the valence electron between the two atoms must be determined by the "effective one-electron" matrix element γ. Thus the residence time on each atom is $|\hbar/\gamma|$. (In the ionized system γ is replaced by $\gamma-\rho$, but this is likely to be a rather small correction). If the time scale picture is correct, therefore, the appearance of the core PE signals should be determined by the relative magnitudes of γ and J. With $|\gamma|>>J$ we have fast exchange, and we should observe a single peak, whereas $|\gamma|<<J$ implies slow exchange, and should give two peaks. This is indeed borne out by the quantitative predictions of the Hush model, illustrated in Figure 18.

Although the principal conclusions of the time-scale picture are correct, Figure 18 shows an important difference between X-PES and a low energy spectroscopic method such as NMR. In the latter case, faster exchange leads to a symmetrical merging of the signals into a single "exchange-narrowed" line (29). In X-PES, the spectrum with exchange is asymmetric, with low intensity, even for rapid exchange. The difference comes from the fact that the energy difference of 2 eV or so in core

PES must seriously perturb the electron exchange phenomena which is under observation. In other techniques, the energy differences are much smaller -10^{-7} eV or less in Mössbauer spectroscopy, much less in NMR. This of course leads to a longer time-scale; but it also ensures that the spectroscopic measurement cannot sensibly affect the exchange process.

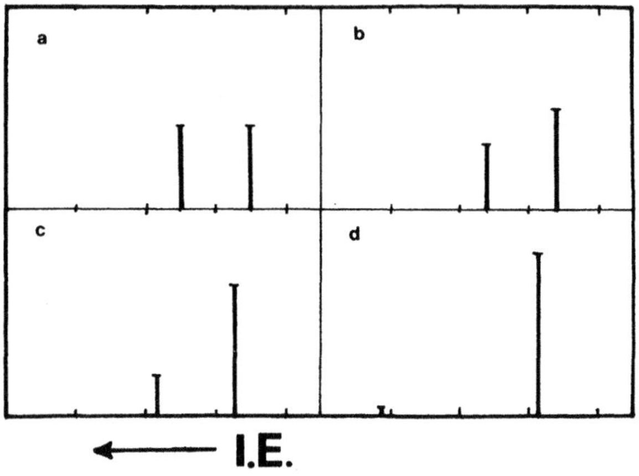

Fig. 18 The form of the core PE signals predicted by the model described in the text for different values of γ/J : (a) 0; (b) -0.1; (c) -0.3; (d) -1.0.

The large perturbation produced in the X-PES experiment, and the associated short time-scale, can lead to serious difficulties in interpretation of the core PE spectra of mixed valency compounds. There certainly may be cases where distinct oxidation states appear in the spectrum of a compound otherwise known to be of class III (23). It would be possible to describe this, as has been done for the tungsten bronzes, in terms of valence trapping in the final, ionized state produced (24). The relevance of the time-scale discussion is then to show that such trapping will only occur if the valence electron exchange (as determined, for example, by the band-width) is slow enough. The limitation of this argument, however, is shown in cases where final-state electron trapping can cause satellite structure in compounds which are not mixed valency (25).

In conclusion, it does appear to be useful to associate a time-scale with X-PES measurements. Such a time-scale is given by $\hbar/\Delta E$, and may typically be around 3×10^{-16} s for a mixed-valency compound. We must bear in mind, however, that there are many other causes of satellites in core PES than mixed-valency character, and even in mixed-valency compounds there are other ways of describing the satellite structure which may appear.

REFERENCES

1. A.D. Baker and C.R. Brundle (eds.), 'Electron Spectroscopy : Theory, Techniques and Applications', vol 1, Academic Press, London, 1977.

D. Briggs (ed.), 'Handbook of X-ray and UV photoelectron spectroscopy', Heyden, London 1977.

2. T. Koopmans, Physica, 1934, 1, 104.
3. W.G. Richards, Internat. J. Mass Spectrosc. Ion Phys., 1969, 2, 419.
4. K. Siegbahn et al., 'ESCA: Atomic, Molecular and Solid State Structure studied by means of Electron Spectroscopy', Nova Act. Reg. Soc. Sci. Upsaliensis, ser. IV, 1967, vol 20 .
5. K. Siegbahn et al., 'ESCA Applied to Free Molecules', North Holland, Amsterdam, 1969.
6. A.F. Orchard et al., unpublished observations.
7. e.g. P. Burroughs, A. Hamnett, J.F. McGilp and A.F. Orchard, J. Chem. Soc., Faraday Trans II, 1975, 71, 177
8. J.M. Thomas and M.J. Tricker, J. Chem. Soc., Faraday Trans. II, 1975, 71, 336 ; S. Evans and J.M. Thomas, ibid, 1974, 70, 313.
9. M.B. Robin and P. Day, Advan. Inorg. Chem. Radiochem., 1967, 10, 247.
10. J.C. Helmer, J. Electron Spectroscopy, 1972, 1, 259.
11. G. Thornton, D. Phil. thesis, University of Oxford, 1976.
12. D. Chadwick and J. Graham, Nature, Phys. Sci., 1972, 237,127.
13. P. Burroughs, A. Hamnett and A.F. Orchard, J. Chem. Soc., Dalton Trans., 1974, 565; M.J. Tricker, I. Adams and J.M. Thomas, Inorg. Nucl. Chem. Letters, 1972, 8, 633.
14. P. Gerard, A. Deneuville, G. Hollinger and Tran Minh Duc, J. Appl. Phys., 1977, 48, 4252.
15. A.F. Orchard and G. Thornton, J. Chem. Soc., Dalton Trans., 1977, 1238.
16. R.G. Egdell, A.F. Orchard and E.G. Scott, unpublished work.
17. A. Hamnett and A.F. Orchard, unpublished work.
18. P. Burroughs, A. Hamnett, A.F. Orchard and G. Thornton, J. Chem. Soc., Dalton Trans., 1976, 1686.
19. T.A. Carlson, M.O. Krauze and W.E. Moddeman, J. Phys. (Paris) Colloq., 1971, 32, C4-76.

20. D.C. Frost, C.A. McDowell and A. Ishitani, Mol. Phys.,1972, 24, 862.

21. C. Creutz and H. Taube, J. Amer. Chem. Soc., 1969, 91 , 3988.

22. P.H. Citrin, J. Amer. Chem. Soc., 1973, 95 , 6472.

23. N.S. Hush, Chem. Phys., 1975, 10, 361.

24. M. Campagna, G.K. Wertheim, H.R. Shanks, F. Zumsteg and E. Banks, Phys. Rev. Letters, 1975, 34, 738 ; J.N. Chazalviel, M. Campagna, G.K. Wertheim and H.R. Shanks, Phys. Rev., 1977, B 16, 697.

25. N. Beatham, P.A. Cox, R.G. Egdell, and A.F. Orchard, Chem. Phys. Letters, 1980, 69, 479.

26. A. Kotani and Y. Toyazawa, J. Phys. Soc. Japan, 1974, 37 , 912.

27. N. Beatham, R.G. Egdell and A.F. Orchard, unpublished results.

28. B.A. De Angelis and M. Schiavello, Chem. Phys. Letters, 1976, 38, 155 ; 1978, 58, 249.

29. J.W. Emsley, J. Feeney and L.H. Sutcliffe, 'High Resolution Nuclear Magnetic Resonance Spectroscopy', Pergamon Press, Oxford 1965, ch. 9.

30. W.L. Jolly, in Baker and Brundle, ref. 1.

LIST OF CONTRIBUTORS TO THE DISCUSSION SESSIONS

G. Amthauer, California Institute of Technology, USA	"Spectroscopic Studies on Mixed Valence Fe Bearing Silicate and Phosphate Minerals"
J. Barnes, University of Dundee, Scotland	"A Possible Mixed Valence Terbium Iodate"
H. Clark, University of Texas, USA	"Electron Transfer in Mixed Valence Solids"
S. Cooper, Harvard University, USA	"Mixed Valence Complexes of Higher Valent Manganese"
T. Cox, Inorganic Chemistry Laboratory, England	"The Time-Scale Problem in Spectroscopy of Mixed-Valence Compounds"
C. Dismukes, Princeton University, USA	"Low-lying Excited Spin States and the Ground Electronic Structure of Iron Sulfur Proteins"
R.G. Egdell, Wolfson College, England	"Photoemission Studies of Mixed Valence Spinel Phase $Li_{1+X}Ti_{2+X}O_4$"
B.J. Evans, University of Michigan, USA	"Influence of Composition on Electron Transfer in Some Synthetic Oxides and Natural Silicates"

O. Evrard, Universite de Nancy I, France	"Properties of the New Compound $LuFe^{II}Fe^{III}O_4$"
L. Ferris, University College London, England	"Resonance Raman Spectra of Te_4^{2+} and Se_4^{2+} Ions"
H. Fritzer, Technical University Graz, Austria	"Magnetic Studies on Some Molybdenum Oxides"
R. Gagne, California Institute of Technology, USA	"Mixed Valence Complexes Derived from Binucleating Macrocyclic Ligands"
C. Gleitzer, Universite de Nancy I, France	"Iron Oxidation Number and Properties in Solid Solutions Derived from Iron Monoxide (Wustite)"
D. Guthrie, Iowa State University, USA	"Two New Zirconium Cluster Compounds"
D. Hendrickson, University of Illinois, USA	"Crystal Structure and Physical Properties of Mixed Valent $[(n^5\text{-}C_5H_5)Fe(n^5\text{-}C_5H_4\text{-}Se\text{-}n^5\text{-}C_5H_4)Fe(n^5\text{-}C_5H_5)]I_3\text{-}I_2\text{-}1/2CH_2Cl_2$"
A. Hewson, Imperial College, England	"The Possibility of Polaronic Effects in Mixed and Intermediate Valence Systems"
M. Kizilyalli, Middle East Technical University, Turkey	"Synthesis and Structure of Mixed Valence Copper Iron Sulphides"
E. Kober, University of North Carolina, USA	"Relationships Between Charge-Transfer and Intervalence-Transfer Bands in Osmium Polypyridal Complexes"
M. Kurmoo, University College, London, England	"Resonance Raman Spectra of $[Pt(tn)_2][Pt(tn)_2X_2](ClO_4)_4$; X = Cl, Br or I

B. McQuillan, University of California, USA	"Investigation of the Arsenic Species in Graphite/AsF_5 Using Synchroton Radiation"
J. Merrick, Iowa State University, USA	"XPS Results Suggesting Possible Mixed Valence States in Scandium Monosulfide"
U. Mueller-Westerhoff, IBM Research Laboratory, USA	"Mixed Metal and Mixed Valence Metallocenophanes"
M.J. Ondrechen, University of Chicago, USA	"Bridged Ruthenium Mixed Valence Compounds: I. Hartree-Fock-Slater Results; II. Some Formal Considerations"
G. Papavassiliou, National Hellenic Research Foundation, Greece	"Preparation and Properties of Some Simple New Mixed-Valence Compounds of Pt and Pd"
H. Patterson, University of Maine, USA	"Electronic Structure of Platinum Blues"
C. Pierpont, University of Colorado, USA	"Tautomeric Catecholate-Semiquinone Interconversion via Metal-Ligand Electron Transfer. Structural and Magnetic Properties of (3,5-Di-*t*-butyl-catecholates)(3,5-di-*t*-butylsemiquinone)(bipyridyl)cobalt(III), A Complex Containing 'Mixed Valence' Ligands"
D. Richardson, Stanford University, USA	"Electronic Coupling in Ruthenium (II,III) Mixed-Valence Ions"
G. Smith, Technische Universitat Berlin, West Germany	"Evidence for Absorption by Exchange-Coupled Fe^{+2} - Fe^{+3} Pairs in the Near Infra-red Spectra of Minerals"

R. Strens, Department of Mineralogy and Petrology, England	"Pressure Effects (to 70 kbar) on Charge-Transfer and d-d Spectra of Exchange-Coupled Ion Pairs in Silicates and Oxide Minerals"
B. Sullivan, University of North Carolina, USA	"Investigations of Intramolecular Outer Sphere Intervalence Transfer in a Homologous Series of Diphosphine Bridged Ruthenium Dimers"
H. Toftlund, Odense University, Denmark	"Partially Oxidized Bis (squarato) Platinum (II) Salts"
C. Torardi, Iowa State University, USA	"Sodium Tetramolybdenum Hexoxide ($NaMo_4O_6$). A Metallic Infinite Chain Polymer Derived by Condensation of Octahedral Clusters"
U. Traeger, Anorganisch-Chemisches Institut der Universitat, West Germany	"Structure and Chemistry of Wolffram's Salt Analogues"
R. Weller, University of North Carolina, USA	"Magnetic Properties of Mixed Valence Charge Transfer Complexes of Copper"
K. Yamaji, Universite Paris-Sud, France	"New Lattice Softening Mechanism in the Quasi-One-Dimensional TCNP Conductors"
A. Zdetsis, N.R.C. Demokritos, Greece	"Optical Properties of Mixed-Valence Compounds of Pt and Pd"

INDEX

GPSR Compliance

The European Union's (EU) General Product Safety Regulation (GPSR) is a set of rules that requires consumer products to be safe and our obligations to ensure this.

If you have any concerns about our products, you can contact us on ProductSafety@springernature.com

In case Publisher is established outside the EU, the EU authorized representative is:

Springer Nature Customer Service Center GmbH
Europaplatz 3
69115 Heidelberg, Germany

Batch number: 10816188

Printed by Printforce, the Netherlands